SO-AKY-106

STEREOCHEMISTRY OF
HETEROCYCLIC COMPOUNDS

GENERAL HETEROCYCLIC CHEMISTRY SERIES

Edward C. Taylor and Arnold Weissberger, Editors

MASS SPECTROMETRY OF HETEROCYCLIC COMPOUNDS
by Q. N. Porter and J. Baldas

NMR SPECTRA OF SIMPLE HETEROCYCLES
by T. J. Batterham

HETEROCYCLES IN ORGANIC SYNTHESIS
by A. I. Meyers

PHOTOCHEMISTRY OF HETEROCYCLIC COMPOUNDS
by Ole Buchardt

STEREOCHEMISTRY OF HETEROCYCLIC COMPOUNDS
by W. L. F. Armarego

Part I, Nitrogen Heterocycles
Part II, Oxygen; Sulfur; Mixed N, O, and S; and Phosphorus Heterocycles

STEREOCHEMISTRY OF HETEROCYCLIC COMPOUNDS

Part I Nitrogen Heterocycles

W. L. F. ARMAREGO

The Australian National University
Canberra, Australia

A Wiley-Interscience Publication

JOHN WILEY & SONS, New York · London · Sydney · Toronto

CHEMISTRY

Copyright © 1977 by John Wiley & Sons, Inc.

All rights reserved. Published simultaneously in Canada.

No part of this book may be reproduced by any means, nor transmitted, nor translated into a machine language without the written permission of the publisher.

Library of Congress Cataloging in Publication Data:

Armarego, W L F
    Stereochemistry of heterocyclic compounds.

    (General heterocyclic chemistry series)
    Includes bibliographical references.
    CONTENTS: pt. 1. Nitrogen heterocycles.
    1. Heterocyclic compounds.   2. Stereochemistry.

I. Title.
QD400.3.A75   547'.59   76-26023
ISBN 0-471-01892-9

Printed in the United States of America

10 9 8 7 6 5 4 3 2 1

QD 400
.3
A 751
v. 1

CHEMISTRY
LIBRARY

To the memory of

E. E. Turner, F.R.S.

INTRODUCTION TO THE SERIES

General Heterocylic Chemistry

The series, "The Chemistry of Heterocyclic Compounds," published since 1950 by Wiley-Interscience, is organized according to classes of compounds. Each volume deals with syntheses, reactions, properties, structure, physical chemistry, etc., of compounds belonging to a specific class, such as pyridines, thiophenes, and pyrimidines, three-membered ring systems. This series has become the basic reference collection for information on heterocyclic compounds.

Many aspects of heterocyclic chemistry have been established as disciplines of *general* significance and application. Furthermore, many reactions, transformations, and uses of heterocyclic compounds have specific significance. We plan, therefore, to publish monographs that will treat such topics as nuclear magnetic resonance of heterocyclic compounds, mass spectra of heterocyclic compounds, photochemistry of heterocyclic compounds, X-Ray structure determination of heterocyclic compounds, UV and IR spectroscopy of heterocyclic compounds, and the utility of heterocyclic compounds in organic synthesis. These treatises should be of interest to *all* organic chemists as well as to those whose particular concern is heterocyclic chemistry. The new series, organized as described above, will survey under each title *the whole field of heterocyclic chemistry* and is entitled "General Heterocyclic Chemistry." The editors express their profound gratitude to Dr. D. J. Brown of Canberra for his invaluable help in establishing the new series.

Department of Chemistry   Edward C. Taylor
Princeton University
Princeton, New Jersey
Research Laboratories   Arnold Weissberger
Eastman Kodak Company
Rochester, New York

13688

PREFACE

In 1974, the periodical *Tetrahedron* commemorated the centenary of van't Hoff and Le Bel's proposal of the tetrahedral carbon atom by devoting 500 pages to articles on stereochemistry. Over the years many aspects of the stereochemistry of organic compounds have been reviewed, but the stereochemistry of heterocyclic compounds has never been considered separately and described in its entirety. It is therefore timely that the stereochemistry of heterocyclic compounds should be collected in a systematic form in two volumes.

The availability of nuclear magnetic resonance spectrometers in the early sixties is mainly responsible for the explosion in the number of publications containing stereochemical data on heterocycles. Not only has nuclear magnetic resonance been useful in deducing relative configurations at chiral centers and conformational preferences with a high degree of certainty, but it has also provided thermodynamic and kinetic data for a variety of equilibria from which stereochemical data have been evaluated. For this reason a large proportion of the literature surveyed in these volumes is post-1960. The literature has been covered until the end of 1974 and incompletely for 1975. More than 4500 references on the stereochemistry of nitrogen, oxygen, sulfur, and phosphorus heterocycles are included in the two volumes, and to keep the books to a reasonable size it was necessary to limit the discussions for many references. Only a few leading references to the stereochemistry of natural products are given to identify them with the ring systems under discussion. These monographs take the form of "guides" into the literature and at the same time give a panoramic view of the stereochemistry of nitrogen heterocycles (in Part I) and oxygen; sulfur; mixed N, O, and S; and phosphorus heterocyclic compounds (in Part II).

No words can express my gratitude to Dr. D. J. Brown for his continued and inspiring guidance, and for his provision of every facility possible during the months of writing. The manuscript would have taken much longer to produce but for the efficient and accurate work of my research assistant, Mrs. Beverly A. Milloy, B.Sc., who carried out the painstaking job of checking the references and the typescript, and whose artistic talent has turned all the formulas into proper drawings that show the three-dimensional structures. My wife assisted immensely in reading and proof-reading the whole manuscript. Finally, I am thankful to Mrs. A. Sirr for carrying out the arduous task of typing the manuscript.

W. L. F. Armarego

Canberra, Australia
June 1976

CONTENTS

STEREOCHEMISTRY OF
HETEROCYCLIC COMPOUNDS

1 INTRODUCTION

One generally associates heterocyclic compounds with planar molecules because heterocycles are not usually regarded in three-dimensional terms. A glance at the reviews in the well established series *Advances in Heterocyclic Chemistry*[1] will confirm this statement. There are, however, a large number and variety of heterocyclic compounds that possess stereochemical properties. Many (but not all) of these heterocycles belong to the known classes of heterocyclic compounds, and in a large number of cases are related to or derived from them by reduction. It is the purpose of this monograph to describe systematically the stereochemical aspects of nitrogen heterocycles (the oxygen; sulfur; mixed N, O, and S; and phosphorus heterocycles are in Part II).

I.   GENERAL STEREOCHEMICAL PROPERTIES OF HETEROCYCLIC COMPOUNDS

The stereochemical properties of heterocyclic compounds arise in a variety of ways. The first to be considered is the transition state in the electrophilic and nucleophilic substitution reactions of "aromatic" heterocycles. Here the stereochemistry of the approach of the reagent and the departure of the substituent involved, together with the nonplanar structure of the transition state, have to be accounted for in a full understanding of the processes. Cycloaddition reactions must also be considered in the above terms.

Reduced heterocyclic compounds possess many of the stereochemical features of their carbocyclic analogues. The presence of substituents on

1

the carbon atoms in the ring can introduce chiral centers and display cis and trans isomerism. In addition to these properties, the heteroatom also alters the geometry of the ring with respect to the carbocyclic analogue by changing the bond distances and bond angles adjacent to the heteroatom. The alterations may be small or large depending on the heteroatom, the number of heteroatoms, and the relative positions of the heteroatoms in the ring. These changes are not relatively large when one heteroatom is involved, and the general structure of the reduced heterocycle can be extrapolated from the known structure of the analogous carbocycle. The differences are, however, subtle and show up in the nonbonded interactions and consequently in the conformational properties of the molecules.

The heteroatom in reduced heterocycles introduces properties which are characteristic of the heteroatom itself. Pyramidal atomic inversion[2] is a property which distinguishes nitrogen, oxygen, sulfur, and phosphorus atoms from tetravalent carbon. Trivalent nitrogen in a conformationally flexible ring, for example, inserts another conformational property to the ring. The hydrogen atom or substituent on the nitrogen atom in a ring can attain two equilibrating conformations by virtue of the inverting nitrogen atom. This points out the conformational property of the nitrogen lone pair of electrons, and considerable attention has been devoted in several laboratories to the "size" (space demand) of the nitrogen lone pair. The rate of atomic inversion is affected by the size of the ring in which the nitrogen atom is inserted and by the substituents. An oxygen atom directly attached to the ring nitrogen atom can decrease its inversion rate to the extent that the nitrogen atom is almost "locked" in a chiral configuration, and it introduces a source of optical activity. The inversion rate, and the effect of substituents on it, varies from one heteroatom to another. The oxygen heteroatom also undergoes inversion, but it is not as interesting as the nitrogen atom because it is divalent. Inversion of the oxygen atom does not alter the situation because it has two lone pairs of electrons which exchange place during the inversion without effecting a serious alteration. Oxonium ions derived from saturated oxygen heterocycles, on the other hand, will exhibit pyramidal inversion, but these present experimental difficulties because of their chemical reactivity. The properties of some oxonium ions, however, have been reported. The oxygen atom in reduced heterocycles produces strong dipolar effects with respect to polar substituents particularly on the adjacent carbon atoms. It tends to force the substituents into an axial conformation (anomeric effect) which affects the conformational properties of the molecule as a whole. Sulfur heterocycles are similar to oxygen heterocycles, but in addition the sulfur atom can expand its valency shell. Oxidation of the sulfur atom yields sulfoxides which have very high barriers to atomic inversion and are therefore possible centers of asymmetry in the molecule. Sulfur forms stable S–S bonds, and these adjacent sulfur atoms in a heterocyclic ring are a source of chirality in their own right by virtue of the helical nature of the C–S–S–C bonding arrangement. The energies involved in altering the torsion angles in this bonding arrangement are greater than in the corresponding carbocyclic systems and therefore have a large effect on the conformational properties of these sulfur containing molecules. Unlike the above heterocycles, many phosphorus heterocycles are known in which the valency state of the phosphorus atom is II, III, IV, V or VI. The phosphorus atom has a particular stereochemistry in each valency state. The heteroatom in phosphorus(III) heterocycles is similar to a nitrogen atom in its pyramidal atomic inversion but is generally much slower. The most interesting examples are found among the P(V) compounds. The trigonal bipyramid P(V) compounds undergo permutational isomerism (not observed in the other heteroatoms) in which the energy required for substituents to exchange places, for example, in inversion of configuration, without breaking and making bonds may be quite low, and the process may occur very readily. Not only have many of these P(V) structures been postulated as intermediates or transition states in reactions, but a host of them have been isolated and characterized.

Several dissymmetric heterocyclic molecules are known which exhibit optical activity. Biheteroaromatic compounds which owe their dissymmetry to restricted rotation about the bond which joins the two heteroaromatic rings have been resolved into their optical antipodes. Their optical stabilities, when compared with their carbocyclic counterparts, demonstrate the steric and other effects attributed to the heteroatoms. Rigid dissymmetric molecules such as heterotwistanes, heterohelicenes, and heteroadamantanes can also exhibit optical activity if the heteroatoms are placed in appropriate positions. A number of examples are known in which the heteroaromatic ring has a serious effect on the chiroptical properties of a chiral center in a side chain when there is enough interaction between the heteroaromatic ring and the asymmetric center.

The three-dimensional properties of many bi- and polycyclic reduced heterocycles, for example, heterobicyclo[x.y.z]alkanes, may not have special stereochemical features but are discussed in this monograph because a knowledge of the relative arrangement of all the atoms in space is necessary for a complete understanding of the properties and reactions of the systems. Transannular interactions due to the spatial proximity of a heteroatom and another distant atom in the same ring have been observed and have led to interesting polycyclic systems.

Short introductions to chapters and to several sections in chapters are provided throughout, and should be consulted for further general stereochemical properties of the relevant classes of compounds.

II. ORGANIZATION OF STEREOCHEMICAL DATA

The general arrangement of data for each class or group of compounds is in four basic parts. In the first part the stereochemical course of the syntheses of heterocyclic compounds is described. This is followed by a discussion of the configuration and then the conformation of heterocyclic molecules. The stereochemical reactions of heterocycles are reviewed in the fourth part.

A clear distinction between configuration and conformation is made in these sections, and these two terms have been defined in several excellent texts.[3-5] Although a detailed explanation of these terms is not warranted, a brief statement is necessary because they are major issues under discussion. The *configuration* of a molecule or a substituted atom is the relative arrangement of the atoms in a molecule, and this *relative arrangement is unaltered by bond rotation*. Except for pyramidal atomic inversion and permutational isomerism (see above) the configuration can only be altered by bond cleavage and the reformation of a new bond. Thus the configuration of an asymmetric carbon atom, and the cis and trans relationship of substituents in cyclic compounds can only be altered by breaking a bond(s) and reforming another bond(s) as in substitution and rearrangement reactions. The energy required to invert the configuration of the tetrahedral bonds of a carbon atom by way of a square planar arrangement of the bonds was calculated. It was found that the process required about 1046 kJ mol^{-1} and was about 2.5 times the C–C bond energy.[6] The *conformation* of a molecule is also a description of the relative arrangement of the atom and groups of atoms in space in a molecule. However, unlike the configuration, it is altered only by rotating one or more bonds. The essential difference between the two terms is that the configuration of a molecule describes the rigid arrangement of atoms or groups of atoms, whereas the conformation of a molecule describes the flexible arrangement of the atoms or groups of atoms. This is where the difference between the two terms becomes very fine. Thus the fixed arrangement of atoms in the crystal of 4-methylpiperidine (**1**) is in fact its configuration, but the same arrangement of atoms in solution is only the preferred conformation because it is in dynamic equilibrium with conformer **2,** among other conformers, and the hydrogen atoms in the methyl group are also rotating about the bond between the methyl carbon atom and C–4 of the piperidine ring.

For absolute configuration see the following section.

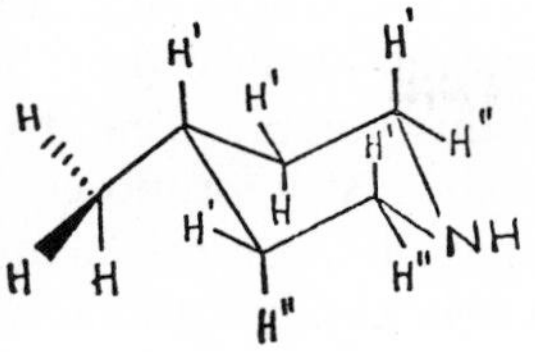

1

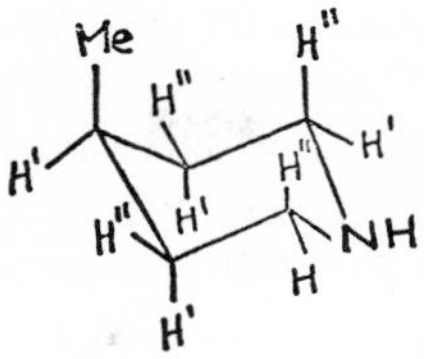

2

III.   NOMENCLATURE

The substitutive naming system is adopted as much as possible in accordance with the International Union of Pure and Applied Chemistry (IUPAC) nomenclature for organic chemistry[7]; the monograph by Fletcher, Dermer, and Fox[8]; and the *Handbook for Chemical Society Authors*.[9] The *Ring Index* nomenclature[10] is used for the basic aromatic ring systems, and for the highly reduced bi- and polycyclic compounds the Baeyer system[7-9] is adopted throughout. In this system the heterocyclic compound is named after the parent alkane. The number of atoms between the bridgehead atoms, in decreasing number, is placed in square brackets before the name of the alkane. The numbering of the atoms begins at the bridgehead atom and proceeds by way of the longest chain to the second bridgehead atom, then through the second longest chain to the first bridgehead atom, and finally through the smallest chain to the second bridgehead atom. Whenever possible the double bond, the heteroatom, and the substituents (in this order) are given the smallest number. The most abundant examples in the present monograph are tricyclic systems, for example, bicyclo[$x.y.z$]alkanes (**3**). They are called *bicyclo* because the first two rings define the

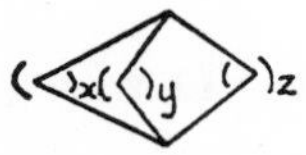

3

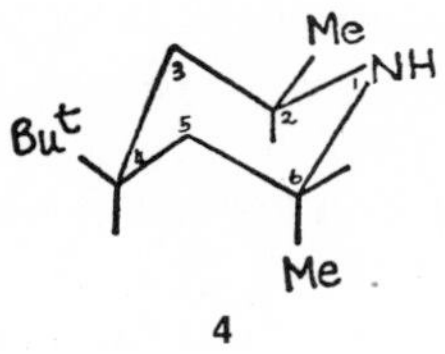

4

third ring. When x, y, or z is zero the molecule becomes a true bicyclic compound; for example, quinolizidine is 1-azabicyclo[4.4.0]decane. There are a few more complicated rings in the text, but these have been named and their formulas have been adequately numbered.

The rules for the notation of *absolute configuration* of asymmetric centers and for dissymmetric molecules proposed by Cahn, Ingold, and Prelog[11] have gained universal acceptance in the last decade. These rules and definitions of stereochemical properties were admirably described by Bentley[3] and Mislow[4] and have been used throughout this volume. Tentative rules for the nomenclature in fundamental stereochemistry proposed by the IUPAC[12] in 1969 and published in the Journal of Organic Chemistry in 1970[13] (see also ref. 8) for absolute and relative configurations, and for conformations are used as far as possible. The relative configurations of asymmetric centers are given in the usual R and S notation but are distinguished from absolute configurations by an asterisk, R^* and S^*. These are used in examples where the absolute configurations are not known. In a second system which is also used, one asymmetric center is taken as reference and denoted by r (reference) while the other centers are cis or trans with respect to it. This is particularly useful in rigid or cyclic molecules, for example, r-2-*trans*-6-dimethyl-*cis*-4-t-butylpiperidine (4).

A proposal for the designation of conformations of small rings was made by Shaw,[14] but most of the generally accepted terminology in conformational analysis is adopted in this work. Also, for the pseudorotating systems, as in pyrrolidine (see Part II, Chapter 2, Scheme 5), the designations proposed by Shaw were not used; instead the conformers were denoted as V for "envelope" conformations **5** or **5a**, and T for the "twist" conformations **6** or **6a**. The superscript numbers denote atoms above the plane of at least three atoms of the ring, and the subscript numbers denote the atom below the plane of at least three atoms of the ring. The numbering in the ring begins at the heteroatom and is counterclockwise.

IV. ENERGY TERMS AND OPTICAL ACTIVITY

The main energy terms $\triangle G$, $\triangle H$, and E_a are expressed in kilojoules per mole, and for conversion into kilocalories per mole they should be divided by 4.185. In most cases the solvents used for these measurements are not indicated in the text because the intention was only to give some indication of the energies involved, and usually the solvent did not alter the magnitude of these values.

A rule of thumb relating optical stability with the free energy in the interconversion of enantiomers can be proposed. If the free energy of interconversion of enantiomers is approximately greater than 80 kJ mol^{-1}, then they can be separated into their optical antipodes by the usual methods of resolution.[17] When the free energy is between approximately 25 and 80 kJ mol^{-1} the molecules are optically labile and racemize very readily, but the enantiomers can be shown to possess optical activity by dynamic methods requiring kinetic resolution or asymmetric transformations.[18,19] For enantiomers with interconversion energies of approximately less than 25 kJ mol^{-1} it is not possible by the present methods to demonstrate optical activity, and the enantiomers are truly conformational isomers. However, they can be observed by nmr methods from which the energy parameters can be evaluated.

Although signs of rotation have been indicated in the text, sometimes together with the absolute configuration, they are truly meaningless because the solvent in which they were measured is not stated. They have, however, been included because they indicate that the compound mentioned is optically active and, in examples where the absolute configuration is provided, they demonstrate that the optical activity was correlated with the absolute configuration.

V.   REFERENCES TO OTHER WORKS

A conscious effort has been made in this monograph to avoid repetition of many discussions that have been treated adequately in reviews or texts, and many of the relevant references are cited below. A basic knowledge of both heterocyclic chemistry and stereochemistry is assumed. Excellent texts and reviews are available on the basic principles of heterocyclic chemistry,[20-24] on the chemistry of separate heterocyclic systems,[25,26] and on advances in heterocyclic chemistry.[1] Specialist reports on the chemistry of saturated heterocyclic and heteroaromatic compounds by the London Chemical Society[27] and the *M.T.P. International Review of Science*[28] are continuously updating the field and contain much stereochemical data. Similarly, reviews and texts on optical rotatory power,[29] parity and optical rotation,[30] theory of optical activity,[31] optical rotatory dispersion[32-34] and circular dichroism,[33,34] general stereochemistry,[3-5,35,36] and conformational analysis[15,16,37-39] have been written by prominent authors. The continuing series entitled *Progress in Stereochemistry*[40] and *Topics in Stereochemistry*[41] are providing and updating stereochemical knowledge, and they are written by authorities on the various aspects of the subject. Fifty papers and reviews were published in *Tetrahedron*[42] in 1974 in commemoration of the centenary of the proposal of the tetrahedral carbon atom by van't Hoff[43] and Le Bel.[44] Three of these articles are particularly interesting to read because they are of historical interest,[45–47] and a short review on principal developments in stereochemistry during the last hundred years was also published in 1974.[48]

The relative and absolute configurations have been collated in an *Atlas of Stereochemistry*.[49] Compilations of interatomic distances and configurations of molecules and ions,[50] and molecular structures and dimensions of organic molecules in the crystalline state[51] contain useful reference material for a very large number of compounds, and they hopefully will be kept up to date.

VI.   REFERENCES

1. A. R. Katritzky and A. J. Boulton, Eds., *Advances in Heterocyclic Chemistry*, Vol. 1 (1963) to Vol. 17 (1974), continuing series, Academic, New York.

2. J. B. Lambert, *Topics Stereochem.*, **6**, 19 (1971); A. Rauk, L. C. Allen, and K. Mislow, *Angew. Chem. Internat. Edn.*, **9**, 400 (1970).

3. R. Bentley, *Molecular Asymmetry in Biology*, Vol. 1, Academic, New York, 1969.

4. K. Mislow, *Introduction to Stereochemistry*, Benjamin, New York, 1966.

5. E. L. Eliel, *Stereochemistry of Carbon Compounds*, McGraw-Hill, New York, 1962.

6. H. J. Monkhorst, *Chem. Comm.*, 1111 (1968).

7. IUPAC, *Nomenclature for Organic Chemistry*, Sects. A, B, and C, Butterworths, London, 1965 and 1966.

8. J. H. Fletcher, O. C. Dermer, and R. B. Fox, *Nomenclature of Organic Chemistry*, Advances in Chemistry Series, Vol. 126, American Chemical Society, Washington, 1974.

9. The Chemical Society, *Handbook for Chemical Society Authors*, Burlington House, W.1., London, 1960.

10. A. M. Patterson, L. T. Capell, and D. F. Walker, *The Ring Index*, 2nd ed., American Chemical Society, Washington, 1960; Suppl. I, pp. 7728–9734; Suppl. II, pp. 9735–11524 (1964); and Suppl. III, p. 11525 (1965).

11. R. S. Cahn, C. K. Ingold, and V. Prelog, *Experientia*, **12**, 81 (1956) and *Angew. Chem. Internat. Edn.*, **4**, 385 (1966); R. S. Cahn and C. K. Ingold, *J. Chem. Soc.*, 612 (1951).

12. *IUPAC Information Bulletin*, **35**, 36 (1969).

13. *J. Org. Chem.*, **35**, 2849 (1970).

14. D. F. Shaw, *Tetrahedron Letters*, 1 (1965).

15. D. H. R. Barton and R. C. Cookson, *Quart. Rev.*, **10**, 44 (1956).

16. E. L. Eliel, N. L. Allinger, S. J. Angyal, and G. A. Morrison, *Conformational Analysis*, Interscience, New York, 1965.

17. S. H. Wilen, *Topics Stereochem.*, **6**, 107 (1971); P. H. Boyle, *Quart. Rev.*, **25**, 323 (1971).

18. M. M. Harris, *Progr. Stereochem.*, **2**, 157 (1958).

19. E. E. Turner and M. M. Harris, *Quart. Rev.*, **1**, 299 (1947).

20. A. Albert, *Heterocyclic Chemistry*, 2nd ed., Melbourne University Press, 1968.

21. A. R. Katritzky and J. M. Lagowski, *The Principles of Heterocyclic Chemistry*, Academic, New York, 1968.

22. J. A. Joule and G. F. Smith, *Heterocyclic Chemistry*, Van Nostrand Reinhold, London, 1972.

23. L. A. Paquette, *Principles of Modern Heterocyclic Chemistry*, Benjamin, New York, 1968.

24. R. M. Acheson, *An Introduction to the Chemistry of Heterocyclic Compounds*, 2nd ed., Interscience, New York, 1967.

25. A. Weissberger and E. C. Taylor, Eds., *The Chemistry of Heterocyclic Compounds*, Vol. 1 (1950) to Vol. 39 (1974), continuing series, Interscience, New York.

26. R. C. Elderfield, Ed., *Heterocyclic Compounds*, Vol. 1 (1950) to Vol. 9 (1967), series terminated, Wiley, New York.

27. W. Parker, Senior Reporter, *Saturated Heterocyclic Chemistry*, Vols. 1, 2, and 3, Specialist Periodical Reports, The Chemical Society, 1970–1975; C. W. Bird and G. W. H. Cheeseman, Senior Reporters, *Aromatic and Heteroaromatic Chemistry*, Vols. 1, 2, and 3, Specialist Periodical Reports, 1973–1975.

28. K. Schofield, Ed., *Heterocyclic Compounds, M.T.P. International Review of Science*, Vol. 4, Butterworths, London, 1973 and 1975.

29. S. F. Mason, *Quart. Rev.*, **17**, 20 (1963).

30. T. L. V. Ulbricht, *Quart. Rev.*, **13**, 48 (1959).

31. Symposium on the Theory of Optical Activity in memőry of J. G. Kirkwood and W. Moffit, *Tetrahedron*, **13**, 1 (1961).

32. C. Djerassi, *Optical Rotatory Dispersion*, McGraw-Hill, New York, 1960.

33. P. Crabbé, *Optical Rotatory Dispersion and Circular Dichroism in Organic Chemistry*, Holden-Day, San Francisco, 1965; P. Crabbé, *An Introduction to the Chiroptical Methods in Chemistry*, Syntex, S.A. and Universidad Nacional Autonoma de Mexico, Universidad ibero americana, Mexico, 1971.

34. J. A. Schellman, *Chem. Rev.*, **75**, 323 (1975).

35. E. L. Eliel, *Elements of Stereochemistry*, Verlag Birkhaüser, Basle-Stuttgart, 1972.

36. M. S. Newman, Ed., *Steric Effects in Organic Chemistry*, Wiley, New York, 1956; J. D. Morrison and H. S. Mosher, *Asymmetric Organic Reactions*, Prentice-Hall, New Jersey, 1971.

37. E. L. Eliel, *Angew. Chem. Internat. Edn.*, **11**, 739 (1972).

38. F. G. Riddell, *Quart. Rev.*, **21**, 364 (1967).

39. R. Rahman, S. Safe, and A. Taylor, *Quart. Rev.*, **24**, 208 (1970).

40. W. Klyne, P. B. D. de la Mare, B. J. Aylett, and M. M. Harris, Eds., *Progress in Stereochemistry*, Vol. 1 (1954) to Vol. 4 (1968), Butterworths, London.

41. N. L. Allinger and E. L. Eliel, Eds., *Topics in Stereochemistry*, Vol. 1 (1967) to Vol. 8 (1974), continuing series, Interscience, New York.

42. *Tetrahedron*, **30**, 1477–2007 (1974).

43. J. H. van't Hoff, *Arch. neer.*, **9**, 445 (1874); *Bull. Soc. chim. France*, **23**, 295 (1875).

44. J. A. Le Bel, *Bull. Soc. chim. France*, **22**, 337 (1874).

45. R. Robinson, *Tetrahedron*, **30**, 1477 (1974).

46. M. J. T. Robinson, *Tetrahedron*, **30**, 1499 (1974).

47. F. G. Riddell and M. J. T. Robinson, *Tetrahedron*, **30**, 2001 (1974).

48. J. Weyer, *Angew. Chem. Internat. Edn.*, **13**, 591 (1974).

49. W. Klyne, *Atlas of Stereochemistry*, Chapman and Hall, London, 1974.

50. L. E. Sutton, Ed., *Tables of Interatomic Distances*, *Chem. Soc. Special Publ.*, The Chemical Society, Burlington House, W.1., London, No. 11 (1958) and No. 18 (1959).

51. O. Kennard, D. G. Watson, F. H. Allen, N. W. Isaacs, W. D. S. Motherwell, R. C. Petterson, and W. G. Town, Eds., *Molecular Structures and Dimensions*, Vol. 1 to 4 (1935–1972) and Vol. 1A (1960–1965, 1972), Crystallographic Data Centre, Cambridge and Interatomic Union of Crystallography.

2 NITROGEN HETEROCYCLES: THREE-, FOUR- AND FIVE-MEMBERED RINGS

I.   THREE-MEMBERED RINGS

A.   Aziridines

Aziridines are dihydro derivatives of the parent azirines. Two azirines are possible, the 1*H*- and 2*H*-azirines, which have a C–C and C–N double bond respectively. 1*H*-Azirines are apparently unknown despite several attempts to synthesize them, but many examples of the isomeric 2*H*-azirines have been prepared, and their properties have been studied in detail.[1] The molecule is planar and possesses one possible source of chirality—the 2-carbon atom. However, no optically active 2*H*-azirine has as yet been synthesized.

The aziridines have per se been the most explored group of three-membered ring heterocyclic compounds. The three-membered ring is planar and under strain. This is reflected in the two stereochemical properties of the molecule: (a) the spatial arrangement of substituents on C–2 and C–3 (cis and trans groups are slightly further apart than they are in ethane, with the cis substituents eclipsed) and (b) the pyramidal inversion rate of the nitrogen atom is decreased (in comparison with aliphatic amines) to values which are amenable to dynamic nmr studies. Lambert has discussed the "pyramidal atomic inversion" of nitrogen in various systems together with those of other elements which show a similar phenomenon.[2]

1.   Synthesis of Aziridines

Some syntheses of aziridines that were devised mainly for the purpose of obtaining a particular stereoisomer or for the study of the steric course of the reaction are described below. This is not exhaustive and the reader is referred to papers in later sections which contain further examples of the relevant aziridines.

a.   *Wenker and Related Syntheses.*   One of the earliest methods for preparing aziridines is the Wenker synthesis. It involves the successive reaction of β-hydroxyamines with sulfuric acid to form the intermediate *O*-sulfuric ester and cyclization of this with alkali. Lucas and collaborators[3] showed that the formation of the sulfuric ester occurred with retention of configuration, but that ring closure involved a Walden inversion at the β-carbon atom. Optically active *erythro*-3-aminobutan-2-ol (**1**, R = H) gave the ester **1** (R = SO₃H) which cyclized to optically active *N*-methyl *trans*-2,3-dimethylaziridine, whereas the optically active *threo*-alcohol gave optically inactive, and therefore *meso-N*-methyl *cis*-2,3-dimethylaziridine. Similarly *trans*-2-hydroxycyclopentyl-,[4] cyclohexyl-,[5] cycloheptyl-,[6] and cy-

clooctyl amines[7] (**3**) were cyclized to the respective *cis*-bicyclo[*n*.1.0]alkanes (**4**). The limit is *n* = 10 because Fanta and co-workers[8] synthesized *trans*-13-azabicyclo[10.1.0]tridecane, but they showed that although alkali converted *trans*-2-aminocyclodecyl hydrogen sulfate into *cis*-11-azabicyclo-

[8.1.0]undecane (4, $n = 8$) together with cyclododecanone and *trans*-2-cyclo-decen-1-ylamine (for conformational reasons), only the last two compounds were formed from similar treatment of *cis*-2-aminocyclodecyl hydrogen sulfate. The stereochemical purity of the aziridine is dependent on the S_N2 nature of the displacement reaction, and in examples where the hydroxy group is on a benzylic carbon atom, as in ephedrines, some retention of configuration has been observed when chlorosulfonic acid was used as reagent.[9] The main course of the reaction, however, proceeds with inversion of configuration.

More recently Brois and Beardsley[10] succeeded in converting (−)-ephedrine (erythro) and (+)-pseudoephedrine, and their nor isomers, into the respective *trans*-(+)- and *cis*-(−)-1,2-dimethyl-3-phenylaziridines with very high stereospecificity by using sulfuric acid in the first step. The mono-substituted $S(−)$-2-isopropyl and $S(−)$-2-isobutyl aziridines were obtained in a similar way from $S(+)$-valinol and $S(+)$-leucinol.[11] The cyclization of (−)-*N*-β-hydroxyethylephedrine, also with concentrated sulfuric acid, gave optically active 1,2-dimethyl-3-phenylaziridine[12] and not a morpholine as originally proposed.[13]

β-Chloroamines undergo ring closure to aziridines with inversion at the carbon atom bearing the halogen atom. This was first demonstrated by Weissberger and Bach[14] who obtained optically active 2,3-diphenylaziridine from *erythro*-(−)-α-amino-α′-chlorodibenzyl, but optically inactive *meso*-2,3-diphenylaziridine from the *threo*-(−)-chloroamine. Displacement of halogen by a weakly basic nitrogen atom is also possible, for example, *trans*-3-iodo-4-methoxycarbonylamino-1-methylcyclohex-1-ene cyclizes to 7-aza-7-methoxycarbonyl-3-methylbicyclo[4.1.0]hept-3-ene under the influence of a base.[15] In a similar stereospecific reaction olefins are converted into β-azidobromides by reaction with *N*-bromosuccinimide and sodium azide. Reduction of these with lithium aluminum hydride afforded the aziridines with the same configuration as the original olefins, that is, *trans*-olefins gave *trans*-aziridines[16] (Equation 1), with the initial addition taking place according to Markownikoff's rule.

$$(1)$$

In a somewhat similar series of reactions the all-*cis*-1,2,3,4,5,6-triepoxy-cyclohexane was transformed into 1,2,4-triazido-3,5,6-trihydroxycyclohex-ane with sodium azide and its tri-*O*-tosyl derivative, whose structure was confirmed by ^{1}H and ^{13}C nmr spectroscopy, was reduced with lithium aluminum hydride, and the triamine cyclized to the all-*cis*-benzenetriimine apparently with inversion of configuration at all the carbon atoms[17] (Equation 2).

$$(2)$$

Benzal aniline condenses with ethyl α-chloroacetate in the presence of a base, in an extension of Darzens glycidic ester synthesis, to form only *trans*-2-ethoxycarbonyl-1,3-diphenylaziridine (**6**). *N,N*-Diethyl α-chloroacetamide, on the other hand, gives an 8:1 mixture of *cis*- and *trans*-2-*N,N*-diethylcarbamoyl-1,3-diphenylaziridine. Assuming that the internal displacement of halogen was purely S_N2, then the reactive intermediates in the conformations **5** and **7** should predominate for the ester **6** and the amide **8** respectively.[18] The equilibration of **6** and **8** in CCl_4 gives a 50:50 and 52:48 mixture of *cis*- and *trans*-aziridine esters and amides respectively, demonstrating that the stereospecificity observed in the synthesis is due to the preferred conformations **5** and **7** in the transition state.[18] Analogous selectivity is found in β-haloamines prepared from α-haloacrylic esters or nitriles and primary amines.[19]

$$(3)$$

The synthesis of 2-acyl and 2-aroyl aziridines from α,β-unsaturated ketones by reaction with primary amines in the presence of iodine, or alternatively from the *N*-halo derivative of the primary amine obtained from the addition, or from primary amines and α-bromochalcones, proceeds with some stereoselectivity.[20-23] The configuration of the aziridines formed is dependent on the nature of the substituents on the carbon and nitrogen atoms and on the solvent used. For example, altering the solvent from methanol to benzene greatly increases the proportion of the trans isomer of aziridine ketones in the reactions of *trans*-chalcone and *trans*-4-nitro-chalcone with cyclohexylamine and iodine, while the proportion of *cis*-aziridine is increased in the reactions of cyclohexylamine with *trans*-α-bromochalcones or *trans*-α-bromo-4-nitrochalcones.[20] The cis to trans ratio of aziridine ketones was shown to depend on the relative bulk of the substituents and their influence on the preferred conformations in the transition state.[22]

An intramolecular addition of an NH group to a double bond to form an aziridine ring, for example, 1-azatricyclo[3.2.1.0²,⁷]octane, is brought about by *N*-bromosuccinimide or lead tetraacetate[24] (Equation 3). The structure of the original amine dictates the stereochemistry of the product. The potential of this reaction for the synthesis of alkaloids, for example, ibogamine, has been realized.[25] A nitrene may well be involved in this addition reaction (see Equation 4).

b.   *By Intermolecular Addition of a Nitrogen or Carbon Unit.*   Nitrogen addition on to a double bond to yield aziridines is achieved in various ways. Nitrenes generated from *N*-aminophthalimide,[26,27] alkoxyamines[28] (by lead tetraacetate), benzoyl[29] and benzenesulfonyl azides,[30,30a] ethoxy-carbonyl azide[31] (by thermolysis or photolysis), and pentafluoronitroso-benzene[32] (by triethyl phosphite or photolysis) add to olefins and give aziridines which have the same cis or trans configuration as in the original olefin, that is, a stereoselective cis addition. The selectivity with penta-fluorophenyl nitrene is the one expected from the addition of singlet nitrene,[32] but with ethoxycarbonyl nitrene both singlet and triplet nitrenes are involved.[31] In such cases the stereoselectivity is poor; whereas the former (singlet) is highly stereospecific, the latter (triplet) is completely nonspecific and although the nitrene adds more rapidly to one carbon atom, the second carbon atom, now a radical, may have time to rotate before cyclizing to provide the inverted aziridine.[31] These are depicted in Equations 4 and 5.

$$
R'-N: \; (singlet) \;\; + \;\; \Big[\begin{array}{c} H \,\diagup\, R^3 \\ H \,\diagdown\, R^2 \end{array} \Big] \;\; \longrightarrow \;\; RN \Big\langle \begin{array}{c} H \,\diagup\, R^3 \\ H \,\diagdown\, R^2 \end{array} \tag{4}
$$

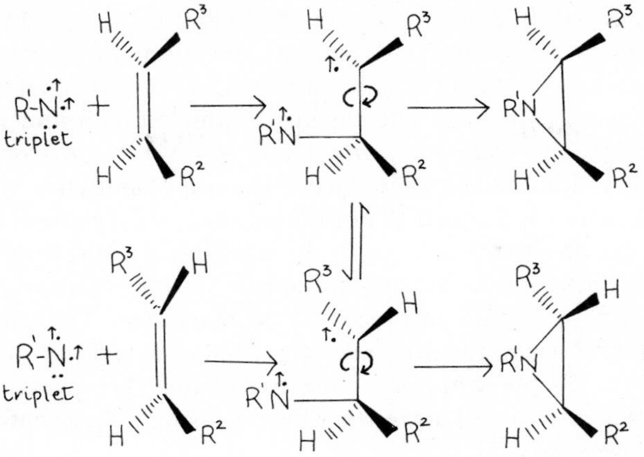

(5)

Another specificity is encountered when the olefin is a rigid molecule. Benzenesulphonyl azide adds to *cis-endo-* and *cis-exo*-norbornene-5,6-dicarboxylic anhydride and dimethyl *cis-exo*-norbornene-5,6-dicarboxylate from the endo side of the molecule (9). Dimethyl *cis-endo*-norbornene-5,6-dicarboxylate, under the same conditions of refluxing CCl$_4$, on the other hand, yields exclusively the *endo*-aziridine. The anhydrides are attacked from the exo side in the presence of ultraviolet light.[30] It is also shown that the initial product from norbornadiene and benzoyl azide is the *exo*-aziridine (10) free from the endo isomer.[29] It appears that the stereochemistry of the products depends on both steric factors and the affinity of the nitrene for the olefin.

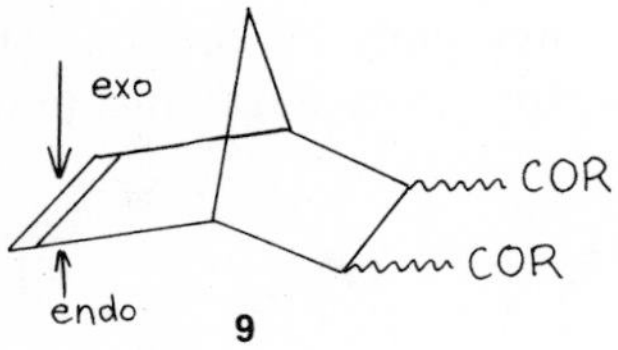

9

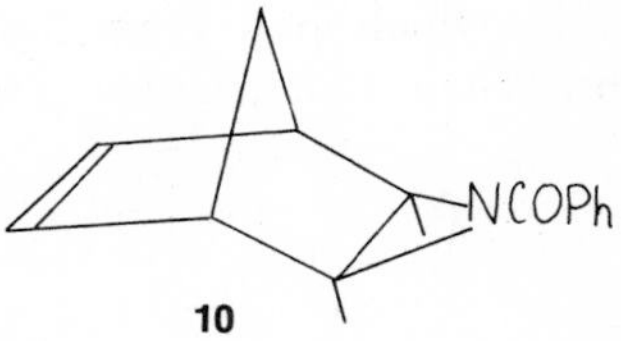

10

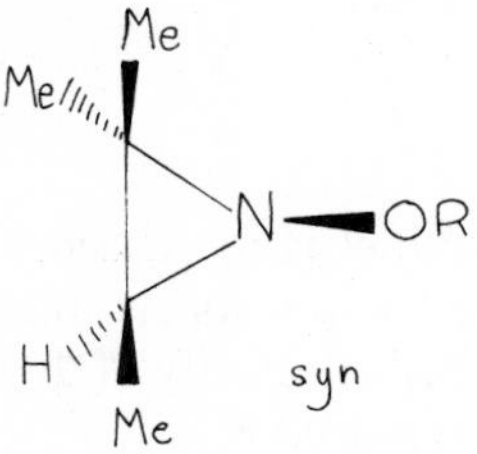

11

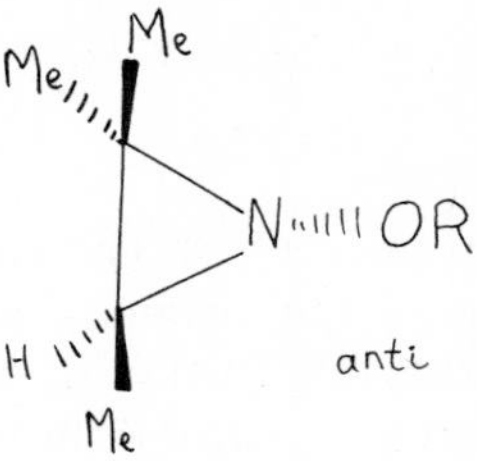

12

The *N*-alkoxyaziridines derived from alkoxyamines and lead tetraacetate do not undergo rapid nitrogen (pyramidal) inversion. Two invertomers are formed, for example, syn (11) and anti (12), in this type of isomerism. These can be separated on tlc and the syn to anti ratios found varies from approximately 1.1 to 1.5 on going from R = Me to R = isoPr.[28] The 1-phthalimidoaziridines behave similarly—the ethyl 1-phthalimidoaziridine-2-carboxylates initially formed (from phthalimido nitrene and α,β-unsaturated esters) are the thermodynamically less stable invertomers in which the ester and phthalimido groups are syn.[27] Nitrene addition to olefins has been shown in certain instances to be reversible. The photolysis of 1-phthalimidonitrenes in the presence of cyclohexene yields 7-phthalimido-7-azabicyclo[4.1.0]heptane and the respective olefin. This reversibility, however, is not general for all nitrenes which possess a heteroatom on the aziridine nitrogen atom.[33]

The synthesis of aziridinium salts from iminium salts by the addition of one carbon atom with diazomethane was developed by Leonard and his school,[34] and it is particularly useful when the double bond is tetrasubstituted. In an analogous way 2,2-dichloro-1,3-diphenylaziridine is formed when dichlorocarbene is reacted with benzal aniline.[35] The stereochemical aspects of these additions, however, have not been explored.

   c.   *Hoch-Campbell and Related Syntheses.*   The Hoch-Campbell synthesis of aziridines from ketoximes and Grignard reagents was found to be stereospecific and regiospecific. The specificity is affected by the type and concentration of the Grignard reagent,[36,36a] by geometrical isomerism in the oxime (syn and anti),[37] and it is subject to asymmetric induction if the oxime possesses a chiral substituent.[38] The 2- and 3-substituents in the aziridine, which arise from the ketone, are invariably cis oriented (Equation 6). When the Grignard reagent is MeMgBr a methylaziridine (e.g., 13)

$$R = Me, Ph \qquad (6)$$

is also formed.[36,37,39] The amount of this product is increased if the concentration of the Grignard reagent is high and supports the intermediacy of an azirine (e.g., 14) in the mechanism. Higher selectivity and yields have been recently obtained if the =NOH group is replaced by an =N–NMe$_3^+$I$^-$ group, that is, a hydrazone.[39a] The steric effect of the substituents is clearly demonstrated in the reactions of cyclohexanone oximes (Equation 7). The

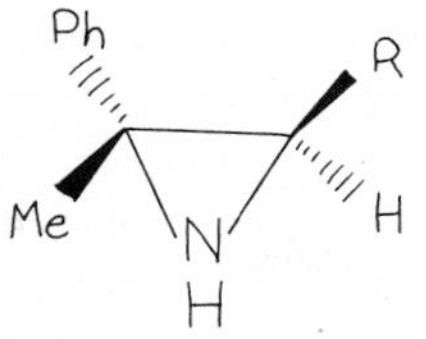

13

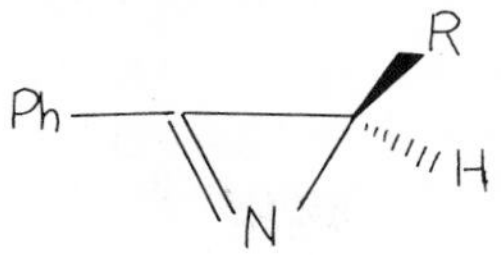

14

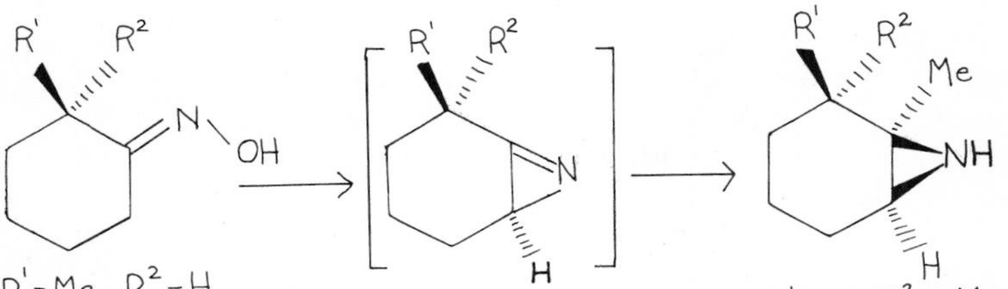

(7)

15 $R' = Me, R^2 = H$

15a $R' = Ph, R^2 = H$

16 $R' = H, R^2 = Me$

16a $R' = Ph, R^2 = H$

methyl derivative **15** gives 60% of the cis isomer **16** (and 40% of the trans isomer), and the phenyl oxime **15a** gives 75% of the trans isomer **16a** (and 25% of the cis isomer).[39] Steric control by the substituent is therefore incomplete.

Lithium aluminum hydride (LAH) also brings about the conversion of ketoximes, and their *O*-tosylates, into aziridines stereoselectively and regio-selectively. It gives mainly the *cis*-2,3-substituted aziridines,[40,41] and does not suffer from further attack by a carbon nucleophile (i.e., the Grignard reagent). Syn and anti oximes have been separated, and the aziridines formed depend on the proximity of the OH group to the α-methylene group which is involved in the reaction.[41,42] A probable mechanism has been discussed;[41] the reaction being essentially the attack of the oximino nitrogen atom on the α-CH$_2$ group of the substituent. In order to avoid the problem of syn-anti isomerism, 2-oxazolines are reduced with LAH instead, and aziridines are thus formed stereospecifically (Equation 8). The yields, however, are variable and the products are sometimes contaminated

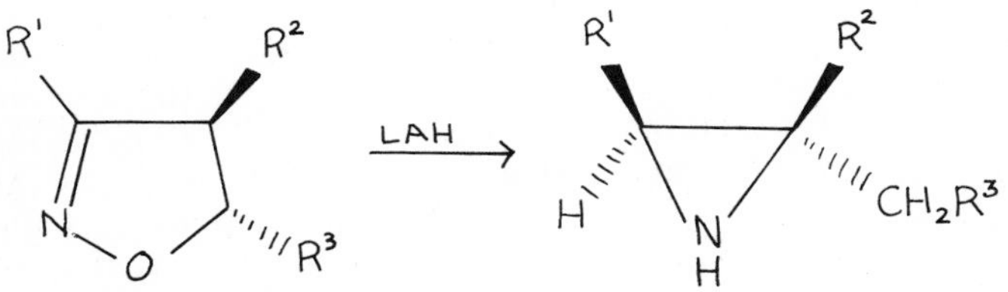

(8)

with amino alcohols formed by cleavage of the N–O bond after reduction of the C–N double bond.[43] cis(Z)- and trans(E)-Nitronic (methyl) esters condense with benzoylacetylene in a stereoselective manner to form the intermediate N-methoxy-4-isoxazoline which rearranges into 2-benzoyl-1-methoxyaziridine. The synthesis is under kinetic control and the Z- and E-nitronic esters yield predominantly the syn- and anti-2-benzoyl-1-methoxyaziridines, respectively.[43a]

   d.   From △²-1,2,3-Triazolines.   △²-1,2,3-Triazolines, readily obtained from the cycloaddition of azides onto olefins, yield aziridines spontaneously, by thermolysis or photolysis.[30,44] The elimination of nitrogen occurs with some retention of configuration at the carbon atoms. This selectivity is not necessarily high, varies with the substituent and depends on whether the reaction is induced thermally or photochemically.[30] In the photolysis of trans-4-methyl-1,5-diphenyl-△²-1,2,3-triazoline, the ratio of trans- to cis-2-methyl-1,3-diphenylaziridines is 66:22 whereas with the cis-triazoline the ratio is 17:65. Both reactions are unaffected by oxygen quenchers, and a 1,3-biradical intermediate was postulated. In the singlet biradical cyclization occurs prior to C–C bond rotation, but in the triplet biradical the molecule would have time to undergo C–C bond rotation to produce the diastereoisomers (Equation 9). Spin inversion in the triplet biradical is relatively slower than bond rotation.[45] There is no isomerization in the original triazolines.

(9)

   endo- and exo-△²-1,2,3-Triazolines are possible when the double bond of the olefin is in bicyclic systems such as norbornene (compare with 9). exo-Triazolines derived from bicyclo[2.2.1]hept-2-enes[30] and 7-oxabicyclo-[2.2.1]hept-2-enes[46] are found to yield the corresponding aziridines, with loss of N₂, on photolysis and thermolysis respectively without change in configuration at C-2 and C-3. The stereospecificity, however, may not always be high.[30] This reaction is particularly useful for preparing aziridines of required cis configuration as in Equation 10.[47] For further information on triazolines see Section III.I.

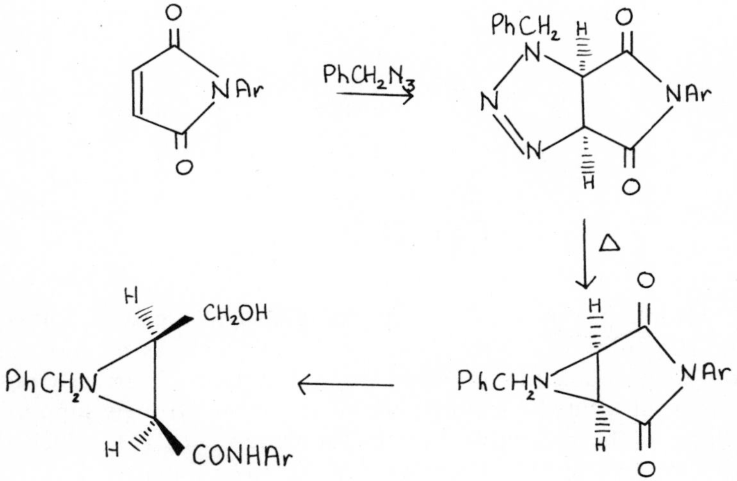

(10)

e. *By Intramolecular Rearrangement.* 4,5-Bis(methoxycarbonyl)-3,6-diphenyl-1-*p*-tosyl-1*H*-azepine is in equilibrium with 3% of the benzimine (**17**) in acetone solution. The stereochemistry follows from the transformation.[48] Such rearrangements sometimes require much energy as in the conversions of 3- and 4-methyl-(but not 2,7-dimethyl)1-ethoxycarbonyl-1*H*-azepines into 2- and 3-methyl-7-azabicyclo[4.1.0]hepta-2,4-diene (i.e., **18**) which require heating to 200°C.[49] 1-Azonine is transformed into 9-azabicyclo[6.1.0]nona-2,4,6-triene (**19**, R = H) by ultraviolet light.[50] The bicyclo compounds (**19**, R = CONMe₂, CO₂Me, or CO₂Et) are also obtained by heating the 4-azabicyclo[5.2.0]nona-2,5,8-trienes (**20**).[50,51]

A very interesting isomerization is the photolytic rearrangement of *N*-methylpyridinium hydroxides into 6-methyl-6-azabicyclo[3.1.0]hex-3-en-

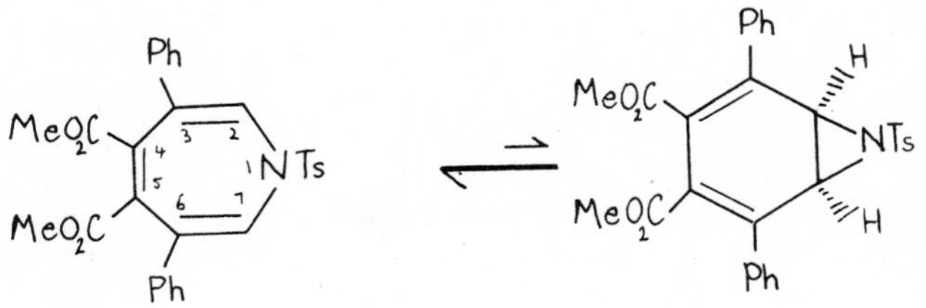

17

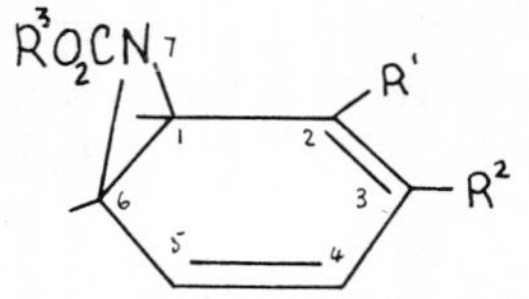

18

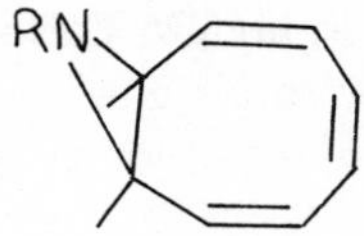

19

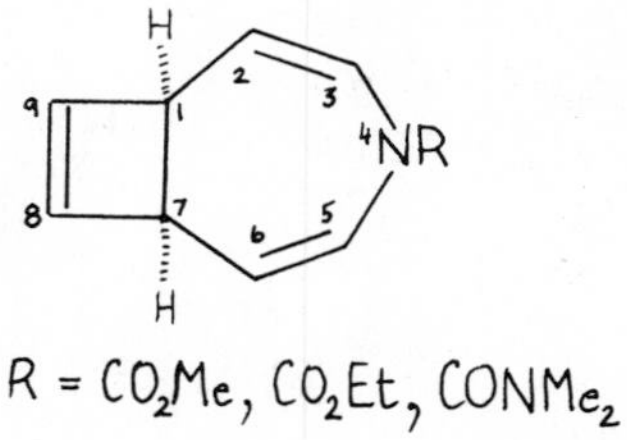

$$R = CO_2Me, \ CO_2Et, \ CONMe_2$$

20

2-*exo*-ol (**22**) in aqueous solutions containing a slight excess of potassium hydroxide (Equation 11). The respective 2-methyl ethers are formed in methanol. The alcohols (**22**) obtained from photolysis of variously mono- and dimethyl substituted N-methylpyridinium cations strongly support the intermediacy of the respective 1-methylazoniabulvalene (**21**) which then reacts with the solvent by the pathways indicated in Equation 11.[52]

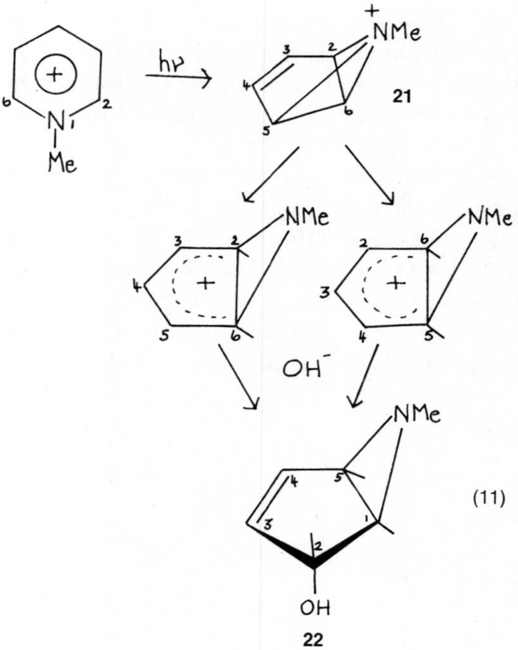

(11)

22

f.   *From Azirines.*   2-Methyl-3-phenyl-2*H*-azirine (**23**) dimerizes in a stereospecific manner on irradiation in benzene solution to a 2:1 mixture of 2-*exo*-6-*exo*- (**24**) and 2-*endo*-6-*exo*-dimethyl-4,5-diphenyl-1,3-diazabicyclo[3.1.0]hex-3-ene (**25**). Both hexenes are hydrolyzed to the same *trans*-3-benzoyl-*r*-2-methyl-*cis*-3-phenylaziridine (Equation 12).[53] In the dimeriza-

tion the azirine acts as a dipole and a 1,3-dipolarophile after ring opening. Other examples in which azirines act as dipoles are the reactions of 2-phenyl- and 2,3-diphenyl-2*H*-azirine with the 1,3-dipolar reagent derived from the iminochloride of benzoyl *p*-nitrobenzylamide and triethylamine to give bicyclohexenes related to **24** and **25**. The bridgehead phenyl group and the phenyl group in the aziridine ring of these products are cis oriented.[54] Compounds with active methylene groups, for example, dimedone, also add across the polar double bond of 3-phenyl-2*H*-azirine in boiling xylene to form the trans adducts (e.g., **26**).[54]

For reviews on the general syntheses and properties of aziridines see Fanta, and Dermer and Mum.[55]

2.   *Configuration of Aziridines*

Several examples of cis and trans aziridines were described in the previous section. Most of the structures have been deduced from the chemical shifts of substituents and from coupling constants of protons on C–2 and C–3. Before nuclear magnetic resonance spectroscopy (nmr) was available cis and trans isomers possessing similar substituents at C–2 and C–3 were distinguished by the absence (cis) or presence (trans) of optical activity and this necessitated the study of optically active diastereomers.[3,14] Cromwell and co-workers extensively examined a variety of *racemic* 2-acyl and 2-aroyl aziridines in an endeavor to find physical and chemical methods to distinguish between cis and trans isomers. They found uv and ir differences which showed in certain instances, but not always,[56] that the higher melting isomer was trans substituted.[57,58] These methods also gave some information regarding the conformation of the RCO– group. In the solid ground state, *cis*-2-aryl-3-aroylaziridines have a nonconjugated *gauche* conformation (27), but in CCl₄ solution they exist as mixtures of nonconjugated *gauche* and conjugated *cisoid* (28) rotational isomers. The trans

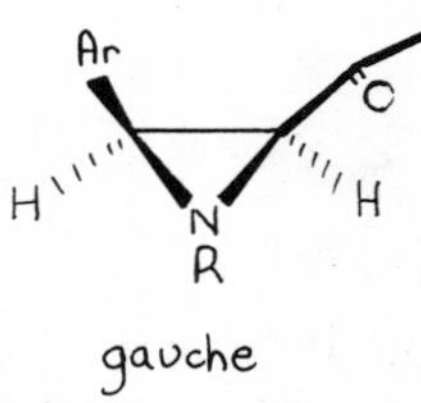

gauche
27

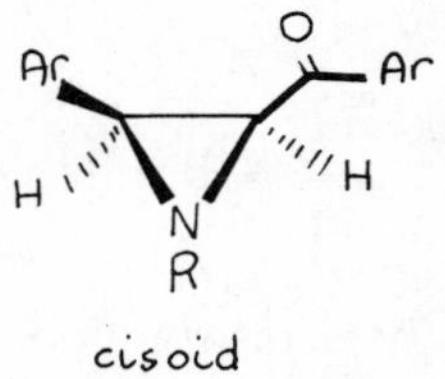

cisoid
28

isomers on the other hand, have a conjugated cisoid conformation in the solid state and in solution. The aziridine ring transmitted the electrical effects of the 2-phenyl substituent to the carbonyl group in the trans isomer, in the excited state, with consequent uv shifts, but *p*-substituents in the phenyl group did not appear to affect the conjugation further.[59] The reactions with phenyl hydrazine have been very useful in distinguishing between cis and trans acyl and aroyl aziridines. The cis isomers yield a pyrazole because of the ready trans amine elimination, but the trans isomers give a 2-pyrazoline (Equations 13 and 14). More recently nmr methods have been used to assign the configuration of aroylaziridines,[60] and the mass spectra of the isomers differ enough to distinguish between them.[61] For nmr spec-

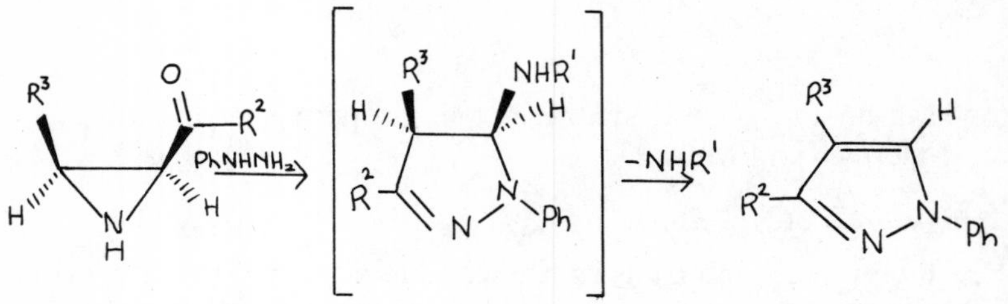

(13)

$$(14)$$

tral data of three-membered ring nitrogen heterocycles see Batterham's monograph.[61a]

The two carbon atoms of aziridines are potentially chiral and when unsymmetrically substituted should exist in four optical isomers—two cis and two trans. If the carbon atoms bear the same substituents then the cis isomer is meso and optically inactive because of internal compensation of optical rotation.

It had been realized that compounds of the type $NR^1R^2R^3$ with an asymmetric nitrogen atom should be capable of optical isomerism, but the activation energy between the isomers is too small for resolution because of rapid nitrogen inversion.[62] The activation energy in aziridines is higher than in acyclic amines, but attempts to prepare two optically active amides derived from 2,2-dimethylaziridine and camphorsulfonyl chloride or α-bromocamphorsulfonyl chloride failed.[63]

In 1968 Brois separated syn and anti isomers of *R*,*S*-2-methyl-1-chloro-aziridine (racemates of **29** and **30**) by gas chromatography and showed that they did not interconvert below 135°C.[64] He also found that the barrier of nitrogen inversion of *N*-chloro- and *N*-bromo-2,2-dimethylaziridines was the highest observed at the time for aziridines. Their coalescence temperatures were above 120°C and should therefore be resolvable into stable optical antipodes at room temperature.[65] At the same time, but independently, Felix and Eschenmoser obtained a 4:1 mixture of 7-*exo*-chloro- (**31**) and 7-*endo*-chloro-7-azabicyclo[4.1.0]heptane (**32**) by chlorination with sodium hypochlorite. They separated these by tlc and were able to distill them in a vacuum without isomerization.[66] In this case the nitrogen atom was not asymmetric but examples with a chiral nitrogen atom have been prepared.

Optically active *S*(−)-2-methylaziridine, derived from natural *S*-alanine,[67] was oxidized with sodium hypochlorite into a mixture of syn and

29   **30**

31   32

anti *N*-chloro derivatives. These two diastereomers were separated by glc into the syn isomer 1*R*,2*S*-1-chloro-2-methylaziridine $[\alpha]_D^{20}$ − 81° (in C_9H_{20}) (29) and the anti isomer 1*S*,2*S*-1-chloro-2-methylaziridine $[\alpha]_D^{20}$ + 94° (in C_9H_{20}) (30) (the nitrogen lone pair of electrons is taken as the smallest substituent). 1*S*,2*S*-1-Bromo-2-methyl-*(anti)*, 1*S*,2*S*-1-chloro-2-propyl-, and 1*S*,2*S*-1-bromo-2-propylaziridine were also described.[68] In examples where C–3 and C–2 are chiral, then a very slow inverting nitrogen atom is also chiral and eight (2^3) optical isomers are theoretically possible.

Optical resolution of aziridines by salt formation is not satisfactory because they are usually acid labile and tend to polymerize under these conditions. In order to circumvent this difficulty, Fujita, Imamura, and Nozaki[69] have prepared and resolved menthyl α-(iodomethyl)benzyl carbamate (obtained from styrene, iodocyanate, and (−)-menthol). The resolved ester gave *R*(−)- and *S*(+)-2-phenylaziridine on alkaline treatment. The absolute configuration was deduced by reduction, with lithium aluminum hydride, of the diastereomeric menthyl esters to *S*(−)- and *R*(+)-*N*-methyl-α-phenyl ethylamine of known absolute configuration.[69]

Ord and cd curves of *S*(−)-2-methylaziridine have been measured and compared with those of *S*(+)-2-methylazetidine and *R*(−)-2-methylpyrrolidine.[67] Also the *N*-halo derivatives were examined and a quadrant rule was postulated for the absolute configuration of the 2-substituent.[67,70] A similar rule was proposed for the *N*–CN chromophore.[67]

Trefonas and collaborators determined the three-dimensional structure of a number of aziridines by X-ray crystallography. Some of these included the cis fused 6-(*p*-bromobenzoyl)-6-azabicyclo[3.1.0]hexane,[71] 7-(*p*-iodobenzenesulfonyl) - 7 - azabicyclo[4.1.0]heptane,[72] 8,8 - dimethyl - 8 - azabicyclo-[5.1.0]octane iodide,[73] 13,13-dimethyl-13-azabicyclo[10.1.0]tridecane iodide (33),[74] and 6 - (*p*-iodobenzenesulfonyl) - 3 - oxa - 6 - azabicyclo[3.1.0]hexane (34).[75] The structure of the last named compound 34 has been confirmed[76] by nmr spectroscopy and is similar to that of the nitrogen analogue 3-(*p*-methoxyphenyl)-6-benzyl-3,6-diazabicyclo[3.1.0]hexane.[77] The cyclo-

33   34

hexane ring of 1-(*p*-iodobenzene sulfonyl)-1-azaspiro(2,5)octane was also shown by an X-ray study to be in an undistorted chair conformation.[78]

The X-ray structure of 1,3-bisadamantylaziridin-2-one is of particular interest. The adamantyl substituents are oriented trans to each other, and the configuration at the nitrogen atom is pyramidal. It infers that very little, if any, amide resonance is present in the molecule.[78a]

3. Conformation of Aziridines

Studies of the conformations of aziridines have been very fruitful partly because the changes in conformation are readily observable by variable temperature nmr spectroscopy (dynamic nmr or dnmr) and partly because of the refinements in the methods for obtaining reliable values of the coalescence temperatures and thermodynamic parameters from the spectra.[79-81] The conformational changes arise mainly from inversion of the nitrogen atom because the aziridine ring is planar and rigid. It is not the intention here to give a comprehensive survey of the nmr spectra and conformational analysis of aziridines, but only to point out briefly the effect of various substituents on the inversion barriers and on the preferred orientation of the *N*-substituent.

Nmr spin coupling between $^1H-C-^{14}N$ is observed in some aziridines and is found to be dependent on the conformation of the nitrogen lone pair of electrons. This information shows that in 1-amino-2-phenylaziridine the anti conformation 35 is preferred.[82] Proton coupling constants and benzene induced chemical shifts in 2-(1-naphthyl)-aziridine indicate that it is much more populated with the anti isomer 36 even though rapid N–H exchange is taking place. Ohtsuru and Tori[83] were thus able to determine the three $^1H-C-^{15}N$ coupling constants and showed that they were between *J*, $^{15}N{=}CH$ and *J*, $^{15}N{-}CH$ values suggesting some π bond character for the N–C bond, and this sheds light on the mechanism of inversion. While working with 2,2- and *trans*-2,3-diphenylaziridines, Martino and collaborators[84] devised conditions in which N—H exchange was excluded, and they were able to observe *H*—C—N*H* coupling constants after coalescence. The coalescence temperatures were 73 and 32°C for the diphenylaziridines respectively which revealed the influence of the C-phenyl substituent on the inversion barrier. It was also shown by nmr that in *cis*-2,3-diphenylaziridines the anti invertomers 37 were predominant, and it was confirmed by an ir study.[85] Some interesting examples discovered by Roberts and co-workers[86-88] are 1-methyl- and 1-ethyl-2-ethylideneaziridines because their inversion rates are much faster than those of the corresponding saturated derivatives. The zwitterionic intermediate 38 was postulated to account for the low barrier. Similarly the coalescence temperature of 2,3-dichloro-1-phthalimidoaziridine was below 0°C, as compared with several other 2- and 3-substituted (with Me, Ph, or CO_2Me) 1-phthalimidoaziridines which were in the range 75–125°C. The higher inversion rate in the chloro compound was attributed to a structure such as 39.[89] Cromwell and collaborators[90] found that in several 1-alkyl-2-aryl-3-benzoylaziridines and methyl 1-alkyl-2-arylaziridine-3-carboxylates the cis isomers displayed slight temperature dependent nmr spectra, whereas the trans isomers showed large changes in the same temperature range (70 to −40°C). The reason given for these results was that the trans isomers existed predominantly in a conformation in which the N-substituent was syn to the carbonyl group.

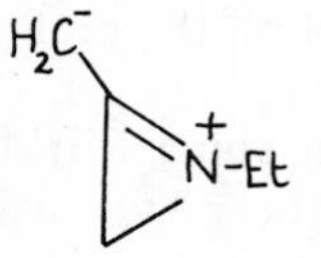

38

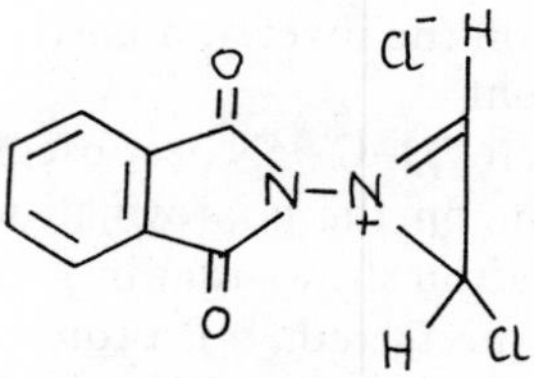

39

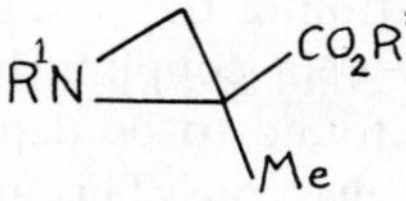

40

In a series of 2-alkoxycarbonyl-2-methylaziridines derived from 40 with R^1 = phthalimido, 2-methylquinazolin-4-on-3-yl, benzoxazolid-2-on-3-yl, and quinol-2-on-1-yl the ratios of syn to anti invertomers favored increasing amounts of the syn form on varying R^2 from methyl to t-butyl. The ratio of syn to anti was 3.5 for the ester 40 (R^1 = phthalimido, $R^2 = t$ = butyl) and shows an unexpected attraction between the heterocyclic substituent R^1 and the increased size of the ester alkyl group.[91] In N-alkyl or aryl aziridine carboxylic esters the equilibria lie toward the invertomer that is predicted from steric interactions (e.g., trans-1-phenyl-2-ethoxycarbonylaziridine [41] and r-2-alkyl-trans-1-phenyl-cis-3-ethoxycarbonylaziridine [42]).[92]

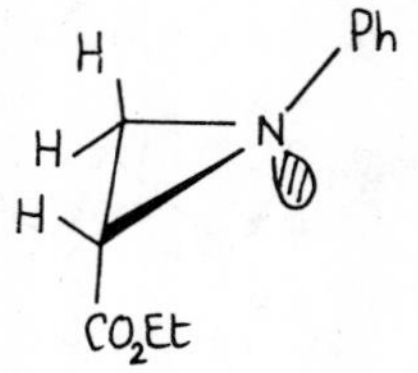

41

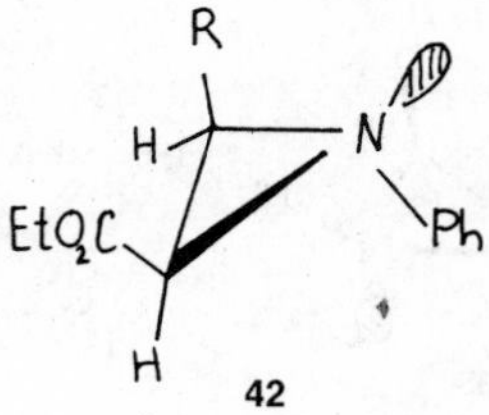

42

The effect of the solvent on inversion barriers is critical. Thus inversion rates are decreased in hydroxylic solvents due to stabilization of the separate invertomers by H-bonding with the nitrogen atom.[88] The ratio of invertomers is also affected by temperature and the addition of polar compounds to the solvent. The equilibrium for *N*-methyl *trans*-2-isopropyl-3-methylaziridine, **43** ⇌ **44**, is 4.1 (at −60°C) and 3.5 (at 20°C) in favor of **43** in cyclopentane[93] (and a somewhat similar ratio in CDCl₃).[94] When

phenol is added to a cyclopentane solution containing the aziridine, the equilibrium is shifted towards the thermodynamically less stable isomer, that is, 2.1 (at −60°C) and 3.2 (at 20°C) in favor of invertomer **43**. The shift in equilibrium is larger at lower temperatures and when bulky groups are substituted in the phenol. Ir spectroscopy indicates strong hydrogen bonding. The change in equilibria is attributed to selective hydrogen bonding of the phenol with the less stable invertomer because it has more available space around the nitrogen atom. Compare the space around the nitrogen atom in **44** with that in **43**.[93]

An empirical scheme for correlating inversion barriers at the nitrogen atom with the ground state geometries has been devised and the inversion energy barriers were plotted against the out-of-plane angle θ in the pyramidal ground state or the valence bond angle ϕ. It was found that the energy barrier increases steeply as the angle θ approaches 60°. Thus the energy increases rapidly from dimethylamine to aziridine, fluoramine (H_2NF), oxaziridine, and 2-azirine.[95]

The free energy of activation of aziridine and (N—²H)-aziridine were compared in the gas phase by FT (fourier transform) nmr spectroscopy, and the signals from the *N*-deuterio derivative had a higher coalescence temperature (79°C, $\Delta G_c^{\ddagger} = 74.9$ kJ mol⁻¹) compared with the *N*-protio-aziridine (68°C, 72.4 kJ mol⁻¹), that is, a higher free energy of activation.[96] Similar results were obtained for 2,2,3,3-tetramethylaziridine and its *N*-deuterio derivative in CCl₄ solution.[97] The effect of increasing the bulk of the *N*-alkyl substituent is to increase the inversion rates.[93,98] These can be deduced by ¹H nmr spectroscopy from the chemical shifts of the protons on C–2 and C–3. In the case of *cis*-2-methyl-3-phenylaziridine[99] and aziridine,[100] the C–2 and C–3 protons syn to the magnetically anisotropic

N-alkyl group are shielded. This shielding decreases in the order *N*-Me > *N*-Et > *N*-Pr^i ≫ *N*-Bu^t, and is probably caused by van der Waals (dispersion) interactions between the protons and the *N*-alkyl groups.[99] In *N*-(2-cyanoethyl) and *N*-(2-methoxycarbonylethyl) aziridines the syn and anti hydrogens on C–2 and C–3 were assigned to the high-field and low-field signals respectively.[101]

Baret, Pierre, and Perraud[101a] examined the inter- and intramolecular hydrogen bonding in *N*-alkyl 2-hydroxymethylaziridine by ir spectroscopy. They found that this data can give valuable information about the conformation of the alkyl groups on the nitrogen atom. Alkyl groups anti with respect to the CH$_2$OH exhibit a preponderance of the intramolecular hydrogen bonded species.

Generally, an increase in the size and number of alkyl groups on C–2, C–3, and N–1 decreases the free energy between the invertomers, and the introduction of an aryl group decreases this energy even further.[102] A plot of the Hammett substituent constant against inversion barrier in *N*-substituted phenyl aziridines exhibited a linear correlation indicating that the lone pair of electrons on the inverting nitrogen atom are conjugated with the π system of the aryl group.[103] The rates of N-inversion in *N*-acylaziridines are much higher (i.e., lower inversion barrier) than in 1-alkyl and 1-aryl derivatives, and follow the order 1-acetyl > 1-benzenesulfonyl > 1-(2,4-dinitrobenzenesulfenyl) > 1-benzenesulfinyl aziridines.[104,105]

The spectra of *N*-acyl-aziridines can be further complicated by rotational barriers about the N–CO bond.[105] The barriers for N-inversion in N-SR compounds have also been measured and are greater than those in the corresponding four-membered rings (azetidines).[106,107] Some *N*-phosphorus compounds have also been studied,[108] and in the case of 2,2-dimethyl-1-triphenylphosphinoaziridine an inversion barrier could not be observed invoking a steric deceleration because of the torsion about the P–N bond.[109] An interesting example is diethyl 2-aziridinylphosphonate (**45**) in which intramolecular hydrogen bonding is the cause for slow N-inversion.[110]

N-Halogen substitution considerably raises the barrier for N-inversion[111,112]—to the extent that in some examples the invertomers have been isolated and characterized (Chapter 2, Section I.A.2).[64,66] This marked effect is more pronounced in three-membered than in four-membered[111,112] and five-membered rings.[111,113] The most stable compound observed to date is 1-fluoro-2,2-tris(trifluoromethyl)aziridine (**46**) which is unchanged when heated up to 190°C.[114]

The electron withdrawing nitrogen and oxygen atoms also raise the N-inversion barriers as has been observed in *N*-amino, *N*-phthalimido-,[27,91,115] 1-(benzoxazol-2-on-3-yl)-,[116] and *N*-alkoxy-aziridine.[28] 1-(Benzoxazol-2-on-3-yl)-*cis*- and *trans*-2,3-dimethylaziridines are of par-

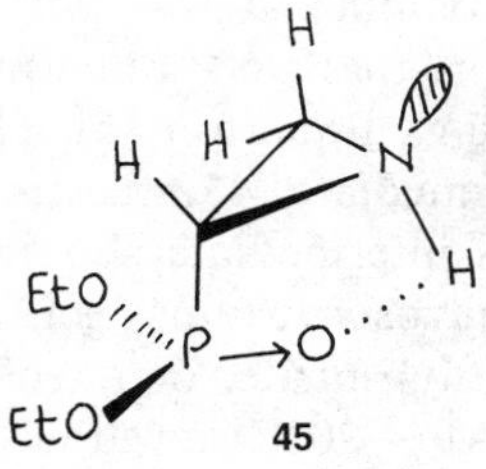

45

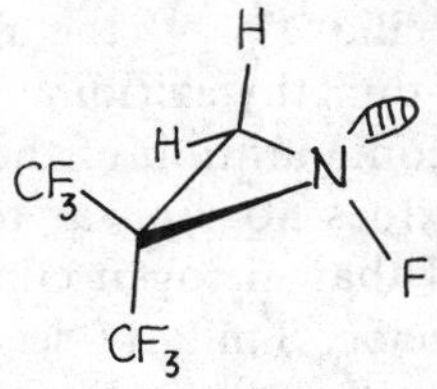

46

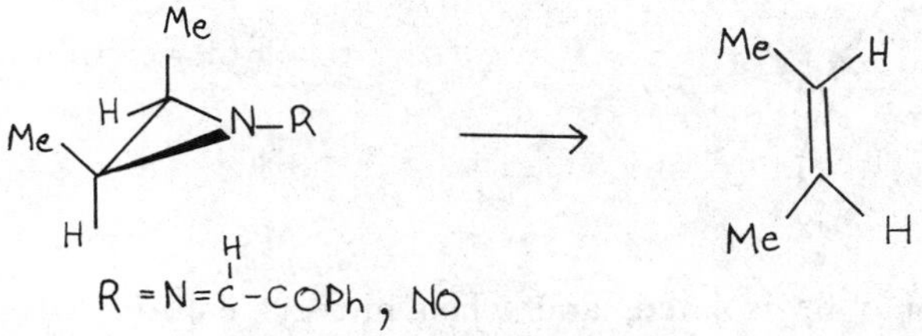

(15)

$$R = N = \overset{H}{\underset{|}{C}} - COPh \, , \; NO$$

ticular interest because the trans isomer exhibits a coalescence temperature around 140°C, but the cis isomer shows no change in the nmr spectrum at 180°C.[116] Some *N*-alkoxyaziridines are so stable that the invertomers have been separated by glc (see Chapter 2, Section I.A.1).[28] This increased stability is also conferred on the aziridine when the nitrogen or oxygen atoms are part of the ring as in diaziridine[102] and oxaziridine.[117]

For tables of activation parameters and for the mechanism of pyramidal atomic inversion the reader is referred to the excellent review by Lambert[2] and references therein.

4.   Reactions of Aziridines

a.   *Nitrogen Extrusion Reactions.*   The removal of the aziridine nitrogen atom is achieved in a variety of ways. Stereospecific deamination of *cis*-1-benzyl and *cis*-1-cyclohexyl-2,3-dibenzoyl aziridine was brought about by *m*-chloroperbenzoic acid in the dark. *cis*-Aziridines and *trans*-aziridines gave *cis*- and *trans*-olefins in 85–90% yields, and the *N*-alkyl groups were transformed to the respective alkyl nitroso dimers. An unstable *N*-oxide intermediate was postulated.[118] A similar reaction with *cis*-1,2,3-triphenyl-aziridine, however, was not stereospecific and the products were *cis*- and *trans*-stilbene, *cis*- and *trans*-stilbene oxide, nitrosobenzene, and benzalde-hyde.[118,119] Benzaldehyde was probably formed from an unstable 2,3,4-triphenyloxazetane which would result from the rearrangements of the nitroxide.[119]

Thermal fragmentation of the phenylglyoxal hydrazones of the diastereomeric 1-amino-2,3-methyl- and 1-amino-2,3-diphenylaziridines were highly stereospecific,[120] and the reaction of difluoroamine at 0°C with *cis*- or *trans*-2,3-dimethylaziridine proceeded with more than 96% retention of the relative configuration of the alkyl groups[121] (e.g., Equation 15). The latter reaction does not appear to involve the intermediate, **47**, because theory predicted that nitrogen elimination from this intermediate should not be stereospecific. Similarly decomposition of *N*-nitrosoaziridines generated by reaction with 3-nitro-*N*-nitrosocarbazole, methyl nitrite, or nitrosyl chloride were better than 99% stereoselective. At −20°C a yellow nitroso-

$$PhCH = CHCOPh$$

48

47

aziridine could be isolated, and when nitrosyl chloride was used some aziridine hydrochloride (stable at −78°C) was also isolated.[122] *n*-Butyl nitrite, together with a little triethylamine, similarly brought about this reaction in high yields and stereospecificity (e.g., Equation 15).[40]

Photolysis of *trans*-1-cyclohexyl-2-benzoyl-3-phenylaziridine in ethanol was not as clean as the above reactions and produced the *cis*- and *trans*-chalcones **48** and cyclohexylhydroxylamine. The *cis*-aziridine decomposed in a different way yielding *N*-benzalcyclohexylamine, acetophenone, and benzaldehyde.[123,124] Triplet $n \rightarrow \pi^*$ excited states were postulated for the phototransfer in both these reactions.[124]

Carbenes generated in a low intensity carbon arc at high vacuum reacted with *cis*-2,3-dimethylaziridine to form *cis*- and *trans*-2-butene (48:52) in low yield (8%). The loss of stereochemical identity was explained by the reaction scheme in Equation 16.[125]

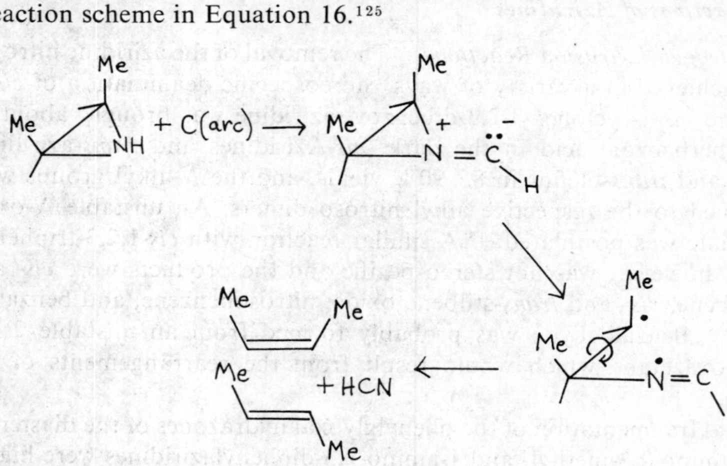

(16)

b. *C–N Bond Cleavage Reactions.* Hydrolytic cleavage of the aziridine ring takes place between N–1 and C–2, or N–1 and C–3 depending on the substituent. The nucleophile attacks the carbon atom involved in the cleavage usually with inversion of configuration[5,10,126,127] and retention of configuration at the carbon atom bearing the amino group (Equation 17). The

specificity is very high when the reaction is clearly S_N2. The situation may be more complicated if cleavage occurs at either of the two C–N bonds simultaneously, or when an S_N1 process is favored over an S_N2 process as in the case of 1-substituted 2-aroyl-3-arylaziridines.[128] Dimethyl sulfoxide is clearly not a good solvent to use in these reactions because it attacks the aziridine ring. For example 1-ethoxycarbonyl-2-methyl-3-phenylaziridine is cleaved exclusively at the 1,3- bond in the trans-isomer (Equation 18), but it occurs along the 1,3- and 1,2- bonds (1:1) in the cis isomer.[129] Bond cleavage in 2,3-dialkylaziridines is caused by carbon disulfide and is fol-

lowed by ring closure to 4,5-dialkylthiazolidin-2-thiones. These have a geometry opposite to that of the aziridines, but whereas the *cis*-aziridines react stereospecifically, the trans isomers are only stereoselective.[130] In an analogous reaction, thiocyanic acid yields the respective 2-aminothiazolidine with inversion of configuration, but it is 100% stereospecific with both isomers.[131] 1-Azoniatricyclo[4.4.1.0^{1,6}]undecane perchlorate (**49**) and related compounds reacted with methyl nitrile to yield 11-methyl-12-aza-1-azoniatricyclo[4.4.3.0^{1,6}]tridec-11-ene perchlorate (**50**). 5,5-Diethyl-$\triangle^1$-pyrrolidine 1-oxide also reacted by insertion and produced 5,5-dimethyl-6-aza-7-oxa-1-azoniatetracyclo[7.4.4.0^{1,9},0^{2,6}] heptadecane perchlorate (**51**).[132]

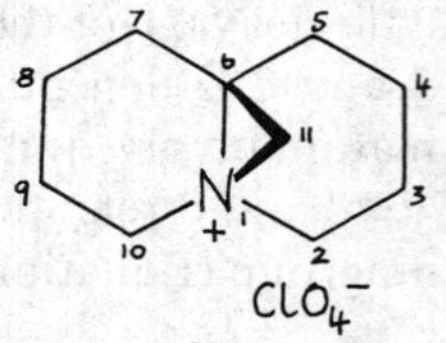

49

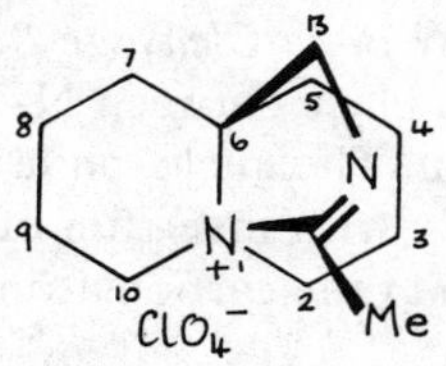

50

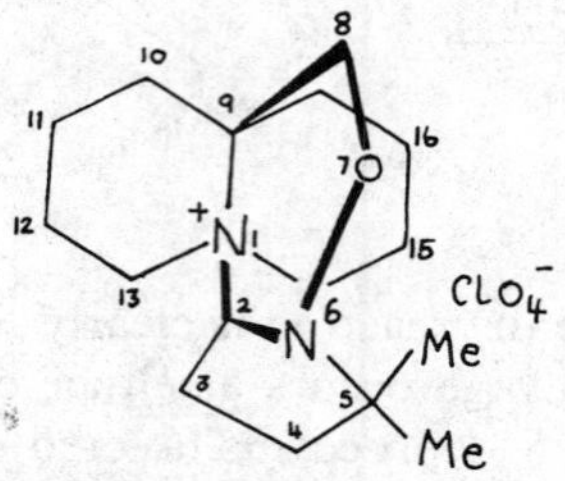

51

These reactions most probably proceed with inversion of configuration at the aziridine carbon atom (e.g., C–11 in **49**) because they are similar to the reaction of *cis*-7-ethoxycarbonyl-7-azabicyclo[4.1.0]heptane with methyl nitrile in the presence of $BF_3 \cdot Et_2O$ which yields 1-ethoxycarbonyl-2-methyl-*trans*-4,5-tetramethylene imidazol-2-ine as an intermediate.[133]

Catalytic hydrogenolysis of the C–N bond in aziridine does not always take place with high stereospecificity and depends on the nature of the catalyst.[11,134] Hydrogenation of 2-methyl-2-phenylaziridine yields mainly 2-phenylpropylamine, but small amounts of 1-phenylpropylamine may also be formed. 2-Phenylpropylamine obtained by hydrogenating optically active 2-methyl-2-phenylaziridine with $Pd(OH)_2$, PtO_2, Pt black, Raney nickel, and Raney cobalt exhibited 72% inversion and 95, 73, 23, and 14% retention of configuration with the respective catalysts.[134] In the hydrogenolysis of alkylaziridines over Raney nickel, on the other hand, the less hindered C–N bond is cleaved with retention of configuration at both carbon atoms to yield alkylethylamines.[134a]

Aziridine *N*-oxides displayed a dual pathway of decomposition. Normally nitrogen is extruded as in the decomposition to form ethylenes already mentioned (Chapter 2, Section I.A.4.a). If the geometry is favorable, however, as in 1-*t*-butyl-2-methylaziridine 1-oxide (prepared by oxidation with ozone at −75°C) in which the *transoid* structure **52** is preferred, 1,2- bond cleavage readily occurs and the hydroxylamine **53** is obtained.[135] Pyrolysis of *trans*-13-*p*-nitrobenzoyl-13-azabicyclo[10.1.0]tridecane (**54**) in boiling toluene gives a 2:1 mixture of *trans*- and *cis*-N-(2-cy-

52

53

54

clodododecenyl)-*p*-nitrobenzamide, but the cis isomer in boiling xylene gives a 7:1 mixture of *trans-N*-(2-cyclododecenyl)-*p*-nitrobenzamide and *trans*-2-*p*-nitrophenylcyclododecane[4,5-*d*]oxazoline.[136] In boiling toluene, the simpler 2-benzyl-1-(*p*-nitrobenzoyl)aziridine cleaved in a similar manner to *trans*-3-*p*-nitrobenzamido-1-phenylprop-1-ene.[137]

    c.   *C–C Bond Cleavage Reactions.*   Azomethine ylides are formed when aziridines undergo photolytic or thermolytic C–2, C–3 bond cleavage. The cleavage is disrotatory for photolysis and conrotatory for thermolysis—in agreement with the Woodward-Hoffmann orbital symmetry rules.[138] The kinetics of some of these cleavages have been studied in detail mainly by Huisgen and his school and were found to be first order and reversible. Internal rotation of the ylide is possible and cis-trans equilibration may occur. The product composition from reaction with a dipolarophile depends on the relative rates of cycloaddition compared with cis-trans equilibration.[139] The activation entropies for the recyclization of the ylides derived from *cis*- and *trans*-1-aryl-2,3-bis(methoxycarbonyl)aziridine produced by flash photolysis were -134 kJ mol^{-1} and -88 kJ mol^{-1} inferring that the transition states are cyclic.[140] The equilibria for thermal and photolytic cleavages are shown in Scheme 1. A variety of aziridines were successfully used in such cycloadditions and include *N*-aryl-2,3-bis(alkoxycarbonyl)-,[139–143] *N*-alkyl- or *N*-aryl-,[144] 2-alkoxycarbonyl-3-aryl-,[144,145] 2-aroyl-1,3-diaryl-,[146,147] and 1,2,3-triphenylaziridines.[148] Among the dipolarophiles used were olefins such as maleic and fumaric esters,[140,145,146,148] acetylene carboxylic esters,[141,144] and even naphthalenes, anthracene, phenanthrene,[142] benzaldehyde,[143] and diphenylcyclopropenone.[149] The synthetic potential of these reactions is enormous because reduced nitrogen and mixed heterocycles with centers of asymmetry can be prepared. An unsym-

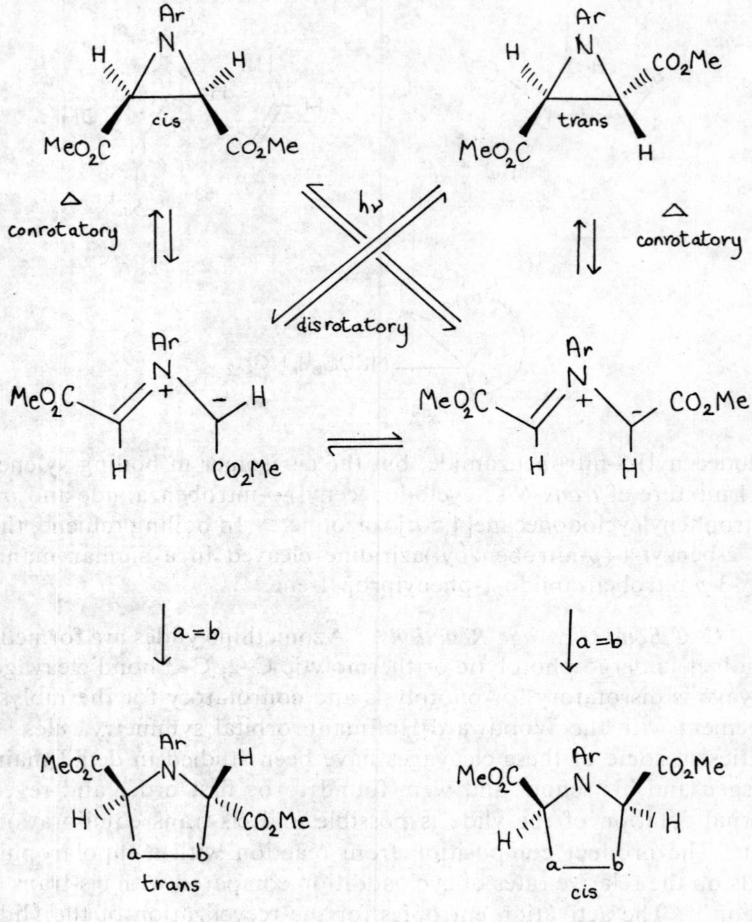

Scheme 1

metrical polarophile can add to the ylide in two ways. The same product is obtained if the substituents on C_1 and C_2 of the aziridines are the same, but if they are different the number of possibilities are doubled (e.g., **55** and **56**. 1,3-Dipolar cycloadditions are also possible when the aziridine is fused to another ring as in the dicarboximide **57** which can be photolytically induced to undergo cycloaddition to dimethyl acetylene dicarboxylate and yield the 1:1 adduct **58**.[150] The reader is referred to Huisgen's articles for further examples.[151]

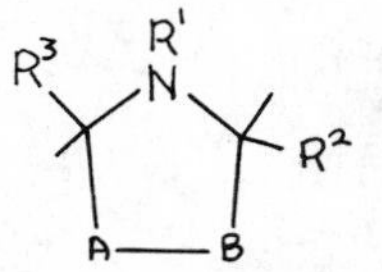

55

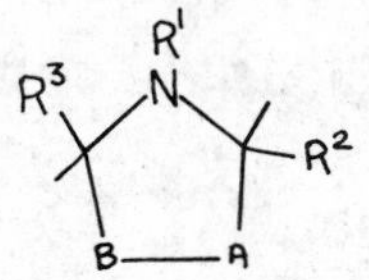

56

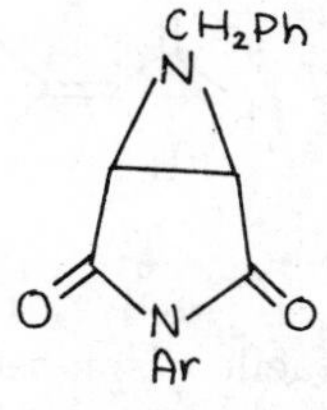

57

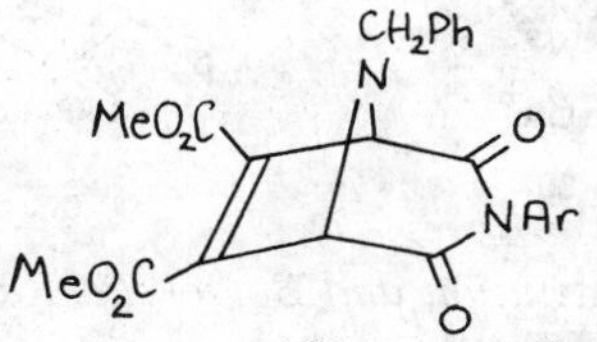

58

In the absence of a dipolarophile the aziridine may undergo isomerization. *cis*-1,2,3-Triphenylaziridine, for example, when heated at 150°C equilibrates to a 22:78 trans-cis mixture.[152] Under uv irradiation, equilibration is accompanied by C–N bond cleavage, the extent of which depends on the solvent used. The position of equilibria in 1-cyano-, 1-ethoxycarbonyl-, and 1-*N,N*-dimethylcarbamoyl-2,3-diphenylaziridines depends on the *N*-substituent, but favors the cis isomer in all cases. A singlet rather than a triplet state was said to be involved and the isomeric preponderance may be due to stronger π interaction in the cis form (Equation 19).[153] Equilibration of *cis*- and *trans*-1-alkyl-2-benzoyl-3-phenyl- and 1-alkyl-2,3-dibenzoylaziridines by base catalysis was studied in detail and depends markedly

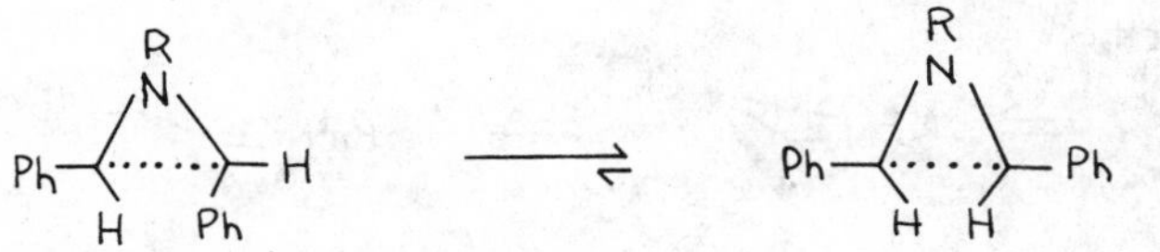

(19)

on the solvent used. It does not appear to involve C–C cleavage and is probably brought about by exchange of the benzylic H or enolization of the carbonyl group.[154]

Although *trans*-1,3-dibenzoyl-2-phenylaziridine rearranges thermally to 1,4-diphenyl-4-benzoyloxy-2-azabutadiene (**59**) clearly by way of a C–C bond cleavage, a reactive intermediate could not be trapped with a dipolarphile.[155] In the photolysis of *trans*-N-*t*-butyl-2-benzoyl-3-phenylaziridine (**60**), 2,5-diphenyloxazole and β-*t*-butylamino-*trans*-benzalacetophenone (**61**) are formed indicating that both C–C and C–N bond cleavage occurred. The cis isomer, on the other hand, gave only the oxazole.[156]

$$\underset{\underset{\underset{Ph}{|}}{Ph\ \overset{\overset{O}{\|}}{C}O\ C}}{=CH-N=CHPh}$$

59

60   **61**

d.   Substitution and Solvolytic Reactions.   The facile displacement of a halogen on a carbon atom in aziridines first demonstrated by Deyrup and Greenwald,[35] was shown to proceed with inversion of configuration. Treatment of 2,2-dichloro-1,3-diphenylaziridine with methyl lithium yields 1-phenyl-*r*-2-chloro-2-methyl-*cis*-3-phenylaziridine in which the second halogen atom is displaced by a phenylthio or cyano group with the respective nucleophiles, and is reduced with LAH to *cis*-3-methyl-1,2-diphenylaziridine (Scheme 2). A mechanism invoking an aziridinyl cation, and not a ring opened structure, as intermediate is proposed. The equilibrium between 1-*t*-butyl-2-aziridinylmethylchloride and 1-*t*-butyl-3-chloroazetidine is solvent dependent and makes the halogen particularly unreactive, but it will undergo substitution under vigorous conditions. The equilibrium involves a tricyclic intermediate (Equation 20).[157] Further related examples

are in Chapter 2, Section II.A.3. 1-*t*-Butyl-2-aziridinylmethanol reacts with thionyl chloride to give mainly a 2:3 mixture of *syn*- and *anti*-3-chloromethyl-1,2,3-oxathiazolidine S-oxides in which the configuration differs at the sulfur atom. The complete mechanism of this reaction is still not clear.[158]

The kinetics of solvolysis of *N*-chloroaziridines was studied in some detail. The relative rates in water at 60°C for 1-chloro-, *syn*-1-chloro-2-methyl-, *anti*-1-chloro-2-methyl-, 1-chloro-*trans*-2,3-dimethyl-, 1-chloro-2,2-dimethyl-, and *anti*-1-chloro-*cis*-2,3-dimethyl-(in MeOH) were 1, 15, 210, 1500, 1800, and 150,000 respectively.[159,160] The nitrogen atom is extruded and two carbonyl fragments are formed. Semiempirical calculations are in agreement with the solvolytic results and the intermediacy of a linear 2-aza-alkyl cation (Equation 21).[159,161]

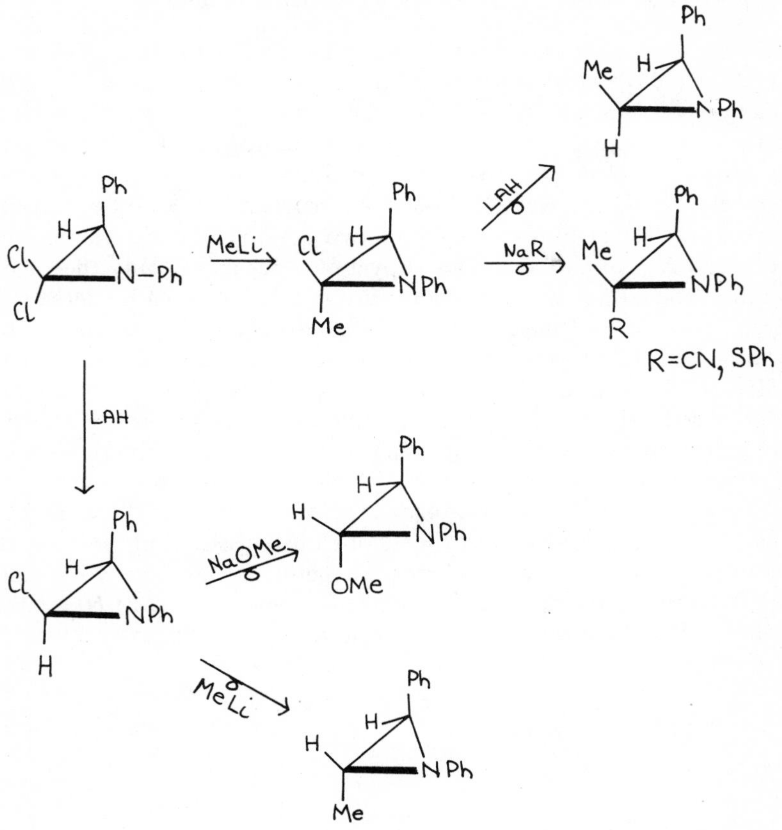

Scheme 2

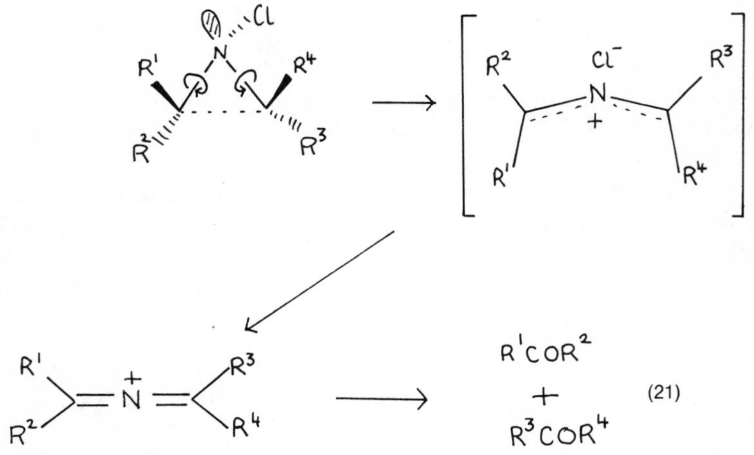

$$R^1COR^2$$
$$+ \qquad (21)$$
$$R^3COR^4$$

e. *Reaction with Asymmetric Induction at 2-α-Carbonyl Groups.* In trying to find ways of distinguishing between *cis*- and *trans*-2-aroyl-1,3-diphenylaziridines prior to the nmr era, Cromwell and co-workers[162] studied the reactions towards organometallic compounds and found some differences in stereoselectivity between them. Later they studied the reduction of *cis*- and *trans*-1-cyclohexyl-2-phenyl-3-aroylaziridines and observed that whereas NaBH$_4$ gave mixtures of the two possible diastereomeric racemic alcohols, LAH or lithium diisopropylamide furnished only the racemate resulting from attack on the carbonyl group from the least hindered side.[163] Deyrup and Moyer[164] also found that the reduction of 2-benzoyl-1-*t*-butyl-aziridine with LAH gave a 94:6 ratio of diastereomeric alcohols, but with NaBH$_4$ a 30:70 ratio was obtained. These ratios were dependent on the solvent used. Pierre, Handel, and Baret[165] further confirmed the higher stereoselectivity of LAH in obtaining only one diastereomeric alcohol by reducing 2-acetyl-1-*t*-butylaziridine with LAH and the other diastereomeric alcohol by reacting 1-*t*-butyl-2-formylaziridine with methyl lithium (Equations 22 and 22a). Bulky substituents obviously assisted in the stereodirection of the reaction because whereas reduction of *cis*-2-acetyl-3-*t*-butyl-1-methylaziridine gave one alcohol, its trans isomer gave a 40:60 mixture of alcohols. The stereoselectivity is sensitive to temperature, solvent, and

reagent. It is apparently assisted by partial coordination of the metal lithium with the carbonyl oxygen and the nitrogen lone pair; the *t*-butyl group assumed a preferred anti conformation.[166] Reduction of *t*-butyl-2-formylaziridine with LAD (and LiI) gave a 75:25 mixture of monodeuterated alcohols **62** and **63,** and in *t*-butyl-2-(^{2}H$_1$)-formylaziridine with LAH (and LiI) the ratio of **62** to **63** was 25:75. Lithium iodide was added to the reaction mixtures because it optimized the stereoselectivity.[167] Similar ster-

62

63

eoselectivity was observed in the reduction of 2-acetylaziridine with $NaBH_4$ in the presence of Na^+ and Li^+ ions.[167a] The asymmetric induction recorded in the syntheses of several 2-aziridine methanols from methylalkyl and methylaryl-α-diketone monooximes and Grignard reagents (Hoch-Campbell synthesis) may well involve a mechanism in which metal coordination between the oxygen and nitrogen atoms assists in the asymmetric induction as in the above reactions.[168]

f.   *N-Alkylation Reactions.*   cis- and trans-1-Methyl- or ethyl-2,3-dimethylaziridines gave methiodides which polymerized on attempted purification. In order to show that a quaternized aziridine was formed and not a dimeric compound (e.g., a piperazine), optically active *trans*-1,2,3-trimethylaziridine was methylated. The product was optically active ($[\alpha]_D^{22}$ + 31.5°) and was the quaternized aziridine because it reacted with ammonia to give an optically active 2-amino-3-dimethylaminobutane and optically inactive *meso*-2,3-bis(dimethylamino)butane.[169] Bottini and co-workers[170] examined the methylation, with CD_3I, of cis and trans (with respect to C–3 and C–2) 1,2-dimethyl-3-isopropylaziridines in which the ratio of syn and anti (with respect to N–Me and C–Pr^i) is 1:99 and 20:80 respectively. The methiodides in both cases were a 1:1 mixture of syn and anti trideuteriomethyl isomers. It was suspected that the methiodides might have equilibrated after the initial alkylation. Indeed, this is the case because when 50% of the reaction of the 2,3-*trans* isomer with CD_3I had proceeded, the ratio of syn and anti, with respect to the isopropyl group, trideuteromethylated salts was 2.4:1, but at 97% completion of the reaction the ratio was 1:1. Equilibration was catalyzed by iodide ions and a reasonable mechanism was proposed (Equation 23). In the case of the benzyl- and p-methyl-,

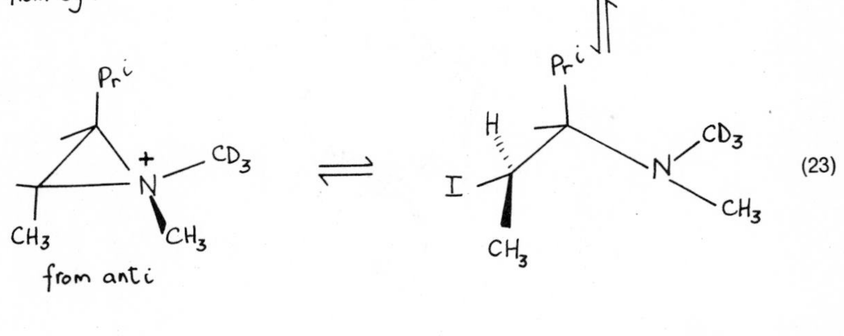

(23)

m-chloro- and *p*-nitrobenzyl quaternary salts of *trans*-1,2-dimethyl-3-iso-propylaziridine the anti to syn (benzyl, isopropyl) ratio is 1.69 for all four diastereomeric pairs.[171] In the alkylation with trideuteromethyl benzene-sulfonate in methanol and benzene the relative rates of quaternization of syn and anti aziridines were obtained. This was possible because the ben-zenesulfonate ion, being a poorer nucleophile than the iodide ion, did not catalyze the equilibration of the quaternary salts. The ratio of the quater-nary salts from syn and anti aziridines obtained in this way is a fairly good estimate of the conformational equilibrium in the parent aziridine.[172]

g.   *Intramolecular Condensations Involving a 2-α-Carbonyl Group and the Ring N-Atom.*   Heine and collaborators[173] showed that *trans*-2-aroyl-3-arylaziridines condensed with ketones in the presence of ammonia to give the monomeric 2,2-disubstituted 4,6-diaryl-1,3-diazabicyclo[3.1.0]hex-3-enes (**64**). The reaction is reversed by acid and bicyclo compounds are photochromic (Equation 24). Irradiation caused a C–C bond cleavage to form the colored azomethine ylide intermediate which could be trapped by dipolarophiles such as tetracyanoethylene[174] or dimethyl acetylene dicarboxylate.[175] Padwa and co-workers[176] isolated and separated *endo-* (64 $R^1 = Ar^1 = Ar^2 = Ph$, $R^2 = H$) and *exo*-2,6,4-triphenyl-1,3-diazabicyclo[3.1.0]-hex-3-enes (64, $R^2 = Ar^1 = Ar^2 = Ph$, $R^1 = H$) from the reaction of *trans*-2-benzoyl-3-phenylaziridine, benzaldehyde, and ammonia, and characterized them by the NOE (Nuclear Overhauser Effect) between the two endo protons on C–2 and C–6 in **64** ($R^2 = Ar^1 = Ar^2 = Ph, R^1 = H$). 2,6,8-Triphenyl-1,5-diazabicyclo[5.1.0]octa-3,5-diene (**65**) was obtained in 25% yield[177] when benzaldehyde was replaced by cinnamaldehyde. Photolytic C–C cleavage of these bicyclo compounds, as shown by trapping experiments, implied that it occurred by a conrotatory motion, whereas a disrotatory motion is predicted for the excited state from orbital symmetry considerations (Equation 24).[178] A conrotatory movement for the aziridine is a symmetry-allowed process in the ground state.[175] Photolysis of the bicyclohexenes **64**[176,179] and the octane **65**[177] in the absence of dipolarophiles gave a variety of phenylpyrazines.

h. *Ring Expansion Reactions.* The interconversion of 2-aziridinecarboxylates to azetidinones (β-lactams) by a variety of carboxylate activating groups proceeds stereospecifically (Equation 25). The intermediate 1-azabicyclo[1.1.0]butan-2-one cation was observed in the nmr spectra of aziridine carboxylic anhydrides in liquid SO_2.[180]

Isomerization of *N*-acyl-2,3-disubstituted aziridines to $\triangle^2$-oxazolines is catalyzed by iodide ions. The trans isomers form *trans*-oxazolines in high yields, but the stereoselectivity is lower for the cis isomers and varies with the iodide ion concentration and the solvent system. The mechanism for the retention of configuration is displayed in Equation 26.[181] The isomerization of *cis*-1-(*N*-phenylcarbamyl or thiocarbamyl)-2,3-dimethylaziridine to the respective oxazolidine (or thiazolidine) is catalyzed by BF_3–Et_2O and proceeds with complete retention of configuration. The mechanism in this case involves coordination of the nitrogen atom with BF_3 and front side attack of the oxygen (or sulfur) atom on the ring carbon atom. Under the variety of conditions of solvents and organic acids as catalysts the oxygen compound gave 100% retention of configuration but the sulfur analogue gave on occasions the trans isomers. The proportion of the latter, however, was never higher than 50% (i.e., complete epimerization).[182]

Hydriodic acid catalyzes the conversion of 2-amino-3-substituted aziridinyl-1,4-naphthaquinones (66) into 1,2,3,4,5,10-hexahydrobenzo[g]quinoxalines (68) by way of the ring opened intermediate iodo compounds 67. When R is phenyl, the quinoxalines 68 are formed directly, otherwise the

66   67   68

reaction stops at the intermediate stage **67**. If the original aziridine ring is monosubstituted or trans substituted then this configuration is retained in **68**. The free bases **67** prepared from *cis*-aziridines (**66**) which can be isolated, however, cyclize with excess of iodide to the trans isomers of **68**.[183]

A useful synthetic entry into $1H$-azepines is the facile thermal isomerization of *cis*-3-methyl-7-methoxycarbonyl-7-azabicyclo[4.1.0]hept-2,4-diene and related compounds, into 4-methyl-1-methoxycarbonyl-$1H$-azepine,[15] and thermolysis of the norbonadiene-azido formate condensation product (**69**) to the dihydro-$1H$-azepine (**70**).[184] Addition of *trans*-2,3-divinylaziridine across the triple bond of hexafluoro-2-butyne at low temperature yields 1,1,1,4,4,4-hexafluoro-2-(*trans*-2,3-divinylaziridin-1-yl)but-2-ene which isomerizes to 4,5-dihydro-2,3-bis(trifluoromethyl)-7-vinyl-$3H$-azepine on heating.[185] *cis*-9-Methoxycarbonyl(phenylsulfonyl or N,N-dimethylcarbamoyl)-9-azabicyclo[6.1.0]nona-2,4,6-triene (**71**) rearranges on warming to *cis*-4-methoxycarbonyl(phenylsulfonyl, or N,N dimethylcarbamoyl)-4-azabicyclo[5.2.0]nona-2,5,8-triene (**72**).[50,186] Further heating of the latter (**72**,R = CO_2Me) yields a mixture of heterocycles which include *cis*- and *trans*-1-methoxycarbonyl-8,9-dihydroindoles and the 6-aza analogue of **72**.[186] If R in the bicyclic compound **71** is phthalimido then the rearrangement takes another course and 9-phthalimido-9-azabicyclo[4.2.1]nona-2,4,7-triene (**73**) is formed.[26]

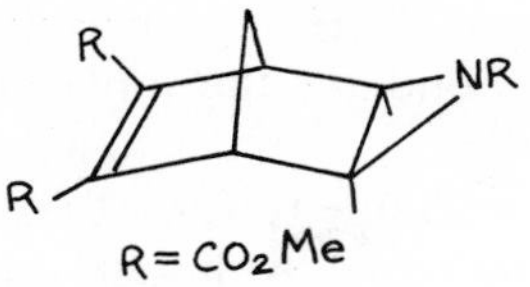
R
R
NR
R= CO2Me

69

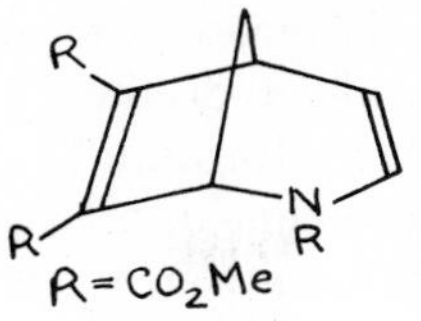
R
R
N
R
R=CO2Me

70

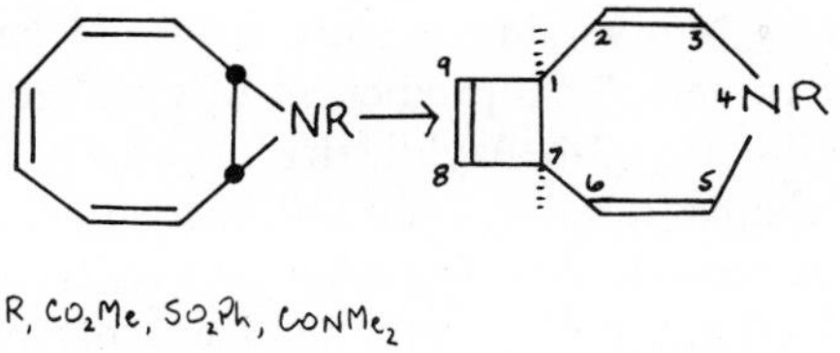
NR
9
1
2
3
4NR
5
6
7
8
R, CO2Me, SO2Ph, CONMe2

71

72

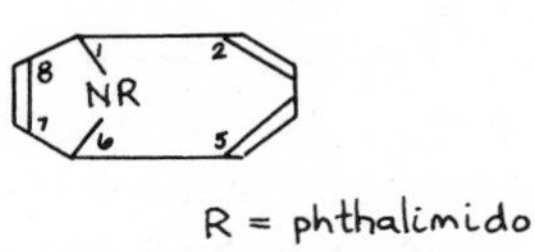
8
1
2
NR
7
6
5
R = phthalimido

73

The rearrangement reactions of aziridines were reviewed by Heine[186a] in 1971.

B. Diaziridines

Information on the stereochemistry of 3*H*-diazirines is scanty, and although the molecule is symmetrical there are two points worth mentioning. The first concerns the thermal extrusion of nitrogen (a chelotropic reaction) in which theoretical arguments on conservation of orbital symmetry predict that concerted decomposition must proceed in an unsymmetrical manner.[187] The second is the magnetic effect of a spirodiazirine as in 8-thiabicyclo-[3.2.1]octane-3-spiro-3′-diazirine and 1,2-diaza-6-methylspiro[2.5]oct-1-ene (**74**). The difference in the nmr chemical shift between the adjacent axial

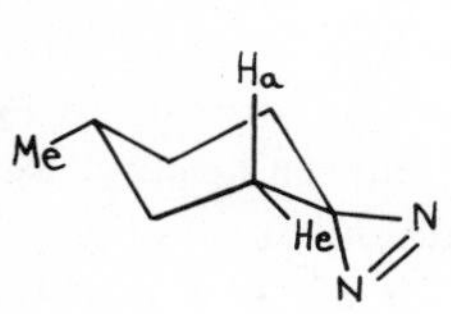
Ha
Me
He
N
N

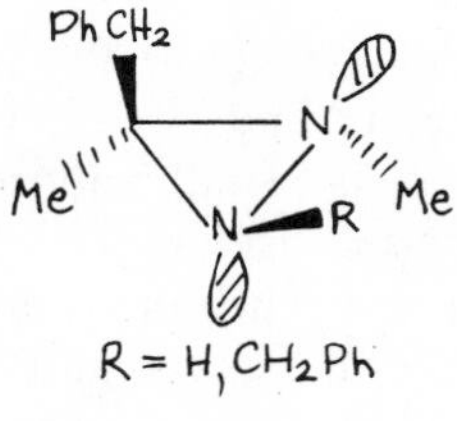
PhCH2
Me
N
Me
N
R
R = H, CH2Ph

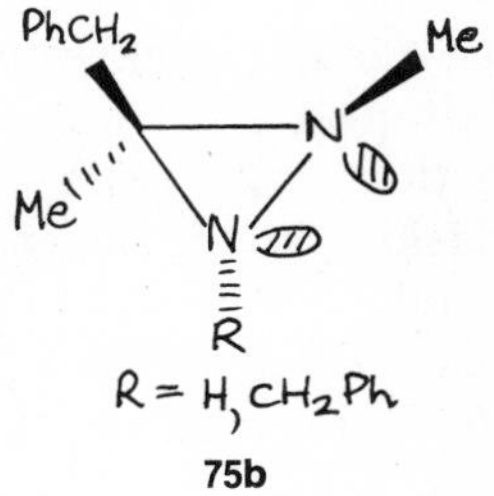
PhCH2
Me
Me
N
N
R
R = H, CH2Ph

74

75a

75b

and equatorial protons may be as large as 94 Hz and may be due to the anisotropic effect of the N–N double bond on the adjacent equatorial protons H_e (see formula **74**).[188]

The replacement of a CH_2 by NH in aziridine to form diaziridine raises the barrier for nitrogen inversion considerably. Although diaziridine has only one ring carbon atom it cannot be asymmetric unless one of the nitrogen atoms carries a substituent. Such a chiral diaziridine has not been prepared, and because N-inversion is slow it should be possible to show the presence of invertomers. This prediction was realized by Mannschreck and Seitz[189] for 1,3-dimethyl-3-benzyldiaziridine and its 2-benzyl derivative (**75a** and **b**). The trisubstituted compound, which was prepared by the reaction of the Schiff base derived from benzyl methyl ketonebenzylamine and *N*-methyl-hydroxylamine-*O*-sulfonic acid was diastereomeric and was separated by tlc. The $\triangle G^{\ddagger}$ value being 112.5 kJ mol^{-1} (at 70°C) with about 50% of each diastereomer at equilibrium. The benzyl derivative **75a** (R = CH$_2$Ph) crystallized out of the mixture at -10°C leaving the invertomer **75b** (R = CH$_2$Ph) in solution. At equilibrium the ratio of **75a** to **75b** was 0.75:1 with a slightly lower $\triangle G^{\ddagger}$ value of 108.3 kJ mol^{-1} (at 70°C).[189] These are obviously more stable than the aziridines.[102] *N*-Acyldiaziridines have also been examined, but they interconverted more rapidly than **75a** and **b** with the N-inversion barrier of the monoacyl derivative (e.g., 3,3-dimethyl-1-(*N*-phenyl)carbamoyldiaziridine) higher than in the diacyl derivatives [e.g., 1-benzoyl(or methoxycarbonyl)-2-methoxycarbonyl-3,3-dimethyldiaziridine].[190]

II.   FOUR-MEMBERED RINGS

A.   Azetidines and Azetidin-2-ones

The stereochemical aspects of azetidines (trimethyleneimine, azacyclobutane) and the β-lactams azetidin-2-ones have recently been the most actively studied group of compounds of the four-membered ring aza heterocycles. 1-Azetines, 2-azetines, and azetidin-3-ones have attracted very little attention. Azetidines have three carbon atoms two of which, if appropriately substituted, can be chiral centers and two in turn can be subject to cis and trans isomerism. The ring is usually nonplanar and undergoes ring inversion. The nitrogen atom also is capable of pyramidal inversion.[2,191] Azetidin-2-ones with two saturated carbon atoms are essentially planar but could theoretically have two chiral centers, and cis and trans arrangements of the substituents on C–3 and C–4 are possible.

The stereochemistry of some syntheses, the configuration, conformation, and reactions will be discussed for azetidines and azetidin-2-ones together, in the following sections.

1.   Syntheses of Azetidines and Azetidin-2-ones

Syntheses of stereochemical interest are described below, but for general methods of preparation reference should be made to the review by Moore.[192]

The synthesis of azetidines by *intramolecular ring closure* has been used to make derivatives of azetidine-2-carboxylic acid. Fowden[193] had isolated the (−)-acid from natural sources and synthesized its enantiomer from S-1,4-diaminobutyric acid. The diamino acid was converted to 4-amino-1-chlorobutyric acid with nitrous acid in the presence of hydrochloric acid, a reaction which occurs with retention of configuration, and this cyclized to (+)-azetidine-2-carboxylic acid with a Walden inversion at C–2. More recently Phillips and Cromwell[194] synthesized the racemic acid by reacting benzyl 1,4-dibromopropionate with diphenylmethylamine in methyl nitrile which cyclized to benzyl N-diphenylmethylazetidine-2-carboxylate and was hydrogenolyzed to the amino acid. The N-methyl derivative was obtained by methylating the intermediate before hydrogenolysis.[195] The racemic acid was resolved by Rodebaugh and Cromwell.[196] Moriconi and Mazzocchi[197] showed that an inversion of configuration takes place at the carbon atom bearing the oxygen atom in the cyclization of N,O-ditosyl 1-hydroxypropyl-amines. When the trans ester **76** was treated with a base it cyclized to cis-7-tosyl-7-azabicyclo[4.2.0]octane (**77**) which is also obtained from the cis ester **78**. Similarly, basificiation of the N,O-ditosyl derivative of trans-2-hydroxymethylcyclohexylamine produced the trans isomer of the bicycle **77**. An analogous intramolecular cyclization of the tosyl derivative of lupinine establishes the relative stereochemistries of the hydroxymethyl group and the bridgehead carbon atom (Equation 27).[198] γ-Aminoolefins (e.g., **79**), which are prepared by the addition of primary amines to α-bromomethyl chalcones, cyclize on treatment with hydrobromic acid, followed by an amine, to *cis*-azetidines **(80)** with high stereospecificity (>95%). The cis isomers are thermodynamically less stable than the trans isomers **(81)** and can be completely converted into the latter by boiling in

methanolic sodium methoxide.[199,200] The amine **79** can be rearranged to the α-alkyl aminomethyl chalcone and with the above treatment yielded the trans isomer **81** directly.[201] β-Substituted γ-aminopropionyl chloride hydrochlorides readily cyclize to 3-substituted azetidin-2-ones. Although these products have a chiral center none of them have been obtained which are optically active.[202] Under photolytic conditions *N,N*-dibenzyl α,β-unsaturated amides cyclize to *N*-benzyl-*trans*-3-alkyl-4-phenylazetidin-2-ones. This takes place by hydrogen abstraction from one of the *N*-benzyl methylene groups by the β-carbon atom, and cyclization of the α-carbon atom on to the same benzylic carbon atom to form the four-membered ring.[202a] Photolysis of α-*N*-alkylamidoacetophenones affords a direct route to azetidin-3-ols. Comparable amounts of the epimeric alcohols (3-hydroxy-

2-methyl-3-phenylazetidines) are formed but are separable by crystallization.[203] Fused azetidin-2-ones are obtained when *N*-(alkoxycarbonyl)diazoacetyl piperidines **(82)** are photolyzed. Nitrogen is lost and a carbene insertion occurs α to the ring nitrogen atom. This reaction was originally developed by Corey and Felix[204] for the synthesis of the penicillin and cephalosporin skeletons and has been used by others for a similar purpose. In the simplest case (Equation 28) and in more elaborate systems, substantial amounts of cis and trans isomers (2:1 from **82**) were formed but their separation presented no major difficulties.[205,206] The cyclization does not take place with the pyrrolidine analogue and the mixture of stereoisomers derived from cyclization of benzyl N-(t-butoxycarbonyl)diazoacetylthiazolidine-4-carboxylate was contaminated with a monocyclic lactone.[205] The photochemical, intramolecular cyclization of 5,5-dimethyl-3,7-dioxocyclooctane carbonyl azide (83) gave the azetidin-2-one (apparently trans) in low yield, whereas cycloheptane carbonyl azide gave a cyclic γ-lactam. Cyclohexane carbonyl azide, cyclopentane carbonyl azide, and pivaloyl azide did not cyclize[207] under these conditions.

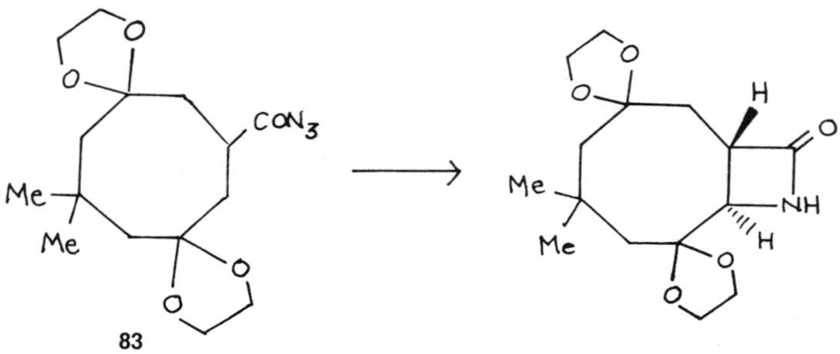

83

Molecular orbital theory predicts that the thermal and photochemical cyclization of 2-azabutadiene (Z isomer shown in Scheme 3) should proceed in a conrotatory and disrotatory manner respectively, and the reverse is predicted for monoaza- and diazahexatrienes.[208]

The *addition reactions* of ketenes to Schiff bases is a very good method for making azetidin-2-ones. The reaction is highly regiospecific in that the C–N and C=C=O double bonds always align themselves in the same way. The reaction between aldoketenes and benzal aniline is usually highly stereospecific in giving mainly the trans isomer.[209–211] In the addition of chloroketene and benzal aniline the major product has a trans structure, but the proportion of the cis isomer is increased by introducing o-substituents in the benzal ring (Equation 29).[212] The reaction is less stereoselective

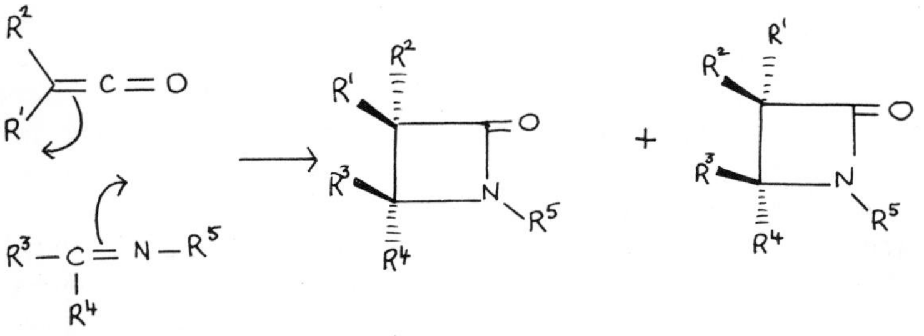

(29)

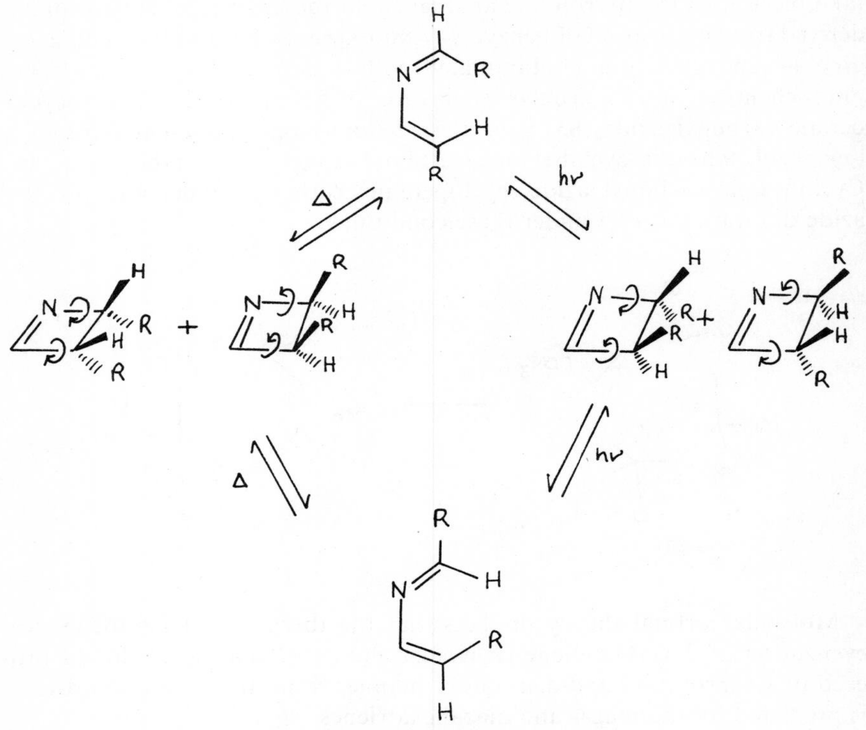

Scheme 3

when substituted ketenes are used.[211,213] In its original form an acetyl chloride was condensed with benzal aniline in the presence of a base,[214] and it is not certain whether the reactions of ketenes are $[2 + 2]\pi$ cyclo-additions as shown in Equation 29, or whether they involve an initial condensation to form the *N*-acyl derivative which cyclizes to an azetidin-2-one.[209] The synthesis can also be brought about by a Reformatzky reaction in which the ketene is generated in position, for example, α-bromo ester and zinc with imines. However, this reaction is not as stereoselective as with ketenes,[210,215] but the selectivity may be improved by lowering the temperature[216] or altering the solvent.[215] In a variant of these reactions α-acetoxyphenylacetyl chloride was condensed with benzal aniline in the presence of triethylamine and gave only 3-acetoxy-1,3,4-triphenylazetidin-2-one in which the 3- and 4-phenyl groups were trans. The same acid chloride reacted with *S*-methyl *N*-phenylbenzthioimidate (PhC(=NPh)SMe) but gave 3-acetoxy-4-methylthio-1,3,4-triphenylazetidin-2-one in which the 3- and 4-phenyl groups were cis, and which gave the cis isomer of 3-acetoxy-

1,3,4-triphenylazetidin-2-one on desulfurizing with Raney nickel.[217] Phenyl-malonyl chloride reacted similarly with benzal anilines to yield azetidin-2-ones in which the aryl groups are cis, and the presence of a phenyl group caused dechlorocarbonylation of the 3-COCl group in the intermediate. If a thioimidate is used instead of benzal aniline the *cis*-3-chlorocarbonyl-4-alkylthioazetidin-2-ones are formed (i.e., no loss of COCl). These are useful intermediates for the synthesis of penicillins.[218]

A one carbon insertion between C–2 and C–3 of 2,2-bis(methoxycar-bonyl)-1,2-diphenylaziridine has been achieved by condensation with sulfur ylides in a stereoselective manner—the cis to trans isomer ratio was better than 79:21 (Equation 30). The aziridine could be replaced by a 1,2-dihydro-oxazoline; and sulfoxide ylides (e.g., $Me_2\overset{+}{S}O-\overset{-}{C}R_2$) are also effective.[219]

The "uniparticulate electrophile" chlorosulfonyl isocyanate (CSI) adds its N and C atoms to a C–C double bond exclusively in a cis configura-

$$ (30) $$

R = CO_2Me, CO_2Et, CO_2Ph

$$ (31) $$

tion to yield *N*-chlorosulfonylazetidin-2-ones. The addition is regiospecific in a Markownikov orientation with the reagent adding to the terminal double bond in a conjugated system.[197,220,221] Under hydrolytic conditions the chlorosulfonyl group can be removed, and reduction with LAH furnishes the corresponding azetidine (Equation 31). CSI adds across the C–C double bond of cyclopentene,[222] norbornene (exo addition),[223] and cyclohexa-1,4-diene.[224,225] Paquette and co-workers extended this study and showed that in examples of the latter type (i.e., **84**) addition occurred on the double bond common to the two rings (i.e., to **85**)[224–226] and succeeded in effecting this addition on to a double bond in bullvalene[227,228] and *cis*-bicyclo[6.1.0]nona-2,4,6-triene (**86**) with most interesting results.[229,230] The addition to the bicyclononatriene(**86**,$R^1 = R^2 = H$) occurred with concurrent cleavage of the C–C bond between the 8- and 3-membered ring leading to a trans fusion between the four- and nine-membered rings in the product 4*H*-10-azabicyclo[7,2.0]undecan-11-one (**87**,R = H). The structure was confirmed by an X-ray study and impressive steric control was observed in this reaction. When R^1 and R^2 were methyl groups and when $R^1 =$ Me and $R^2 =$ H (i.e., syn) in **86** no reaction with CSI was observed, but in the anti compound **86** ($R^1 =$ H, R = Me) the reaction proceeded smoothly

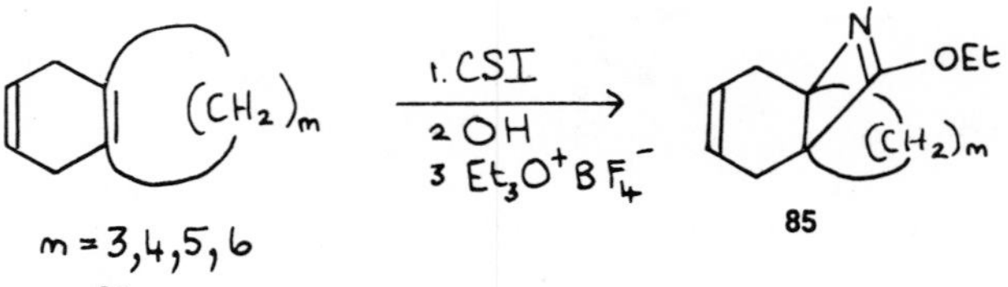

$m = 3,4,5,6$

84

85

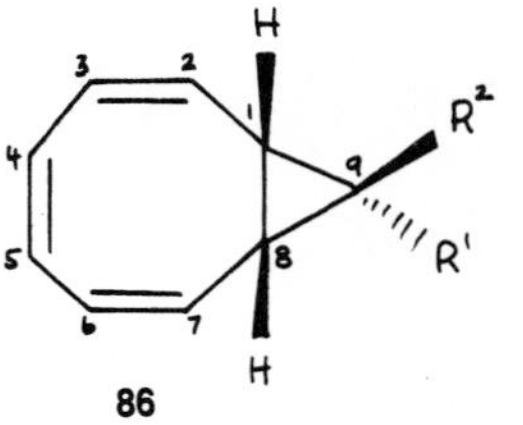

86

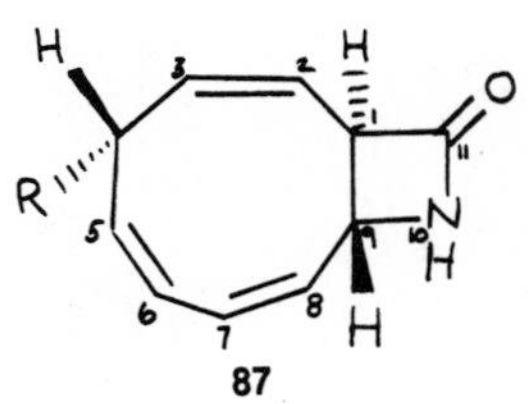

87

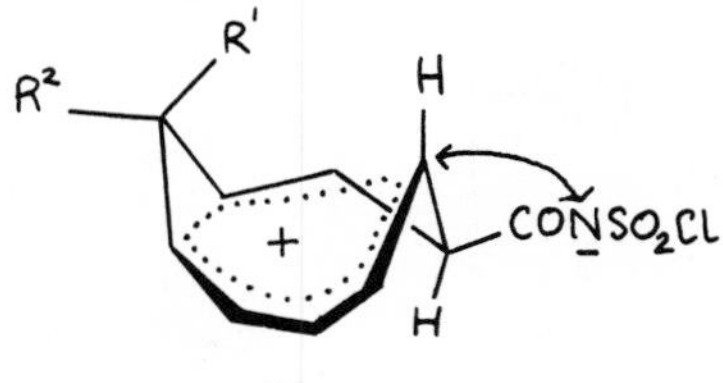

88

and the product had the stereochemistry depicted in **87** (R = Me).[229] The products from the condensation of derivatives of **86** (R^1 = R^2 = H), in which one methyl group was in turn at positions 1, 2, and 4, gave the bicyclo compounds **87**, in which the methyl group appeared at C–3, C–2, and C–9 respectively. These reactions were rationalized by invoking electrophilic attack at C–3 in **86** with formation of the intermediate *trans*-1,3-bishomotropilium ions (**88**). However, initial attack at C–6 was not excluded as an alternative.[230]

Effenberger and co-workers[231] observed [2 + 2] cycloadditions between alkyl enolethers and toluene-*p*-sulfonyl isocyanate with the formation of 4-alkoxyazetidin-2-ones. After the initial cis addition equilibration occurs in solution (and to a slower extent in the solid state) yielding predominantly trans products. There is also concurrently a slower irreversible ring opening to the respective β-alkoxyacrylamides (Scheme 4). The kinetics of

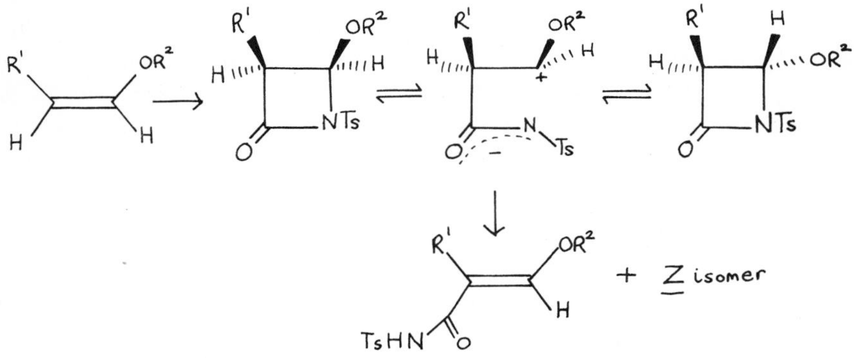

Scheme 4

these transformations have also been examined.[232] Trichloroacetyl isocyanate reacts with alkyl enolethers with the same regiospecificity to yield similar *trans*-3-substituted 4-alkoxyazetidin-2-ones within one hour. On standing for 23 hr, however, the azetidin-2-ones which are in equilibrium with the noncyclic form are transformed into 5-alkyl-6-alkoxy-5,6-dihydro-1,3-oxazin-4-ones.[233] In an interesting ring contraction, Lowe and co-workers[234,235] have carried out a photolytic Wolff rearrangement of 3-diazopyrrolidin-2,4-diones and have obtained azetidin-2-ones (Equation 32). Close to equal amounts of *cis*- and *trans*-azetidin-2-ones were formed, but they were separated and used for the synthesis of nuclear analogues of penicillins and cephalosporins.

7-Hydroxy-7-methoxy-*cis*-bicyclo[4.1.0]heptane (**89**, R^1 = OH, R^2 = OMe) and sodium azide in acetone rearranged, on boiling, to 7-aza-*cis*-bicyclo-

(32)

[4.2.0]octan-8-one (**90**, X = O)[236]; and in a related rearrangement 7-azido-*cis*-bicyclo[4.1.0]heptane (**89**, R^1 = N$_3$, R^2 = H) lost molecular nitrogen and gave 7-aza-*cis*-bicyclo[4.2.0]oct-7-ene which was reduced with NaBH$_4$ to the parent 7-aza-*cis*-bicyclo[4.2.0]octane (**90**, X = H$_2$).[237] All these reactions did not upset the configuration at the bridgehead carbon atoms.

2.   Configuration and Confirmation of Azetidines and Azetidin-2-ones

The absolute configuration of the naturally occurring S(−)-azetidine-2-carboxylic acid was deduced by the synthesis of its enantiomer,[193] and the optical resolution of the racemate has been achieved (see Chapter 2, Section II.A.1). Its dimensions were obtained from an X-ray study.[238] The ord curves of (+)-2-methylazetidine was compared with that of R(+)-coniin[70,239] and assigned the R configuration. The N-chloro derivative was prepared and separated into 1S,2R-(*cis*)- and 1R,2R-(*trans*)-1-chloro-2-methylazetidines which exhibited − ve and + ve Cotton effects respectively. After comparison with chloroderivatives of 2S-2-methylaziridine an Octant rule was postulated.[70] Fontanella and Testa[240] resolved racemic ethyl α-aminomethyl-α-phenyl butyrate into its enantiomers and cyclized these to the respective 3-ethyl-3-phenylazetidin-2-ones with ethyl magnesium bromide. This cyclization, which did not involve the chiral center, was accompanied by a change in the sign of rotation.

The assignment of cis and trans structures is mainly done by nmr[199,217—219] (see references in this and other sections on four-membered rings). The size of the coupling constant for cis and trans vicinal protons as a guide for the structures occasionally becomes unreliable, and this has been ascribed to unusual flattening of the four-membered ring by the presence of large substituents on the nitrogen atom.[199] This has been confirmed by X-ray crystallographic studies of the azetidinium ions **91**,[241] **92**,[238] **93**,[242] **94**,[243] and **95**.[244] The dihedral angles are inscribed in the formulas and

indicate that the degree of puckering varies from one compound to another. The *N,N*-dibenzyl derivative **95** is most interesting because the ring is planar. Vigevani and Gallo[245] studied the ^{1}H nmr chemical shifts, vicinal and geminal coupling constants of azetidin-2-ones, azetidin-2-thiones and 1-azetines, and proposed planar structures for these compounds. The lanthanide shift reagent Eu(dpm)$_3$ was used to assign the cis and trans structures to 1-cyclohexyl-2-phenyl-3-hydroxyazetidines[246] and this approach could be useful in deciding between structures when coupling constants are unreliable.

Studies of coupling constants and chemicals of a series of 1-alkylazetidines, 1-alkylazetidin-3-ols, 1-alkyl-3-methylazetidin-3-ols, 3-aroylazetidines, and C-methylazetidines led Cromwell and co-workers[247,248] to conclude that the ring is nonplanar and conformationally labile; and the nitrogen atom undergoes pyramidal inversion.[2] The four-membered ring is less strained than the aziridine ring and the barriers for inversion are generally lower by about 38 kJ mol^{-1} ($\sim$9 kcal mol^{-1}).[98,102,106,107,111—113] The effect of substituents on the inversion barriers is uniformly similar to those observed in the aziridines (Section II.A.3), for example, alkyl groups[112] on C–3 slow inversion down, and aryl[102] and acyl[106] groups on the nitrogen atom increase it.[106] N-chloro,[191] N-amino,[2,249] and N-nitroso[250] substituents all raise the inversion barrier. Tables of the thermodynamic parameters of inversion are cited in reference 2.

The ir spectra of cis- and trans-1-alkyl-3-hydroxy-2-methylazetidines revealed that inter- and intramolecular hydrogen bonding was taking place. Intramolecular hydrogen bonding is present in structures such as **96**[251] and can be used to distinguish them from their epimers.

96

3.   Reactions of Azetidines and Azetidin-2-ones

Extrusion of the nitrogen atom of azetidines is achieved by treatment with difluoroamine, and by reaction of N-nitroso or N-amino azetidines with sodium dithionite or mercuric oxide respectively. Application of these three methods to cis-2,4-dimethylazetidine gave almost identical products, namely, cis- (16%) and trans-1,2-dimethylcyclopropanes (84%), with the trans isomer predominating. Conversely trans-2,4-dimethylazetidine gave approximately a 68:32 ratio of cis to trans dimethylcyclopropanes. The products of the reaction are accounted for by a planar trimethylene intermediate and incomplete stereospecificity is attributed to "electronic leakage."[252] Pyrolysis of cis- and trans-3,4-dimethyl (and diethyl) azetidin-2-ones at 600°C is virtually stereospecific yielding the corresponding butadiene with almost complete retention of configuration in a concerted $[\sigma2_s + \sigma2_a]$ retrogression. Thermolysis of the cis- and trans-iminoether derivatives, on the other hand, gave the same three ring-opened products in about the same proportion (Scheme 5).[253] N-Chlorosulfonyl azetidin-2-ones undergo ring cleavage on heating in dimethyl formide or hexamethyl phosphorotriamide to yield α,β- and/or β,γ-unsaturated nitriles depending on whether or not a hydrogen atom is present α- to the carbonyl group. The spatial relationship between the substituents on C–3 and C–4 appears to be retained,[254] although this is not always the case.[255] The formation of ω-alkoxyacrylamides from 4-alkoxyazetidin-2-ones after long standing in solution has been mentioned earlier (see Section II.A.1).[231,232]

1-Azabicyclo[1.1.0]butanes are best discussed at this point because they lead to 3-substituted azetidines by nucleophilic attack and the stereochem-

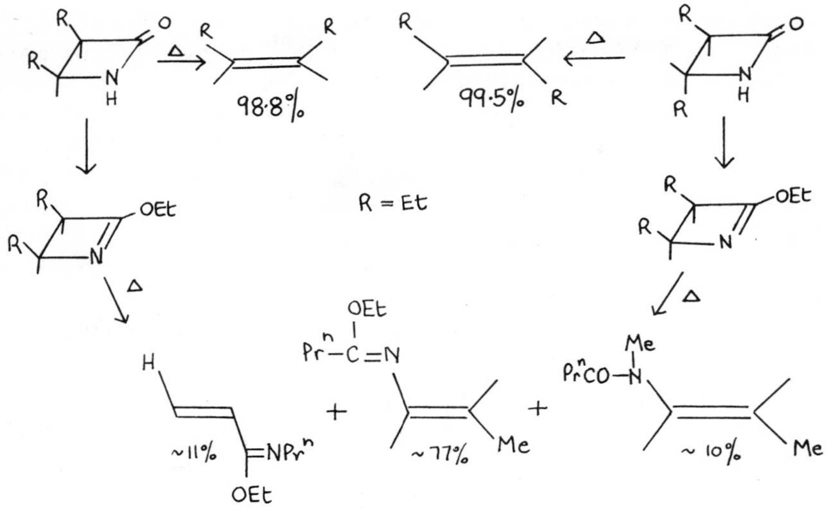

Scheme 5

istry of the nucleophilic displacements at C–3 can be explained by such intermediates. 2-Alkyl-3-aryl-1-azabicyclo[1.1.0]butanes (**97**) are best made from the reaction of 2-alkyl-3-aryl-2*H*-azirines, with dimethyl sulfonium ylide. The N–1,C–3 bond is cleaved by water, methanol, or dry HCl/Et$_2$O to yield 3-hydroxy-, 3-methoxy-, or 3-chloro-2-alkyl-3-phenyl azetidines respectively (Equation 33).[256] Only one diastereomer of **98** is produced—

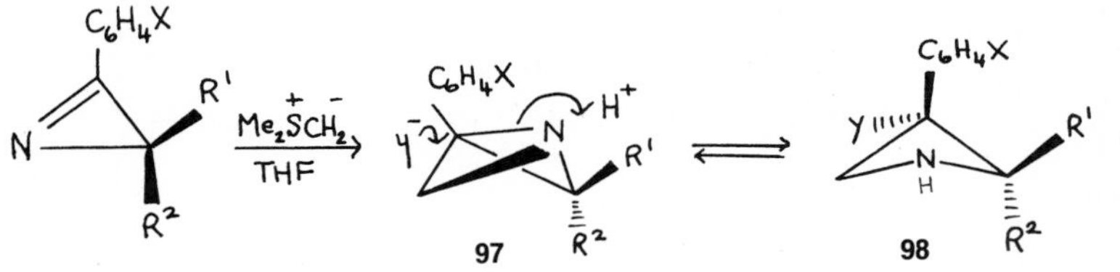

$$\text{X = H, pOMe, pCF}_3 \, ; \quad \text{R=Me,H} \, ; \quad \text{R}_2 = \text{H,Me} \, ; \quad \text{y = OH,OMe,Cl}$$

demonstrating that the ring opening is stereospecific and that the nucleophile attack is from the concave face of the protonated molecule. The nucleophilic displacement of the substituent Y in a number of azetidines (**98**) is shown to occur with retention of configuration as is observed in *cis*- and *trans*-1-cyclohexyl-3-mesyloxy-2-phenylazetidines,[257] and 3-chloro- and 3-hydroxy-3-*p*-substituted-phenylazetidines.[258] The rates of solvolysis of the

tosyl group follows the order *cis*-1-*t*-butyl-2-methyl-3-tosyloxyazetidine >
1-*t*-butyl-3-tosyloxyazetidine > *trans*-1-*t*-butyl-2-methyl-3-tosyloxyazeti-
dine, and the Arrhenius plots for the trans isomer are nonlinear. This is
evidence for postulating anchimeric assistance, by way of a bicyclo intermediate such as **97** exclusively for the cis isomer, and to a lesser extent for
the trans isomer.[259] The deuterium isotope effect in substituting $Y = Cl$
for $Y = OH$ in **98** ($X = H$, $R^1 = R^2 = H$) and **98** ($X = H$, $R^1 = R^2 = D$) is
$K_H/K_D = 0.948$. This value is consistent with a completely inductive or
field transmission effect of the β-deuterium atom. In the intermediate there
is therefore little or no hyperconjugation overlap of the β-deuterium and
the orbitals of the carbonium carbon atom (depicted in **99**) which are directed

99

towards the nitrogen atom.[260] *r*-2-Methyl-*t*-3-hydroxy-*c*-3-phenylazetidine
cannot be directly converted into the diastereomeric *cis*-3-hydroxy derivative because of retention of configuration of C–3 on nucleophilic displacement. However, this could be overcome by decreasing the ability of forming
a bicyclo intermediate by making the *N*-tosyl derivative, converting it into
the 3-*O*-mesylester, and displacing the latter group by OH in a normal
S_N2 reaction (Walden inversion) (Equation 34).[261]

2-Methyl-3-phenyl-1-azabicyclo[1.1.0]butane (**100**) is converted into a 2-phenyl-2-vinylaziridine by lithium diisopropylamine in an unusual intermolecular bond displacement.[262]

R = H, Me

100

Potassium in liquid ammonia equilibrates *cis*- and *trans*-3,4-diphenyl-azetidin-2-ones into a 1:1 mixture by removing H-3 and forming the anion.[221] This is assisted by both the properties of the carbonyl and the phenyl groups, but an aroyl group on C-3 alone is sufficient to make the hydrogen atom on C-3 acidic enough. Thus Doomes and Cromwell[263] showed that *cis*-1-*t*-butyl-3-aroyl-2-phenylazetidine formed a potassium salt which gave a mixture of *cis*- and *trans*-(with respect to the phenyl and aroyl groups)C-3-methylated azetidines on addition of methyl iodide. The cis isomer predominated but the proportion was sensitive to the solvent used.

The displacement of a 3-thio substituent in azetidin-2-one by nucleophiles normally occurs with inversion of configuration[264] even when the S group is part of a fused ring as in penicillanic acid.[265] Cleavage between the oxygen atom and the bridgehead carbon atom takes place in the *cis*-oxazoloazetidine (**101**) on reaction with various alkyl thiols, with formation of *cis*-(**103**) and *trans*-(**104**)4-alkylthio-1-(2-methylpropenyl)-3-phenylaceta-mido-azetidin-2-ones in a 1.5–2.0:1 ratio in favor of the cis isomer, but this depends on the alkyl thiol used. In order to explain this stereoselectivity an azetin-2-one cation intermediate (**102**) was proposed.[266]

Rearrangement of *cis*-1-*t*-butyl-3-aroyl-2-phenylazetidine occurs in the presence of uv light to give 4-aryl-1-*t*-butyl-2-phenylpyrrole, whereas the trans isomer forms a mixture of 3-aryl-2-phenyl- and 5-aryl-2-phenyl-1-*t*-butylpyrroles.[267] A mechanism has been proposed from results with deuterated analogues.[268]

Paquette and co-workers have studied in detail the ring-enlargement equilibrium between fused 2-alkoxyazetines and fused azetidin-2-ones with multimembered ring heterocycles. They showed that the equilibrium in Equation 35 favors the azocine valence tautomer.[224–226,269,270] However, if a constraint is placed on the system, such as replacing R^1 and R^2 in

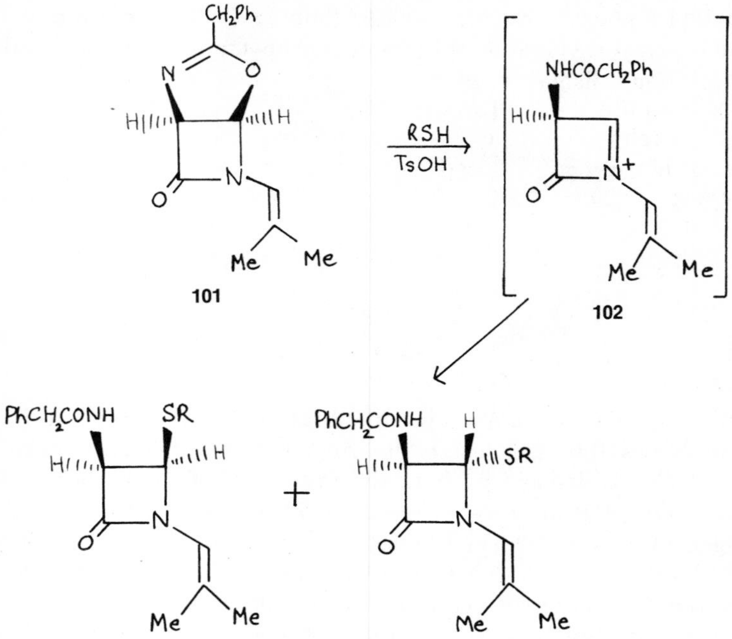

101

102

103   +   104

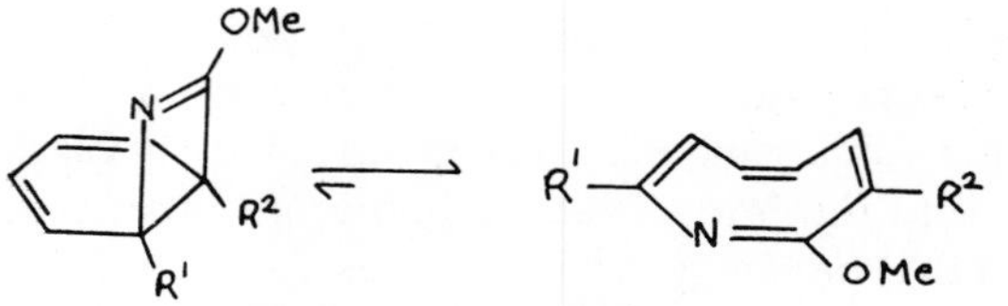

(35)

Equation 35 by a tri-, tetra-, or pentamethylene bridge, then the bicyclic form is favored, but with a hexamethylene bridge the equilibrium is transferred towards the azocine form.[225,226] In the case of the fused azetidines (105, X = H₂) or azetidin-2-ones (105, X = O), the bicyclic structures 106 are preferred. The ratio of 105 to 106 alters with the substituents, for ex-

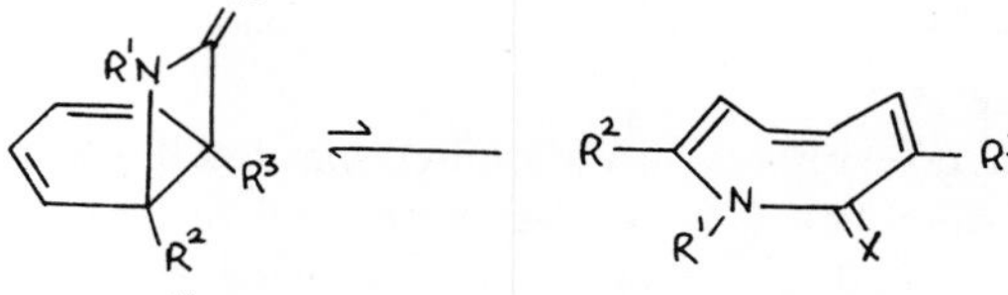

X = O, H₂   X = O, H₂

105   106

ample in **105** $R^1 = H$, $R^2 = R^3 = Me$, $R^1 = R^2 = R^3 = H$, and $R^1 = R^2 = R^3 = Me$ and the percentages of bicyclic form are 97.6, 80.5, and greater than 98 respectively. A rise in temperature favors the azocine but a change of solvent has little effect on the tautomerism.[269,270]

cis-1,4-Diphenyl-3-azidoazetidin-2-one rearranges at 156°C, with elimination of nitrogen, to 1,5-diphenyl-5*H*-imidazolin-2-one in 70% yield, but only 20% of the same imidazolinone is obtained from the *trans*-azido isomer.[271] This difference remains unexplained.

B. Diazetidines

1,2- and 1,3-Diazetidines and their oxo derivatives are known,[192] but only a meager interest in the stereochemistry of the former is displayed in the literature.

Photochemical and thermal addition of dialkyl azodicarboxylates to olefins is a source of 1,2-diazetidines, in which the two nitrogen atoms add in a cis fashion. If any isomerization occurs it is probably due to an alteration in the structure of the olefin.[272] Cycloaddition of quadricyclane yields the exo product because it is the result of the best alignment of the reacting orbitals (Equation 36).[273]

The presence of a second nitrogen atom in the azetidine ring should be expected to raise the barrier for pyramidal nitrogen inversion. The examples studied (**107**, $R = CO_2Et$ or CF_3) cannot give the required information because fluoro groups have a large effect on the inversion param-

107	**108**	**109**	**110**

eters.[274,275] The coalescence temperatures and the free energies for a number of 1,2-diaryldiazetidin-3-ones (**108**) have been observed by the dynamic nmr technique and the process is probably related to inversion of the amide nitrogen[276]—N–1 is expected to invert less rapidly than N–2. In the interconverting conformers of 1,2-dibenzyl-4,4-dimethyldiazetidin-3-one,

109 and **110,** two signals for the prochiral methyl groups are observed at low temperatures only when optically active dibenzoyl tartaric acid is added to the solution. This is taken as proof for the presence of a chiral nitrogen atom.[277]

III.   FIVE-MEMBERED RINGS

A.   Introduction

The five-membered ring nitrogen heterocyclic compounds which are discussed in this section include mainly reduced structures because these have three-dimensional properties of interest. The pyrrolines and pyrrolidines have two, three and four carbon atoms which are potentially chiral, if adequately substituted, and the rings are conformationally labile. The nitrogen atom introduces additional stereochemical interest by virtue of its pyramidal inversion. The research activity in this group of compounds is not surprising when it is considered that the ring system is present in several naturally occurring products, with the essential amino acid proline occupying a dominant place.

Spirobipyrrolidines have a fourfold axis of symmetry and are therefore dissymmetric. The discovery of optical activity in the 1,1′-isomers laid the tetrahedral structure of the quaternary nitrogen atom on a firm foundation. Pyrrole itself is planar, but bipyrryls or arylpyrroles become dissymmetric if rotation about the bond joining the two nuclei is restricted by the presence of bulky substituents in the o,o' positions. These are the only "non-reduced" examples of stereochemical interest. In fused pyrrolidines two bridgehead carbon atoms may be present, as in perhydroindole, and the fusion can be cis or trans, and each of these may be chiral. Several examples of these are discussed in the following subsections. Fused pyrrolidines in which nitrogen is one of the bridgehead atoms are described in Chapter 4, Sections VI, A and B.

The stereochemical properties are altered on increasing the number of nitrogen atoms in the five-membered rings of pyrrolines or pyrrolidines to form reduced pyrazoles, imidazoles, and triazoles. The changes in properties depend on whether the added nitrogen atoms are sp^2 or sp^3 hybridized, but the effects of replacing a $=$C$-$ or $-$CH$_2-$ unit by a $=$N$-$ or $-$NH$-$ unit respectively can be predicted qualitatively.

B.   Pyrrolidines

Several compounds discussed in this section are naturally occurring but have been included because they portray some important features of the stereochemistry of this system. A few pyrrolines are also discussed in this section.

1. Synthesis of Pyrrolidines

a. *By Intramolecular Cyclization.* A straight chain consisting of a 1,4-dibromopropane unit cyclizes readily with a unit containing a nitrogen atom to give a five-membered ring without forming appreciable amounts of long chain polymers. *meso*-2,3-Dimethyl-1,4-dibromobutane reacts with *p*-toluenesulfonamide to give *meso*(*cis*)-3,4-dimethyl-1-tosylpyrrolidine which is hydrolyzed to *cis*-3,4-dimethylpyrrolidine (Equation 37). The *d,l*(trans) isomer was obtained by LAH reduction of *d,l*-2,3-dimethylsuccinimide.[278] 4-Amino-2-bromo-3-methoxybutyric acid cyclized to *cis*- and *trans*-*R,S*-2-carboxy-3-methoxypyrrolidine under the influence of a base. The cis and trans isomers are separable by recrystallization of the copper salts, and the cis and trans *S* isomers were isolated after destruction of the *R* isomers with D-amino acid oxidase. The required 3-hydroxyprolines were finally obtained without epimerization by demethylation of the methoxy compounds with hydrobromic acid.[279] The cyclization of ω-amino-butyric esters form pyrrolidones as in the preparation of (−)-5-ethyl-5-phenylpyrrolidin-2-one from methyl(+)-3-amino-3-phenylhexanoate,[280]

and S(−)-2-carboxypyrrolidin-5-one from the mono γ ester of glutamic acid in alcoholic ammonia,[281] or from glutamic acid by heating at 175°C.[282]

Corey and Hertler[283] examined the mechanism of the Hofmann-Löffler reaction as a means of preparing cyclic amines from the open chain *N*–Cl amines and sulfuric acid (e.g., conversion of di-*n*-butyl-*N*-chloroamine into *N*-butylpyrrolidine). Whereas they showed that chiral 4-hydroxypentyla-mine provided S(−)-2-methylpyrrolidine with complete inversion of con-figuration on treatment with TsCl–NaH, the cyclization of chiral 4-^{2}H$_1$-*N*-chloro-*N*-methylpentylamine to a mixture of 2-deutero-1,2-dimethyl- and 1,2-dimethylpyrrolidines ($K_H/K_D = 3.54$) occurred with complete loss of optical activity. They also established that a free radical chain mechanism requiring intramolecular H-transfer as one of the propagation steps was involved.

The Michael adduct (**111**) from the condensation of acrolein (as bisulfite) and α-aminomalonic acid undergoes ring-chain tautomerism with the pyrrolidine dicarboxylic acid which decarboxylates easily to yield a

33:67 mixture of *cis-* and *trans-*3-hydroxy-D,L-prolines (Equation 38).[284] Izumiya and Witkop[285] transformed *d,l*-2-*N*-benzyloxycarbonylamino-pent-4-enoic acid (**112**, benzyloxycarbonylallyl glycine) with NBS into a mixture of cyclic bromolactones **113** and **114** which, after hydrolysis to the amines, rearranged in alkali to a mixture (77:23) of *allo*-4-hydroxy- and 4-hydroxy-D,L-proline.

In a similar approach Equchi and Kakuta[286] started with D-glutamic acid and obtained chiral 5-carboxytetrahydrofuran-2-one on treatment with nitrous acid. The furanone was converted into 5-chloromethyltetrahydro-furan-2-one (**115**) and provided a mixture (3:2) of epimeric 3-chloro derivatives on chlorination. This gave a 3:2 mixture of *cis-* and *trans-*4-hydroxyprolines on reaction with ammonia. During these transformations the original chiral center of D-glutamic acid was preserved and became the chiral center at C–4 of these prolines (Equation 39).

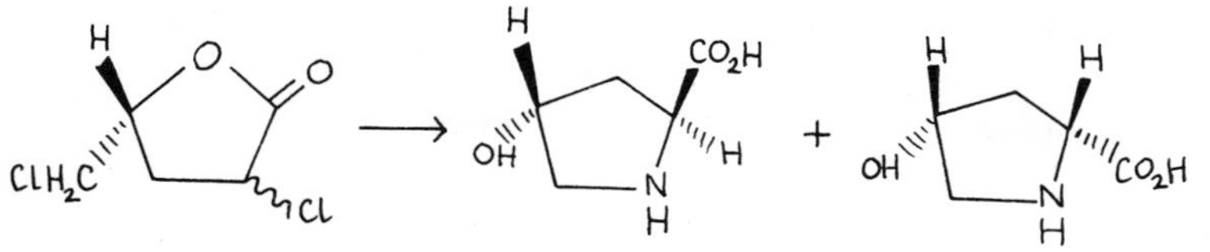

(39)

The thermal intramolecular cyclization of N,N-bis(alk-2-enyl)amines is a new stereoselective method for preparing pyrrolidines and was developed by Oppolzer and co-workers (Equation 40).[287] It was applied to the syn-

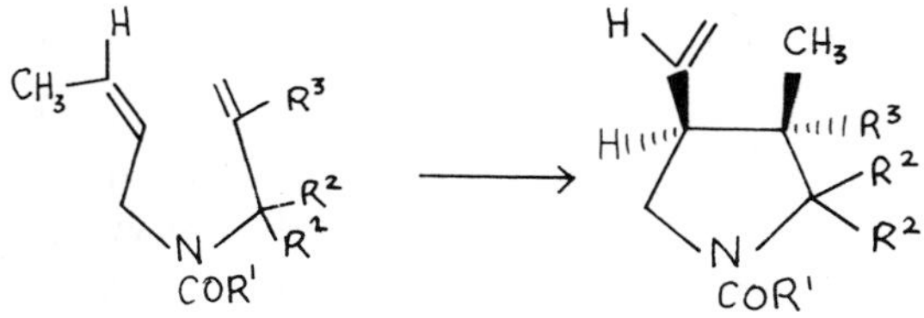

(40)

theses of several other systems and the substituents derived from the olefinic carbon atoms were always cis oriented.

A second synthesis in which the C–3,C–4 bond of pyrrolidine is formed, is the intramolecular base catalyzed cyclization of N-2-cyanoethyl-N-2,3-epoxypropylamines (Equation 40a). When $R^2 = H$, the products formed in good yields are *trans*-3-cyano-4-hydroxymethylpyrrolidine, but when $R^2 =$ aryl, *cis*-iminolactones are formed predominantly.[287a]

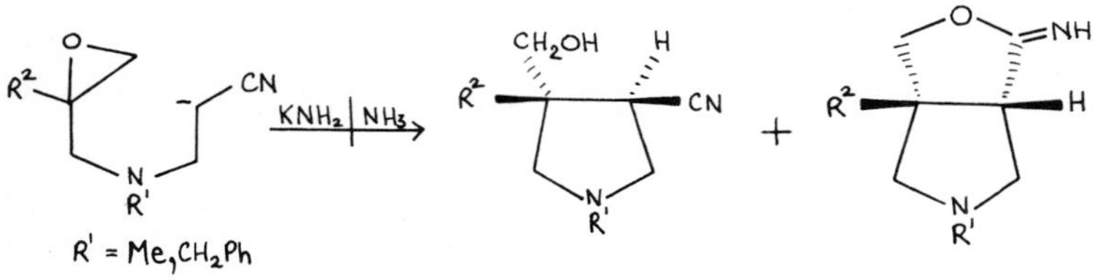

(40a)

Ring cleavage between N–1 and C–4 of 1-benzyl-4-allyl(or 4-phenyl)-azetidin-2-ones is induced by lithium diisopropylamide in THF at $-78°C$, and the intermediate **115a** cyclizes, intramolecularly, to yield *cis*-4-phenyl-5-vinyl(or phenyl)pyrrolidin-2-ones with high stereospecificity.[288]

In order to determine the absolute configuration of $S(+)$-isovaline Yamada and Achiwa[289] converted it into $(+)$-5-methyl-5-ethylpyrrolidin-2-one (**117**) by way of the 2,4-dione formed from an intramolecular cyclization of N-ethoxycarbonylacetylisovaline ethyl ester (**116**) followed by hydrolysis and decarboxylation. The absolute configuration of the pyrrolidinone **117** was deduced by synthesis from the tricarboxylic ester derived from degradation of D($-$)-quinic acid (Scheme 6).

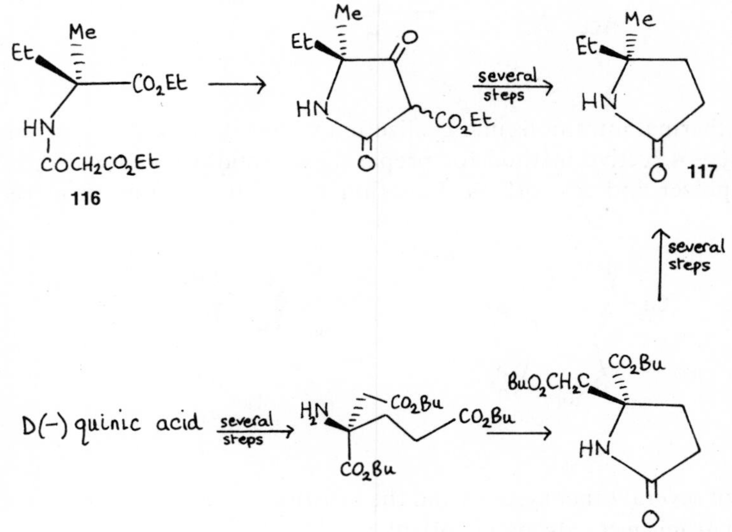

Scheme 6

Intramolecular cyclizations sometimes give products of different configuration depending on the method used. Treatment of 1-phenoxy-3-methyl-4-methylaminopentane with HBr gives *trans*-1,3-dimethyl-2-*n*-propylpyrrolidine whereas catalytic reduction of the imine from 1-phenoxy-3-methylheptan-4-one and methylamine gives the cis isomer. The latter was also obtained by dehydrogenation of the trans isomer followed by hydrogenation in the presence of copper chromite.[290]

b.   *By Cycloaddition.*   The azomethine ylides formed by thermal or photochemical ring opening of aziridines undergo 1,3-dipolar cycloaddition reactions with olefins and acetylenes to give pyrrolidines and pyrrol-3-ines in a stereospecific manner.[291,292] The stereospecificity for C–2 and C–5 depends on the method of generating the ylide and the specificity for C–3 and C–4 depends on whether the olefin is cis or trans. *cis*- or *trans*-Olefins yield 3,4-*cis*- or 3,4-*trans*-substituted pyrrolidines respectively with high specificity, but the specificity for C–2 and C–5, on the other hand, depends entirely on the rate of isomerization of the ylide compared with the rate of cycloaddition (see Section I.C.4). Photolysis of 2,3-diphenyl-2*H*-azirine also gives an ylide which condenses with dipolarophiles to form 1-pyrrolines.

The mechanism has been studied in detail by Padwa and co-workers,[293,294] who showed that the stereospecificity of the cycloaddition is affected by the size of the substituents in the olefin. The cycloaddition with methyl acrylate furnishes cis-4-methoxycarbonyl-2,5-diphenyl-1-pyrroline (Equation 41) which epimerizes to the trans isomer with methoxide. The latter is also formed when 2,4-diphenyl-2-oxazolin-5-one is heated with methyl acrylate.

(41)

c. *By Reduction.*   The presence of one substituent in pyrrolines can have a profound effect on the stereochemistry of reduction. This is observed in the catalytic reduction of 1-ethoxycarbonyl-4-(1-ethoxycarbonyl-1-cyano-methylene)-2-methylpyrrolidine, trans-3,4-dimethoxy-2-p-methoxyphenyl-1-pyrroline, and 3-hydroxyimino-5-methyl-1-phenylpyrrolidin-2-one with a platinum catalyst whereby cis-4-ethoxycarbonylmethyl-2-methylpyrrolidine (after hydrolysis),[295] trans-3-methoxy-trans-4-methoxy-r-2-p-methoxyphen-ylpyrrolidine (mainly),[296] and cis-4-amino-2-methyl-4-phenylpyrrolidin-2-one are formed.[297] The proportion of trans isomer in the latter compound can be increased by altering the catalyst. Sodium borohydride is also stereo-selective and reduces 1,2-diphenyl-,[298] 1-ethoxycarbonyl-2-phenyl-,[299] and 1-benzyloxycarbonyl-2-carboxypyrrolidin-4-ones[300] almost completely to the cis-2,4-disubstituted alcohols. In the last named compound the stereo-selectivity is reduced if the 1-benzyloxycarbonyl group is absent. These re-actions are particularly useful for the conversion of isomers, for example, by oxidizing an alcohol to a ketone and then reducing it back to another alcohol in a specific manner.[300] A remarkably regiospecific reaction is the reduction of 3-substituted 1-methylpyrrolidin-2,6-diones (succinimides) with $NaBH_4$ in which the 2-oxo group only is reduced to an alcohol group in preference to the less sterically hindered 6-oxo group.[301] Reduction of 3-benzylidene-4-hydroxypyrrolidines with LAH gives predominantly trans-3-benzyl-4-hydroxypyrrolidines,[302] and reaction of the diphenylmethyl ester of 1-ben-zyloxycarbonyl-4-methylene-S-proline with di(1,2-dimethylpropyl)borane, then hydrogen peroxide followed by removal of the protecting groups, provides cis-4-hydroxymethyl-S-proline.[303]

Catalytic hydrogenation of 2,5-dimethylpyrrole and its 1-methyl deriva-tive gives exclusively the cis-2,5-dimethylpyrrolidine and cis-1,2,5-trimethyl-pyrrolidine respectively. Reduction with zinc and acetic acid, on the other hand, yields trans-2,5-dimethyl-3-pyrroline which can be reduced catalyti-cally to the trans-pyrrolidine without trans-cis isomerization.[304] Catalytic reduction of cis-1-acetyl-5-methoxycarbonyl-4-methyl-2-pyrroline also gives the respective pyrrolidine without upsetting the stereochemistry.[305]

d.   By Skeletal Rearrangement.   Three interesting examples fall under this heading. The first involves a hydroxide ion induced ring contraction of a pyrid-2-one to a pyrrolidine and has been used extensively by Osugi[306] and Sanno[307] in their synthetic studies of kainic acid. Basically, a 3-halo-geno-4,5-disubstituted pyrid-2-one is converted into a 3,4-disubstituted pyr-rolidine-2-carboxylic acid in which the orientation of the 4- and 5-substitu-ents in the six-membered ring, which become the 3- and 4-substituents in the five-membered ring, is preserved (Equation 42). The second is a con-

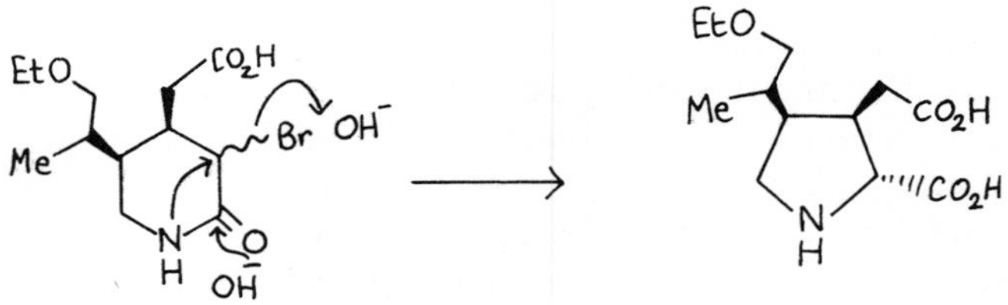

(42)

certed [2,3] anionic sigmatropic rearrangement of a tetrahydropyridine be-taine to a mixture of cis- and trans-2-benzoyl-3-vinylpyrrolidine in which one isomer predominates over the other. The predominance depends on the substituents R^1 and R^2 in Equation 43.[308] In the third example, the cis- and trans-1-benzoyl-2-phenylcyclopropanes, in N-methyl formamide at 180°C, form the intermediate N-formylimino derivative which rearranges

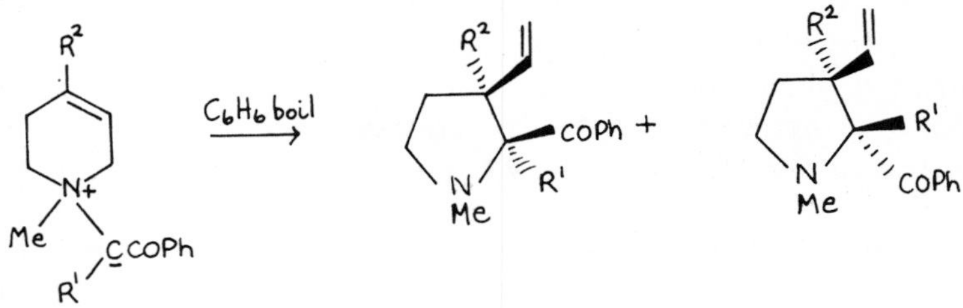

(43)

into a 2:1 mixture of *cis-* and *trans-*1-methyl-2,6-diphenylpyrrolidines irrespective of whether a mixture of *cis-* and *trans-* or only *cis-*cyclopropane is used (Equation 44). Reduction of a double bond must occur at some stage in this reaction and is probably brought about by formic acid.[309]

(44)

e. *From Aldimines and Succinic Anhydrides.* Castagnoli and co-workers[310,311] found that benzaldehyde (and substituted benzaldehydes) *N*-alkylimines react with succinic anhydride by formation of an amide followed by intramolecular addition across the C–N double bond to produce *cis-* and *trans-*4-carboxy-1-methyl-5-phenylpyrrolidin-2-one in which the trans isomer is the major product. This reaction has been extended to pyridine-3-aldehyde *N*-methylimine for the preparation of *trans-*3-methylnicotine (Equation 45).[312]

(45)

f. *Miscellaneous.* Many examples of partial or complete steric control have been observed in the pyrroline and pyrrolidine series (see Section III.C). Oxidation of 1-benzyloxycarbonyl-2-*p*-methoxybenzyl-3-pyrroline furnishes a 5:1 ratio of *trans-* to *cis-*1-benzyloxycarbonyl-3,4-epoxy-2-*p*-methoxybenzylpyrrolidines. The former can be converted into the cis isomer by ring opening of the epoxide group with acetate ions followed by ring closure with methane sulfonyl chloride and a base. Ring opening of the *cis-*epoxide, with trifluoroacetic acid, however, is not stereoselective.[313] Osmium tetroxide oxidation of 1-substituted 2-carboxypyrrolidines yields the *cis-*3,4-dihydroxy derivatives but with no steric control by the carboxyl group,[314] Yamada and collaborators[315] have carried out a large number of transformations starting from (−)-3-hydroxyproline in a determination of the absolute configuration of chiral mercaptosuccinic acid. All their reactions, which involved several pyrrolidines, were carried out with remarkable steric control.

Decarboxylation of 3-phenylpyrrolidin-5-one-2,2-dicarboxylic acid by heating at 150–160°C gave exclusively *cis-*3-phenylpyrrolidin-5-one-2-carboxylic acid. In collidine, on the other hand, a 2:1 ratio of *cis*- to *trans*-monocarboxylic acids[316] was produced. Similar decarboxylations had been reported by Sanno.[317]

2. Configuration of Pyrrolidines

The configuration of many substituted chiral pyrrolidines have been deduced by correlation with *S*-proline. The chemistry and reactions of optically active prolines are admirably described by Greenstein and Winitz.[318] The cd curves of 3- and 4-substituted prolines were examined and compared with those of chiral azetidine-2-carboxylic acid and 4-hydroxypiperidine-2-carboxylic acid.[319] Similarly the curves for *R*-tartrimide and *S*-malimide were measured, and the effects of solvents examined.[320] Prolines, like many amino acids, lack a strong uv absorbing chromophore and give plain ord curves which can make correlations difficult. Djerassi and co-workers[321] succeeded in overcoming this difficulty by obtaining anomalous dispersion curves on introducing a strong chromophore such as a 1-dithio-carbamic ester groups. This makes identification easier and facilitates configurational correlations. The mutarotation of *S*-proline and other chiral α-amino acids in aqueous potassium carbonate observed by Armarego and Milloy[322] is caused by the slow formation of the *N*-carboxy-(carbamate) salts. *R*(+)-3-methylpyrrolidin-2-one associates in *n*-hexane solution because an additional band at 202 nm appears in the cd spectrum as the solution is concentrated.[323] The chiral properties of many more pyrrolidines have been examined for correlation and identification purposes and will be found scattered in several of the references cited in these sections.

Only a few ir studies of pyrrolidines and related compounds have been made and the one that stands out is the effect on the C=N band of a substituent in 3-isoalkylidene-1-pyrrolines (**118**). The band does not shift in the expected manner when R[3] is changed from H to alkyl to aryl; this is explained by the out-of-plane twisting of the exocyclic alkylidene group and/or the R[3] groups.[324]

The absolute configuration of *S*(−)-2-aminoethyl-,[325] *S*(−)-2-methyl- and *S*(−)-1,2-dimethyl-,[283] *S*(+)-2-ethyl-2-methyl-5-oxo-,[289,326] *R*(+)- and *S*(−)-4-methylthio(and methylsulfonyl)-2-oxo-,[316,327] *R*(−)-3-methylthio-

118

119

120

2-oxo-, $R(-)$-3-methylsulfonyl-2-oxo-,[327] $R(+)$-3-methyl-2-oxo-,[323] $S(-)$-2-carboxy-5-oxo-,[282] $R(+)$-2-acetonyl-2-methyl-,[328] and $S(-)$-2-phenylpyrrolidines[328a] with one chiral center were obtained by chemical correlation. Pterolactam, $(+)$-2-methoxypyrrolidin-5-one (119), was isolated from Bracken but its absolute configuration is unknown.[329] It should be optically labile because it is essentially an α-alkanolamine (i.e., methanol adduct of 1-pyrrolin-5-one). The absolute configuration of the α-amino acid, $3R,5S,1'S$-3-(1'-amino-1'-carboxymethyl)-5-carboxypyrrolidin-2-one from *Pentaclethra macrophylla* was deduced by nmr, ord, cd, and X-ray analyses.[329a] Robertson and Witkop[330,331] converted 2-R,S-2-carbamoyl-3-pyrroline into $S(+)$-2-carboxy-3-pyrroline (120) in over 78% yield by an enzymic method. Obviously an asymmetric transformation (kinetic resolution) took place which was mediated by the enzyme. A similar resolution, without amide hydrolysis was achieved with $(+)$-α-bromocamphor-π-sulfonic acid in place of the enzyme. The amide racemized more rapidly than the acid at room temperature in water. The absolute configuration of the acid was deduced by catalytic reduction to S-proline.

The determination of configuration in pyrrolidines which possess two chiral centers is facilitated if the two centers can be related in some way. This is possible between C–2 and C–4 as in the formation of a lactone from *allo*-4-hydroxyproline[300] or a thiolactone from *cis*-4-thio-S-proline,[316,327] and between C–2 and C–5 as in the anhydride of 1-benzyloxycarbonyl-pyrrolidine-2,5-dicarboxylic acid.[332] All these compounds have a cis configuration and are useful starting materials for preparing other pyrrolines of known configuration. The configuration of all the 3-hydroxyprolines have been deduced in the course of the extensive and elegant chemical and biochemical investigations of Witkop and his collaborators.[279,284,333–335] The stereochemistry of all four 4-hydroxyprolines was first studied by Neuberger[336,337] and later in more detail by Witkop and his group.[285,338] *cis*- and *trans*- (i.e., S and R) 4-$^3H_1$-(and 4$^2H_1$)S-Prolines were synthesized by displacement of a 4-tosyl group by $LiAl^3H_4$ or $LiAl^2H_4$ with Walden inversion. The radioactive prolines were then used to show that enzymic hydroxylation of proline to 4-hydroxyproline with molecular oxygen occurred with retention of configuration.[339] If the absolute configuration of one substituent in the pyrrolidine ring is known then the configuration of the second can be determined by chemical bonding (as in the above lactones and anhydride) or from a close examination of the nmr spectra from which the cis or trans relationship of substituents can be assessed. This has been done directly or by comparison with model compounds as in *cis*-4-hydroxymethyl-,[303] *cis*-and *trans*-4-methyl-,[340] *cis*-and *trans*-4-fluoro-,[341] *trans*-4-*n*-propyl-,[342,343] and *cis*-3-methyl-S-prolines.[344] The configuration of *cis*- and *trans*-3-amino-2-carboxypyrrolidines was deduced from a comparison of ord curves, and by reaction with D-amino acid oxidase which destroys R-(D)α-amino acids.[345]

The Clough-Lutz-Jirgenson rule,[346] which states that the rotation of L-α-amino acids is more positive in aqueous acid than in water, predicts that viomycidine (**121**,2-guanidino-5-carboxypyrroline) obtained from the antibiotic viomycin, has the L (or S) configuration.[347]

A few pyrrolidines with three chiral centers are known and are naturally occurring. Several isomers of α-kainic acid (**122**) have been studied intensively by different Japanese workers.[348-353] The original configuration of 3-acetoxy-4-hydroxy-2-p-methoxybenzylpyrrolidine (anisomycin)[354] has been revised to the r-2(R)-cis-3(S)-$trans$-4(S) isomer **123** by an X-ray study of a bromoacetate derivative,[355] and by synthesis.[355a] The atomic parameters for S-proline,[356] 2S,4S-4-hydroxyproline,[357] S-5-iodomethyl- and S-5-carbamoylpyrrolidin-2-one,[358] and 3R-3-carboxy-2,2,5,5-tetramethyl-1-oxo-pyrrolidine (**124**)[359] were obtained from X-ray crystallographic studies. The last mentioned compound (**124**), which can be used in spin labeling experiments, was found to be almost planar with respect to C-2,N-1,C-5,O-6. The X-ray analysis of ($-$)-3,4-methylene-S-proline, obtained by the reac-

tion of carbene with 2S-2-methoxycarbonyl-1-trifluoroacetyl-3-pyrroline, revealed the cis configuration **125** and is identical with the naturally occurring amino acid from chestnuts. The major product from the carbene reaction is the $trans$-($+$) isomer.[360]

Reduction of the tosyl dipyrrolidin-2-one (**126**) with LAH gives a mixture of dipyrrolidines (**127** and **128**). The tartrate salt of isomer **128** gives an optically active base indicating that it is dissymmetric, and the tartrate of the second isomer yields the optically inactive *meso*-1,1'-dicyclohexyl-3,3'-bipyrrolidinyl (**127**).[361]

In their brilliant studies on the total synthesis of Vitamin B$_{12}$, Woodward and Eschenmoser[361a] and their respective schools have prepared a number of chiral reduced pyrroles, reduced bipyrromethines and corrins in which the absolute configurations of the many chiral centers were established.

3. *Conformation of Pyrrolidines*

Studies of the conformations of pyrrolidines have not been very intensive. The five-membered ring is conformationally more labile than the three- and four-membered rings. Inversion barriers of a few pyrrolidines were measured by dnmr methods and compared with those of other systems.[2,98,111,113] The effect of substituents on nitrogen inversion barriers is not very different from those observed in aziridines, but they are not quite as pronounced. Nmr parameters were obtained for 4-hydroxyproline, 4-allohydroxyproline,[362] and 3,4-dehydroprolinamide,[363] and the conformations were deduced from these values. Proton chemical shifts were used to distinguish between invertomers, for example, in the quaternization of 1-alkyl-2-methylpyrrolidines,[364] and in showing that the favorable conformation of 1-methyl-*cis*-2,5-diphenylpyrrolidine is one with the methyl group anti to the phenyl substituents.[365] From nmr studies it was shown that in the crystalline form *N*-nitroso-*S*-proline and 4-hydroxy-*N*-nitroso-*S*-proline existed in the syn conformation—unlike *N*-nitroso-2-*S*-2-carboxyazetidine which is in the anti form.[366] The esr spectra of radicals produced in aqueous solution from pyrrolidine and 1-alkyl derivatives and hydroxyl radicals at low pH values were examined and compared with those of piperidines, morpholines, and 1,4-dioxane. It was deduced that the pyrrolidinyl radicals assume rapidly interconverting half chair structures.[367]

Data regarding barriers to internal rotation in 1-methylpyrrolidin-2-one derived from far ir measurements was explained by almost free rotation of the molecule as opposed to free pseudorotation.[368] Pyrrolidine, on the other hand, like cyclopentane is a free (or only slightly restricted) pseudo-rotator (compare with tetrahydrofuran in Part II, Chapter 2, Scheme 5).[368a] Dipole moments by Lee and Kumler of *N*-acylpyrrolidin-2-one are consistent with a "cis-trans" arrangement with the conformation of the carbonyl groups in opposite direction to each other.[369] They have also determined and discussed the effect of substitution on the dipole moment of pyrrolidin-2-one.[370]

The signs of the cd curves of several substituted pyrrolidin-2-ones were correlated with the conformations which were imposed on the ring by the various substituents.[370a]

4.   Reactions of Pyrrolidines

a.   *Interconversion of Substituents.*   Reactions in which substituents in pyrrolidines are interconverted are numerous and fall into two categories. The first includes examples in which the chiral center is not altered as in the conversion of *S*-proline into *S*(−)-2-methylpyrrolidine by way of *S*-(+)-prolinol, *N,O*-ditosyl-*S*-prolinol, and (−)-1-tosyl-2*S*-2-methylpyr-rolidine.[371] Other examples are the formation of lactones,[315,327,372] esters, amides, and products from the Curtius degradation of the acid azide to an amino group with retention of configuration,[332] the oxidation of a methylthio to a methylsulfonyl group,[315,327] and methylation and esterification of a hydroxy or thiol group.[315,327,373] Examples of the second category are more plentiful and involve nucleophilic substitutions which invariably occur by S_N2 displacements. These displacements are generally very clean and afford products which are relatively free from epimers. *cis*- and *trans*-4-Chloro-, 4-bromo-, 4-amino-,[372] and 4-thio-*S*-pro-lines[315,317,373] and their 1-methyl betaines[374] were prepared in this way. The configuration of a hydroxy group can be inverted by converting it into its tosyl derivative and displacing this group by hydroxide ions.[375] Further examples will be found among the references in Sections 1 and 2 of this chapter.

b.   *Intramolecular Cyclization.*   All the possibilities in which any two atoms of the pyrrolidine nucleus have been joined, by intervening atoms, are known. *S*-Prolinol-*O*-sulfate betaine cyclizes to 1*S*,3*S*(−)-1-azabicyclo-[3.1.0]hexane (**129**) in a boiling alkaline solution.[376] The *N*-ethyl quaternary

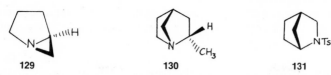

129   130   131

salt of **129** was obtained from both 2-chloromethyl-1-ethylpyrrolidinium chloride and alkali, and from alkaline treatment of 3-chloro-1-ethylpiperidinium chloride.[377,378] The linkage between N–1 and C–3 is brought about by intramolecular reductive cyclization of 4-ethyl-3-hydroxyimino-1-methoxycarbonylmethyl-4-phenylpyrrolidine, which yields 8-ethyl-8-phenyl-1,4-diazabicyclo[3.2.1]octan-3-one,[379] and similar ring closures.[299,379] A cyclization which, for nomenclature reasons only, is described as an N–1,C–4, but which is another example of an N–1,C–3, ring closure is the intramolecular alkylation of cis-4-hydroxyethyl-2-methylpyrrolidine to 2-methyl-1-azabicyclo[2.2.1]heptane **(130)** and is brought about by PBr_5.[295] Several examples in which C–2 and C–3 are linked are discussed separately in Section E.1. The linkage between C–2 and C–4 is achieved by the cyclization of trans-1-tosyl-3-tosyloxy-2-tosyloxymethylpyrrolidine with benzylamine to 5-benzyl-2-tosyl-2,5-diazabicyclo[2.2.1]heptane which is then converted into the parent base by hydrolysis followed by hydrogenolysis.[380] A similar linking is obtained by cyclization of 2-(2,2-dicarboxyethyl)- or 2(2,2-dicarboxyacetyl)-1-tosyl-4-tosyloxypyrrolidine followed by hydrolysis and decarboxylation to 2-tosyl-2-azabicyclo[2.2.1]heptane[381] **(131)**, or its 5-oxo derivative.[382] Examples of pyrrolidines with C–3 and C–4, and C–2 and C–5 bridges are (−)-3,4-methylene-*S*-proline **(125)**[360] and the anhydride of 2,5-dicarboxypyrrolidine.[332]

c.   *Alkylation.*   The methylation of cis-1,2,5-trimethyl- and 3-(diphenylmethylene)1,2,5-trimethyl-pyrrolidines is about 4 and 10 times faster than the methylation of the respective trans isomers as expected from steric considerations.[304] McKenna and collaborators[382a] studied the kinetics of quaternization of 1-alkyl-2-methylpyrrolidine and found that attack of the reagents is preferentially from an axial approach and cis to the 2-methyl group. Alkylation of 1-alkyl-2-phenylpyrrolidines yields a mixture of quaternary salts in which the entering groups are oriented cis and trans to the 2-phenyl group. Product ratios indicate that the 2-phenyl group has some steric control on the direction of alkylation. In the methylation of 1-alkyl-2-phenylpyrrolidines the products in which the quaternizing methyl groups are cis to the 2-phenyl groups predominate, but if the size of the alkylating group is larger than methyl (e.g., $PhCH_2Br$) then the salt in which the quaternizing benzyl group is trans to the 2-phenyl group is predominant.[383]

d.   *Miscellaneous.*   Thermal extrusion of nitrogen from cis and trans-2,5-dimethylpyrrol-3-ine in the presence of nitrohydroxylamine ($NO_2 \cdot NH$-

[OH]) takes place with complete stereospecificity (sigma symmetric) to yield *trans-trans-* and *cis-trans-*2,4-hexadiene respectively in accordance with Woodward-Hoffmann theory.[384]

While correlating *cis-* and *trans-*3-methyl-*S*-proline with alloisoleucine, Witkop and co-workers cleaved the formally enamine bond of 1-acetyl-5-ethoxycarbonyl-4-methyl-2-pyrroline with ethylthiol to yield ethyl 2-acetamido-5,5-bisethylthio-3-methylpentanoate without upsetting the asymmetric centers.[305,385]

Yamada and his school have developed several stereoselective syntheses starting from *S*-proline and its derivatives. Alkylation of the enamine derived the *t*-butyl ester of *S*-proline and 4-methyl(or 4-*t*-butyl)cyclohexanone yielded after hydrolysis a mixture of *cis-* and *trans-*2-alkyl-4-methyl-(or *t*-butyl)cyclohexanone in which the trans isomer predominated.[386] Similar selectivity was observed in the formation of 2-bromocyclohexanone.[387] *S-t-*Butoxycarbonyl-1-(2-phenylprop-1-enyl)pyrrolidine (the enamine from diphenylacetaldehyde) reacts with methyl vinyl ketone and, after hydrolysis and cyclization, yields *R*(+)-4-methyl-4-phenylcyclohex-2-enone in optical yields which vary with the size of the 2-substituent in the pyrrolidine.[388,389] In a third application 1-amino-*S*-proline was used for the asymmetric syntheses of *S*-prolyl-*R*-amino acids.[390]

Chymotrypsin catalyzes the hydrolysis of *p*-nitrophenyl esters of the spin labeled compound 3-carboxy-2,2,5,5-tetramethyl-1-oxypyrrolidine (e.g., **124**) with low enantiomeric specificity.[391]

C.   Spiropyrrolidines

There are two groups of spiropyrrolidines—those in which the quaternary atom is a carbon atom and those in which it is a nitrogen atom. Dissymmetry in these molecules is not because they lack a center or a plane of symmetry but because they lack a rotating axis of symmetry.[392] This simply means that no matter how the dissymmetric molecule is rotated there is at least one position in which it will not superimpose on its mirror image.

The best way of obtaining the absolute configuration of a dissymmetric spiro compound of the first group is to synthesize it in a stepwise sequence. Krow and Hill[393] prepared *S*(−)-2,7-ditosyl-2,7-diazaspiro[4,4]nonane (**132**) starting from *R*(−)-4-carboxymethyl-4-ethoxycarbonylpyrrolidin-2-one (Equation 46). The absolute configuration of the latter was deduced by reduction with LAH and conversion into 2-thia-7-azaspiro[4,4]nonane which was subsequently desulfurized, oxidized, and hydrolyzed to the known *S*(−)-2-ethyl-2-methylsuccinic acid. Steric interactions in the

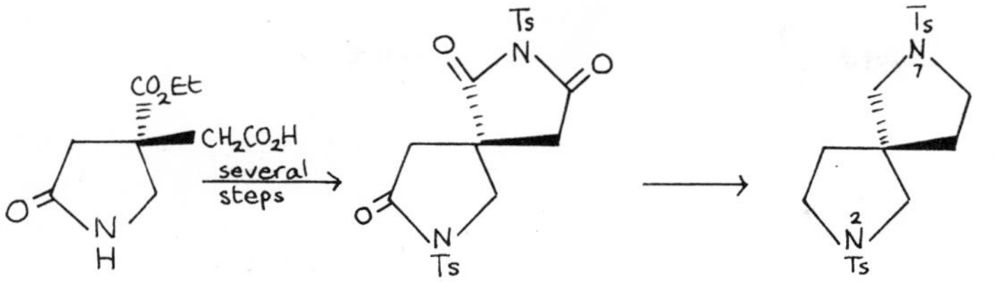

(46)

132

syntheses of spiro compounds can sometimes lead to only one of two possible diastereomers. Such is the formation of 2′-phenyloxindole-3-spiro-3′-pyrrolidine (**133**) from a Mannich reaction of 3-(2′-aminoethyl)oxindole and benzaldehyde. The benzene ring obviously directs the orientation of the 2′-phenyl substituent.[394] A biosynthetic parallel has been observed in the alkaloid rhyncophylline.[395]

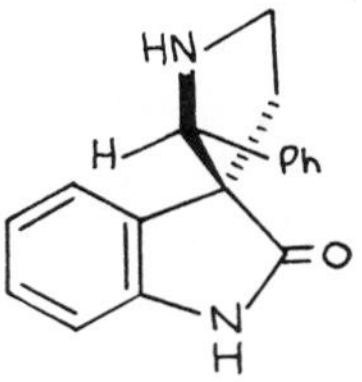

133

In a classical series of papers McCasland and Proskow[396-398] demonstrated experimentally for the first time that the presence of a fourfold alternating axis of symmetry was a sufficient condition for optical inactivity. They prepared all four 3,4,3′,4′-tetramethylspiro[1,1′]bipyrrolidinium (**134-137**) toluene-p-sulfonates from optically active compounds related configu-

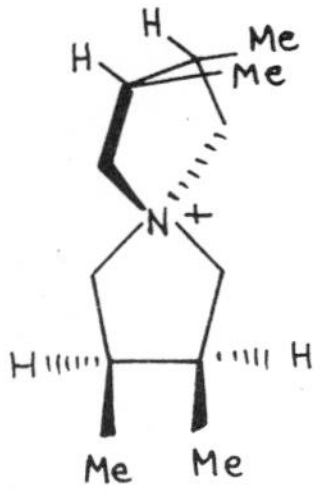

cis/cis
(opt active)

134

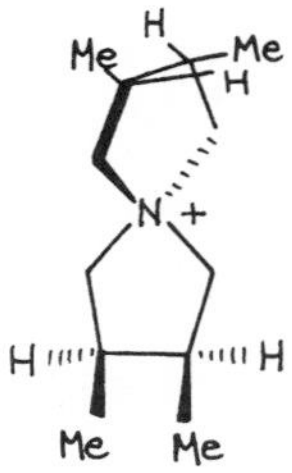

cis/trans
(opt active)

135

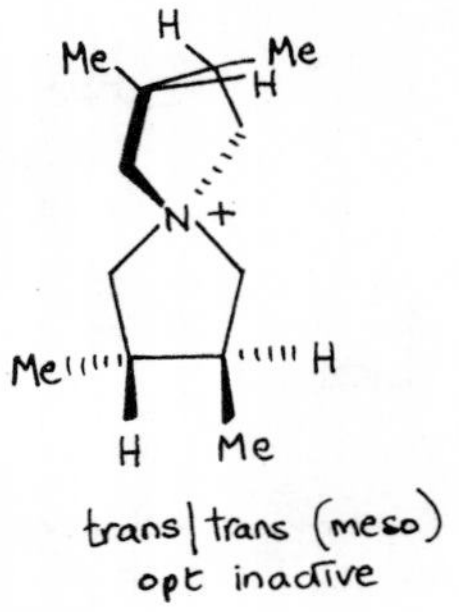

trans | trans (meso)
opt inactive

136

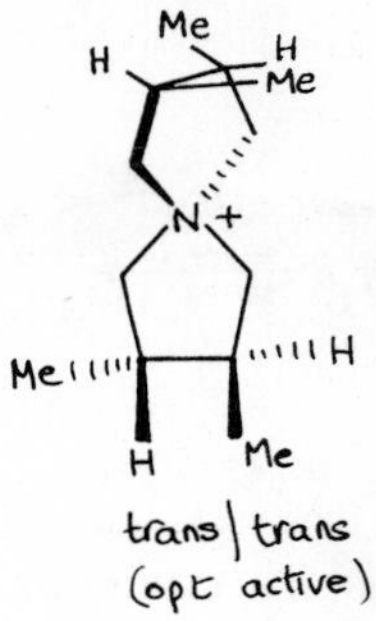

trans | trans
(opt active)

137

rationally to (+)-2,3-dimethylsuccinic acid of now known absolute configuration. (−)-3,4-Dimethylpyrrolidine was condensed with the (−)-di-O-tosyl derivative of butane-2,3-diol and gave the optically inactive meso spirane **136**. The dextro enantiomers gave the same meso compound (**136**). Five-, six-, and seven-membered [1,1'] spiro systems were prepared by the cyclization of suitable tertiary aminocarbinols using toluene-*p*-sulfonyl chloride in the presence of tertiary bases.[399] The absolute configuration of *R*(−)-4,4'-dimethoxy-1,1',3,3'-tetrahydrospiro[isoindole-2,2'-isoindolinium]bromide (**138**) was deduced by conversion with phenyl lithium into *S*(−)-4,8-dimethoxy-5,7,12,12a-tetrahydroisoindolo[2,1-*b*]isoquinoline (**139**) by a Stevens rearrangement. The absolute configuration of **139** was

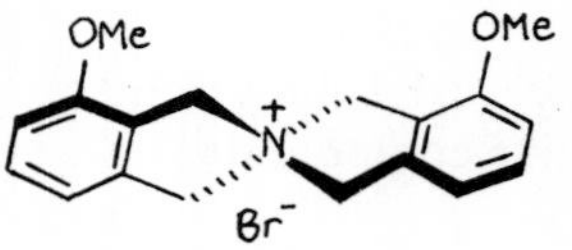

138

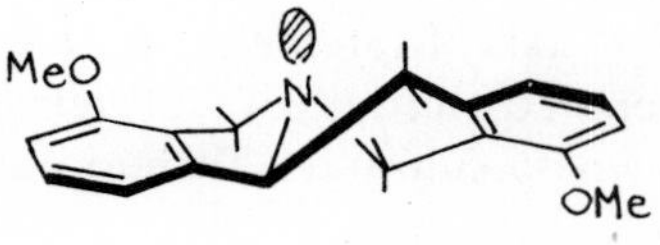

139

obtained by ozonolysis to *N,N*-dicarboxymethyl-*S*-aspartic acid and is in agreement with Lowe's rule, the helix conductor model, and the Eyring-Jones model[400] of optical activity. The spiro compound **132**, however, is an exception to these rules.

D.   Chiral Pyrroles

In 1931 Lions[401] recognized the possibility that 1-phenylpyrroles with bulky substituents in the ortho positions should be chiral and that it should be possible to obtain these in optically active forms. Their chirality is due to restricted rotation about the pivot bond. In these examples inter-conversion of enantiomers is possible only if the substituents ortho to the pivot bond are capable of crossing over each other. Lions, however, was unable to resolve 1-(2-phenyl)phenyl-2,5-diphenylpyrrole-3-carboxylic acid into it optically active enantiomers. Bock and Adams[402] also could not resolve 1-(3-, or 4-carboxyphenyl)-2,5-dimethypyrrole-3-carboxylic acid, but they succeeded in separating the diastereoisomeric brucine salts of 1-(2-car-boxyphenyl)-2,5-dimethylpyrrole-3-carboxylic acid and obtained the enan-tiomeric free acids **140** and **141**. These acids were optically stable in ethanolic solution but racemized in boiling glacial acetic acid or $0.1N$ sodium hy-droxide solutions.[403] Optically active 3,3'-dicarboxy-2,2',4,4',5,5'-hexa-

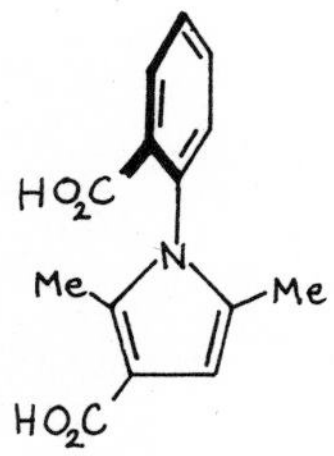

140

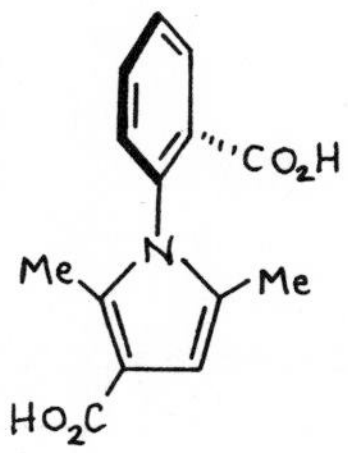

141

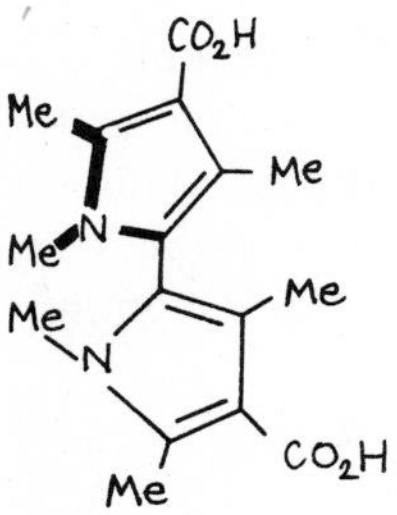

142

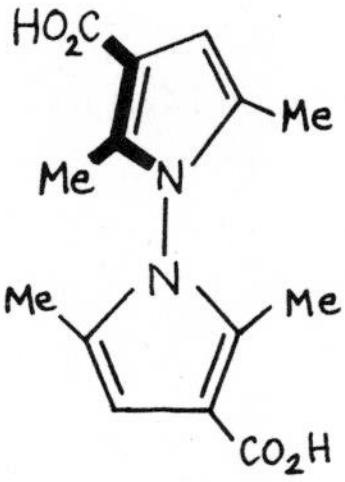

·143

methyl-2,2'-bipyrrolyl (**142**)[404] and 3,3'-dicarboxy-2,2',5,5'-tetramethyl-1,1'-bipyrrolyl (**143**)[405] were also obtained by resolution of the brucine salts. Whereas the free acid **142** racemized on melting, the 1,1'-dipyrrolyl (**143**) was unusually resistant to racemization. Chang and Adams[406] went even further and showed that 4,6-bis(3-carboxy-2,5-dimethyl-1-pyrrolyl)-1,3- xy-lene (**144**) existed in two diastereomeric forms; one was resolved into its enantiomers and the second was the meso form. The dipyrrolylbiphenyl (**145**) behaved similarly.[407]

E.   Fused Pyrrolidines

The fused pyrrolidines described in this section possess at least one C–C bond common to the fused ring, that is, they belong to the azabicyclo-[x.y.0]alkane ring system. Fused pyrrolidines in which the nitrogen atom

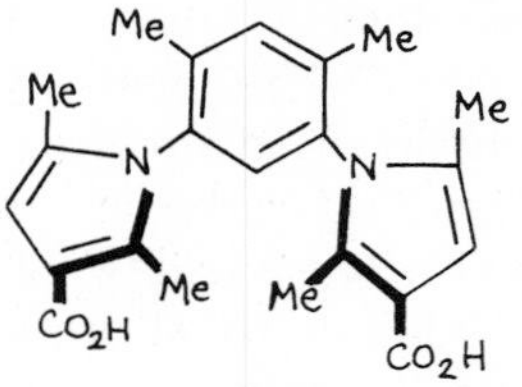

144

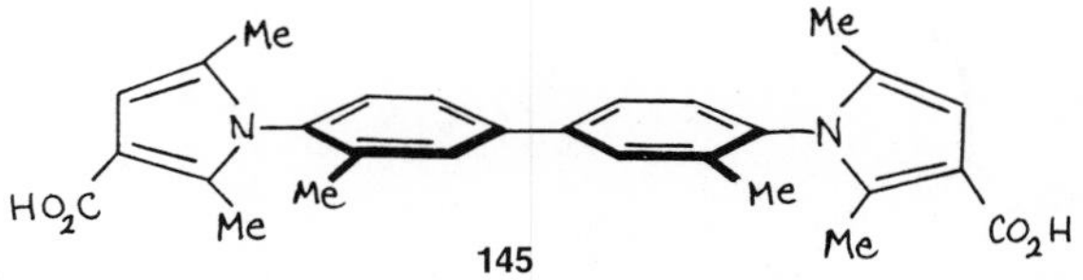

145

is at the bridgehead position between the rings are discussed in Chapter 4, Section VI. A and B.

1.   2- and 3-Azabicyclo [3.1.0] hexanes

The ring fusion in this system is cis because of the small sizes of the rings. Fowler[408] prepared 1-methoxycarbonyl-6-ethoxycarbonyl-2-azabicyclo-[3.1.0]hex-3-ene (146) by decomposing ethyl diazoacetate with Cu_2Cl_2 in the presence of 1-methoxycarbonylpyrrole. The available enamine double bond in the product can react further to yield the two possible tricyclic pyrrolidines: the syn 147 and anti 2-azatricyclo[4.1.0.0^{3,5}]heptane-2,4,7-tricar-

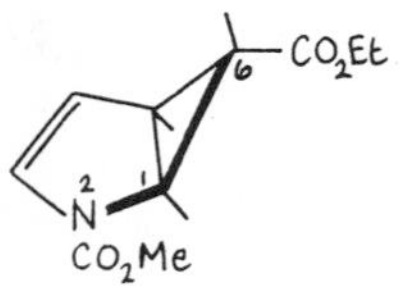

146

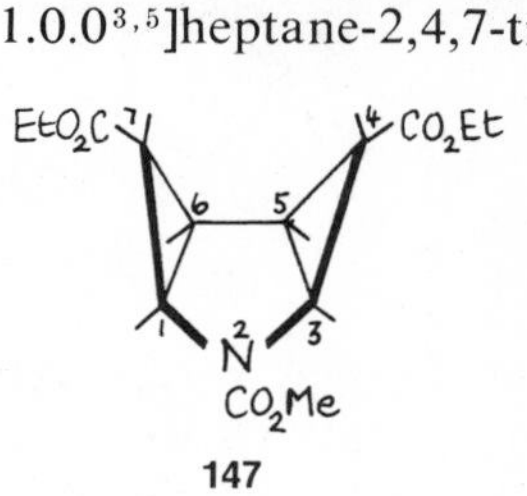

147

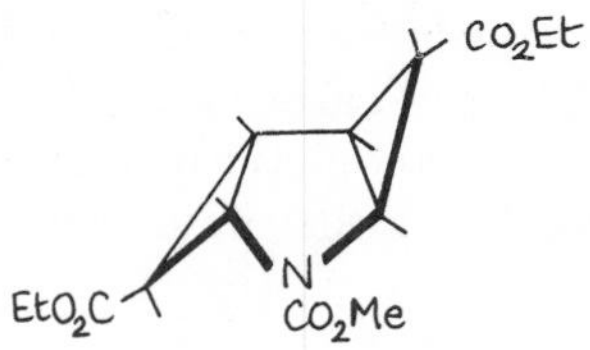

148

boxylic ester (148) ([4.1.0.^{1,6}0^{3,5}] is understood). The carboxylic ester groups in the cyclopropane ring can be removed, and the syn and anti tricyclic compounds rearrange to 1-methoxycarbonyl-2,3-dihydroazepine at 121 and 350°C respectively. The syn isomer also shows greater reactivity by undergoing a cycloaddition reaction with *N*-phenylphthalimide.[409] The bicyclic pyrroline 146 shows dual cycloaddition reactivity. It reacts at the cyclopropane ring with dimethyl acetylene dicarboxylate or *N*-phenylphthalimide to yield 8-azabicyclo[3.2.1]octane derivatives in contrast to ethyl diazoacetate (see 146) and tetracyanoethylene which react at the double bond to give 2-azatricyclo[4.1.0.0^{3,5}]heptane and 2-azatricyclo-[4.2.0.0.^{3,5}]octane derivatives respectively.[410] The ring system 146 can also be obtained by condensation of diphenylcyclopropenthione with enamines as in the preparation of a number of 2-azabicyclo[3.1.0]hex-3-ene-3-thiolates.[411]

One example of the 3-azabicyclo[3.1.0]hexanes has already been described (see 125, Section B.2) but further examples were made by reacting maleic anhydrides with dimethyl sulfoxide methiodide in the presence of sodium hydride[412] or with diaryldiazomethane.[413] The 6,6-diaryl-3-azabicyclo[3.1.0]-hexanes obtained by reduction of the products from the last mentioned reactions were obtained unambiguously by intramolecular cyclization of *cis*-1-aminomethyl-2,2-diaryl-3-hydroxymethylcyclopropane with toluene-*p*-sulfonyl chloride in the presence of triethylamine.[413] In their extensive and comprehensive study of the reactions of chlorosulfonyl isocyanate, Paquette and his collaborators obtained 3-azabicyclo[3.1.0]hexanes by reaction with 1,2,2-trimethyl- and 1,2,3,3-tetramethylbicyclo[1.1.0]butane (Equation 47). The reaction with 1,2-dimethylbicyclo[1.1.0]butane, however, followed a different course.[414]

(47)

2.   2- and 3-Azabicyclo [3.2.0] heptanes

Like the bicyclo[3.1.0]hexane system in the previous section (E.1), the stable configuration of azabicyclo[3.2.0]heptanes has the bridgehead carbon atoms in the cis configuration.

Under the influence of ultraviolet light and in the presence of benzophenone as a sensitizer pyrrole undergoes a cycloaddition reaction with dimethyl acetylene dicarboxylate to yield 6,7-dimethoxycarbonylbicyclo-[3.2.0]hepta-3,6-diene. This rearranges into 3,4-dimethoxycarbonyl-1H-azepine.[415] The reverse reaction, however, is more common, for example, the photochemical isomerization of 2-amino-3H-, 2-dimethylamino-3H-, and 2-ethoxy-3H-azepines into 2-amino-,2-dimethylamino-, and 2-ethoxy-2-azabicyclo[3.2.0]hepta-3,6-dienes[416]; 1-methoxycarbonyl-1H-azepine into 2-methoxycarbonyl-2-azabicyclo[3.2.0]hepta-3,6-diene[417]; 3,7-dimethyl-1,3-dihydro-2H-azepin-2-one and 3,5,7-trimethyl-1,3-dihydro-2H-azepine into 1,4-dimethyl-2-azabicyclo[3.2.0]hept-6-en-3-one[418] and 1,4,6-trimethyl-2-azabicyclo[3.2.0]hept-6-ene[419] respectively (Equation 48). The epimer of

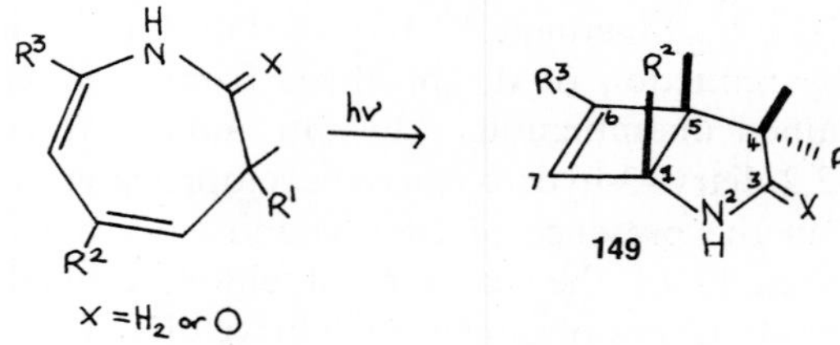

(48)

149

X = H$_2$ or O

149 is sometimes a minor product and both can be reduced to the fully saturated systems. The reverse reaction is thermally allowed. The non-photolytic base catalyzed rearrangement of 4-chloromethyl-3,5-dimethoxy-carbonyl-1,2,6-trimethyl-1,4-dihydropyridine yields a mixture of 4,7-di-methoxycarbonyl-1,2,7-trimethyl-2-azabicyclo[3.2.0]hepta-3,6-diene and the 1H-azepine in varying proportions depending on the concentration of reactants.[420]

The isomeric 3-aza system has been obtained by the [2 + 2]π cycloaddition of ethylene and 2,3-dichloromaleimides under the influence of uv light,[421] and by the cycloaddition of enamines or ynamines and maleimides.[422] The products from the former reactions are a useful source of cis-cyclobutane-1,2-dicarboxylic acid (Equation 49).[421]

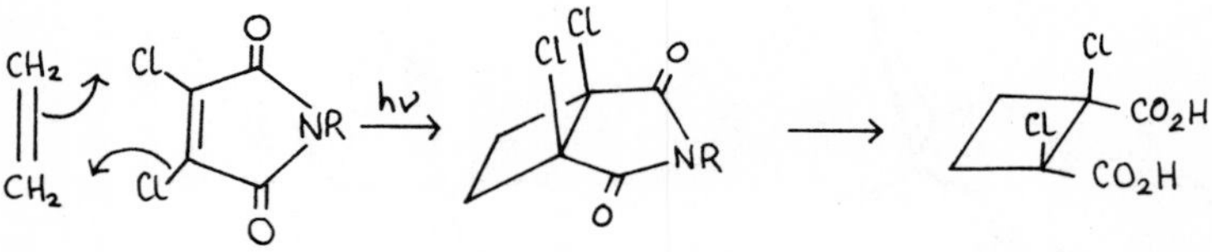

(49)

3. 2-and 3-Azabicyclo [3.3.0] octanes

Barrett and Linstead[423] realized that the strain in cis-fused bicyclo[3.3.0]-octane (150) is less than that in the trans-fused isomer (151), and that it results from tetrahedral distortion at the bridgehead carbon atoms. An inverting nitrogen atom in place of a 2-CH_2 or 3-CH_2 may slightly relieve

150    151

the strain. Both *cis*- and *trans*-2-azabicyclo[3.3.0]octanes have been pre-pared. Prelog and Szpilfogel[424] reduced 2-(2-phenylethyl)cyclopentanone oxime to the amine, and cyclized the derived alcohol to 2-azabicyclo[3.3.0]-octane by way of the 2′-bromoethyl derivative. The trans structure for this product was confirmed by King and his co-workers[425] by a similar synthesis. The experimental difficulties in trying to make the cis isomer were over-come when 2-ethoxycarbonylmethylcyclopentanone oxime was reduced with Raney nickel to a mixture of *trans*-2-ethoxycarbonylmethylcyclo-hexylamine and *cis*-2-azabicyclo[3.3.0]octan-3-one. Reduction of this oxo compound gave the required *cis*-2-azabicyclo[3.3.0]octane. The trans struc-ture for the aminoester was deduced from its reluctance to cyclize to the lactam, but it was successfully reduced to the alcohol which was cyclized to *trans*-2-azabicyclo[3.3.0]octane as before (Scheme 7). The trans isomer has

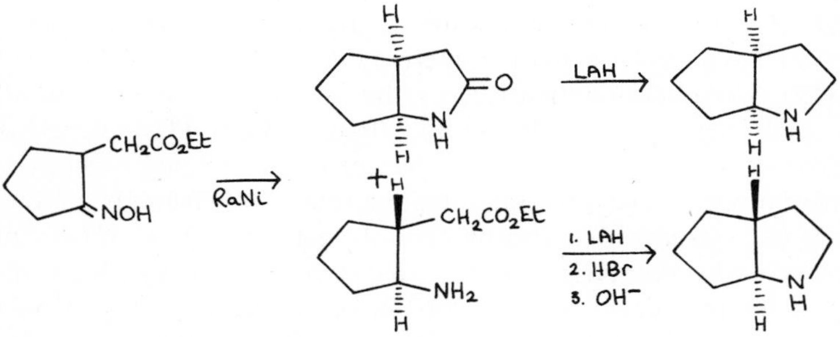

Scheme 7

a higher boiling point and refractive index but lower density than the cis isomer, which is in qualitative agreement with the Auwers-Skita rules which are now expressed more accurately by Allinger's generalization[426] that the higher boiling points, refractive indexes, and densities are attributed to the less stable configuration. Bertho and Rödl[427] obtained *cis*-2-aza-bicyclo[3.3.0]octane and its 2-methyl, 2-benzyl, and 2-phenyl derivatives by LAH reduction of the 3-oxo derivatives which were in turn formed from the Raney nickel reduction of carboxymethylcyclopentanone in the presence of the respective primary amines. In a variant of this method 2-(1,1-dimethylaceton-1-yl)cyclohexanone cyclized in the presence of ammonia to *cis*-1-hydroxy-3,4,4-trimethyl-2-azabicyclo[3.3.0]oct-2-ene which was catalytically reduced into a mixture of diastereomeric *cis*-3,4,4-tri-methyl-2-azabicyclooctanes in which the 3-methyl group was in the exo and endo configurations.[428] The reduction of 2-acetonylcyclohex-1-ene ox-ime with zinc and acetic acid was studied in some detail and found to yield 3-methyl-2-azabicyclo[3.3.0]oct-2-ene and its 8-hydroxy derivative. The latter was reduced to the octane and three isomeric forms were iso-lated. The cis configuration was ascribed to all of these but only one gave a blue-violet color with copper sulfate solution indicating that in two iso-mers the 3-methyl group or 8-hydroxy groups inhibited the reaction of copper ions with the nitrogen atom.[429–432] Dry distillation of *cis*-1-benz-amido-2-carboxymethylcyclopentane, but not the trans isomer, gave *cis*-3-phenyl-2-azabicyclo[3.3.0]oct-2-ene.[433]

N-Methyl cyclopentylethylamine undergoes the Hofmann-Freytag-Löffler reaction on treatment with *N*-chlorosuccinimide and strong acid to give *cis*-2-methyl-2-azabicyclo[3.3.0]octane free from the trans isomer.[434] Photo-chemical cyclization of *N*-nitroso-*N*-methylcyclohex-1-en-3-ylethylamine occurs in methanolic solution yielding a mixture of *cis*-2-methyl-2-aza-bicyclo[3.3.0]octan-8-one and its oxime.[435] 6-*endo*-Phenyl-(and 6-*endo*-methyl)5-*exo*-nitronorborn-2-enes undergo a rearrangement when treated successively with ethanolic potassium hydroxide then hydrochloric acid to give 2-hydroxy-*cis*-4-phenyl-(or methyl)*cis*-2-azabicyclo[3.3.0]oct-7-en-3-one (**152**). The phenyl derivative is reduced and isomerized with alkali to the *trans*-4-phenyl-*cis*-2-azabicyclo[3.3.0]octan-3-one (**153**) which is inde-pendently synthesized by catalytic reduction of 2-(α-phenyl-carboxymethyl)-cyclopentanone in the presence of ammonia. The structure of the inter-mediate **152** was also deduced by ozonolysis to the known *d,l*-erythro-3,4-dicarboxy-4-phenylbutyric acid.[436] The stereospecific synthesis of *cis*-3-methyl-4-vinylpyrrolidines by intramolecular cyclization (Equation 40) has been extended to the synthesis of 2-benzoyl-*trans*-4-methyl-*cis*-2-azabicyclo-[3.3.0]oct-6-ene from *N*-alkyl-*N*-benzoylcyclopent-1-en-3-ylamine.[287]

Four widely different series of reactions have been described for the synthesis of *cis*-3-azabicyclo[3.3.0]octanes. The simplest is the LAH reduc-tion of *cis*-cyclopentane-1,2-dicarboxylic acid imide.[429] In another approach a Dieckmann cyclization of 1-acyl-*cis*-3,4-bismethoxycarbonylmethylpyr-rolidine (prepared by oxidation of the double bond of 8-acyl-*cis*-8-aza-bicyclo[4.3.0]non-3-ene followed by esterification of the acid) produced

152   **153**

3-acyl-8-methoxycarbonyl-*cis*-3-azabicyclo[3.3.0]octan-7-one which was converted into the parent base by hydrolysis, decarboxylation, and Wolff-Kishner reduction of the carbonyl group.[437] 5,5-Dimethyl-1-vinylbicyclo-[2.1.1]hexane undergoes a Ritter reaction with benzonitrile in acid to give a mixture of *cis*-8-ethyl-4,4-dimethyl-2-phenyl-3-azabicyclo[3.3.0]octa-2,7-diene and *cis*-8-ethyl-4,4-dimethyl-2-phenyl-3-azabicyclo[3.3.0]oct-2-ene in low yield.[438] C–1,C–2 bond cleavage in 2-aminoalkyl nitrates is induced under basic conditions and was applied to the cleavage of the norbornane derivative **154**. The stereochemistry of the product *cis*-7-hydroxymethyl-3-methyl-*cis*-3-azabicyclo[3.3.0]octane (**155**), after LAH reduction, follows from that of the starting material.[439]

154   **155**

4. 7-Azabicyclo [4.3.0] nonanes (Perhydroindoles)

The interest in this ring system arose because of its presence in naturally occurring substances, for example, tryptophane, mesembrines, and lycoranes. Willstätter and Jaquet[440] made 7-azabicyclo[4.3.0]nonane by catalytic reduction of indole and described its properties. Reduction of indole with a nickel catalyst also gave 3-ethylcyclohexylamine.[441] This amine was the cis isomer because it was prepared unambiguously, and from this evidence the cis-fused bicyclononane structure (perhydroindole) was proposed.[442] More recently it was shown that although catalytic reduction of indole with Raney nickel in cyclohexane gives perhydroindole, in methanol[442] or ethanol,[442,443] on the other hand, it yields the *N*-methyl and *N*-ethyl derivatives respectively. Catalytic reduction of 7-methoxy-1-methyl-indole in acetic acid using platinum oxide gives a 2:3 mixture of *cis*-7-methyl-7-azabicyclo[4.3.0]nonane and its 5-methoxy derivative. Each of these is probably formed by a different mechanism. The stereochemistry was deduced from the nmr spectrum. The 5-methoxy group is equatorial because demethylation with hydriodic acid provides the alcohol with the same configuration and conformation (156) as the parent which shows very strong intramolecular hydrogen bonding in the ir spectrum.[444] Unlike the trans 7-azabicyclo[4.3.0]nonane which has a rigid conformation (157),

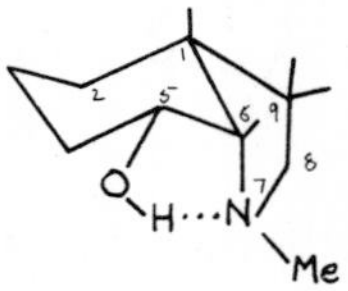

156

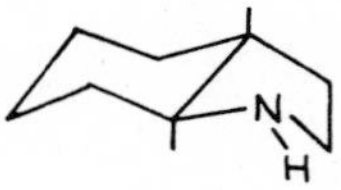

157

the cis isomer is more flexible and can equilibrate between the two extreme conformations of the bicyclic ring, that is, between 156 and 158. Note the relationship between the C–1 and C–9 protons in 156 and 158. Also conformers in which the cyclohexane ring is in the boat form may become important in this system. Catalytic reduction of 7-aryl-7-azabicyclo[4.3.0]-non-1,6-ene[445] and -non-5,6-ene[446] with platinum oxide similarly yields the cis-fused system. In the unusual compound 7-azabicyclo[4.3.0]nona-8,9-dione, which exists predominantly in the enolic form, 9-hydroxy-7-azabi-cyclo[4.3.0]non-9(1)-en-8-one reduction with NaBH₄ yields the 9-hydroxy-nonanone which is in the conformation 159 in aqueous solution.[447]

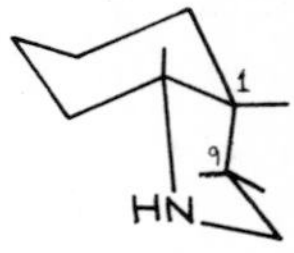

158

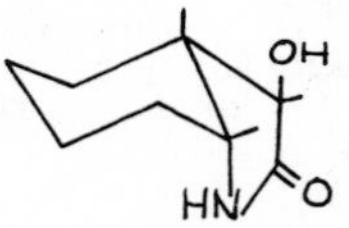

159

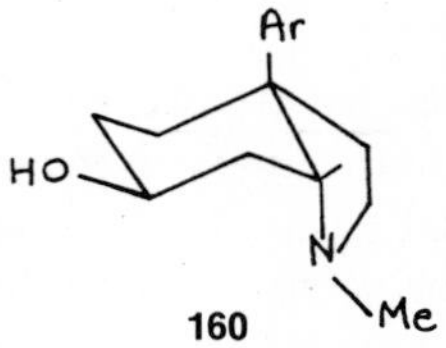

160

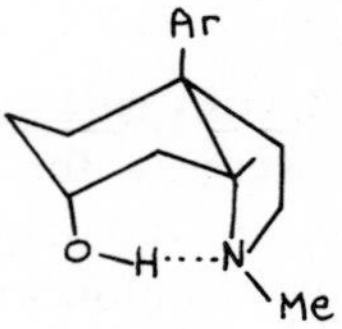

161

King and his school[448,449] synthesized *cis*- and *trans*-7-azabicyclo[4.3.0]-nonanes from authentic *cis*- and *trans*-2-(2-bromoethyl)cyclohexylamines and showed conclusively that the 2-azabicyclo[4.3.0]nonanes of Willstätter and co-workers,[440,441] Metayer,[443] and Adkins and Coonradt[450] had the cis configuration. In this series as in the azabicyclo[3.3.0]octane series (Section III. E.3) the boiling points, refractive indexes, and densities of the cis isomer are fractionally higher than those of the trans isomer in accordance with the Auwers-Skita rule.[448] Unlike the cyclization of cis- and trans-2-(2-bromoethyl)cyclohexylamines, the dry distillation of cis-1-benzamido-2-carboxymethylcyclohexane yields cis-8-phenyl-7-azabicyclo[4.3.0]non-7-ene, but the trans acid does not cyclize.[433] The intramolecular cyclization of 3-N-chloro-N-methylaminoethylcyclohex-1-ene, which is promoted by silver nitrate in methanol, yields a mixture including 5-methoxy-, 5-chloro-, and 5-nitrooxy-cis-7-methyl-7-azabicyclo[4.3.0]nonanes. A mechanism involving a nitrenium ion intermediate was favored.[451] A Claisen rearrangement with cleavage of the N–N bond occurs when acetophenone 2,4-dimethylphenylhydrazone is heated in nitrobenzene and one of the four products has been identified as 1,3-dimethyl-8-phenyl-cis-7-azabicyclo-[4.3.0]nona-3,7-dien-2-one and was reduced to the parent compound.[452]

Several syntheses of 1-aryl-7-azabicyclo[4.3.0]nonanes were developed because of their potential use in the preparation of alkaloids related to mesembrine. X-Ray studies and the determination of the absolute configuration of some of these alkaloids revealed that the five- and six-membered rings are cis fused.[453–455] Because of this and because further cyclization of cis-1-aryl-7-azabicyclo[4.3.0]nonane with formaldehyde yields the crinane skeleton found in the Amaryllidaceae alkaloids, nonanes of this configuration were investigated.[456–457] 1,1-Biscyanomethyl-1-(3,4-dimethoxyphenyl) acetone cyclizes in 65% sulfuric acid to the 1-aryl-7-azabicyclo-[4.3.0]non-5(6)-en-4,8-dione which is a source of cis- and trans-1-aryl-7-azabicyclo[4.3.0]nonane ketones from which d,l-mesembrine and mesembrinols can be obtained.[458,459] A second synthesis developed by Taguchi, Oh-ishi, and Kugita[460] involves Pictet-Spengler cyclization of 3-aryl-3-(butan-3-onyl)-1-pyrroline which gives 1-aryl-cis-7-azabicyclo[4.3.0]nonan-4-one. Reduction of this ketone with NaBH$_4$, after N-methylation, produces a mixture of epimeric alcohols **160** and **161**. The isomer **161** exhibits strong intramolecular hydrogen bonding.[461] Another synthesis is the catalytic reduction of 1-aryl-7-azabicyclo[4.3.0]nonenes with a $\triangle^{6,7}$ or $\triangle^{5,6}$ double bond (obtained by different routes)[456,457,462,463] which yields predominantly the cis-fused nonane. The addition of aryl lithium across a $\triangle^{6,7}$ double bond which gives 1-aryl-7-azabicyclo[4.3.0]nonane may well be stereospecific.[464]

Yamada and Otani[465,466] prepared the antipode of natural (−)-mesembrine by an asymmetric synthesis. They formed the enamine from S-proline pyrrolidide and N-formyl-2(3′,4′-dimethoxyphenyl)-4-methylamino butyraldehyde which was condensed with methyl vinyl ketone, hydrolyzed with acetic acid, and the product was cyclized to (+)-mesembrine (Equation 50). This is yet another application of Yamada's asymmetric syntheses

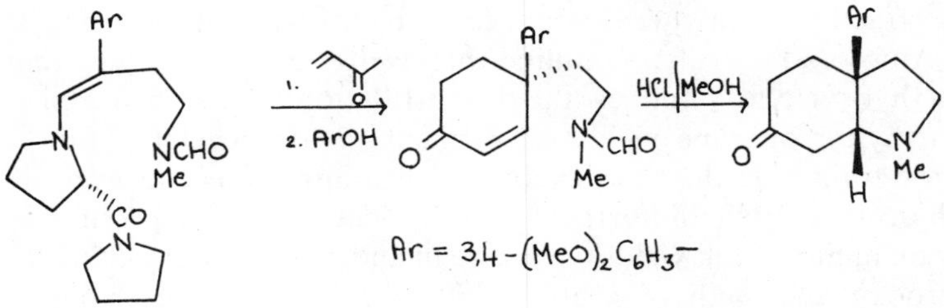

$$Ar = 3,4-(MeO)_2C_6H_3-$$

(50)

involving S-proline enamines (Chapter 2, Section III.B.4.d). ($\pm$)-Mesem-brine and ($\pm$)-desdimethoxymesembrine are formed in approximately 50% yield when 1-methyl-3-aryl-2-pyrrolines react with methyl vinyl ketone and may be potentially useful for the syntheses of related alkaloids because of their simplicity.[467,468] A Diels-Alder reaction has also been used in the total synthesis of ($\pm$)α- and β-lycoranes. For example 3,4-dimethoxy-ω-nitrostyrene cyclizes with methyl hexa-3,5-dienoate and the adduct is then converted in several steps into cis-5-(3,4-dimethoxyphenyl)-cis-7-azabi-cyclo[4.3.0]nonane (Equation 51). A Pictet-Spengler ring closure of this base using formaldehyde gives ($\pm$)-α-lycorine.[469]

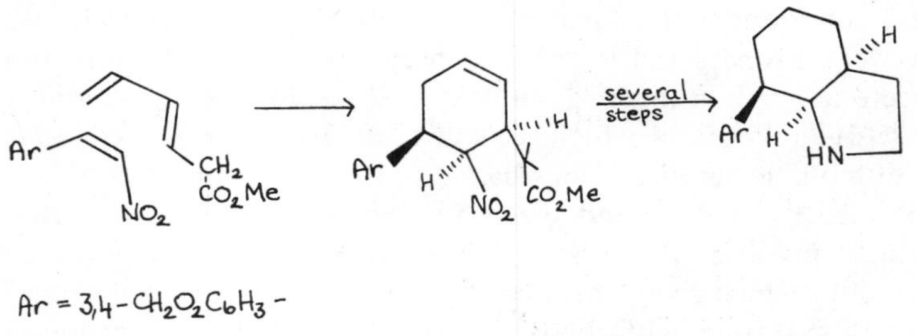

(51)

$$Ar = 3,4-CH_2O_2C_6H_3-$$

Cycloaddition occurs between 3-phenyl-2-pyrrolin-4,5-dione and buta-diene in dimethyl sulfoxide to yield stereospecifically 1-phenyl-cis-7-azabi-cyclo[4.3.0]non-3-en-8,9-dione and is yet another possibility for alkaloid syntheses.[470] N-(1-Benzocyclobutenyl)vinyl acetamide undergoes a thermal rearrangement to a 4.7:1 mixture of trans- and cis-4,5-benzo-7-azabicyclo-[4.3.0]nonen-8-one. Details of the kinetics of this reaction have been studied and the mechanism involves ring opening of the cyclobutene ring followed by intramolecular cyclization (Equation 52).[471]

In studying the differences between cis- and trans-7-azabicyclo[4.3.0]-nonane, King and co-workers[442,449,472] examined the Hofmann elimination reactions. They found that whereas cis-7-methyl-7-azabicyclo[4.3.0]nonane

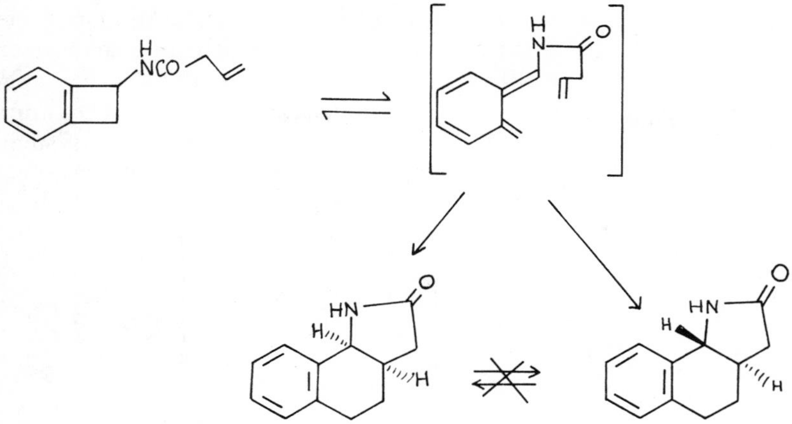

(52)

methiodide underwent C–6, N–7 bond cleavage yielding (after hydrogenation) N,N-dimethylaminoethylcyclohexane, the trans isomer gave *trans*-1-
ethyl-2-N,N-dimethylaminocyclohexane and an ether. The explanation of
these results took into account the necessity for a coplanar H–C–C–N–
arrangement of atoms in facilitating both the removal of the β-hydrogen
atom and the cleavage of the C–C bond.[473] In the cis conformation **162**
H–1, but not H–9 (axial or equatorial), is ideally disposed for the elimination to give the observed products. The equatorial H–9 in the trans con-

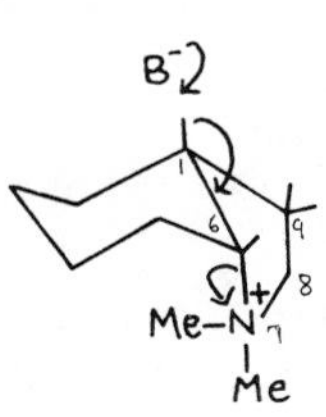

162

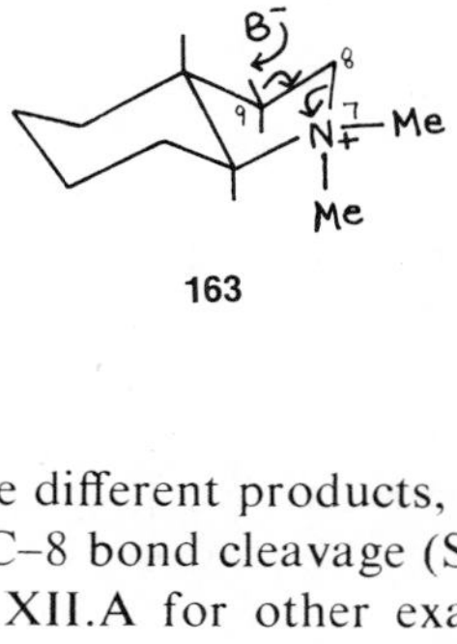

163

formation **163** is similarly oriented, hence the different products, and most
probably H–9 equatorial is lost in the N–7,C–8 bond cleavage (See Chapter 3, Sections VIII.B, IX.B.4.c, X.B, and XII.A for other examples of
Hofmann eliminations in the perhydro-2-pyrindine, decahydroquinoline,
decahydroisoquinoline, and octahydroacridine series).

5. *8-Azabicyclo [4.3.0] nonanes (Perhydroisoindoles)*

The most popular synthesis for this system requires as starting material
the Diels-Alder adduct derived from a conjugated diene and a maleic or
fumaric acid derivative as the dienophile. Maleimide yields the *cis*-8-aza-

bicyclo[4.3.0]nona-7,9-dione (164) system directly whereas fumaric acid forms a *trans*-1,2,3,6-tetrahydrophthalic acid which can be transformed into a *trans*-8-azabicyclo[4.3.0]nonane. The configuration of the ring junction is clearly defined by the dienophile in these reactions and epimerization of a bicyclooxononane generally furnishes the more stable cis system. Dienes used in such condensations included butadienes (or sulfolene),[474–478]

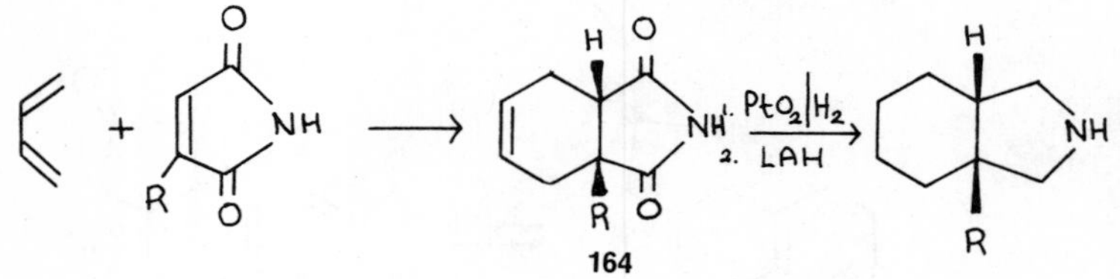

164

(53)

cyclopentadiene,[479–482] cyclohexa-1,3-diene,[483] and furan[484,485] with maleimide, and all yield the cis-fused system as in Equation 53. With the cyclic diene two cis products are possible: an endo (165) and an exo adduct (165a) with the endo isomer usually predominating, but these could be equilibrated.[479,482,486] Maleic anhydride gives different ratios of such ad-

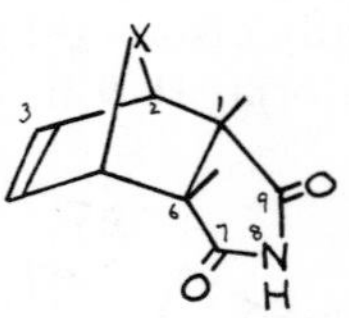

165

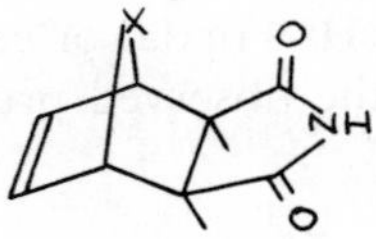

165a

ducts from maleimide, and this can be used to advantage in order to obtain both series.[479,484,486] The equilibrium no doubt occurs by way of a retro Diels-Alder reaction but if the double bonds or carbonyl groups in 165 and 165a are reduced the system becomes fixed.[485] *trans*-8-Azabicyclo-[4.3.0]nonane is best prepared using the synthesis of Christol and his co-workers[487] (Equation 54).

1,2-Benzo-*cis*-3,4-diphenylbutene undergoes a Diels-Alder reaction with *N*-methyl maleimide to yield stereospecifically *d,l*-3,4-benzo-8-methyl-*trans*-

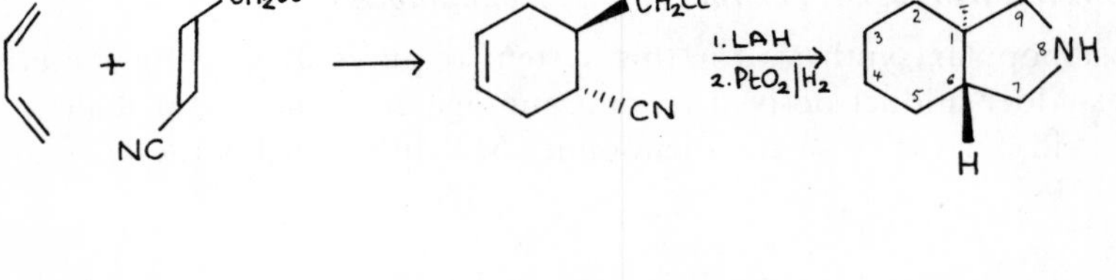

(54)

2,5-diphenyl-*cis*-8-azabicyclo[4.3.0]non-3-en-7,9-dione because the methiodide of the dideoxy base obtained by reduction was resolved into its optical enantiomers with camphor-β-sulfonic acid. The *trans*-butene reacted similarly but gave *meso*-3,4-benzo-8-methyl-*cis*-2,5-diphenyl-*cis*-8-azabicyclo-[4.3.0]nonen-7,9-dione. Thus ring opening of the 1,2-benzo-3,4-diphenyl-butene prior to condensation is stereospecific.[488] A similar reaction, but involving an intramolecular cycloaddition, was developed by Oppolzer[489] and is shown in Equation 55 (compare with Equation 52). The reaction was extended to higher homologues (i.e., **166** where $n = 2$, 3, and 4) with a considerable drop in yield when $n = 3$. *o*-(2-*N*-Ethoxycarbonyl-*N*-alkylaminoethyl)benzaldehyde undergoes a similar rearrangement on photolysis, but the reaction is not as stereospecific as the above.[490] In an analogous,

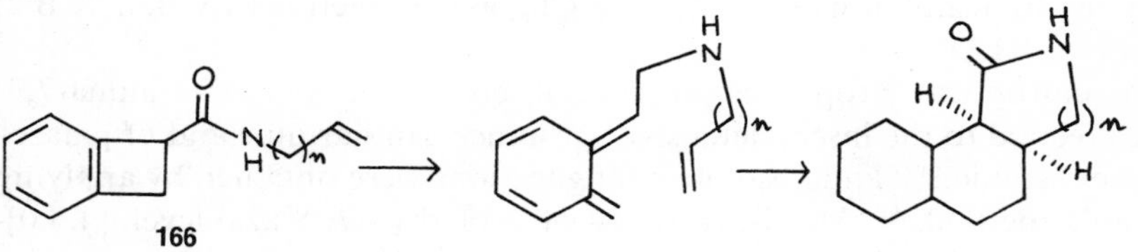

(55)

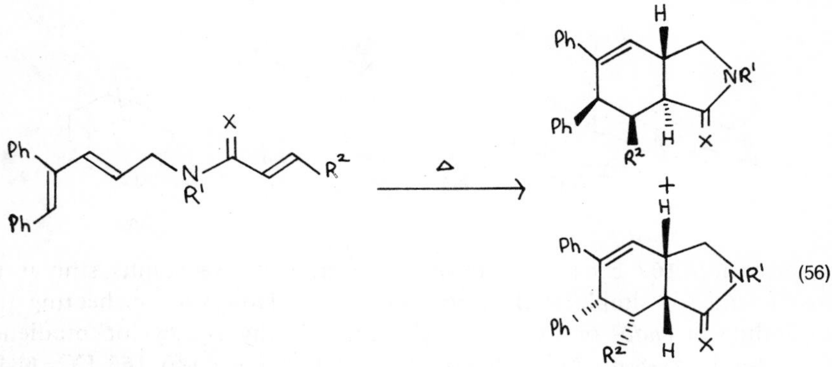

(56)

but thermally induced, intramolecular Diels-Alder reaction of pentadienyl-acrylamides the ratio of cis- to trans-fused bicyclic compounds formed varied with the group R^2 (Equation 56). The trans-fused bicycle was the predominant isomer when $X = O$ and R^2 was CO_2Et or Ph; equal amounts of cis and trans isomers were formed when $X = O$, $R^2 = H$, and the cis isomer was the major product when $X = H_2$ and $R^2 = H$.[491]

A Schmidt reaction of cycloocta-2,4,6-trien-1-one in sulfuric acid, but not in trifluoroacetic acid, gave *cis*-8-azabicyclo[4.3.0]nona-2,4-dien-7-one.

The cis configuration was established by catalytic reduction to the known *cis*-nonan-7-one.[492]

The conformation of *trans*-8-azabicyclo[4.3.0]nonane is rigid like that of the *trans*-7-aza isomer **157**, and although the *cis*-8-azanonane can exist in the two extreme ring conformations **156** and **158** these are equivalent in this particular case. The situation is altered, however, if *cis*-8-azabicyclo-[4.3.0]nonane bears a substituent on one of the ring carbon atoms because C-7 and C-9 are no longer equivalent. Conformers in which the cyclohexane ring is in the boat form may also become important in the system. Armarego and Kobayashi[476] have differentiated between *cis*- and *trans*-8-azabicyclo[4.3.0]nonanes, and their 1-methyl derivatives, by the narrow band envelope observed in the ¹H nmr spectra for the protons on C-2,3,4, and 5 of the cis isomers compared with that of the trans isomers (as in the perhydroazanaphthalenes[476]) (See Chapter 3, Sections IX.B.3, X.B.2, XI.A, and XI.D).

The chiroptical properties of several *cis*-8-azabicyclo[4.3.0]nona-7,9-diones related to the insect defensive substance cantharidine and of palasonine were studied[493] and absolute configurations were obtained by applying Horeau's method.[494] An X-ray study of 8-methyl-*cis*-8-azabicyclo[4.3.0]-non-3-ene methiodide showed that in the crystals it had the conformation **167** and not **168**.[495] 8-Methyl-*cis*-8-azabicyclo[4.3.0]non-3-ene is also in the

167   **168**

169

conformation **167** because quaternization studies give results similar to those found for simple tertiary pyrrolidines.[496] However, on heating the methiodide of *endo*- or *exo*-2-methyl-2-aza-1,2-dihydrodicyclopentadienes (the 7,9-dideoxy-8-methyl derivatives related to the endo **165** [X = CH₂] and exo **165a** [X = CH₂] adducts), the products were mainly the mono-demethylated cyclic bases together with minor quantities of 5-dimethylami-nomethyl-6-hydroxymethyl(or methoxymethyl)bicyclo[2.2.1]hept-2-ene.[480] Thus a Hofmann elimination reaction did not take place as would be predicted, probably because of the lack of a coplanar system containing the four atoms H–C–C–N– (compare Sections III.E.4 and 5). *endo*-(or *exo*-)-2-Aza-1,2-dihydrodicyclopentadiene (the 4,9-dideoxy derivatives of **165** [X = CH₂] and **165a** [X = CH₂]) react with halogen acids to give the 9-halo derivatives, that is, addition across the double bond without exo-endo interconversion. The endo isomer differs from the exo isomer in that it cyclizes to the tertiary base **169** on alkaline treatment.[497,498]

Two intramolecular rearrangements of 8-azabicyclo[4.3.0]nonanes should be mentioned. The first is the acid catalyzed rearrangement of 1-*N'*-phenyl-ureido-*trans*-8-azabicyclo[4.3.0]nonan-7-one (**170**) which gives *cis*-8a-ami-nomethyl-3-phenylperhydroquinazoline-2,4-dione (**171**) in high yield.[476] The

170   **171**   **172**

second is the photochemical equilibration of 8-alkyl-3,4-dimethyl-*cis*-8-azabicyclo[4.3.0]nona-2,4-dien-7,9-dione with *syn*- and *anti*-8-alkyl-3,4-di-methyl-*cis*-8-azatricyclo[4.3.0.0^{2,5}]non-3-ene-7,9-diones. Heat, on the other hand, favors the bicyclic compound.[499]

Epoxidation of 8-benzoyl-*cis*-8-azabicyclo[4.3.0]non-3-ene is stereospe-cific and gives the *syn*-3,4-epoxy derivative **172**. The anti isomer is obtained by the addition of the elements of HOBr followed by dehydrobromina-tion.[500] Epoxidation of 8-(2-phenylethyl)-*cis*-8-azabicyclo[4.3.0]non-3-ene-7,9-dione is less stereospecific and gives a 2:1 mixture of *syn*- and *anti*-3,4-epoxy derivatives. These results are consistent with a bicycle in the conformation **167**.[501]

F.   Indoles, Indolines and Isoindolines

Indole is a planar molecule and any stereochemistry associated with it must involve asymmetry in a side chain. The compound in this group which has been studied most is tryptophan, but its chemistry will not be discussed here because it has been described admirably by Greenstein and Winitz.[318] The other indolyl compounds that have attracted attention are related to the plant growth hormone indolyl-3-acetic acid. α-(3-Indolyl)pro-pionic acid and its 1-methyl derivative have been resolved into their pure enantiomers,[502,503] and the absolute configurations were derived by Fredga's quasi-racemate method by comparison with α-(1-naphthyl)propi-onic acid.[504,505] The absolute configuration of the antibiotic indolomycin was determined by degradation of the (+) enantiomer of α-indolmycenic acid to $(+)$-α-(3-indolyl)propyl alcohol which was identified with the alcohol from the plant hormone $S(+)$-α-(3-indolyl)propionic acid. The configuration of the antibiotic (173) is therefore $5R,6S$.[506]

173

174

175

2,3-Disubstituted indolines are not planar and exhibit cis-trans isomerism. Steche[507] reduced 2,3-dimethylindole with tin and hydrochloric acid and obtained 2,3-dimethyl-2,3-indoline which was thought to be the trans isomer, but was later shown to be a 60:40 mixture of *cis*- and *trans*-2,3-dimethylindolines.[508] Catalytic reduction of 2,3-dimethylindole, on the other hand, yields only the trans indoline.[508,509] The intramolecular cyclization of *N*-crotylaniline (174, $R^1 = H$, $R^2 = Me$) with polyphosphoric acid yields a mixture of 2,3-dimethylindole and *cis*-2,3-dimethylindoline (175, $R^1 = H$, $R^2 = Me$). This reaction involves a Claisen rearrangement followed by cyclization, and it was thought that the indole was formed by dehydrogenation of some *trans*-2,3-dimethylindoline which was formed.[510] Dehydrogenation of *cis*- and *trans*-2,3-dimethylindolines did not readily yield the indole, thus the previous explanation was incorrect and biforcation in the pathway was postulated. Closer examination of this reaction revealed that in fact a 3:1 mixture of *trans*- and *cis*-2,3-dimethylindolines was formed.[508] A similar, but much more stereospecific cyclization is the photolysis of *N*-arylenamines which affords *trans*-2,3-disubstituted indolines (Equation 57). The reaction proceeds by a conrotatory ring closure followed by electron demotion and a thermal suprafacial [1,4]-sigmatropic hydride shift.[511] When R^2 and R^3 form part of a carbocyclic ring, that is, $R^2,R^3 = -(CH_2)_n-$, then if $n = 4$ or 5 the stereospecificity is essentially the same as above (a little of the cis isomer is formed), but when $n = 3$ the cis isomer is the only product. This suggests that in the last case the hydride

$$(57)$$

shift occurs after the two fused five-membered rings rearrange from the trans to a cis orientation as a result of the relief of strain.[511]

Methyl magnesium iodide adds across the 1,2- double bond of 3,3-disubstituted indolenines stereospecifically when one of the substituents is bulky (e.g., isopropyl or phenyl) and the group entering position 2 is oriented trans to the 3-bulky group. The diastereomeric *cis*-indolines were prepared by catalytic reduction of the respective 2,3,3-trisubstituted indolenines.[512] A trans addition of nitric acid across the 2,3 double bond of 1-benzoyl(or acetyl)-2,3-dimethylindole furnishes the *trans*-1-benzoyl-2-hydroxy-(or 3-nitro)-3-nitro-2,3-dimethylindoline which is converted into *trans*-1-benzoyl-2,3-dihydroxy-2,3-dimethylindoline (**176**) by boiling in ethanol containing alumina.[513] *cis*-1-Benzoyl-2,3-dihydroxy-2,3-dimethylindoline, on the other hand, is obtained by reaction of the corresponding indole with osmium tetroxide.[514] The reactivity of these two isomers towards hydroxide ions is quite different. The cis isomer yields 2,2-dimethylindoxyl whereas the trans compound forms a mixture of 2-methyl-2-phenylindoxyl (**177**) and 3-hydroxy-3-methyl-2-phenylindolenine (**178**).[514]

Berti and co-workers[515] evaporated a benzene solution of 2,3-dimethylindole over a period of days and isolated a nonperoxidic oxidation product. They showed that it was 3,3a,4,8b-tetrahydro-3',3a,3b-trimethylspiro[2H-furo[3,2-b]indole-2,2'-indoline]-3'-ol and that the physical data suggested the stereochemistry depicted in formula **179**. McLean and co-workers[516] reexamined this product and confirmed previous data but revised the stereochemistry to structure **180** after an X-ray study.

Indoles undergo photocycloaddition reactions with olefins such as methyl acrylate, acrylonitriles, acrylamides, and ethyl vinyl ether across the

2,3- double bond stereospecifically to form tetrahydro-1H-cyclobut[b]indo-lines (181). The substituent X can be exo or endo, that is, cis or trans to the vicinal bridgehead hydrogen atom, and both isomers are usually obtained in varying amounts depending on the olefin and the substituent R[2,517]

179   180

181

Several optically active indolines have been prepared and show some interesting behavior. Chiral 3-hydroxyindolin-2-one is obtained from o-ni-tromandelic acid. The acid was resolved as the brucine salt, reduced with ferrous sulfate to ($-$)- and ($+$)-o-aminomandelic acids which were then cyclized to ($+$)- and ($-$)-3-hydroxyindolin-2-ones with a change of rotation on ring closure. These racemized very slowly (312 hr at 20°C) in 1N HCl and in pyridine. The chiral N-hydroxy derivatives were formed directly from the optically active o-nitromandelic acids by reduction with zinc dust in ammonia.[518]

Corey and his co-workers[519] have conceived a new and ingenious approach to the asymmetric synthesis of α-amino acids. The method uses chiral 2-hydroxymethylindoline as the reagent—which can be recycled. The indoline is converted into its N-amino derivative and is condensed with an α-keto ester (with the carbon skeleton of the proposed α-amino acid) to form a 3-substituted 2,5,6,7-tetrahydro-1-oxa-4,5-diazepin-2-one. This is first reduced stereospecifically to the hexahydro derivative, then the N–N bond is cleaved by hydrogenolysis, and finally hydrolysis yields the required optically active α-amino acid and the original indoline (Scheme 8, R[1] = H,-Me, R[2] = H,Me). The reduction 183 → 184 could not be effected catalytically but was achieved by "chemical" means using aluminum amalgam under carefully controlled conditions. It should be noted that the S reagent gave the R-amino acid and the converse is true with the R reagent. When R[1] = Me and R[2] = H, the optical purity of the α-amino acids (alanine and butyrine) were 80–90%. However, on increasing the asymmetry in the indoline 182 by making R[1] = H and R[2] = Me, the $S_N R_o$($-$)-indoline (i,e..

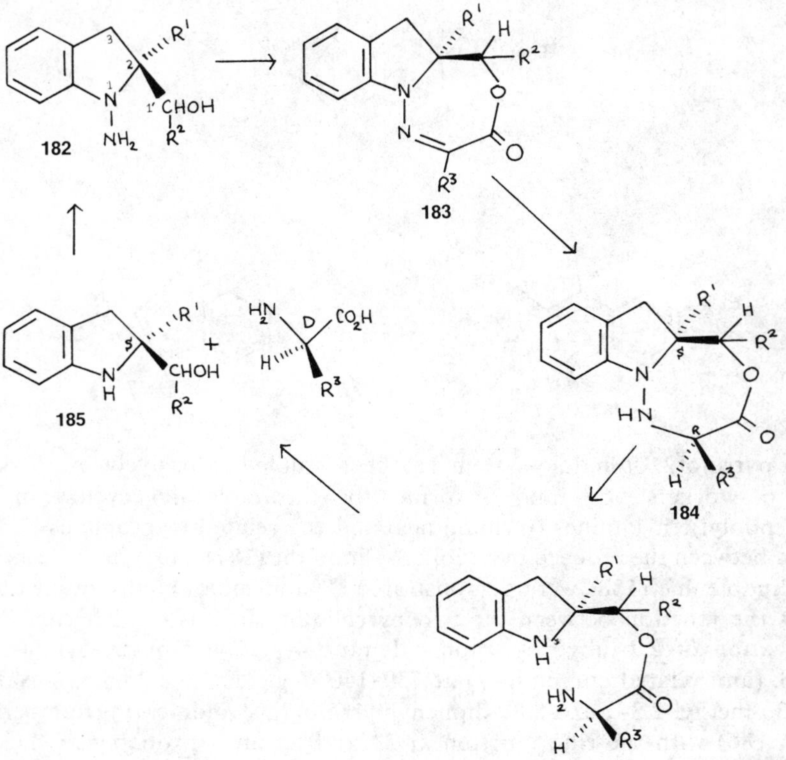

Scheme 8

S at C–2 and R at C–1′), the optical purities of R-alanine, R-butyrine, R-valine, and R-isoleucine are increased to 96, 97, 97, and 99%, and further recrystallization gives 100% purity. The X-ray crystal structure of the reduced oxadiazepine **184** ($R^1 = H, R^2 = R^3 = Me$) is consistent with the observed stereospecificities.[520] In this elegant chemical work the absolute configuration of the (+)-indoline **185** ($R^1 = R^2 = H$) was deduced by N-methylation and ring cleavage of the methiodide with sodium in liquid ammonia to 2-methylamino-3-phenylpropanol methiodide with a configuration that was correlated with $S(\text{L})$ (−)-phenylalanine.[515] The absolute configuration of **185** ($R^1 = Me, R^2 = H$) was similarly derived.

Julian and Pikl resolved 2-(5-ethoxy-1,3-dimethylindolin-2-on-3-yl)ethyl-N-methylamine by successive salt formation with d-camphorsulfonic acid and (+)-tartaric acid in an early synthesis of the alkaloid physostigmine (eserine).[521] The absolute configuration of physostigmine (**186**) was derived from nmr studies and the observation of nuclear Overhauser effects.[522]

186

187   **188**

This pyrrolo[2,3-b]indole system has been studied extensively by Witkop and co-workers[523-525] and is formed by intramolecular cyclization of 2-(2-indolyl)ethylamines (tryptamines) and the related tryptophans. Equilibria between the reduced pyrrolo[2,3-b]indoline (**187**) and 3-methylamino-ethylindolenine (**188**) were demonstrated,[526] and undoubtedly in all these cases the junction between the two pyrrolidine rings is cis. Thermal isomerization of 2,3-dihydro-1-diphenylamino-4-methyl-5-phenylpyrrol-2,3-dione (and related compounds) at 130–140°C yields 1,2,3,3a,8,8a-hexahydro-3a-methyl-2,3-dioxo-8,8a-diphenylpyrrolo[2,3-b]indole (ring numbering as in **186**) with the ring junction at 3a and 8a most probably in the cis configuration.[527] An unusual rearrangement occurs when cis- and trans-2-ethoxycarbonyl-1,2,3,3a,8,8a-hexahydro-3a-(2-hydroxy-4-nitrobenzyl)-pyrrolo[2,3-b]indole (chiral at C–2) is treated with ethanolic hydrochloric acid and gives 2-(2-hydroxy-4-nitrobenzyl)tryptophan ethyl ester hydrochloride. A mixture of cis- and trans-pyrrole[2,3-b]indole (57:43) is formed also by rearrangement of a mixture of epimeric 3-(2-hydroxy-4-nitrobenzyl)-tryptophan ethyl ester hydrochlorides (this is in the indolenine form) prepared in the attempted benzylation of tryptophan ethyl ester hydrochloride with 2-hydroxy-4-nitrobenzyl bromide.[528]

Ollis, Ormand, and Sutherland[529] studied the rearrangement of thermochromic spirans derived from the indolines **189** by variable temperature ¹H nmr spectroscopy and explained the spectral changes by the involvement of electrocyclic reactions (Scheme 9).[529]

Stereochemical studies in the isoindoline series are very scanty and only a few cases of interest are described here. Bonnett and White[530] found that reductive dimerization of 2,5-dimethylpyrrole with tin and hydrochloric acid gives a mixture of cis- and trans-1,3,4,7-tetramethylisoindolines (85:15). The trans isomer is racemic and the cis isomer is meso, and they

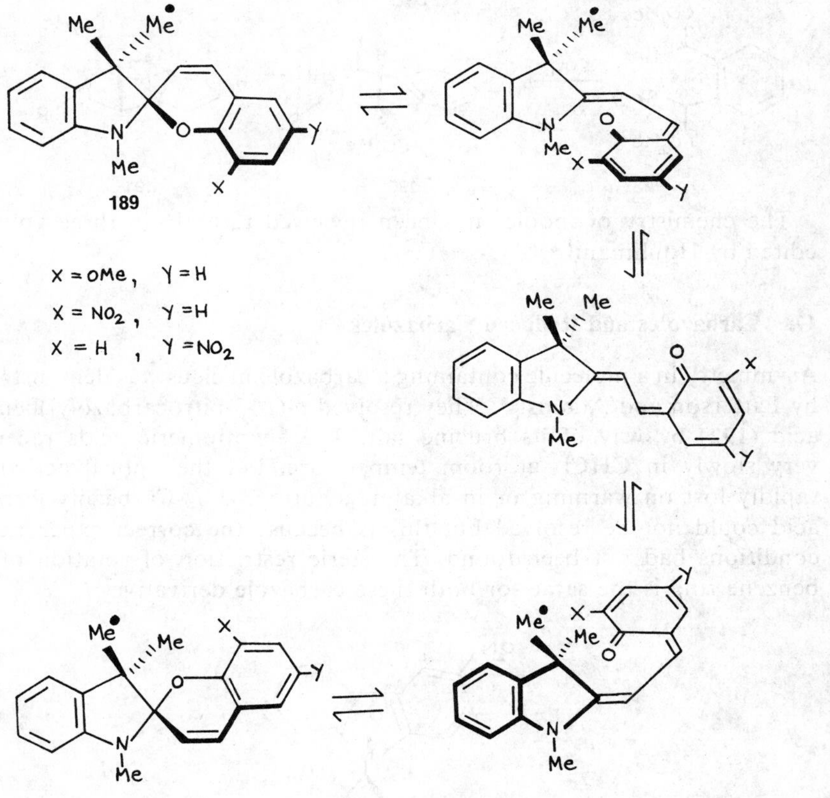

X = OMe, Y = H
X = NO₂, Y = H
X = H, Y = NO₂

Scheme 9

can be distinguished by means of the ¹H nmr spectra of the *N*-benzyl and *N*-neopentyl derivatives.[531] The two isomers are in equilibrium in acid medium and the equilibrium may favor the trans isomer at low temperatures. *cis*- and *trans*-2-Benzyl-1,3-diphenylisoindolines are also distinguishable by ¹H nmr spectroscopy.[532] *trans*-1,3-Diphenylisoindoline[533] is obtained in low yield by reduction of 1-hydroxy-1,3-diphenyl-1-*H*-isoindole with Zn–AcOH and the cis isomer is formed by reduction of the same hydroxyindole with LAH–AlCl₃.[532,534]

Cignarella and Saba[535] reacted *o*-bis(methoxycarbonyl-α-bromomethyl)-benzene with various amines and obtained a mixture of *cis*- and *trans*-1,3-bismethoxycarbonylisoindolines (Equation 58) which were separated. Alkaline hydrolysis of the esters **190** and **191** (R = H) gives a 1:1 mixture of the corresponding acids. However, when R is large (e.g., *p*-tolyl) the *cis*-dicarboxylic acid is the major product from the hydrolysis of either ester, in accordance with the expected greater stability of structure **191**.

190   191   (58)

The chemistry of indoles has been reviewed recently in three volumes edited by Houlimann.[536]

G.   Carbazoles and Reduced Carbazoles

Asymmetry in a molecule containing a carbazole nucleus was demonstrated by Patterson and Adams.[537] They resolved o-(N-3-nitrocarbazolyl)benzoic acid (192) by way of its brucine salt. The enantiomeric acids racemize very slowly in $CHCl_3$ at room temperature, but the optical activity is rapidly lost on warming or in alkaline solution. o-(N-Carbazolyl)benzoic acid could not be resolved but this is because the correct experimental conditions had not been found. The steric restriction of rotation of the benzene ring is the same for both these carbazole derivatives.

192

193   194

Carbazole is a planar molecule and therefore by itself is not stereo-chemically interesting. Its 4a-H tautomer, however, has an asymmetric carbon atom and should be chiral. No examples of these are known, but Fritz and Stock[538] resolved a tetrahydro derivative, 1,2,3,4-tetrahydro-4a-methyl-4a-H-carbazole, with S(+)-tartaric acid. The absolute configuration of the $S(-)$enantiomer **193** was obtained by converting it into (S)-9-formyl-2,3,4,9-tetrahydro-4a-methyl-4a-H-carbazole and comparing the ord curves of its chromophore with that of C-curarin III, a degradation product of strychnine.[539] The free base formed from the 9-methiodide of **193** is 2,3,4,9-tetrahydro-4a,9-dimethyl-4a-H-carbazole (**194,** R = Me). Methanol can add across the 1,9a double bond to produce the trans adduct 1,2,3,4,4a,9a-hexahydro-9a-methoxy-4a,9-dimethylcarbazole in which a second asymmetric center has been introduced.[538] When the $S(-)$-formyl derivative **194** (R = CHO) is treated with POCl$_3$ and then water, hydration occurs across the 1,9a double bond and cis-1-formyl-1,2,3,4,4a,9a-hexahydro-9a-R-hydroxy-4a-S-methylcarbazole is formed. Further treatment of the latter with POCl$_3$ and methanol gave the 9a-S-methoxy-4a-S-methyl derivative whereby the 9a-hydroxy group is replaced by a methoxy group with inversion of configuration.[540] ^{1}H nmr spectra revealed that the rotation of the formyl group is restricted by the 9a-substituent. The absolute configurations of these formyl carbazole adducts were correlated with those of the cyclic ether **196** (R = CHO) obtained from 4a-R,9a-S-echibolin (**195,** R = H) of known absolute configuration (Equation 59).[541] The racemate of the tetrahydrocarbazole **193** reacts with formaldehyde to give

R' = H or Me

195

R = H or CHO

196

(59)

various products. The axial and equatorial 1-hydroxymethyl derivatives obtained were identified and distinguished by the ability of the equatorial isomer to exhibit intramolecular hydrogen bonding.[542]

1,2,3,4,4a,9a-Hexahydrocarbazole exists in cis and trans forms (an example has already been mentioned above), but molecular models demonstrate that there is much more strain associated with the trans structure than with the cis structure. This strain shows up in the chemistry of hexahydrocarbazoles. Reduction of carbazole to hexahydrocarbazole invariably yields the cis isomer (see below).[450] Perkin, Gurney, and Plant[543] reduced 900 g of carbazole with tin and hydrochloric acid and obtained mainly the cis isomer, but they succeeded in separating 12 g of the trans isomer from it. The stereochemistry of the 4a,9a-disubstituted 1,2,3,4,4a,9a-hexahydrocarbazoles, described by Plant and Tomlinson,[544,545] is not clear, but Witkop[546] examined the rearrangement of the dihydroxy derivative.

Catalytic reduction of 4a-ethyl-1,2,3,4-tetrahydro-4a-H-carbazole in acetic anhydride gives the 9-acetyl-cis-hexahydro derivative with structure **197** which was confirmed by an X-ray study.[547] Lithium aluminum hydride reduction of the racemate of **193** also gives the cis-hexahydro derivative.[548] In an application of the Fischer indole synthesis the phenyl hydrazones of cyclohexanone and 2-methylcyclohexanone are cyclized in formic acid to cis-1,2,3,4,4a,9a-hexahydrocarbazole and its 4a-methyl derivative respectively.[549] Yet another synthesis which yields a cis-hexahydrocarbazole is the intramolecular cyclization of 1-benzyl- or 1-methyl-3-(2-oxobutyl)indoles in trifluoroacetic acid to the 2-keto derivatives of 9-benzyl- or 9-methyl-cis-1,2,3,4,4a,9a-hexahydrocarbazole. The 1-unsubstituted indole yields a 2,2′-biindolyl derivative with this reagent but the carbazole can be obtained if $BF_3 \cdot Et_2O$ is used instead.[550] One way of obtaining a trans-

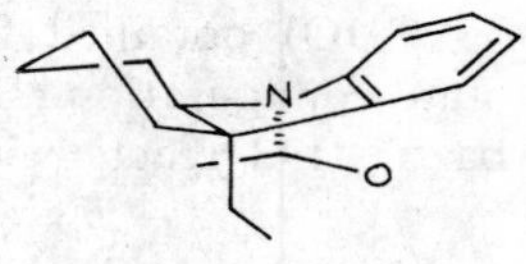

197

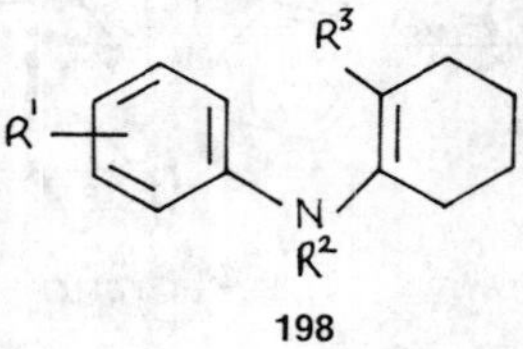

198

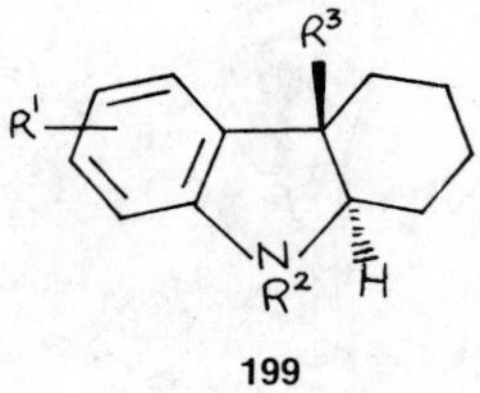

199

hexahydrocarbazole is by the photocyclization of N-aryl enamines **198** → **199**.[511] This cyclization has been described before (Section III.F) in the synthesis of indolines, and it involves a conrotatory cyclization followed by electron demotion to a dipolar species which then undergoes a suprafacial [1,4]-hydride shift. In examples where the cyclohexene ring in **198** is replaced by dihydronaphthalene or cycloheptene some cis-fused indolines are also formed.[551] The trans structure in the case of 6,8-dibromo-1,2,-3,4,4a,9a-hexahydro-4a,9-dimethylcarbazole was confirmed by an X-ray analysis.[551,552]

A substituent on C–3 of cis-1,2,3,4,4a,9a-hexahydro-9-methylcarbazole can be syn or anti to the hydrogen atoms on C–4a and C–9a. The syn isomer is the one in which the hydrogen atoms on C–3, 4a and 9a are all cis. Both isomers with 3-*t*-Bu, 3-Me, or 3-OMe groups have been prepared, separated, and identified according to their ¹H nmr spectra and the theory of conformational analysis.[553] The stereochemistry of the 3-substituent influences the rates of quaternization at N–9 and here again the syn and anti nature of the 3-substituent can be identified by ¹H nmr.[554] The spectra for the syn isomers are consistent with a rigid structure in which the cyclohexane ring is in a slightly flattened boat conformation (**200**).[555] Hofmann degradation of *cis*-1,2,3,4,4a,9a-hexahydro-9,9-dimethylcarbazole

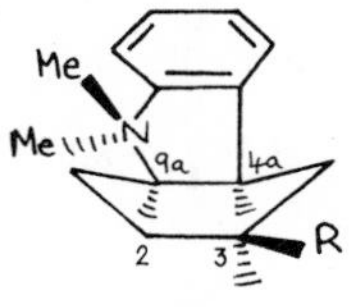

200

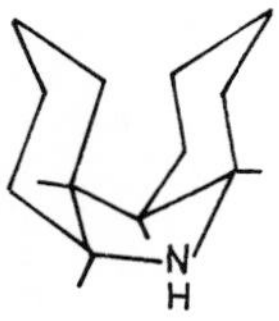

201

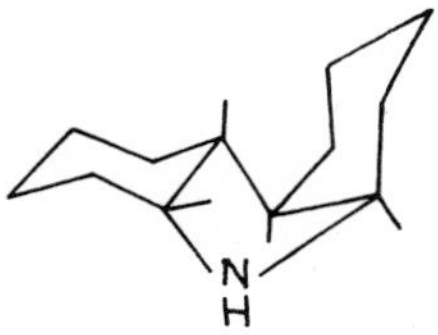

202

hydroxide gives *o*-(cyclohex-1-enyl)-*N*,*N*-dimethylaniline as a result of the loss of H–4a in the elimination. Loss of H–1 is an alternative pathway because the 4a,9,9-trimethyl hydroxide is degraded to *o*-(1-methylcyclohex-2-enyl)-*N*,*N*-dimethylaniline, but a little of the demethylated product *cis*-9,4a-dimethylhexahydrocarbazole is also formed.[556]

Catalytic reduction of carbazole[450] and *cis*-hexahydrocarbazole,[546] or electrolytic reduction of 1,2,3,4,5,6,7,8-octahydrocarbazole (and its 9-methyl and 9-ethyl derivatives)[557] give dodecahydrocarbazole. Witkop stated that the product is the cis,cis isomer, but we still do not know whether it is the syn (**201**) or anti (**202**) cis,cis isomer.

H. Diazolines

1. *Pyrazolines and Pyrazolidines*

The chemistry of pyrazolines and pyrazolidines was reviewed in 1967 by Jarboe[558] and reference should be made to it for information about syntheses and reactions. Here the stereochemistry in the 1-pyrazolines, 2-pyrazolines, and pyrazolidines are treated separately and as little as possible of the discussions in the above review is duplicated.

a.   *1-Pyrazolines*.   The main interest in the syntheses of 1-pyrazolines has been in the steric course of [2 + 3] cycloaddition reactions. The addition itself is cis and highly stereospecific, but the alignment of the reacting species is subject to the direction of polarization in the olefin and the bulk of the substituents near the reacting bonds (Scheme 10). The number of products, however, is severely restricted by the high regiospecificity of the

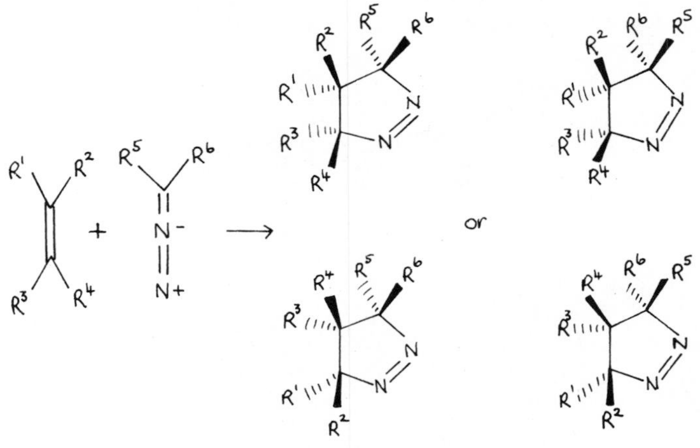

Scheme 10

reactions and by the choice of polarophile. Huisgen summarized the chemistry of these cycloadditions in 1963 and 1965.[151] The most commonly used 1,3-dipole is diazomethane which reacted with the following olefins: crotonic esters[559]; ethylidene malonic ester[559,560]; fumaric ester[559]; maleic ester[559]; ω-phenyl-,[561,562] ω-chloro-,[563] α-acetyl-,[564,565] and α-cyanoacrylic esters[565,566]; isoprenyl methyl ketone[567]; *p*-benzoquinones[568—570]; 2,3,4,5-tetraphenylcyclopentadien-1-one[571]; derivatives of 1-methoxycarbonylcyclohex-1-ene[572,573]; dehydronorcamphor[574]; and 1,2-bischlorocarbonylcyclobutene.[575] Divinyl- and phenyl vinylsulfones give *d-*, *l-*, and *meso*-1-pyrazolin-3-ylsulfones.[576]

Diazopropane reacts with methyl acrylate and its methyl derivatives in the normal way, that is, in accordance with the electronic effects to give the expected 1-pyrazolines. Cycloaddition of diazopropane and diazomethane onto methyl 3-methylbut-2-enoate, however, gives the 1-pyrazoline with the reversed orientation; and methyl 4,4-dimethylpent-2-enoate also undergoes reverse addition with diazopropane but not with diazomethane or diazoethane. This is probably due to unfavorable eclipsing interactions.[577] Methyl but-2-ynoate, but not methyl propiolate, yields a 1-pyrazoline of opposite orientation and is attributed to steric factors in the transition state.[578]

The cycloaddition of diazopropane with allenic esters and nitriles has also been studied and the orientations were discussed in terms of steric and electronic effects.[579] 1-Pyrazolines fused onto a cyclobutane ring are obtained from diazopropane and cyclobutene. Halogen substituents at C–3 and C–4 influence the direction of addition (Equation 60).[580,581]

Aryl diazomethanes react with styrenes to form about equal amounts of *cis*- and *trans*-3,5-diaryl-1-pyrazolines,[582] and with 1,3-butadiene a mixture of three diastereoisomers of 3,3′-diphenyl-5,5′-bi-1-pyrazolines, namely: cis,cis, trans-cis (meso), and trans,trans is obtained.[583] A partial asymmetric synthesis results when menthyl acrylate (or its α-methyl derivative) condenses with diphenyl diazomethane because the methyl 5,5-diphenyl-1-pyrazolin-3-carboxylate formed gives optically active 2,2-diphenylcyclopropane carboxylic acid. The asymmetric synthesis, however, is less than 10% stereospecific.[584]

Other 1,3-dipoles that yield 1-pyrazolines are triphenylsilyl diazoalkanes (in which one of the substituents is a triphenylsilyl group),[585] vinyl diazomethane,[586] and ethyl diazoacetate.[587,588] In the last case isomerization of the 1-pyrazolines formed to 2-pyrazolines can be so rapid, because of the presence of the 3-alkoxycarbonyl group, that the same product may be obtained from both the *cis*- and *trans*-olefins. Thus diethyl fumarate and maleate give the same *trans*-3,4,5-trisethoxycarbonyl-2-pyrazoline.[588]

Identification of cis- and trans-substituted 1-pyrazolines can be achieved by ^1H nmr spectroscopy,[589] and reference should be made to the respective papers (above) and to a recent monograph[61a] for this information. Chiral *trans*-3,5-dimethyl-1-pyrazoline is obtained by optical resolution of its (+)-10-camphorsulfonate salt and the absolute stereochemistry of the (+) isomer was found to be 3R,5R by an independent synthesis from R-2-hydroxypent-4-ene (Equation 61).[590] cis-3,5-Dimethyl-1-pyrazoline is, of course, in the meso configuration.[591] In addition to nmr and optical activity, dipole moment measurements have been used to assign the configuration of 1-pyrazolines (e.g., *trans*-3,5-diaryl-1-pyrazolines[592]).

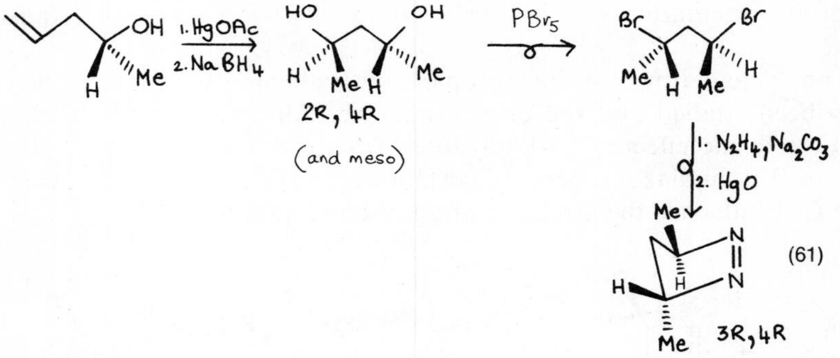

(61)

The decomposition of 1-pyrazolines is a nitrogen extrusion reaction in which cyclopropanes are the main products, but olefins also may be produced. The decomposition is effected thermally or photolytically. The answer to the question of the stereospecificity of this reaction has been the subject of considerable investigation. The reason for this is that the reaction is complex and may well proceed by various pathways, although a common intermediate may be formed initially. The general reaction for a *trans*-1-pyrazoline which yields a mixture of *cis*- and *trans*-cyclopropanes is shown in Equation 62, with the olefins (if any) omitted. The ratio of the

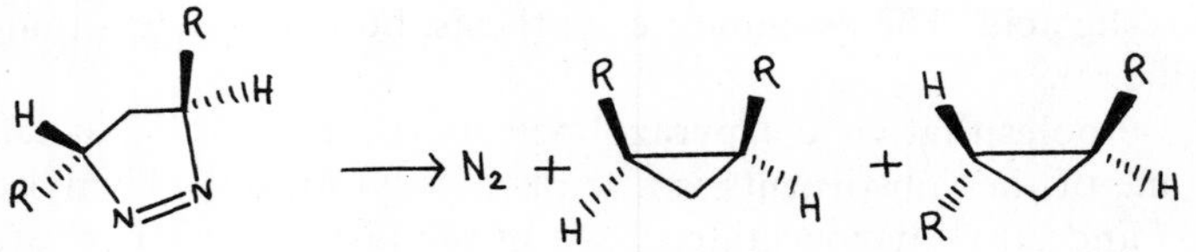

(62)

cyclopropanes formed varies with the nature of the substituents, reaction conditions, and on whether the decomposition was thermally or photochemically induced.

The thermal decomposition will be considered first and a few examples will illustrate that the reaction can only be vaguely predictable. Decomposition of *trans*-3,5-dimethyl-1-pyrazoline gives a 73:25 mixture of *cis*- and *trans*-1,2-dimethylcyclopropanes, and *cis*-3,5-dimethyl-1-pyrazoline gives a 33:66 mixture of *cis*- and *trans*-1,2-dimethylcyclopropanes.[590,593] Decomposition of *trans*-3,4-bismethoxycarbonyl-3-methyl-5-phenyl-1-pyrazoline yields exclusively *trans*-1,2-bismethoxycarbonyl-3-methyl-5-phenylcyclopropane, whereas the cis isomer yields a 75:25 mixture of *cis*- and *trans*-cyclopropanes[594] (with respect to the methoxycarbonyl groups). *cis* or *trans*-3,5-Diaryl-1-pyrazoline furnishes mainly *trans*-1,2-diarylcyclopropane although the small proportion of *cis*-cyclopropane from the cis isomer is much higher than that from the *trans*-1-pyrazoline.[582,595—597] McGreer and McKinley[598] divided their study into three classes: the 3,5-disubstituted, the 3,4-disubstituted, and the 3,4,5-trisubstituted pyrazolines, but with compounds of the last class four possible cyclopropanes are formed (Scheme 11). They found that some stereoselectivity was present and that the transition state retained several of the stereochemical features of the parent pyrazoline. This would necessitate a concerted cleavage of the C–N bonds with the orbitals of the new bonds beginning to overlap in the transition state. The lack of complete stereospecificity indicates that the three-carbon fragment can behave in a syn or anti conformation with energies which cannot differ by more than a few kilojoules per mole.[598] Crawford and collaborators[586,599—601] also agree that the cleavage of the two C–N bonds occurs simultaneously, unlike in the decomposition of aliphatic azo compounds. The general scheme of Overberger and Anselm[595] for the decomposition of 3,5-diaryl-1-pyrazolines holds true and is redrawn in Scheme 11 after the fashion of Crawford and Mishra.[590,593] There is

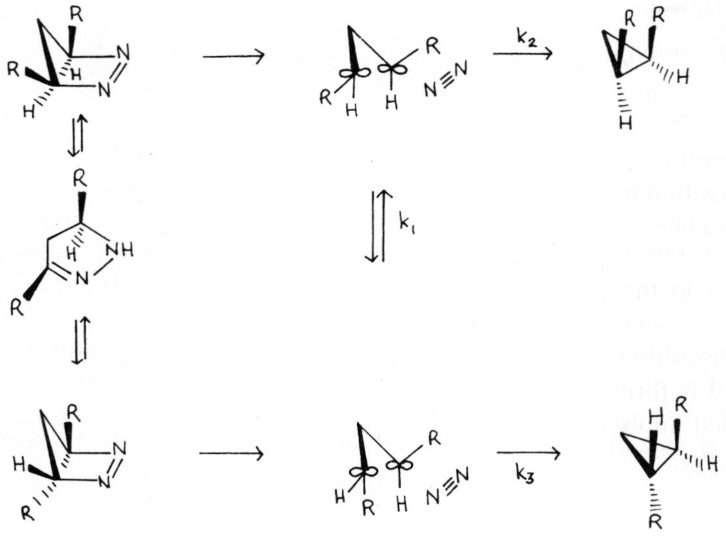

Scheme 11

evidence from esr studies that a free biradical intermediate is involved which cannot always be clearly demonstrated,[595] but there is kinetic evidence in support of a biradical intermediate.[602,603] The stereoselectivity of the reaction depends on the relative rates of inversion k_1 of the biradical compared with the cyclization rates k_2 and k_3 (Scheme 11).

Three other thermal decompositions are worthy of mention, namely, the pyrolysis of *cis*- and *trans*-3-methoxycarbonyl-5-methyl-3,4-diphenyl-1-pyrazolines[604] and *cis*- and *trans*-3-methoxycarbonyl-3,5-dimethyl-1-pyrazolines[605] which yield a mixture of cyclopropanes and olefins, and the thermal decomposition of *cis*- and *trans*-4-bromo-3-methoxycarbonyl-3-methyl-1-pyrazoline which follows a different path by eliminating the elements of HBr followed by rearrangement of the methoxycarbonyl group to give 4-methoxycarbonyl-3-methylpyrazole.[606] A kinetic and product analysis investigation of the thermolysis of 4-aryl-3-cyano-3-methoxycarbonyl-4-methyl-1-pyrazolines was made by McGreer and Wigfield.[607] They found that *cis*-1-pyrazolines (e.g., **203**, $R^1 = Me, R^2 = Ar$) gave predominantly olefins in which the aryl group migrated (i.e., **204**) whereas *trans*-1-pyrazolines (e.g., **203**, $R^1 = Ar, R^2 = Me$) gave olefinic products in which the methyl group migrated (i.e., **205**).

Several comparisons of the products of thermal and photochemical decomposition of 1-pyrazolines were made, and it is generally found that the photochemical reactions are more stereoselective than thermal decompositions.[582,608-613] Also the bias towards the *cis*- or *trans*-cyclopropanes formed in the photochemical reaction is in the opposite sense from the one in the thermal reaction.[567,582,608-610] Quantum yields of photochemical decomposition were measured[614] and it is generally accepted that a biradical is formed which is probably in an excited singlet or triplet state, but that an excited triplet biradical could be formed in the presence of a sensitizer.[615,616]

(63)

Schneider, Erben, and Merz[617] examined the thermolysis and photolysis of the rather more rigid bicyclic systems 4-substituted 2,3-diaza-cis-bicyclo-[3.3.0]octa-2,7-dienes and 7-substituted 8,9-diaza-cis-bicyclo[4.3.0]nona-2,8-dienes. These yield 6-substituted cis-bicyclo[3.1.0]hex-2-enes and 7-substituted cis-bicyclo[4.1.0]hept-2-enes respectively; but whereas in the photolytic extrusion of nitrogen there is retention of configuration at the chiral centers carrying the substituents, in the thermal decompositions inversion of configuration at these centers is generally observed (Equation 63).

b. *2-Pyrazolines.* 2-Pyrazolines possess one tetrahedral carbon less than their isomeric 1-pyrazolines and are consequently less complicated. Also they do not extrude nitrogen quite as easily as the 1-pyrazolines and are therefore more stable.

Diazomethane undergoes a [2 + 3] cycloaddition with olefins to form 1-pyrazolines (Section III.H.1.b), but if the olefin has a strongly electron attracting substituent then the product is the isomeric 2-pyrazoline in which the double bond is between a ring nitrogen atom and a carbon atom bearing the electron withdrawing group. The cycloaddition is cis and is regiospecific. β-Acetyl (or β-benzoyl) acrylates and diazomethane give approximately a 75:25 mixture of 3-acetyl-(or benzoyl)4-methoxycarbonyl and 4-acetyl-(or benzoyl)3-methoxycarbonyl 1-pyrazolines.[618] If the electron withdrawing substituent is on the 1,3-dipole (e.g., ethyl diazoacetate) and on the olefin, then the reactants align themselves with the electron withdrawing substituents opposite to each other. In addition to this the substituents can be syn or anti to each other. The 1-pyrazolines are formed initially but rearrange to the 2-pyrazolines. There are two possibilities for these rearrangements, but generally one is excluded for steric reasons (Scheme 12).[619,620,587] The reactions of ethyl diazoacetate and diazomethane with 2,6-dichlorobenzoquinone are highly regiospecific[621] as are the cycloadditions of vinyl diazomethane[622] and silyl diazoalkanes with olefins.[623]

Huisgen and his school have described the new 1,3-dipole, diphenyl nitrile imine, which is generated in position from 2,5-diphenyltetrazole or from treating the iminochloride of benzoyl phenylhydrazide with triethylamine.[624,625] It reacts with open chain[626] and cyclic olefins,[624–626] and with cyclopropenes[627] stereospecifically (Equation 64). Variable temperature ¹H nmr studies of the cyclopropane adduct can be explained by the ring inversion equilibrium in Equation 64 which probably occurs by way of either a diaziridine or a zwitterionic 1,4-dihydropyrazine intermediate.[627] Cyanogen azide behaves as the above with hexamethylbenzene (Dewar benzene, hexamethylbicyclo[2.2.0]hexa-1,5-diene) to form cis-2-cyanamino-

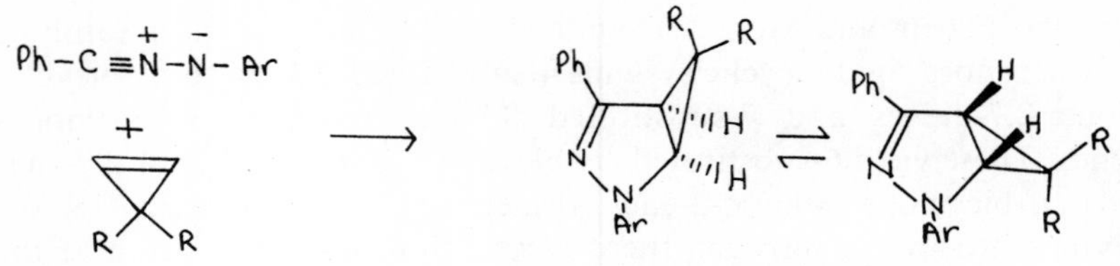

$$\text{Ph}-\text{C}\equiv\overset{+}{\text{N}}-\overset{-}{\text{N}}-\text{Ar}$$

(64)

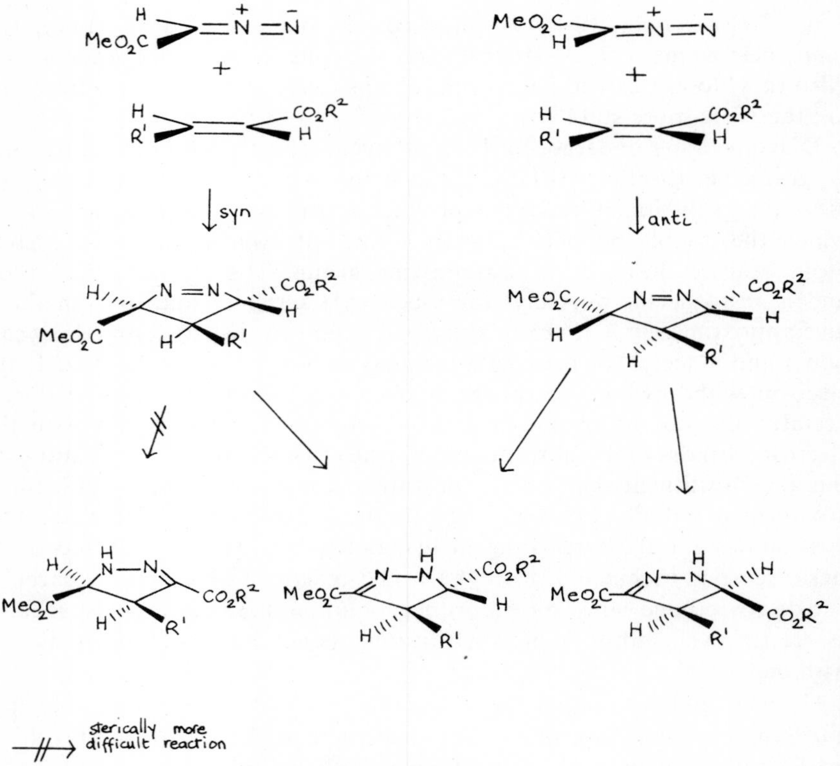

Scheme 12

1,4,5,6,7-pentamethyl-8-methylene-1,2-diazabicyclo[3.3.0]octa-3,6-diene   as one of the complex products.[628]

The intramolecular cyclization of phenylhydrazones of α,β-unsaturated ketones produces 2-pyrazolines stereoselectively. The selectivity depends on the substituents and on the reaction conditions (Equation 65). Phenyl hydrazone (of 1-acetyl-4-t-butylcyclohexene) in acetic acid at 60°C for seven days forms exclusively the *trans*-2-pyrazoline, but reflux for 6 hr gives a cis to trans ratio of 1:4. The phenyl hydrazones of 1-benzoyl- and 1-acetylcyclohex-1-enes gave the *cis*-2-pyrazolines some gummy material and some pyrazole. Although the cis-trans equilibration does occur under the reaction conditions it does not exclude the possibility that the gummy products were formed from the *trans*-pyrazolines.[629,630] In a similar way hydrazine reacts with α,β-unsaturated aldehydes or ketones to yield *cis*- and *trans*-4,5-disubstituted 2-pyrazolines in which the *trans* isomer predominated.[631—633] *trans*-Chalcone epoxides and *threo*-chalcone chlorohydrins react stereospecifically with hydrazine to form only *trans*-3,5-diaryl-4-hydroxy-2-pyrazolines, but the cis isomers are obtained from the *erythro*-chlorohydrins.[634]

Several 4-substituted 3-amino-2-pyrazolin-5-ones were prepared which have an asymmetric carbon atom at C–4, but none of them were obtained in optically active forms.[635—637]

2-Pyrazolines can be distinguished from 1-pyrazolines by ¹H nmr and by ir spectroscopy.[592,638] They are formed by heating[618,619] or by the action of uv light[638] on the 1-isomers. The reverse isomerization, that is, from 2- to 1-pyrazolines is also initiated by heat,[639] and it is believed that the thermal extrusion of nitrogen from 2-pyrazolines proceeds by way of the tautomeric 1-pyrazolines and that the stereochemistry of the cyclopropanes produced is controlled by the intermediate 1-pyrazoline.[640,641] Photolytic decomposition of 5-phenyl-3-methyl(or unsubstituted)-1-methyl-2-pyrazoline in benzene also probably causes the migration of the double bond from N–2,C–3 to N–1,N–2 because a mixture of *cis*- and *trans*-1-methyl(or unsubstituted)-1-methylazo-2-phenylcyclopropanes was formed.[642] This is an example in which the two C–N bonds are not broken simultaneously (see above). For further reading, for tables of compounds,[558] and for ¹H nmr data[61a,643] see the references indicated.

Elguero, Jacquier, and Tizané[644] examined the protonation of 3-pyrazo-lines which behave like enamines and rearrange to 2-pyrazolinium cations. This protonation, which occurs at C–3 in 4,5-disubstituted 3-pyrazolines is thermodynamically controlled and yields predominantly *trans*-4,5-disubstituted 2-pyrazolinium salts.

c. *Pyrazolidines.* Reduction of 2-pyrazolines with LAH produces isomeric pyrazolidines in which the orientation of substituents at C–4 and C–5 is unaltered. However, the proportion of isomeric pyrazolidines (at C–2) varies because the reduction is subject to steric control by the substituents already present. Elguero, Jacquier, and Tizané[645] also studied the reduction of 1,2,3,4-tetrasubstituted pyrazolinium salts with LAH in the presence of iodine which give mixtures of *cis*- and *trans*-1,2,3,4-tetrasubstituted pyrazolidines in which the cis isomer predominates. Similar reduction of 4-dimethylamino-2,3-dimethyl-1-phenyl-3-pyrazolin-5-one[646] and -5-thione,[647] on the other hand, gives only *trans*-4-dimethylamino-2,3-dimethyl-1-phenylpyrazolidin-5-one and -5-thione respectively.

A second method for preparing pyrazolidines stereospecifically is by the cycloaddition of dialkyl azodicarboxylate with olefins, and it proceeds with cis specificity (Equation 66).[648,649] Photolysis of *N*-substituted 2,3-

$$
\text{(illustration of cyclopentadiene + azodicarboxylate} \longrightarrow \text{bicyclic pyrazolidine)} \quad (66)
$$

dihydro-1,2-diazepines causes stereospecific ring contraction to form the bicyclic pyrazolidines (e.g., 2,3-diaza-*cis*-bicyclo[3.2.0]hept-6-enes).[650]

Pyrazolidines are conformationally mobile with both nitrogen atoms capable of undergoing pyramidal inversion. Such inversion was observed by variable temperature ^{1}H nmr spectroscopy in 1,2,4,4-tetramethyl-[2,249] and perfluoro-1,2-dimethylpyrazolidines,[275] and in 1,2-dibenzyl- and 2-benzyl-1,5,5-trimethylpyrazolidin-3-one.[651] The amide nitrogen atom, that is, N–2, is most probably the faster inverting nitrogen atom of the two present in the molecule. Ring inversion and pseudorotation are slowed down by the *ananchomeric* (anchoring group) effect of a *t*-butyl group in 4-*t*-butyl-1,2-diphenylpyrazolidine, and the coalescence of signals observed at −90°C is caused by the equilibrium **206** ⇌ **207** due to nitrogen inversion.[652]

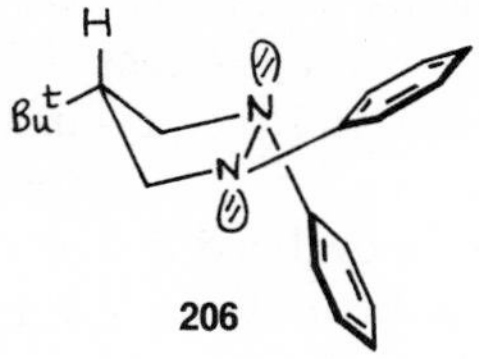

206

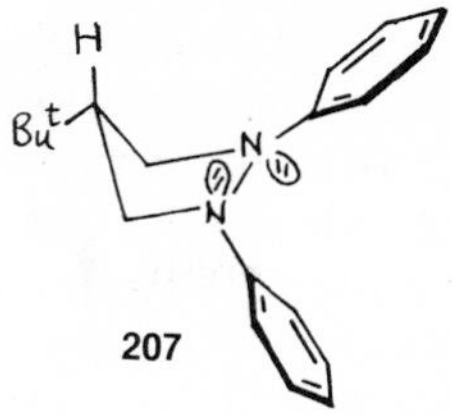

207

2. Imidazolines and Imidazolidines

2-Imidazolines and imidazolinones have attracted the most interest, although studies have not been extensive. 1,3,5-Triphenyl-2,4-diaza-1,4-pentadiene (hydrobenzamide), prepared from benzaldehyde and liquid ammonia, cyclizes by a disrotatory process to cis-1,3,5-triphenyl-2-imidazoline (amarine) on heating in benzene. Equilibration with potassium t-butoxide in t-butanol converted it into a 96:4 mixture of trans- (isoamarine) and cis-1,3,5-triphenyl-2-imidazolines[653] (Equation 67). Isoamarine is the trans isomer because it was obtained in (−) and (+) forms when the (+)-tartrate salt was resolved into its diastereoisomers.[654] Hydrobenzamide gave a deep blue color in tetrahydrofuran containing phenyl lithium which on work-up with acetic acid gave mainly cis-1,3,5-triphenyl-2-imidazoline. Irradiation of the blue solution before work-up increased the proportion of trans isomer. The latter is formed by a conrotatory cycliza-

(67)

tion.[653] The kinetics and isotope effects of the base catalyzed cyclization of hydrobenzamide have also been studied and suggest that in the cyclization and prototropy a common W-shaped carbanion intermediate is formed which cyclizes readily.[655] Another unusual synthesis is that of cis-1-ethoxycarbonyl-3a,8a-dihydro-2-methylindeno[1,2-d]imidazole (208) from a free radical reaction of N,N-dichlorourethane and indene in methyl nitrile.

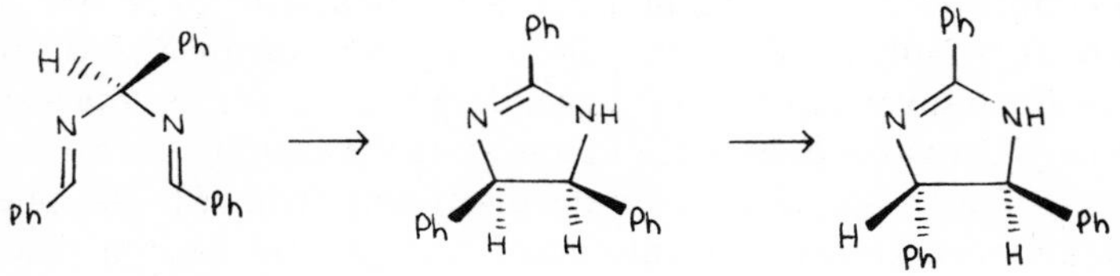

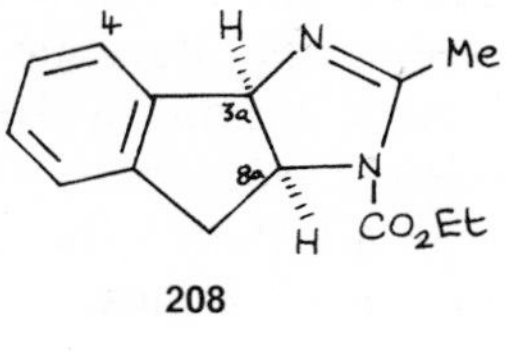

208

The initial reaction is a trans addition of the elements of Cl and NCO_2Et across the C–2,C–3 double bond of indene followed by attack of the halogen atom by methyl nitrile with Walden inversion and then cyclization to the 2-imidazoline **208**.[656]

In a study of α-amino acids Karrer and co-workers cyclized N-acetyl-[657] and N-benzoyl-S(L)-asparagine[658] with alkaline hypobromite into S-4-carboxy-2-imidazolinone (**209**). The reaction obviously proceeds by way of the isocyanate which ring closes to the 3-acetyl- or 3-benzoylimidazolidone followed by hydrolysis, because under similar conditions N-benzyloxycarbonyl asparagine gave the 3-benzyloxycarbonyl derivative of **209**.[659] Similarly, d,l-3-benzamido-3-phenylpropionamide gave d,l-4-phenyl-2-imidazolinone (**210**, R = H,X = H$_2$).[660]

209   210

The conformation of a molecule can have an appreciable influence on the course of a cyclization. This is notable in the fusion of d,l-ephedrine and (+)-pseudoephedrine hydrochlorides with urea. d,l-Ephedrine yields trans-1,5-dimethyl-3-phenyl-2-imidazolinone, by way of an S_N2 displacement with Walden inversion, but (+)-pseudoephedrine forms trans-3,4-dimethyl-5-phenyl-2-oxazolinone without upsetting the asymmetric centers.[661] The synthesis of R(−)-4-phenyl-2-imidazolinthione from R(−)-1-phenylethylene diamine of known absolute configuration and carbon disulfide occurs without affecting the asymmetric centers. The S(+)-enantiomer was also prepared, and each gave the R(+)- and S(−)-enantiomers of the broad spectrum anthelmintic 2,3,5,6-tetrahydro-6-phenylimidazo[2,1-b]-thiazole "tetramisole," on reaction with 1,2-dibromoethane. Inversion of optical rotation was observed in this last step.[662]

A considerable amount of work was done in the study of biotin, and one of the problems that arose was the structure of the desulfurization product—desthiobiotin. Four desthiobiotins are possible, two cis and two trans isomers (i.e., two racemates). Racemic desthiobiotin has the cis structure **211**. Wood and du Vigneaud synthesized it from 7-amino-8-oxononanoic acid and potassium isocyanate which gave the imidazolone, followed by reduction of the 4,5 double bond with Raney nickel[663] at 100°C. d,l-Desthiobiotin obtained in this way was contaminated with

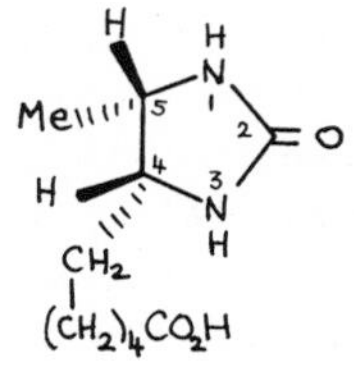

211

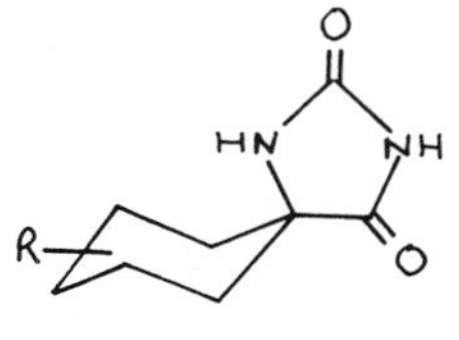

212

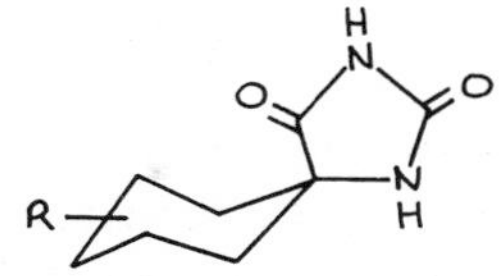

213

some of the trans isomer, *d,l*-allodesthiobiotin, because high pressure hydrogenation isomerizes this compound. Reduction using Adam catalyst in acetic acid produces sterically purer desthiobiotin.[664]

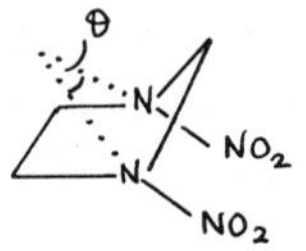

214

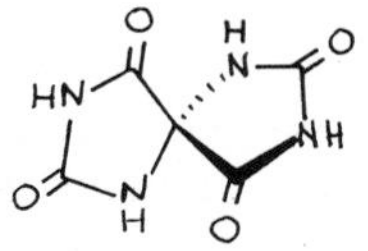

215

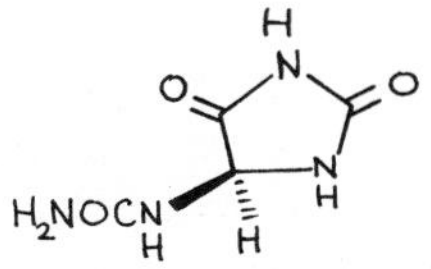

216

Lippich in 1908[665] and Dakin in 1911[666] made several optically active 5-monosubstituted hydantoins from naturally occurring α-amino acids. Dakin found that some racemized spontaneously during synthesis. (−)-Isovaline lacks an α-hydrogen atom, and yields a 5,5-disubstituted hydantoin which does not racemize. Sobotka, Holzman, and Kahn[667] resolved 5-ethyl-5-phenylhydantoin by way of its brucine salt and found that it was as resistant to racemization as the hydantoin from isovaline. The rapid racemization in the presence of *N*-sodium hydroxide is clearly attributed to the ionization of H–5 in 5-monosubstituted hydantoins. The racemization of 5-monosubstituted hydantoins has also been measured in 0.002*N* NaOH in 95% ethanol. The rates were faster if the nitrogen atoms were substituted suggesting that the anion formed by loss of H from N decreases enolization.[668] A number of optically active *R*(−)-5-phenylhydantoins (**210**, R = H,Me,Et) were prepared by cyclodehydration of optically pure *R*(−)-2-phenylhydantoic acid and its N–5 methyl and ethyl derivatives in dilute hydrochloric acid. Detailed kinetics of racemization were studied in dilute hydrochloric acid and in trifluoroacetic anhydride. It was found that with the former acid racemization of the hydantoins was negligible.[669]

Pope and Whitworth[670] resolved spiro-5,5-bihydantoin (**215**) and observed interesting changes in rotation from +12° in the neutral species to

$+59°$ and $-6°$ on increasing the alkali concentration. They explained these changes in terms of "tautomeric" (now mesomeric) mono- and dianions which would have different specific rotations. The preparation of spirohydantoins from monomethyl or dimethyl (not gemminal) cyclohexanones by the Bucherer reaction $(NaCN/(NH_4)_2CO_3)$ yields predominantly one of two stereoisomers **212** or **213**. The other isomer (**213** or **212**) is formed when a modification of the reaction using NH_4CN and CS_2 was followed by desulfurization of the resulting dithiohydantoin with hot chloroacetic acid to the hydantoin.[671] The absolute identity of these two spirohydantoins is not known.

$(-)$-5-Ureidohydantoin (alantoin, **216**) was isolated after incubation of the racemate with *Pseudomonas aeruginosa* whereby the $(+)$-enantiomer is degraded. The *R*-absolute configuration was deduced by comparison of the ord and cd curves of *S*-5-carboxymethylhydantoin and *R*-carbamoyl-alanine.[672]

The structure of streptolidine, from the antibiotic streptothricin, was revised to *trans*-2-amino-4-carboxy-5-(2-amino-1-hydroxyethyl)-2-imidazo-line on ^1H nmr evidence[673]; this structure was originally proposed.[674]

Optically active 2-α-hydroxyalkyl-2-imidazolines possess a chiral center outside the heterocyclic ring and are readily formed from optically active 2-α-hydroxyalkyl formamidine hydrochlorides and ethylene diamine.[675] The most important imidazole with an optically active side chain is the essential α-amino acid *S*-(L)-histidine.[676]

The photodimer of 2-phenyl-1-azirine, previously formulated as 4-phenyl-3-phenylimino-1-azabicyclo[2.1.0]pentane, is now shown to be 4,5-diphenyl-1,3-diazabicyclo[3.1.0]hex-3-ene.[677] It contains a 3-imidazoline ring fused with an aziridine ring. The basic structure, **64** was discussed previously (Section I.A.4.g).

Imidazolidines are the least stable members of this ring system essentially because they are cyclic aldehyde imines. The reaction of 1,2-diaminoalkanes with aldehydes most frequently leads to the *cis*-4-alkyl-2-phenylimidazoli-dines. However when phenyl groups are present on the nitrogen atoms, the trans isomers predominate.[678] Attempted cyanoalkylation of *d,l*- and *meso*-2,3-diphenylethylene diamine with formaldehyde and sodium cyanide in an acidic medium unexpectedly yields *trans*- and *cis*-1,3-bis(cyano-methyl)-4,5-diphenylimidazolidines. The lower yields of the latter may be a reflection of the conformation of the diamine [see $(+)$-pseudoephedrine above].[679]

Dipole moment measurements of 1,3-dinitroimidazolidine are in agree-ment with values calculated for the envelope conformation **214** in which θ is $135°$ $14'$.[680]

For further reading and for tables of imidazolines and imidazolidines see Hofmann's monograph.[681]

I. Triazolines

Stereochemical interest in $\triangle^1$-1,2,3-triazolines is centered around the possible asymmetry at C–4 and C–5 and the orientation of the R–N–N=N– group with respect to it. The more common method of synthesizing 1-triazolines is by a [3 + 2] cycloaddition of aryl azides and olefins, and the general applications of this method were reviewed in 1963.[151] The additions are always cis and usually regiospecific (Equation 68). Because the 1-tri-

azolines are thermally labile (they extrude N_2) it is important that the reactions are carried out at as low a temperature as possible—sometimes this requires weeks to complete. *cis*- and *trans*-β-Methylstyrenes and phenyl azide yield only *cis*- and *trans*-1,5-diphenyl-4-methyl- $\triangle^2$-1,2,3-triazolines, respectively.[682] Rigid alkenes such as norbornene condense with azides on heating to form only the exo adduct of 3-aryl-3,4,5-triazatricyclo-[5.2.1.0^{2,6}]dec-4-ene (**217**).[683–686] The addition of phenyl azide to the bicyclic systems containing two double bonds (e.g., bicyclo[3.2.1]octa-2,6-

217   **218**   **219**

diene, 6-*exo*-methylenebicyclo[2.2.1]hept-2-ene, and tricyclo[5.2.1.0^{2,6}]deca-3,9-diene) occurs readily at the angle-strained double bond to form triazolines with the two possible exo orientations, for example, **218** and **219** are formed in the ratio of 1.3:1. The predominant isomer produced is the one in which there is more stabilization of the dipolar transition state by the proximal unreactive double bond. The orientation and ratio of isomers was derived from their ^{1}H nmr data.[687]

The double bond of enol-ethers will also accept aryl azides as 1,3-dipoles with the *cis*- and *trans*-ethers producing *cis*- and *trans*-5-alkoxy-1-phenyl-$\triangle^2$-1,2,3-triazolines with high stereospecificity.[688] Also enamines under controlled conditions yield 5-amino-1-aryl- $\triangle^2$-1,2,3-triazolines stereo- and regiospecifically. In this case, however, the aminotriazolines derived from *p*-nitro(or *p*-chloro)phenyl azide undergo cis-trans rearrangement involving the amino group, which occurs readily at higher temperatures or on addition of acid catalysts. Several of these were studied and the ratio of cis to trans isomers at equilibrium varied from 28:72 for **220** ($R^1 = R^3 = Me$, $R^2 = Et$) to 0:100 for **220** ($R^1 = Ph, R^2 = Me$ and $R^3 = -[CH_2]_5-$, i.e., morpholino), and the time for the attainment of equilibrium varied from 30 min to 2 hr respectively (Equation 69). A ring opened intermediate was suggested.[689] The triazoline from the enamine, 2-morpholinonorbornene, like norbornene furnished the exo-fused triazoline in which the morpholino group was in the endo position.[690] The cycloaddition of cyclic enamines

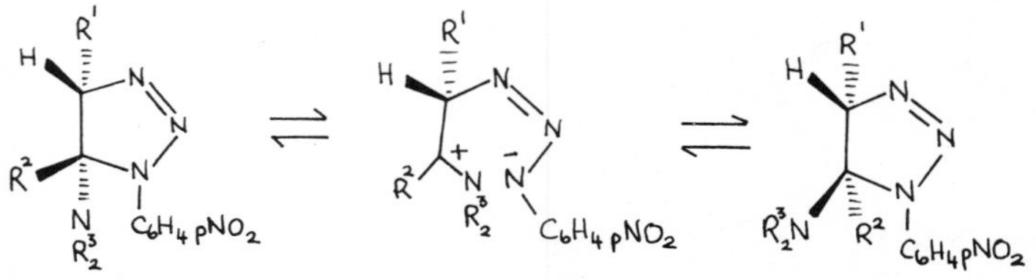

(69)

220

(e.g., 3-methoxycarbonyl-2,3-dehydroquinuclidine, and phenyl azide) are similar to morpholinonorbornene both in stereo- and regiospecificity.[691]

Δ^1-1,2,3-Triazolines decompose thermally or photochemically to aziridines with loss of molecular nitrogen.[682] A full discussion of similar reactions has already been made (see Section I.A.1). In the attempt to trap the intermediate from the pyrolysis of 3-phenyl-*exo*-3,4,5-triazatricyclo-[5.2.1.0^{2,6}]dec-4-ene (**217**), Baldwin and co-workers[692] added phenyl isocyanate to the system. The diaryl product isolated, which was thought to be the 2-imidazolin-one **221**, turned out to have a structure in which the norbornane ring was cleaved.[693] The product from the reaction with

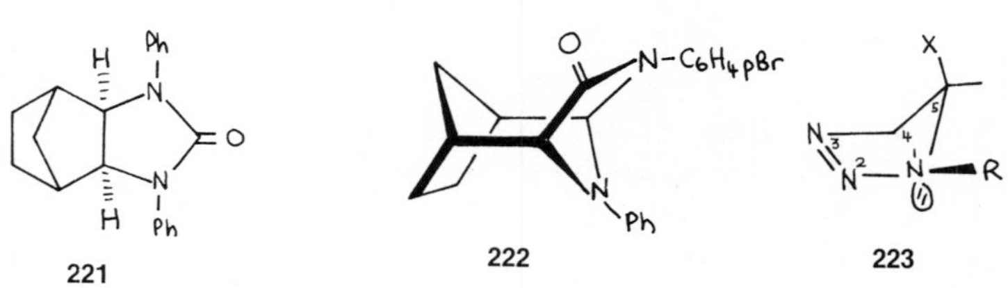

221   **222**   **223**

p-bromophenyl isocyanate (**222**) was similar and an X-ray study confirmed that the norbornane ring was cleaved.[686]

Olsen examined the ^1H nmr spectra of 5-alkoxy- and 5-hydroxy-Δ^2-1,2,3-triazolines and explained the results in terms of an envelope conformation with the 5-substituent in a pseudoaxial position at the flap and with the 1-substituent pseudoequatorial (**223**, X = OH, OEt, OBu, OPr).[694]

This preferred conformation may well be due to a type of anomeric effect. An X-ray investigation of 5-ethyl-1,4-dimethyl-5-hydroxy-$\triangle^2$-1,2,3-triazoline revealed that an analogous situation exists in the crystal.[695] 5-Amino-1-(p-nitrophenyl)-$\triangle^2$-1,2,3-triazolines also prefer the envelope conformation (**223,** X = NH$_2$,R = 4–NO$_2$C$_6$H$_4$–) only when the other substituent at C–5 is a hydrogen atom.[694]

Phenyl osotriazoles derived from aldo-sugars are among the few triazoles known in which the potential source of asymmetry is outside the five-membered rings. Ord and cd studies have shown that the phenyl osotriazole chromophore is optically anisotropic and produces a Cotton effect. A correlation was observed between the Cotton effects of the phenyl osotriazoles and the configuration of the corresponding sugars, and a sector rule was proposed.[695]

IV. REFERENCES

1. F. W. Fowler, "Synthesis and Reactions of 1-Azirines," *Adv. Heterocyclic Chem.*, **13,** 45 (1971).

2. J. B. Lambert, *Topics Stereochem.*, **6,** 19 (1971).

3. F. H. Dickey, W. Fickett, and H. J. Lucas, *J. Amer. Chem. Soc.*, **74,** 944 (1952).

4. P. E. Fanta, *J. Chem. Soc.*, 1441 (1957).

5. O. E. Paris and P. E. Fanta, *J. Amer. Chem. Soc.*, **74,** 3007 (1952).

6. P. B. Talukdar and P. E. Fanta, *J. Org. Chem.*, **24,** 555 (1959).

7. D. V. Kashelikar and P. E. Fanta, *J. Amer. Chem. Soc.*, **82,** 4927 (1960).

8. P. E. Fanta, L. J. Pandya, W. R. Groskopf, and H-J. Su, *J. Org. Chem.*, **28,** 413 (1963); P. E. Fanta, R. Golda, and H-J. Hsu, *J. Chem. and Eng. Data*, **9,** 24 (1964); *Chem. Abs.*, **73,** 14814 (1970).

9. T. Taguchi and M. Kojima, *Chem. and Pharm. Bull. (Japan)*, **7,** 103 (1959); K. Tanaka, *J. Pharm. Soc. Japan*, **70,** 212, 220 (1950); K. Takaka and T. Sugawa, *J. Pharm. Soc. Japan*, **72,** 1548, 1551 (1952).

10. S. J. Brois and G. P. Beardsley, *Tetrahedron Letters*, 5113 (1966).

11. H. Rubinstein, B. Feibush, and E. Gil-Av, *J.C.S. Perkin I*, 2094 (1973).

12. R. Haberl, *Monatsh.*, **89,** 814 (1958).

13. W. G. Otto, *Angew. Chem.*, **68,** 181 (1956).

14. A. Weissberger and H. Bach, *Ber.*, **64,** 1095 (1931); **65,** 631 (1932).

15. L. A. Paquette, D. E. Kuhla, J. H. Barrett, and R. J. Haluska, *J. Org. Chem.*, **34,** 2866 (1969).

16. D. VanEnde and A. Krief, *Angew. Chem. Internat. Edn.*, **13,** 279 (1974).

17. R. Schwesinger and H. Prinzbach, *Angew. Chem. Internat. Edn.*, **12,** 989 (1973).

18. J. A. Deyrup, *J. Org. Chem.*, **34,** 2724 (1969).

19. G. Bouteville, Y. Gelas-Mialhe, and R. Vessière, *Bull. Soc. chim. France*, 3264 (1971).

20. P. L. Southwick and R. J. Shozda, *J. Amer. Chem. Soc.*, **82,** 2888 (1960).

21. P. L. Southwick and R. J. Shozda, *J. Amer. Chem. Soc.*, **81**, 5435 (1959).

22. N. H. Cromwell, R. P. Cahoy, W. E. Franklin, and G. D. Mercer, *J. Amer. Chem. Soc.*, **79**, 922 (1957).

23. N. H. Cromwell and R. P. Cahoy, *J. Amer. Chem. Soc.*, **80**, 5524 (1958).

24. W. Nagata, S. Hirai, K. Kawata, and T. Aoki, *J. Amer. Chem. Soc.*, **89**, 5045 (1967).

25. W. Nagata, S. Hirai, T. Okumura, and K. Kawata, *J. Amer. Chem. Soc.*, **90**, 1650 (1968).

26. G. C. Tustin, C. E. Monken, and W. H. Okamura, *J. Amer. Chem. Soc.*, **94**, 5112 (1972).

27. R. S. Atkinson and R. Martin, *J.C.S. Chem. Comm.*, 386 (1974).

28. B. V. Ioffe and E. V. Koroleva, *Tetrahedron Letters*, 619 (1973).

29. A. C. Oehlschlager and L. H. Zalkow, *Chem. Comm.*, 70 (1965).

30. R. L. Hale and L. H. Zalkow, *Tetrahedron*, **25**, 1393 (1969).

30a. A. C. Oehlschlager and L. H. Zalkow, *Canad. J. Chem.*, **47**, 461 (1969).

31. J. S. McConaghy, Jr. and W. Lwowski, *J. Amer. Chem. Soc.*, **89**, 2357 (1967).

32. R. A. Abramovitch and S. R. Challand, *J.C.S. Chem. Comm.*, 1160 (1972); R. A. Abramovitch, S. R. Challand, and Y. Yamada, *J. Org. Chem.*, **40**, 1541 (1975).

33. T. L. Gilchrist, C. W. Rees, and E. Stanton, *J. Chem. Soc. (C)*, 988 (1971).

34. N. J. Leonard and K. Jann, *J. Amer. Chem. Soc.*, **84**, 4806 (1962).

35. J. A. Deyrup and R. B. Greenwald, *J. Amer. Chem. Soc.*, **87**, 4538 (1965).

36. Y. Diab, A. Laurent, and P. Mison, *Tetrahedron Letters*, 1605 (1974).

36a. A. Bartnik and A. Laurent, *Bull Soc. chim. France*, 173 (1975).

37. A. Laurent and A. Muller, *Tetrahedron Letters*, 759 (1969).

38. G. Alvernhe and A. Laurent, *Tetrahedron Letters*, 1913 (1971).

39. R. Chaabouni and A. Laurent, *Bull. Soc. chim. France*, 2680 (1973).

39a. G. Alvernhe, S. Arsenyiadis, R. Chaabouni, and A. Laurent, *Tetrahedron Letters*, 355 (1975).

40. R. M. Carlson and S. Y. Lee, *Tetrahedron Letters*, 4001 (1969).

41. K. Kotera, T. Okada, and S. Miyazaki, *Tetrahedron*, **24**, 5677 (1968).

42. K. Kitahonoki, Y. Takano, A. Matsuura, and K. Kotera, *Tetrahedron*, **25**, 335 (1969).

43. K. Kotera, Y. Takano, A. Matsuura, and K. Kitahonoki, *Tetrahedron*, **25**, 539 (1970).

43a. R. Crée and R. Carrié, *J.C.S. Chem. Comm.*, 112 (1975).

44. K. Alder, G. Stein, and W. Friedrichsen, *Annalen*, **1**, 501 (1933).

45. P. Scheiner, *J. Amer. Chem. Soc.*, **88**, 4759 (1966).

46. N. S. Zefirov, P. P. Kadzyauskas, and Y. Kur'ev, *J. Gen. Chem. (U.S.S.R.)*, **35**, 262 (1965).

47. S. Oida and E. Ohki, *Chem. and Pharm. Bull. (Japan)*, **17**, 980 (1969).

48. H. Günther, J. B. Pawliczek, B. D. Tunggal, H. Prinzbach, and R. H. Levin, *Chem. Ber.*, **106**, 984 (1973).

49. L. A. Paquette, D. E. Kuhla, and J. A. Barrett, *J. Org. Chem.*, **34**, 2879 (1969).

50. A. G. Anastassiou, S. W. Eachus, R. L. Elliott, and E. Yakali, *J.C.S. Chem. Comm.*, 531 (1972).

51. A. G. Anastassiou and R. L. Elliott, *J.C.S. Chem. Comm.*, 601 (1973).

52. L. Kaplan, J. W. Pavlik, and K. E. Wilzbach, *J. Amer. Chem. Soc.*, **94**, 3283 (1972).

53. M. Gakis, M. Märky, H-J. Hansen, and H. Schmid, *Helv. Chim. Acta*, **55**, 748 (1972).

54. N. S. Narasimhan, H. Heimgartner, H-J. Hansen, and H. Schmid, *Helv. Chim. Acta*, **56**, 1351 (1973).

55. P. E. Fanta, *Heterocyclic Compounds with Three and Four-Membered Rings*, Part 1, A. Weissberger, Ed., Interscience, New York, 1964, pp. 524–575; O. C. Dermer and G. E. Mum, *Ethyleneimine and Other Aziridines*, Academic Press, New York, 1969.

56. N. H. Cromwell and G. D. Mercer, *J. Amer. Chem. Soc.*, **79**, 3815 (1957).

57. N. H. Cromwell and R. J. Mohrbacher, *J. Amer. Chem. Soc.*, **75**, 6252 (1953).

58. N. H. Cromwell, N. G. Barker, R. A. Wankel, P. J. Vanderhorst, F. W. Olson, and J. H. Anglin, Jr., *J. Amer. Chem. Soc.*, **73**, 1044 (1951).

59. N. H. Cromwell, R. E. Bambury, and J. L. Adelfang, *J. Amer. Chem. Soc.*, **82**, 4241 (1960).

60. A. E. Pohland, R. C. Badger, and N. H. Cromwell, *Tetrahedron Letters*, 4369 (1965).

61. N. H. Cromwell, P. B. Woller, H. E. Baumgarten, R. G. Parker, and D. L. von Minden, *J. Heterocyclic Chem.*, **9**, 587 (1972).

61a. T. J. Batterham, *N.M.R. Spectra of Simple Heterocycles*, Wiley-Interscience, New York, 1972.

62. J. F. Kincaid and F. C. Henriques, Jr., *J. Amer. Chem. Soc.*, **62**, 1474 (1940).

63. R. Adams and T. L. Cairns, *J. Amer. Chem. Soc.*, **61**, 2464 (1939).

64. S. J. Brois, *J. Amer. Chem. Soc.*, **90**, 506 (1968).

65. S. J. Brois, *J. Amer. Chem. Soc.*, **90**, 508 (1968).

66. D. Felix and A. Eschenmoser, *Angew. Chem. Internat. Edn.*, **7**, 224 (1968).

67. R. G. Kostyanovsky, I. M. Gella, V. I. Markov, and Ž. E. Samojlova, *Tetrahedron*, **30**, 39 (1974).

68. R. G. Kostyanovsky, Z. E. Samojlova, and I. I. Tchervin, *Tetrahedron Letters*, 719 (1969).

69. S. Fujita, K. Imamura, and H. Nozaki, *Bull. Chem. Soc. Japan*, **44**, 1975 (1971).

70. R. G. Kostyanovsky, V. I. Markov, and I. M. Gella, *Tetrahedron Letters*, 1301 (1972).

71. H. M. Zacharis and L. M. Trefonas, *J. Heterocyclic Chem.*, **5**, 343 (1968).

72. L. M. Trefonas and R. Majeste, *J. Heterocyclic Chem.*, **2**, 80 (1965).

73. L. M. Trefonas and R. Towns, *J. Heterocyclic Chem.*, **1**, 19 (1964).

74. L. M. Trefonas, R. Towns, and R. Majeste, *J. Hererocyclic Chem.*, **4**, 511 (1967).

75. L. M. Trefonas and T. Sato, *J. Heterocyclic Chem.*, **3**, 404 (1966).

76. P. E. Fanta and E. N. Walsh, *J. Org. Chem.*, **31**, 59 (1966).

77. S. Oida, H. Kuwano, Y. Ohashi, and E. Ohki, *Chem. and Parm. Bull. (Japan)*, **18**, 2478 (1970).

78. H. M. Zacharis and L. M. Trefonas, *J. Heterocyclic Chem.*, **7**, 1301 (1970).

78a. A. H-J. Wang, I. C. Paul, E. R. Talaty, and A. E. Dupuy, Jr., *J.C.S. Chem. Comm.*, 43 (1972).

79. H. S. Gutowsky and C. H. Holm, *J. Chem. Phys.*, **25**, 1228 (1956).

80. S. Alexander, *J. Chem. Phys.*, **37**, 967 and 974 (1962); **38**, 1787 (1963); **40**, 2741 (1964).

81. A. Allerhand, H. S. Gutowsky, J. Jonas, and R. A. Meinzer, *J. Amer. Chem. Soc.*, **88**, 3185 (1966).

82. H. Paulsen and W. Greve, *Chem. Ber.*, **103**, 486 (1970).

83. M. Ohtsuru and K. Tori, *Tetrahedron Letters*, 4043 (1970).

84. R. Martino, J. Abeba, and A. Lattes, *J. Heterocyclic Chem.*, **10**, 91 (1973).

85. R. Martino, J. Abeba, and A. Lattes, *Tetrahedron Letters*, 433 (1973).

86. A. T. Bottini and J. D. Roberts, *J. Amer. Chem. Soc.*, **78**, 5126 (1956).

87. A. Loewenstein, J. F. Neumer, and J. D. Roberts, *J. Amer. Chem. Soc.*, **82**, 3599. (1960).

88. A. T. Bottini and J. D. Roberts, *J. Amer. Chem. Soc.*, **80**, 5203 (1958).

89. D. J. Anderson and T. L. Gilchrist, *J. Chem. Soc.* (*C*), 2273 (1971).

90. D. L. Nagel, P. B. Woller, and N. H. Cromwell, *J. Org. Chem.*, **36**, 3911 (1971).

91. D. J. Anderson, D. C. Horwell, and R. S. Atkinson, *Chem. Comm.*, 1189 (1969).

92. J-L. Pierre, P. Baret, and P. Arnaud, *Bull. Soc. chim. France*, 3619 (1971).

93. R. L. VanEtten and J. J. Dolhun, *J. Org. Chem.*, **33**, 3904 (1968).

94. A. T. Bottini, R. L. VanEtten, and A. J. Davidson, *J. Amer. Chem. Soc.*, **87**, 755 (1965).

95. J. Stackhouse, R. Baechler, and K. Mislow, *Tetrahedron Letters*, 3437 (1971).

96. R. E. Carter and T. Drakenberg, *J.C.S. Chem. Comm.*, 582 (1972).

97. T. J. Bardos, C. Szantay, and C. K. Navada, *J. Amer. Chem. Soc.*, **87**, 5796 (1965).

98. H. Kessler, *Angew. Chem. Internat. Edn.*, **9**, 219 (1970).

99. S. J. Brois, *Tetrahedron*, **26**, 227 (1970).

100. S. J. Brois, *J. Amer. Chem. Soc.*, **89**, 4242 (1967).

101. H. Saitô, K. Nukada, T. Kobayashi, and K. Morita, *J. Amer. Chem. Soc.*, **89**, 6605 (1967).

101a. P. Baret, J-L. Pierre, and R. Perraud, *Bull. Soc. chim. France*, 707 (1975).

102. H. Kessler and D. Leibfritz, *Tetrahedron Letters*, 4297 (1970).

103. J. D. Andose, J. M. Lehn, K. Mislow, and J. Wagner, *J. Amer. Chem. Soc.*, **92**, 4050 (1970).

104. F. A. L. Anet, R. D. Trepka, and D. J. Cram, *J. Amer. Chem. Soc.*, **89**, 357 (1967).

105. F. A. L. Anet and J. M. Osyany, *J. Amer. Chem. Soc.*, **89**, 352 (1967).

106. J. B. Lambert, B. S. Packard, and W. L. Oliver, Jr., *J. Org. Chem.*, **36**, 1309 (1971).

107. J. M. Lehn and J. Wagner, *Chem. Comm.*, 1298 (1968).

108. A. Hassner and J. E. Galle, *J. Amer. Chem. Soc.*, **92**, 3733 (1970).

109. A. H. Cowley, M. J. S. Dewar, W. R. Jackson, and W. B. Jennings, *J. Amer. Chem. Soc.*, **92**, 5206 (1970).

110. S. Rengaraju and K. D. Berlin, *J. Org. Chem.*, **37**, 3304 (1972).

111. H. Kessler and D. Leibfritz, *Tetrahedron Letters*, 4289 (1970).

112. J. M. Lehn and J. Wagner, *Chem. Comm.*, 148 (1968).

113. J. M. Lehn and J. Wagner, *Tetrahedron*, **26**, 4227 (1970).

114. R. G. Kostyanovsky, I. I. Tchervin, A. A. Fomichov, Z. E. Samojlova, C. N. Makarov, Yu. V. Zeifman, and B. L. Dyatkin, *Tetrahedron Letters*, 4021 (1969).

115. S. J. Brois, *Tetrahedron Letters*, 5997 (1968).

116. R. S. Atkinson, *Chem. Comm.*, 676 (1968).

117. H. Kessler and D. Leibfritz, *Tetrahedron Letters*, 4293 (1970).

118. H. W. Heine, J. D. Myers, and E. T. Peltzer, *Angew. Chem. Internat. Edn.*, **9**, 374 (1970).

119. A. Padwa and L. Hamilton, *J. Org. Chem.*, **31**, 1995 (1966).

120. R. K. Müller, D. Felix, J. Schreiber, and A. Eschenmoser, *Helv. Chim. Acta*, **53**, 1479 (1970).

121. J. P. Freeman and W. H. Graham, *J. Amer. Chem. Soc.*, **89**, 1761 (1967).

122. R. D. Clark and G. K. Helmkamp, *J. Org. Chem.*, **29**, 1316 (1964).

123. A. Padwa and L. Hamilton, *J. Amer. Chem. Soc.*, **87**, 1821 (1965).

124. A. Padwa and L. Hamilton, *J. Amer. Chem. Soc.*, **89**, 102 (1967).

125. J. Villaume and P. S. Skell, *J. Amer. Chem. Soc.*, **94**, 3455 (1972).

126. R. Ghirardelli and H. J. Lucas, *J. Amer. Chem. Soc.*, **79**, 734 (1975).

127. K. Tanaka, *J. Pharm. Soc. Japan*, **70**, 220 (1950); K. Tanaka and T. Sugawa, *J. Pharm. Soc. Japan*, **72**, 1551 (1952).

128. N. H. Cromwell, G. V. Hudson, R. A. Wankel, and P. J. Vanderhorst, *J. Amer. Chem. Soc.*, **75**, 5384 (1953).

129. S. Fujita, T. Hiyama, and H. Nozaki, *Tetrahedron*, **26**, 4347 (1970).

130. T. A. Foglia, L. M. Gregory, G. Maerker, and S. F. Osman, *J. Org. Chem.*, **36**, 1068 (1971).

131. R. A. Wohl and D. F. Headley, *J. Org. Chem.*, **37**, 4401 (1972).

132. N. J. Leonard, D. B. Dixon, and T. R. Keenan, *J. Org. Chem.*, **35**, 3488 (1970).

133. T. Hiyama, H. Koide, S. Fukita, and H. Nozaki, *Tetrahedron*, **29**, 3137 (1973).

134. S. Mitsui and Y. Sugi, *Tetrahedron Letters*, 1287 (1969).

134a. Y. Sugi, M. Nagata, and S. Mitsui, *Bull. Chem. Soc. Japan*, **48**, 1663 (1975).

135. J. E. Baldwin, A. K. Bhatnagar, S. C. Choi, and T. J. Shortbridge, *J. Amer. Chem. Soc.*, **93**, 4082 (1971).

136. I. J. Burnstein, P. E. Fanta, and B. S. Green, *J. Org. Chem.*, **35**, 4084 (1970).

137. D. V. Kashelikar and P. E. Fanta, *J. Amer. Chem. Soc.*, **82**, 4930 (1960).

138. R. B. Woodward and R. Hoffmann, *The Conservation of Orbital Symmetry*, Verlag. Chemie-Academic, New York, 1970; R. B. Woodward and R. Hoffmann, *Angew. Chem. Internat. Edn.*, **8**, 781 (1969).

139. R. Huisgen and H. Mäder, *J. Amer. Chem. Soc.*, **93**, 1777 (1971).

140. H. Hermann, R. Huisgen, and H. Mäder, *J. Amer. Chem. Soc.*, **93**, 1779 (1971).

141. R. Huisgen, W. Scheer, and H. Huber, *J. Amer. Chem. Soc.*, **89**, 1753 (1967).

142. R. Huisgen and W. Scheer, *Tetrahedron Letters*, 481 (1971).

143. R. Huisgen, V. Martin-Ramos, and W. Scheer, *Tetrahedron Letters*, 477 (1971).

144. P. B. Woller and N. H. Cromwell, *J. Org. Chem.*, **35**, 888 (1970).

145. F. Texier and R. Carrié, *Bull. Soc. chim. France*, 310 (1974).

146. J. W. Lown, J. P. Moser, and R. Westwood, *Canad. J. Chem.*, **47**, 4335 (1969).

147. A. Padwa and L. Hamilton, *J. Heterocyclic Chem.*, **4**, 118 (1967).

148. J. H. Hall and R. Huisgen, *Chem. Comm.*, 1187 (1971).

149. J. W. Lown, R. Westwood, and J. P. Moser, *Canad. J. Chem.*, **48**, 1682 (1970); J. M. Lown, R. K. S. Malley, G. Dallas, and T. W. Maloney, *Canad. J. Chem.*, **48**, 89 (1970).

150. S. Oida and E. Ohki, *Chem. and Pharm. Bull. (Japan)*, **17**, 2461 (1969).

151. R. Huisgen, *Angew. Chem. Internat. Edn.*, **2**, 565, 633 (1963); **7**, 321 (1968); R. Huisgen, W. Scheer, and H. Mäder, *Angew. Chem. Internat. Edn.*, **8**, 602, 604 (1969); R. 'Huisgen, *Bull. Soc. chim. France*, 3431 (1965).

152. J. H. Hall, R. Huisgen, C. H. Ross, and W. Scheer, *Chem. Comm.*, 1188 (1971).

153. A. G. Anastassiou and R. B. Hammer, *J. Amer. Chem. Soc.*, **94**, 303 (1972).

154. R. E. Lutz and A. B. Turner, *J. Org. Chem.*, **33**, 516 (1968).

155. A. Padwa and W. Eisenhardt, *J. Org. Chem.*, **35**, 2472 (1970).

156. A. Padwa and W. Eisenhardt, *J. Amer. Chem. Soc.*, **93**, 1400 (1971).

157. V. R. Gaertner, *J. Org. Chem.*, **35**, 3952 (1970).

158. J. A. Deyrup, C. L. Moyer, and P. S. Dreifus, *J. Org. Chem.*, **35**, 3428 (1970).

159. P. G. Gassman, D. K. Dygos, and J. E. Trent, *J. Amer. Chem. Soc.*, **92**, 2084 (1970).

160. P. G. Gassman and D. K. Dygos, *J. Amer. Chem. Soc.*, **91**, 1543 (1969).

161. R. G. Weiss, *Tetrahedron*, **27**, 271 (1971).

162. N. H. Cromwell, J. H. Anglin, Jr., F. W. Olsen, and N. G. Barker, *J. Amer. Chem. Soc.*, **73**, 2803 (1951).

163. D. K. Wall, J.-L. Imbach, A. E. Pohland, R. C. Badger, and N. H. Cromwell, *J. Heterocyclic Chem.*, **5**, 77 (1968).

164. J. A. Deyrup and C. L. Moyer, *J. Org. Chem.*, **35**, 3424 (1970).

165. J. L. Pierre, H. Handel, and P. Baret, *J.C.S. Chem. Comm.*, 551 (1972).

166. J. L. Pierre, H. Handel, and P. Baret, *Tetrahedron*, **30**, 3213 (1974).

167. J. L. Pierre and H. Handel, *Tetrahedron Letters*, 4431 (1974).

167a. H. Handel and J. L. Pierre, *Tetrahedron*, **31**, 997 (1975).

168. R. Bartnik and A. Laurent, *Tetrahedron Letters*, 3869 (1974).

169. G. K. Helmkamp, R. D. Clark, and J. R. Koskinen, *J. Org. Chem.*, **30**, 666 (1965).

170. A. T. Bottini and R. L. VanEtten, *J. Org. Chem.*, **30**, 575 (1965).

171. A. T. Bottini, B. F. Dowden, and L. Sousa, *J. Amer. Chem. Soc.*, **87**, 3249 (1965).

172. A. T. Bottini, B. F. Dowden, and R. L. VanEtten, *J. Amer. Chem. Soc.*, **87**, 3250 (1965).

173. H. W. Heine, R. H. Weese, R. A. Cooper, and A. J. Durbetaki, *J. Org. Chem.*, **32**, 2708 (1967).

174. T. DoMinh and A. M. Trozzolo, *J. Amer. Chem. Soc.*, **92**, 6997 (1970).

175. T. DoMinh and A. M. Trozzolo, *J. Amer. Chem. Soc.*, **94**, 4047 (1972).

176. A. Padwa, S. Clough, and E. Glazer, *J. Amer. Chem. Soc.*, **92**, 1778 (1970).

177. A. Padwa and L. Gehrlein, *J. Amer. Chem. Soc.*, **94**, 4933 (1972).

178. A. Padwa and E. Glazer, *J. Org. Chem.*, **38**, 284 (1973).

179. A. Padwa and E. Glazer, *J. Amer. Chem. Soc.*, **94**, 7788 (1972).

180. J. A. Deyrup and S. C. Clough, *J. Org. Chem.*, **39**, 902 (1974)

181. T. A. Foglia, L. M. Gregory, and G. Maerker, *J. Org. Chem.*, **35**, 3779 (1970).

182. T. Nishiguchi, H. Tochio, A. Nabeya, and Y. Iwakura, *J. Amer. Chem. Soc.*, **91**, 5835 (1969).

183. G. Casini, F. Claudi, M. Grifantini, and S. Martelli, *J. Heterocyclic Chem.*, **11**, 377 (1974).

184. M. Klaus and H. Prinzbach, *Angew. Chem. Internat. Edn.*, **10**, 273 (1971).

185. E. L. Stogryn and S. J. Brois, *J. Amer. Chem. Soc.*, **89**, 605 (1967).

186. A. G. Anastassiou, R. L. Elliott, and A. Lichtenfeld, *Tetrahedron Letters*, 4569 (1972).

186a. H. W. Heine, in *Mechanisms of Molecular Migrations*, Vol. 3, B. S. Thyagarajan, Ed., Wiley-Interscience, New York, 1971, p. 145.

187. J. P. Snyder, R. J. Boyd, and M. A. Whitehead, *Tetrahedron Letters*, **5**, 4347 (1972).

188. J. J. Uebel and J. C. Martin, *J. Amer. Chem. Soc.*, **86**, 4618 (1964).

189. A. Mannschreck and W. Seitz, *Angew. Chem.*, **81**, 224 (1969); *Angew. Chem. Internat. Edn.*, **8**, 212 (1969).

190. R. G. Kostyanovsky, K. S. Zakharov, M. Zaripova, and V. F. Rudtchenko, *Tetrahedron Letters*, 4207 (1974).

191. J. B. Lambert, W. L. Oliver, Jr., and B. S. Packard, *J. Amer. Chem. Soc.*, **93**, 933 (1971).

192. J. A. Moore, *Heterocyclic Compounds with Three- and Four-Membered Rings*, Vol. 19, Part 2, A. Weissberger, Ed., Interscience, New York, 1964, Chap. 7, pp. 885–983.

193. L. Fowden, *Biochem. J.*, **64**, 323 (1956).

194. B. A. Phillips and N. H. Cromwell, *J. Heterocyclic Chem.*, **10**, 795 (1973).

195. R. M. Rodebaugh and N. H. Cromwell, *J. Heterocyclic Chem.*, **8**, 421 (1971).

196. R. M. Rodebaugh and N. H. Cromwell, *J. Heterocyclic Chem.*, **6**, 993 (1969).

197. E. J. Moriconi and P. H. Mazzocchi, *J. Org. Chem.*, **31**, 1372 (1966).

198. F. Galinovsky and H. Nesvadba, *Monatsh.*, **85**, 1300 (1954).

199. E. Doomes and N. H. Cromwell, *J. Org. Chem.*, **34**, 310 (1969).

200. M. C. Eagen, R. H. Higgins, and N. H. Cromwell, *J. Heterocyclic Chem.*, **8**, 851 (1971).

201. J-L. Imbach, E. Doomes, R. P. Rebman, and N. H. Cromwell, *J. Org. Chem.*, **32**, 78 (1967).

202. G. Cignarella, G. F. Cristiani, and E. Testa, *Annalen*, **661**, 181 (1963).

202a. H. Hagesawa, H. Aoyama, and Y. Omote, *Tetrahedron Letters*, 1901 (1975).

203. E. H. Gold, *J. Amer. Chem. Soc.*, **93**, 2793 (1971).

204. E. J. Corey and A. M. Felix, *J. Amer. Chem. Soc.*, **87**, 2518 (1965).

205. G. Lowe and J. Parker, *Chem. Comm.*, 577 (1971).

206. D. M. Brunwin, G. Lowe, and J. Parker, *J. Chem. Soc.* (C), 3756 (1971).

207. K. Kawashima and I. Agata, *J. Pharm. Soc. Japan*, **89**, 1426 (1969).

208. Z. Neiman, *J.C.S. Perkin II*, 1746 (1972).

209. A. K. Bose, G. Spiegelman, and M. S. Manhas, *Tetrahedron Letters*, 3167 (1971).

210. J. L. Luche and H. B. Kagan, *Bull. Soc. chim. France*, 2450 (1968).

211. J. Decazes, J. L. Luche, and H. B. Kagan, *Tetrahedron Letters*, 3665 (1970).

212. D. A. Nelson, *Tetrahedron Letters*, 2543 (1971).

213. J. Decazes, J. L. Luche, and H. B. Kagan, *Tetrahedron Letters*, 3661 (1970).

214. J. C. Sheehan and J. J. Ryan, *J. Amer. Chem. Soc.*, **73**, 1204 (1951).

215. J. L. Luche and H. B. Kagan, *Bull. Soc. chim. France*, 3500 (1969).

216. F. Dardoize, J. L. Moreau, and M. Gaudemar, *Bull. Soc. chim. France*, 1668 (1973).

217. A. K. Bose, B. Lai, B. Dayal, and M. S. Manhas, *Tetrahedron Letters*, 2633 (1974).

218. A. K. Bose, J. C. Kapur, B. Dayal, and M. S. Manhas, *J. Org. Chem.*, **39**, 312 (1974).

219. M. Vaultier, R. Danion-Bougot, D. Danion, J. Hamelin, and R. Carrié, *Tetrahedron Letters*, 1923 (1973).

220. E. J. Moriconi and W. C. Meyer, *J. Org. Chem.*, **36**, 2841 (1971).

221. H. J. Freidrich, *Tetrahedron Letters*, 2981 (1971).

222. E. Nativ and P. Rona, *Israel J. Chem.*, **10**, 55 (1972).

223. E. J. Moriconi and W. C. Crawford, *J. Org. Chem.*, **33**, 370 (1968).

224. L. A. Paquette and T. Kakihana, *J. Amer. Chem. Soc.*, **90**, 3897 (1968).

225. L. A. Paquette, T. Kakihana, J. F. Hansen, and J. C. Philips, *J. Amer. Chem. Soc.*, **93**, 152 (1971).

226. L. A. Paquette and J. C. Philips, *J. Amer. Chem. Soc.*, **90**, 3898 (1968).

227. L. A. Paquette, S. Kirschner, and J. R. Malpass, *J. Amer. Chem. Soc.*, **92**, 4330 (1970).

228. L. A. Paquette, S. Kirschner, and J. R. Malpass, *J. Amer. Chem. Soc.*, **91**, 3970 (1969).

229. L. A. Paquette, M. J. Broadhurst, C. Lee, and J. Clardy, *J. Amer. Chem. Soc.*, **94**, 630 (1972).

230. L. A. Paquette and M. J. Broadhurst, *J. Amer. Chem. Soc.*, **94**, 632 (1972).

231. F. Effenberger, G. Prossel, and P. Fischer, *Chem. Ber.*, **104**, 2002 (1971).

232. F. Effenberger, P. Fischer, G. Prossel, and G. Kiefer, *Chem. Ber.*, **104**, 1987 (1971).

233. A. B. Arbuzov, N. M. Zobova, and I. I. Andronova, *Bull. Acad. Sci.*, *U.S.S.R.* (*div. Chem. Sci.*), **23**, 1484 (1975)

234. G. Lowe and D. D. Ridley, *J.C.S. Chem. Comm.*, 328 (1973); G. Lowe and D. D. Ridley, *J.C.S. Perkin I*, 2024 (1973).

235. J. R. Hlubucek and G. Lowe, *J.C.S. Chem. Comm.*, 419 (1974).

236. H. H. Wassermann and M. S. Baird, *Tetrahedron Letters*, 3721 (1971).

237. D. S. Wulfman and T. R. Steinheimer, *Tetrahedron Letters*, 3933 (1972).

238. H. M. Berman, E. L. McGandy, J. W. Burgner II, and R. L. VanEtten, *J. Amer. Chem. Soc.*, **91**, 6177 (1969).

239. M. S. Toy and C. C. Price, *J. Amer. Chem. Soc.*, **82**, 2613 (1960).

240. L. Fontanella and E. Testa, *Annalen*, **616**, 148 (1958).

241. E. L. McGandy, H. M. Berman, J. W. Burgner II, and R. L. VanEtten, *J. Amer. Chem. Soc.*, **91**, 6173 (1969).

242. R. L. Towns and L. M. Trefonas, *J. Amer. Chem. Soc.*, **93**, 1761 (1971).

243. C. L. Moret and L. M. Trefonas, *J. Heterocyclic Chem.*, **5**, 549 (1968).

244. R. L. Snyder, E. L. McGandy, R. L. VanEtten, L. M. Trefonas, and R. L. Towns, *J. Amer. Chem. Soc.*, **91**, 6187 (1969).

245. A. Vigevani and G. G. Gallo, *J. Heterocyclic Chem.*, **4**, 583 (1967).

246. T. Okutani, A. Morimoto, T. Kaneko, and K. Masuda, *Tetrahedron Letters*, 1115 (1971).

247. R. H. Higgins, N. H. Cromwell, and W. W. Paudler, *J. Heterocyclic Chem.*, **8**, 961 (1971).

248. R. H. Higgins, E. Doomes, and N. H. Cromwell, *J. Heterocyclic Chem.*, **8**, 1063 (1971).

249. J. M. Lehn, *Fortschr. Chem. Forsch.*, **15**, 311 (1970).

250. C. L. Bumgardner, K. S. McCallum, and J. P. Freeman, *J. Amer. Chem. Soc.*, **83**, 4417 (1961).

251. R. H. Higgins and N. H. Cromwell, *J. Heterocyclic Chem.*, **8**, 1059 (1971).

252. J. P. Freeman, D. G. Pucci, and G. Binsch, *J. Org. Chem.*, **37**, 1894 (1972).

253. L. A. Paquette, M. J. Wyvratt, and G. R. Allen, Jr., *J. Amer. Chem. Soc.*, **92**, 1763 (1970).

254. E. J. Moriconi and C. C. Jalandoni, *J. Org. Chem.*, **35**, 3796 (1970).

255. K. Hirai, H. Matsuda, and Y. Kishida, *Chem. and Pharm. Bull. (Japan)*, **21**, 1305 (1973).

256. A. G. Hortmann and D. A. Robertson, *J. Amer. Chem. Soc.*, **94**, 2758 (1972).

257. T. Okutani and K. Masuda, *Chem. and Pharm. Bull. (Japan)*, **22**, 1498 (1974).

258. J. L. Kurz, B. K. Gillard, D. A. Robertson, and A. G. Hortmann, *J. Amer. Chem. Soc.*, **92**, 5008 (1970).

259. R. H. Higgins and N. H. Cromwell, *J. Amer. Chem. Soc.*, **95**, 120 (1973).

260. B. K. Gillard and J. L. Kurz, *J. Amer. Chem. Soc.*, **94**, 7199 (1972).

261. D. D. Miller, J. Fowble, and P. N. Patil, *J. Medicin. Chem.*, **16**, 177 (1973).

262. A. G. Hortmann and J. Koo, *J. Org. Chem.*, **39**, 3781 (1974).

263. E. Doomes and N. H. Cromwell, *J. Heterocyclic Chem.*, **6**, 153 (1969).

264. J. C. Sheehan, D. Ben-Ishai, and J. U. Piper, *J. Amer. Chem. Soc.*, **95**, 3064 (1973).

265. R. J. Stoodley and N. R. Whitehouse, *J.C.S. Perkin I*, 32 (1973).

266. D. F. Corbett and R. J. Stoodley, *J.C.S. Chem. Comm.*, 438 (1974).

267. A. Padwa and R. Gruber, *J. Amer. Chem. Soc.*, **92**, 100 (1970).

268. A. Padwa and R. Gruber, *J. Amer. Chem. Soc.*, **92**, 107 (1970).

269. L. A. Paquette, T. Kakihana, J. F. Kelly, and J. R. Malpass, *Tetrahedron Letters*, 1455 (1969).

270. L. A. Paquette, T. Kakihana, and J. F. Kelly, *J. Org. Chem.*, **36**, 435 (1971).

271. J. S. Millership and H. Suschitzky, *Chem. Comm.*, 1496 (1971).

272. E. K. von Gustorf, D. V. White, B. Kim, D. Hess, and J. Leitich, *J. Org. Chem.*, **35**, 1155 (1970).

273. N. Rieber, J. Alberts, J. A. Lipsky, and D. M. Lemal, *J. Amer. Chem. Soc.*, **91**, 5668 (1969).

274. W. D. Phillips, *Determination of Organic Structures by Physical Methods*, Vol. 2, F. C. Nachod and W. D. Phillips, Eds., Academic, New York, 1962, pp. 450, 453.

275. P. Ogden, *Chem. Comm.*, 1084 (1969).

276. E. Fahr, W. Fischer, A. Jung, L. Sauer, and A. Mannschreck, *Tetrahedron Letters*, 161 (1967).

277. A. Mannschreck, V. Jonas, and B. Kolb, *Angew. Chem. Internat. Edn.*, **12**, 583 (1973).

278. G. E. McCasland and S. Proskow, *J. Amer. Chem. Soc.*, **76**, 6087 (1954).

279. J. C. Sheehan and J. G. Whitney, *J. Amer. Chem. Soc.*, **85**, 3863 (1963).

280. P. M. G. Bavin, *J. Medicin. Chem.*, **9**, 52 (1966).

281. A. F. Beecham, *J. Amer. Chem. Soc.*, **76**, 4615 (1954).

282. A. F. Beecham, *J. Amer. Chem. Soc.*, **76**, 4613 (1954).

283. E. J. Corey and W. R. Hertler, *J. Amer. Chem. Soc.*, **82**, 1657 (1960).

284. K. Morita, F. Irreverre, F. Sakiyama, and B. Witkop, *J. Amer. Chem. Soc.*, **85**, 2832 (1963).

285. N. Izumiya and B. Witkop, *J. Amer. Chem. Soc.*, **85**, 1835 (1963).

286. C. Eguchi and A. Katuka, *Bull. Chem. Soc. Japan*, **47**, 1704 (1974).

287. W. Oppolzer, E. Pfenninger, and K. Keller, *Helv. Chim. Acta*, **56**, 1807 (1973).

287a. R. Achini and W. Oppolzer, *Tetrahedron Letters*, 369 (1975).

288. T. Durst, R. V. D. Elzen, and M. J. LeBelle, *J. Amer. Chem. Soc.*, **94**, 9261 (1972).

289. S-i. Yamada and K. Achiwa, *Chem. and Pharm. Bull. (Japan)*, **12**, 1525 (1964).

290. R. Adams and E. F. Rogers, *J. Amer. Chem. Soc.*, **63**, 228 (1941).

291. F. Texier and R. Carrié, *Bull. Soc. chim. France*, 2373 (1972).

292. F. Texier and R. Carrié, *Bull. Soc. chim. France*, 2381 (1972).

293. A. Padwa and J. Smolanoff, *J. Amer. Chem. Soc.*, **93**, 548 (1971).

294. A. Padwa, M. Dharan, J. Smolanoff, and S. I. Wetmore, Jr., *J. Amer. Chem. Soc.*, **95**, 1945 (1973).

295. H. Pracejus and M. Grass, *J. prakt. Chem.*, **8**, 362 (1959).

296. I. Felner and K. Schenker, *Helv. Chim. Acta*, **53**, 754 (1970).

297. K. Okumura, K. Kotera, and I. Inoue, *Chem. and Pharm. Bull. (Japan)*, **12**, 725 (1964).

298. P. E. Hanna and A. E. Ahmed, *J. Medicin. Chem.*, **16**, 963 (1973).

299. P. E. Hanna, *J. Heterocyclic Chem.*, **10**, 747 (1973).

300. A. A. Patchett and B. Witkop, *J. Amer. Chem. Soc.*, **79**, 188 (1957).

301. J. B. P. A. Wijngerg, W. N. Speckamp, and H. E. Schoemaker, *Tetrahedron Letters*, 4073 (1974).

302. P. L. Southwick, N. Latif, B. M. Fitzgerald, and N. H. Zlacek, *J. Org. Chem.*, **31**, 1 (1966).

303. M. Bethell and G. W. Kenner, *J. Chem. Soc.*, 3850 (1965).

304. S. Ohki and M. Yoshino, *Chem. and Pharm. Bull. (Japan)*, **16**, 269 (1968).

305. A. B. Mauger, F. Irreverre, and B. Witkop, *J. Amer. Chem. Soc.*, **87**, 4975 (1965).

306. K. Osugi, *J. Pharm. Soc. Japan*, **78**, 1348, 1353 (1958) and earlier papers.

307. Y. Sanno, *J. Pharm. Soc. Japan*, **80**, 603 (1960).

308. S. Mageswaran, W. D. Ollis, and I. O. Sutherland, *J.C.S. Chem. Comm.*, 656 (1973).

309. E. Breuer and D. Melumad, *J. Org. Chem.*, **37**, 3949 (1972).

310. N. Castagnoli, Jr., *J. Org. Chem.*, **34**, 3187 (1969).

311. M. Cushman and N. Castagnoli, Jr., *J. Org. Chem.*, **36**, 3404 (1971).

312. M. Cushman and N. Castagnoli, Jr., *J. Org. Chem.*, **37**, 1268 (1972).

313. S. Oida and E. Ohki, *Chem. and Pharm. Bull. (Japan)*, **17**, 1405 (1969).

314. C. B. Hudson, A. V. Robertson, and W. R. J. Simpson, *Austral. J. Chem.*, **21**, 769 (1968).

315. S-i. Yamada, Y. Murakami, and K. Koga, *Tetrahedron Letters*, 1501 (1968).

316. F. Zymalkowski and P. Pachaly, *Chem. Ber.*, **100**, 1137 (1967).

317. Y. Sanno, *J. Pharm. Soc. Japan*, **78**, 1113 (1958).

318. J. P. Greenstein and M. Winitz, *The Chemistry of the Amino Acids*, Wiley, New York, 1961, Vol. 1, pp. 191–199 and Vol. 3, p. 2178.

319. L. Fowden, P. M. Scopes, and R. N. Thomas, *J. Chem. Soc.* (*C*), 833 (1971).

320. H. R. Dave and M. K. Hargreaves, *Chem. Comm.*, 743 (1967).

321. C. Djerassi, H. Wolf, and E. Bunnenberg, *J. Amer. Chem. Soc.*, **84**, 4552 (1962).

322. W. L. F. Armarego and B. A. Milloy, *J.C.S. Perkin I*, 1905 (1974).

323. N. J. Greenfield and G. D. Fasman, *J. Amer. Chem. Soc.*, **92**, 177 (1970); A. P. Volosov, V. A. Zubkov, and T. M. Birshtein, *Tetrahedron*, **31**, 1259 (1975).

324. A. I. Meyers, *J. Org. Chem.*, **24**, 1233 (1959).

325. S. Schnell and P. Karrer, *Helv. Chim. Acta*, **39**, 2036 (1955).

326. K. Achiwa and S-i. Yamada, *Chem. and Pharm. Bull.* (*Japan*), **14**, 537 (1966).

327. Y. Murakami, K. Koga, H. Matsuo, and S-i. Yamada, *Chem. and Pharm. Bull.* (*Japan*), **20**, 543 (1972).

328. R. Lukeš, J. Kovár, J. Kloubek, and K. Bláha, *Coll. Czech. Chem. Comm.*, **25**, 483 (1960).

328a. R. Morlacchi, V. Losacco, and V. Tortorella, *Gazzetta*, **105**, 349 (1975).

329. K. Takatori, S. Nakano, S. Nagata, K. Okumura, I. Hirono, and M. Shimizu, *Chem. and Pharm. Bull.* (*Japan*), **20**, 1087 (1972).

329a. A. Welter, M. Marlier, and G. Dardenne, *Bull. Soc. chim. belges*, **84**, 243 (1975).

330. A. V. Robertson and B. Witkop, *J. Amer. Chem. Soc.*, **82**, 5008 (1960).

331. A. V. Robertson and B. Witkop, *J. Amer. Chem. Soc.*, **84**, 1697 (1962).

332. C. Gallina, F. Petrini, and A. Romeo, *J. Org. Chem.*, **35**, 2425 (1970).

333. F. Irreverre, K. Morita, A. V. Robertson, and B. Witkop, *J. Amer. Chem. Soc.*, **85**, 2824 (1963).

334. F. Sakiyama, F. Irreverre, S. L. Friess, and B. Witkop, *J. Amer. Chem. Soc.*, **86**, 1842 (1964).

335. B. Witkop, *Recent Advances on Naturally Occurring Nitrogen Heterocyclic Compounds*, Special Publication of the Chemical Society No. 3, 60 (1955); A. B. Mauger and B. Witkop, *Chem. Rev.*, **66**, 47 (1966).

336. A. Neuberger, *J. Chem. Soc.*, 429 (1945).

337. C. S. Hudson and A. Neuberger, *J. Org. Chem.*, **15**, 24 (1950).

338. B. Witkop and T. Beiler, *J. Amer. Chem. Soc.*, **78**, 2882 (1956).

339. Y. Fujita, A. Gottlieb, B. Peterkofsky, S. Udenfriend, and B. Witkop, *J. Amer. Chem. Soc.*, **86**, 4709 (1964).

340. R. J. Abraham, K. A. McLaughlan, S. Dalby, G. W. Kenner, R. C. Sheppard, and L. F. Burroughs, *Nature*, **192**, 1150 (1961).

341. J. T. Gerig and R. S. McLeod, *J. Amer. Chem. Soc.*, **95**, 5725 (1973).

342. B. J. Magerlein, R. D. Birkenmeyer, R. R. Herr, and F. Kagan, *J. Amer. Chem. Soc.*, **89**, 2454 (1967).

343. G. Slomp and F. A. MacKellar, *J. Amer. Chem. Soc.*, **89**, 2459 (1967).

344. S. Nakamura, T. Chikaike, H. Yonehara, and H. Umezawa, *Chem. and Pharm. Bull.* (*Japan*), **13**, 599 (1965).

345. C. Gallina, C. Marta, C. Colombo, and A. Romeo, *Tetrahedron*, **27**, 4681 (1971).

346. J. P. Greenstein, *Adv. Protein Chem.*, **9**, 121 (1954).

347. J. R. Dyer, H. B. Hayes, E. G. Miller, Jr., and R. F. Nassar, *J. Amer. Chem. Soc.*, **86**, 5363 (1964).

348. T. Sugawa, Y. Sanno, and A. Kurita, *J. Pharm. Soc. Japan*, **75**, 845, 856 (1955).

349. S. Murakami, T. Takemoto, Z. Tei, and K. Daigo, *J. Pharm. Soc. Japan*, **75**, 866, 869 (1955).

350. H. Morimoto, *J. Pharm. Soc. Japan*, **75**, 901 (1955) and following papers.

351. Y. Ueno, H. Nawa, J. Ueyanagi, H. Morimoto, R. Nakamori, and T. Matsuoka, *J. Pharm. Soc. Japan*, **75**, 807 (1955) and following papers.

352. H. Watase and I. Nitta, *Bull. Chem. Soc. Japan*, **30**, 889 (1957).

353. K. Murayama, S. Morimura, and G. Sunagawa, *J. Pharm. Soc. Japan*, **85**, 757 (1965).

354. J. J. Beereboom, K. Butler, F. C. Pennington, and I. A. Solomons, *J. Org. Chem.*, **30**, 2334 (1965).

355. J. P. Schaefer and P. J. Wheatley, *J. Org. Chem.*, **33**, 166 (1968).

355a. C. M. Wong, *Canad. J. Chem.*, **46**, 1101 (1968); S. Oida and E. Ohiki, *Chem. and Pharm. Bull. (Japan)*, **16**, 2087 (1968).

356. A. McL. Mathieson and H. K. Welsh, *Acta Cryst.*, **5**, 599 (1952).

357. I. Donohue and K. N. Trueblood, *Acta Cryst.*, **5**, 414, 419 (1952); J. Zussman, *Acta Cryst.*, **4**, 73, 493 (1951).

358. J. A. Molin-Case, E. Fleischer, and D. W. Urry, *J. Amer. Chem. Soc.*, **92**, 4728 (1970).

359. S. S. Ament, J. B. Wetherington, J. W. Moncrief, K. Flohr, M. Mochizuki, and E. T. Kaiser, *J. Amer. Chem. Soc.*, **95**, 7896 (1973).

360. Y. Fugimoto, F. Irreverre, J. M. Karle, I. L. Karle, and B. Witkop, *J. Amer. Chem. Soc.*, **93**, 3471 (1971).

361. P. L. Southwick, J. A. Vida, and D. P. Mayer, *J. Org. Chem.*, **29**, 1492 (1964).

361a. R. B. Woodward, *Pure Appl. Chem.*, **25**, 283 (1971); **17**, 519 (1968); A. Eschenmoser, *Quart. Rev.*, **24**, 366 (1970).

362. R. J. Abraham and W. A. Thomas, *J. Chem. Soc.*, 3739 (1964).

363. T. J. Batterham, N. V. Riggs, A. V. Robertson, and W. R. J. Simpson, *Austral. J. Chem.*, **22**, 725 (1969).

364. J. E. Becconsall and R. A. Y. Jones, *Tetrahedron Letters*, 1103 (1962).

365. E. Breuer and D. Melumad, *J. Org. Chem.*, **38**, 1601 (1973).

366. W. Lijinsky, L. Keefer, and J. Loo, *Tetrahedron*, **26**, 5137 (1970).

367. B. C. Gilbert, R. O. C. Norman, and M. Trenwith, *J.C.S. Perkin II*, 1033 (1974).

368. D. P. McDermott and H. L. Strauss, *J. Amer. Chem. Soc.*, **94**, 5124 (1972).

368a. P. J. McCullough, *J. Chem. Phys.*, **29**, 966 (1958).

369. C. M. Lee and W. D. Kumler, *J. Amer. Chem. Soc.*, **84**, 565, 571 (1962).

370. C. M. Lee and W. D. Kumler, *J. Amer. Chem. Soc.*, **83**, 4593 (1961).

370a. T. Konno, H. Meguro, and K. Tuzimura, *Tetrahedron Letters*, 1305 (1975).

371. P. Karrer and K. Ehrhardt, *Helv. Chim. Acta*, **34**, 2202 (1951).

372. R. H. Andreatta, V. Nair, A. V. Robertson, and W. R. J. Simpson, *Austral J. Chem.*, **20**, 1493 (1967).

373. A. J. Verbiscar and B. Witkop, *J. Org. Chem.*, **35**, 1924 (1970).

374. A. A. Patchett and B. Witkop, *J. Amer. Chem. Soc.*, **79**, 185 (1957).

375. R. Gaudry and C. Godin, *J. Amer. Chem. Soc.*, **76**, 139 (1954).

376. P. G. Gassman and A. Fentiman, *J. Org. Chem.*, **32**, 2388 (1967).

377. C. F. Hammer and S. R. Heller, *Chem. Comm.*, 919 (1966).

378. C. F. Hammer, S. R. Heller, and J. H. Craig, *Tetrahedron*, **28**, 239 (1972).

379. A. Hirshfeld, W. Taub, and E. Glotter, *Tetrahedron*, **28**, 1275 (1972).

380. P. S. Portoghese and A. A. Mikhail, *J. Org. Chem.*, **31**, 1059 (1966).

381. P. S. Portoghese, D. L. Lattin, and V. G. Telang, *J. Heterocyclic Chem.*, **8**, 993 (1971).

382. P. S. Portoghese and D. T. Sepp, *J. Heterocyclic Chem.*, **8**, 531 (1971).

382a. J. McKenna, J. M. McKenna, A. Tulley, and J. White, *J. Chem. Soc.*, 1711 (1965), and following papers.

383. A. Solladié-Cavallo and G. Solladié, *Tetrahedron Letters*, 4237 (1972).

384. D. M. Lemal and S. D. McGregor, *J. Amer. Chem. Soc.*, **88**, 1335 (1966).

385. A. B. Mauger, F. Irreverre, and B. Witkop, *J. Amer. Chem. Soc.*, **87**, 4975 (1965).

386. K. Hiroi and S-i. Yamada, *Chem. and Pharm. Bull.* (*Japan*), **21**, 4721 (1973).

387. K. Hiroi and S-i. Yamada, *Chem. and Pharm. Bull.* (*Japan*), **21**, 54 (1973).

388. G. Otani and S-i. Yamada, *Chem. and Pharm. Bull.* (*Japan*), **21**, 2112 (1973).

389. T. Sone, K. Hiroi and S-i. Yamada, *Chem. and Pharm. Bull.* (*Japan*), **21**, 2331 (1973).

390. K. Achiwa and S-i. Yamada, *Tetrahedron Letters*, 1799 (1974).

391. K. Flohr and E. T. Kaiser, *J. Amer. Chem. Soc.*, **94**, 3675 (1972).

392. K. Mislow, *Introduction to Stereochemistry*, Benjamin, New York, 1966.

393. G. Krow and R. K. Hill, *Chem. Comm.*, 430 (1968).

394. J. Harley-Mason and R. F. J. Ingleby, *J. Chem. Soc.*, 3639 (1958).

395. J. C. Seaton and L. Marion, *Canad. J. Chem.*, **35**, 1102 (1957).

396. G. E. McCasland and S. Proskow, *J. Amer. Chem. Soc.*, **77**, 4688 (1955).

397. G. E. McCasland and S. Proskow, *J. Amer. Chem. Soc.*, **78**, 5646 (1956).

398. G. E. McCasland and S. Proskow, *J. Org. Chem.*, **22**, 122 (1957).

399. N. B. Mehta, R. E. Brooks, J. Z. Strelitz, and J. W. Horodniak, *J. Org. Chem.*, **28**, 2843 (1963).

400. J. H. Brewster and R. S. Jones, Jr., *J. Org. Chem.*, **34**, 354 (1969).

401. F. Lions, *J. Amer. Chem. Soc.*, **53**, 1176 (1931).

402. L. H. Bock and R. Adams, *J. Amer. Chem. Soc.*, **53**, 3519 (1931).

403. L. H. Bock and R. Adams, *J. Amer. Chem. Soc.*, **53**, 374 (1931).

404. J. L. A. Webb, *J. Org. Chem.*, **18**, 1413 (1953).

405. C. Chang and R. Adams, *J. Amer. Chem. Soc.*, **53**, 2353 (1931).

406. C. Chang and R. Adams, *J. Amer. Chem. Soc.*, **56**, 2089 (1934).

407. R. Adams and R. M. Joyce, *J. Amer. Chem. Soc.*, **60**, 1491 (1938).

408. F. W. Fowler, *Chem. Comm.*, 1359 (1969).

409. S. R. Tanny and F. W. Fowler, *J. Amer. Chem. Soc.*, **95**, 7320 (1973).

410. S. R. Tanny and F. W. Fowler, *J. Org. Chem.*, **39**, 2715 (1974).

411. T. Eicher and S. Böhm, *Chem. Ber.*, **107**, 2238 (1974).

412. P. T. Izzo, *J. Org. Chem.*, **28**, 1713 (1963).

413. R. Baltzly, N. B. Mehta, P. B. Russell, R. E. Brooks, E. M. Grivsky, and A. M. Steinberg, *J. Org. Chem.*, **27**, 213 (1962).

414. L. A. Paquette, G. R. Allen, Jr., and M. J. Broadhurst, *J. Amer. Chem. Soc.*, **93**, 4503 (1971).

415. R. P. Gandhi and V. K. Chadha, *Indian J. Chem.*, **9**, 305 (1971).

416. R. A. Odum and B. Schmall, *Chem. Comm.*, 1299 (1969).

417. L. A. Paquette and J. H. Barrett, *J. Amer. Chem. Soc.*, **88**, 1718 (1966).

418. L. A. Paquette and W. C. Farley, *J. Org. Chem.*, **32**, 2725 (1967).

419. L. A. Paquette, *J. Amer. Chem. Soc.*, **86**, 4092 (1964).

420. R. F. Childs and A. W. Johnson, *Chem. Comm.*, 95 (1965).

421. H. D. Scharf and F. Korte, *Chem. Ber.*, **98**, 764 (1965).

422. R. H. Rynbrandt, *J. Heterocyclic Chem.*, **11**, 787 (1974).

423. J. W. Barrett and R. P. Linstead, *J. Chem. Soc.*, 436 (1935).

424. V. Prelog and S. Szpilfogel, *Helv. Chim. Acta*, **28**, 178 (1945).

425. H. Booth, F. E. King, K. G. Mason, J. Parrick, and R. L. St. D. Whitehead, *J. Chem. Soc.*, 1050 (1959).

426. N. L. Allinger, *Experientia*, **10**, 328 (1954).

427. A. Bertho and G. Rödl, *Chem. Ber.*, **92**, 2218 (1959).

428. G. F. Hennion and F. X. Quinn, *J. Org. Chem.*, **35**, 3054 (1970).

429. R. Griot, *Helv. Chim. Acta*, **42**, 67 (1959).

430. T. Wagner-Jauregg and M. Roth, *Chem. Ber.*, **93**, 3036 (1960).

431. R. Griot and T. Wagner-Jauregg, *Helv. Chim. Acta*, **42**, 605 (1959).

432. R. Griot and T. Wagner-Jauregg, *Helv. Chim. Acta*, **42**, 121 (1959).

433. I. Murakoshi, W. Murata, and J. Haginiwa, *J. Pharm. Soc., Japan*, **84**, 674 (1964).

434. S. V. Kessar, A. L. Rampal, and K. P. Mahajan, *J. Chem. Soc.*, 4703 (1962).

435. Y. L. Chow, R. A. Perry, and B. C. Menon, *Tetrahedron Letters*, 1549 (1971).

436. W. E. Noland, R. B. Hart, W. A. Joern, and R. G. Simon, *J. Org. Chem.*, **34**, 2058 (1969).

437. K. Murayama, S. Morimura, Y. Nakamura, and G. Sunagawa, *J. Pharm. Soc. Japan*, **85**, 130 (1965).

438. T. Sasaki, S. Eguchi, and T. Ishii, *J. Org. Chem.*, **35**, 2257 (1970).

439. K. S. Pillay, R. N. Lockhart, T. Tezuka, and Y. L. Chow, *J.C.S. Chem. Comm.*, 80 (1974).

440. R. Willstätter and D. Jaquet, *Ber.*, **51**, 767 (1918).

441. R. Willstätter, F. Seitz, and J. von Braun, *Ber.*, **58**, 385 (1925).

442. F. E. King, J. A. Baltrop, and R. J. Walley, *J. Chem. Soc.*, 277 (1945).

443. M. Metayer, *Bull. Soc. chim. France*, 1093 (1948).

444. M. Mokotoff, *J. Heterocyclic Chem.*, **10**, 1063 (1973).

445. A. Bertho and H. Kurzmann, *Chem. Ber.*, **90**, 2319 (1957).

446. A. Bertho and J. F. Schmidt, *Chem. Ber.*, **97**, 3284 (1964).

447. W. L. F. Armarego and B. A. Milloy, *J.C.S. Perkin I*, 2814 (1973).

448. F. E. King, D. M. Bovey, K. G. Mason, and R. L. St. D. Whitehead, *J. Chem. Soc.*, 250 (1953).

449. H. Booth and F. E. King, *J. Chem. Soc.*, 2688 (1958).

450. H. Adkins and H. L. Coonradt, *J. Amer. Chem. Soc.*, **63**, 1563 (1941).

451. M. Mokotoff and R. F. Sprecher, *Tetrahedron*, **30**, 2623 (1974).

452. R. B. Carlin and D. P. Carlson, *J. Amer. Chem. Soc.*, **81**, 4673 (1959).

453. P. Coggon, D. S. Farrier, P. W. Jeffs, and A. T. McPhail, *J. Chem. Soc.*, (*B*), 1267 (1970).

454. P. A. Luhan, A. T. McPhail, and P. M. Gross, *J.C.S. Perkin II*, 51 (1973).

455. P. W. Jeffs, R. L. Hawks, and D. S. Farrier, *J. Amer. Chem. Soc.*, **91**, 3831 (1969).

456. K. Okada, *Chem. and Pharm. Bull.* (*Japan*), **10**, 398 (1962).

457. K. Okada, *Chem. and Pharm. Bull.* (*Japan*), **10**, 401 (1962).

458. T. Oh-ishi and H. Kugita, *Chem. and Pharm. Bull.* (*Japan*), **18**, 299 (1970).

459. T. Oh-ishi and H. Kugita, *Chem. and Pharm. Bull.* (*Japan*), **18**, 291 (1970).

460. H. Taguchi, T. Oh-ishi, and H Kugita, *Chem. and Pharm. Bull.* (*Japan*), **18**, 1008 (1970)

461. H. Taguchi, T. Oh-ishi, and H. Kugita, *Tetrahedron Letters*, 5763 (1968).

462. K. Okada, *Chem. and Pharm. Bull.* (*Japan*), **10**, 852 (1962).

463. G. L. Smith and H. W. Whitlock, Jr., *Tetrahedron Letters*, 2711 (1966).

464. E. F. Godefroi and L. H. Simanyi, *J. Org. Chem.*, **27**, 3882 (1962).

465. S-i. Yamada and G. Otani, *Tetrahedron Letters*, 1133 (1971).

466. G. Otani and S-i. Yamada, *Chem. and Pharm. Bull.* (*Japan*), **21**, 2130 (1973).

467. R. V. Stevens and M. P. Wentland, *J. Amer. Chem. Soc.*, **90**, 5580 (1968).

468. S. L. Keely and F. C. Tahk, *J. Amer. Chem. Soc.*, **90**, 5584 (1968).

469. R. K. Hill, J. A. Joule, and L. J. Loeffler, *J. Amer. Chem. Soc.*, **84**, 4951 (1962).

470. Y. Tsuda, K. Isobe, and A. Ukai, *Chem. Comm.*, 1554 (1971).

471. W. Oppolzer, *J. Amer. Chem. Soc.*, **93**, 3834 (1971).

472. F. E. King and H. Booth, *J. Chem. Soc.*, 3798 (1954).

473. J. McKenna, *Chem. and Ind.*, 406 (1954).

474. I. N. Nazarov and V. F. Kucherov, *Izvest. Akad. Nauk. S.S.S.R.*, *Ser. khim.*, 289 (1952); *Chem. Abstr.*, **47**, 5363 (1953); T. E. Sample, Jr. and L. F. Hatch, *J. Chem. Educ.*, **45**, 55 (1968); L. E. Miller and D. J. Mann, *J. Amer. Chem. Soc.*, **72**, 1484 (1950).

475. L. M. Rice, E. E. Reid, and C. H. Grogan, *J. Org. Chem.*, **19**, 884 (1954).

476. W. L. F. Armarego and T. Kobayashi, *J. Chem. Soc.* (*C*), 3222 (1971).

477. K. Murayama, N. Yoshida, S. Morimura, Y. Nakamura, and G. Sunagawa, *J. Pharm. Soc. Japan*, **85**, 765 (1965).

478. L. M. Rice and C. H. Grogan, *J. Org. Chem.*, **20**, 1687 (1955).

479. C. F. Culberson and P. Wilder, Jr., *J. Org. Chem.*, **25**, 1358 (1960).

480. C. F. Culberson and P. Wilder, Jr., *J. Org. Chem.*, **27**, 4205 (1962).

481. H. M. Walton, *J. Org. Chem.*, **22**, 315 (1957).

482. D. Craig, *J. Amer. Chem. Soc.*, **73**, 4889 (1951); S. C. Harvey, *J. Amer. Chem. Soc.*, **71**, 1121 (1949); K. Adler and G. Stein, *Angew. Chem.*, **50**, 514 (1937).

483. M. Fujimoto and K. Okabe, *Chem. and Pharm. Bull.* (*Japan*), **10**, 714 (1962).

484. J. A. Berson and R. Swidler, *J. Amer. Chem. Soc.*, **76**, 4060 (1954).

485. L. M. Rice and C. H. Grogan, *J. Org. Chem.*, **25**, 393 (1960).

486. H. Stockmann, *J. Org. Chem.*, **26**, 2025 (1961).

487. H. Christol, A. Donche, and F. Plénat, *Bull. Soc. chim. France*, 1315 (1966).

488. G. Quinkert, K. Opitz, W. W. Wiersdoff, and M. Finke, *Annalen*, **693**, 44 (1966).

489. W. Oppolzer, *J. Amer. Chem. Soc.*, **93**, 3833 (1971).

490. W. Oppolzer and K. Keller, *Angew. Chem. Internat. Edn.*, **11**, 728 (1972).

491. H. W. Gschwend and A. O. Lee, *J. Org. Chem.*, **38**, 2169 (1973).

492. A-H. Khuthier and J. C. Robertson, *J. Org. Chem.*, **35**, 3760 (1970).

493. M. G. Peter, G. Snatzke, F. Snatzke, K. N. Nagarajan, and N. Schmid, *Helv. Chim. Acta*, **57**, 32 (1974).

494. R. Weidmann and A. Horeau, *Tetrahedron Letters*, 2979 (1973), and references cited therein.

495. C. D. Smith, R. D. Otzenberger, B. P. Mundy, and C. N. Caughlan, *J. Org. Chem.*, **39**, 321 (1974).

496. R. D. Otzenberger, K. B. Lipkowitz, and B. P. Mundy, *J. Org. Chem.*, **39**, 319 (1974).

497. P. Wilder, Jr. and C. F. Culberson, *J. Amer. Chem. Soc.*, **81**, 2027 (1959).

498. C. F. Culberson and P. Wilder, Jr., *J. Amer. Chem. Soc.*, **82**, 4939 (1960).

499. J. B. Bremner and R. N. Warrener, *Chem. Comm.*, 926 (1967).

500. R. Achini, H. R. Loosli, and F. Troxler, *Helv. Chim. Acta*, **57**, 572 (1973).

501. A. P. Gray, D. E. Heitmeier, and H. Kraus, *J. Amer. Chem. Soc.*, **84**, 89 (1962).

502. B. Sjöberg, *Arkiv Kemi*, **12**, 251 (1958).

503. A. Fredga and L-B. Agenäs, *Arkiv Kemi*, **15**, 327 (1960).

504. A. Fredga, *Tetrahedron*, **8**, 126 (1960).

505. B. Sjöberg, *Arkiv Kemi*, **15**, 419 (1960).

506. T. H. Chan and R. K. Hill, *J. Org. Chem.*, **35**, 3519 (1970).

507. A. Steche, *Annalen*, **242**, 367 (1887).

508. A. R. Bader, R. J. Bridgwater, and P. R. Freeman, *J. Amer. Chem. Soc.*, **83**, 3319 (1961).

509. H. Adkins and R. E. Burks, Jr., *J. Amer. Chem. Soc.*, **70**, 4174 (1948).

510. J. E. Hyre and A. R. Bader, *J. Amer. Chem. Soc.*, **80**, 437 (1958).

511. O. L. Chapman and G. L. Eian, *J. Amer. Chem. Soc.*, **90**, 5329 (1968).

512. A. H. Jackson and P. Smith, *J. Chem. Soc.* (*C*), 1667 (1968).

513. C. M. Atkinson, J. W. Kershaw, and A. Taylor, *J. Chem. Soc.*, 4426 (1962).

514. J. W. Kershaw and A. Taylor, *J. Chem. Soc.*, 4320 (1964).

515. G. Berti, A. Da Settimo, G. Di Colo, and E. Nannipieri, *J. Chem. Soc.* (*C*), 2703 (1969).

516. S. McLean, E. K. Strom-Gundersen, K. S. Dichmann, J. K. Fawcett, and S. C. Nyburg, *Tetrahedron Letters*, 2645 (1970).

517. D. R. Julian and R. Foster, *J.C.S. Chem. Comm.*, 311 (1973).

518. A. McKenzie and P. A. Stewart, *J. Chem. Soc.*, 104 (1935).

519. E. J. Corey, R. J. McCaully, and H. S. Sachdev, *J. Amer. Chem. Soc.*, **92**, 2476 (1970).

520. E. J. Corey, H. S. Sachdev, J. Z. Gougoutas, and W. Saenger, *J. Amer. Chem. Soc.*, **92**, 2488 (1970).

521. P. L. Julian and J. Pikl, *J. Amer. Chem. Soc.*, **57**, 755 (1935).

522. G. R. Newkome and N. S. Bhacca, *Chem. Comm.*, 385 (1969).

523. M. Ohno, T. F. Spande, and B. Witkop, *J. Amer. Chem. Soc.*, **90**, 6521 (1968).

524. M. Ohno, T. F. Spande, and B. Witkop, *J. Amer. Chem. Soc.*, **92**, 343 (1970).

525. B. Witkop and R. K. Hill, *J. Amer. Chem. Soc.*, **77**, 6592 (1955).

526. A. H. Jackson and A. E. Smith, *J. Chem. Soc.*, 5510 (1964).

527. G. Kollenz, E. Ziegler, M. Eder, and E. Prewedourakis, *Monatsh.*, **101**, 1597 (1970).

528. G. M. Loudon, D. Portsmouth, A. Lukton, and D. E. Koshland, Jr., *J. Amer. Chem. Soc.*, **91**, 2792 (1969).

529. W. D. Ollis, K. L. Ormand, and I. O. Sutherland, *Chem. Comm.*, 1697 (1968).

530. R. Bonnett and J. D. White, *J. Chem. Soc.* (*C*), 1648 (1963).

531. C. O. Bender and R. Bonnett, *J. Chem. Soc.* (*C*), 2186 (1968).

532. L. A. Caprino, *Chem. Comm.*, 494 (1966).

533. D. R. Boyd and D. E. Ladhams, *J. Chem. Soc.*, 2089 (1928).

534. L. A. Caprino, *J. Amer. Chem. Soc.*, **84**, 2196 (1962).

535. G. Cignarella and A. Saba, *Gazzetta*, **99**, 450 (1969).

536. W. J. Houlimann, Ed., *Indoles*, Vols. 1, 11, 111, Interscience, New York, 1972.

537. W. I. Patterson and R. Adams, *J. Amer. Chem. Soc.*, **55**, 1069 (1933).

538. H. Fritz and W. Stock, *Annalen*, **721**, 82 (1969).

539. H. Fritz, R. Oehl, and E. Besch, *Annalen*, **721**, 87 (1969).

540. H. Fritz and R. Oehl, *Annalen*, **756**, 79 (1972).

541. H. Fritz and P. Uhrhan, *Annalen*, **756**, 87 (1972).

542. K. Bernauer, *Helv. Chim. Acta*, **46**, 211 (1963).

543. J. Gurney, W. H. Perkin, Jr., and S. G. P. Plant, *J. Chem. Soc.*, 2676 (1927).

544. S. G. P. Plant and M. L. Tomlinson, *J. Chem. Soc.*, 298 (1933).

545. S. G. P. Plant and M. L. Tomlinson, *J. Chem. Soc.*, 3324 (1931).

546. B. Witkop, *J. Amer. Chem. Soc.*, **72**, 614 (1950).

547. S. McLean, U. O. Trotz, K. S. Dichmann, J. K. Fawcett, and S. C. Nyburg, *Tetrahedron Letters*, 4561 (1970).

548. Y. Ban, H. Kinoshita, S. Murakami, and T. Oishi, *Tetrahedron Letters*, 3687 (1971).

549. J. Shimizu, S. Murakami, T. Oishi, and Y. Ban, *Chem. and Pharm. Bull.* (*Japan*), **12**, 2561 (1971).

550. R. R. Wittekind and S. Lazarus, *J. Heterocyclic Chem.*, **7**, 1241 (1970).

551. O. L. Chapman, G. L. Eian, A. Bloom, and J. Clardy, *J. Amer. Chem. Soc.*, **93**, 2918 (1971).

552. A. Bloom and J. Clardy, *Chem. Comm.*, 531 (1970).

553. A. Smith and J. H. P. Utley, *J. Chem. Soc.* (*C*), 1 (1970).

554. A. Smith and J. H. P. Utley, *J. Chem. Soc.* (*B*), 1201 (1971).

555. D. Shaw, A. Smith, and J. H. P. Utley, *J. Chem. Soc.* (*B*), 1161 (1970).

556. H. Booth, F. E. King, and J. Parrick, *J. Chem. Soc.*, 2302 (1958).

557. W. H. Perkin, Jr. and S. G. P. Plant, *J. Chem. Soc.*, 1503 (1924).

558. C. H. Jarboe, *Pyrazoles, Pyrazolines, Pyrazolidines and Condensed Rings*, R. H. Wiley, Ed., Intersciecne, New York, 1967, Ch. 6, pp. 177–285.

559. K. v. Auwers and F. König, *Annalen*, **496**, 26 (1932).

560. R. Danion-Bougot and R. Carrié, *Bull. Soc. chim. France*, 313 (1969).

561. J. P. Deleux, G. Leroy, and J. Weiler, *Tetrahedron*, **29**, 1135 (1973).

562. A. Nabeya, F. B. Culp, and J. A. Moore, *J. Org. Chem.*, **35**, 2015 (1970).

563. D. T. Witiak and M. C. Lu, *J. Org. Chem.*, **23**, 4451 (1968); D. T. Witiak and B. K. Sinha, *J. Org. Chem.*, **35**, 501 (1970).

564. R. Danion-Bougot and R. Carrié, *Bull. Soc. chim. France*, 2526 (1968).

565. R. Danion-Bougot and R. Carrié, *Bull. Soc. chim. France*, 3511 (1972).

566. J. Hamelin and R. Carrié, *Bull. Soc. chim. France*, 2162 (1968).

567. D. E. McGreer, N. W. K. Chiu and M. G. Vinje, *Canad. J. Chem.*, **43**, 1398 (1965).

568. F. M. Dean, P. G. Jones, R. B. Morton, and P. Sidisunthorn, *J. Chem. Soc.*, 5336 (1963).

569. F. M. Dean, P. G. Jones, and P. Sidisunthorn, *J. Chem. Soc.*, 5186 (1962).

570. W. Rundel and P. Kästner, *Annalen*, **737**, 87 (1970).

571. B. Eistert and A. Langbein, *Annalen*, **678**, 78 (1964).

572. J. F. W. Keana and C. U. Kim, *J. Org. Chem.*, **36**, 118 (1971).

573. J. F. W. Keana and C. U. Kim, *J. Org. Chem.*, **35**, 1093 (1970).

574. R. S. Bly, F. B. Culp, Jr., and R. K. Bly, *J. Org. Chem.*, **35**, 2235 (1970).

575. F. B. Kipping and J. J. Wren, *J. Chem. Soc.*, 1733 (1957).

576. R. Helder, T. Doornbos, J. Strating, and B. Zwanenburg, *Tetrahedron*, **29**, 1375 (1973).

577. S. D. Andrews, A. C. Day, and A. N. McDonald, *J. Chem. Soc.* (*C*), 787 (1969).

578. A. C. Day and R. N. Inwood, *J. Chem. Soc.* (*C*), 1065 (1969).

579. S. D. Andrews, A. C. Day, and R. N. Inwood, *J. Chem. Soc.* (*C*), 2443 (1969).

580. M. Franck-Neumann, *Tetrahedron Letters*, 2979 (1968).

581. W. Ried, W. Kuhn, and A. H. Schmidt, *Chem. Ber.*, **107**, 759 (1974).

582. C. G. Overberger, N. Weishenker, and J.-P. Anselme, *J. Amer. Chem. Soc.*, **86**, 5364 (1964).

583. C. G. Overberger, R. E. Zangaro, R. E. K. Winter, and J.-P. Anselme, *J. Org. Chem.*, **36**, 975 (1971).

584. F. J. Impastato, L. Barash, and W. M. Walborsky, *J. Amer. Chem. Soc.*, **81**, 1514 (1959).

585. A. G. Brook and P. F. Jones, *Canad. J. Chem.*, **49**, 1841 (1971).

586. R. J. Crawford and M. Ohno, *Canad. J. Chem.*, **53**, 3134 (1974).

587. A. N. Pudovik and R. D. Gareev, *J. Gen. Chem.* (*U.S.S.R.*), **44**, 1407 (1974).

588. A. D. Forbes and J. Wood, *J. Chem. Soc.* (*B*), 646 (1971).

589. R. Jacquier and G. Maury, *Bull. Soc. chim. France*, 306 (1967).

590. A. Mishra and R. J. Crawford, *Canad. J. Chem.*, **47**, 1515 (1969).

591. R. J. Crawford, A. Mishra, and R. J. Dummel, *J. Amer. Chem. Soc.*, **88**, 3959 (1966).

592. C. G. Overberger, J.-P. Anselme, and J. R. Hall, *J. Amer. Chem. Soc.*, **85**, 2752 (1963).

593. R. J. Crawford and A. Mishra, *J. Amer. Chem. Soc.*, **87**, 3768 (1965).

594. W. M. Jones and W.-T. Tai, *J. Org. Chem.*, **27**, 1324 (1962).

595. C. G. Overberger and J.-P. Anselme, *J. Amer. Chem. Soc.*, **86**, 658 (1964).

596. C. G. Overberger and J.-P. Anselme, *J. Amer. Chem. Soc.*, **84**, 869 (1962).

597. J. W. Timberlake and B. K. Bandlish, *Tetrahedron Letters*, 1393 (1971); R. C. Neuman, Jr. and E. W. Ertley, *J. Amer. Chem. Soc.*, **97**, 3130 (1975).

598. D. E. McGreer and J. W. McKinley, *Canad. J. Chem.*, **49**, 105 (1971).

599. R. J. Crawford, D. M. Cameron, and H. Tokunaga, *Canad. J. Chem.*, **52**, 4025 (1974).

600. R. J. Crawford and H. Tokunaga, *Canad. J. Chem.*, **52**, 4033 (1974).

601. R. J. Crawford and A. Mishra, *J. Amer. Chem. Soc.*, **88**, 3963 (1966).

602. R. J. Crawford, R. J. Dummel, and A. Mishra, *J. Amer. Chem. Soc.*, **87**, 3023 (1965).

603. M. P. Schneider and R. J. Crawford, *Canad. J. Chem.*, **48**, 628 (1970); R. J. Crawford and D. M. Cameron, *Canad. J. Chem.*, **45**, 691 (1967).

604. J. P. Deleux, G. Leroy, M. Sana, and J. Weiler, *Bull. Soc. chim. belges*, **82**, 423 (1973).

605. D. E. McGreer, P. Morris, and G. Carmichael, *Canad. J. Chem.*, **41**, 726 (1963).

606. D. E. McGreer and Y. Y. Wigfield, *Canad. J. Chem.*, **47**, 2095 (1969).

607. D. E. McGreer and Y. Y. Wigfield, *Canad. J. Chem.*, **47**, 3965 (1969).

608. K. L. Rinehart, Jr., and T. V. Van Auken, *J. Amer. Chem. Soc.*, **82**, 5251 (1960).

609. C. G. Overberger, N. Weinshenker, and J.-P. Anselme, *J. Amer. Chem. Soc.*, **87**, 4119 (1965).

610. T. Sanjiki, M. Ohta, and H. Kato, *Chem. Comm.*, 638 (1969).

611. T. V. Van Auken and K. L. Rinehart, Jr., *J. Amer. Chem. Soc.*, **84**, 3736 (1962).

612. S. Inagaki and K. Fukui, *Bull. Chem. Soc. (Japan)*, **45**, 824 (1972).

613. T. Sasaki, S. Eguchi, and F. Hibi, *J.C.S. Chem. Comm.*, 227 (1974).

614. P. S. Engel and L. Shen, *Canad. J. Chem.*, **52**, 4040 (1974).

615. R. Moore, A. Mishra, and R. J. Crawford, *Canad. J. Chem.*, **46**, 3305 (1968).

616. E. B. Klunder and R. W. Carr, Jr., *Chem. Comm.*, 742 (1971); S. D. Nowacki, P. B. Do and F. H. Dorer, *J.C.S. Chem. Comm.*, 273 (1972).

617. M. Schneider, A. Erben, and I. Merz, *Chem. Ber.*, **108**, 1271 (1975).

618. N. El-Ghandour, O. Henri-Rousseau, and J. Soulier, *Bull. Soc. chim. France*, 2817 (1972).

619. P. Eberhard and R. Huisgen, *Tetrahedron Letters*, 4337 (1971).

620. R. Huisgen and P. Eberhard, *Tetrahedron Letters*, 4343 (1971).

621. B. Eistert, J. Riedinger, G. Küffner, and W. Lazik, *Chem. Ber.*, **106**, 727 (1973).

622. A. R. Bassindale and A. G. Brook, *Canad. J. Chem.*, **52**, 3474 (1974).

623. I. Tabushi, K. Takagi, M. Okano, and R. Oda, *Tetrahedron*, **23**, 2621 (1967).

624. R. Huisgen, M. Seidel, G. Wallbillich, and H. Knupfer, *Tetrahedron*, **17**, 3 (1962).

625. R. Huisgen, K. Adelsberger, E. Aufderhaar, H. Knupfer, and G. Wallbillich, *Monatsh.*, **98**, 1618 (1967).

626. R. Huisgen, H. Knupfer, R. Sustmann, G. Wallbillich, and V. Weberndorfer, *Chem. Ber.*, **100**, 1580 (1967).

627. J. P. Visser and P. Smael, *Tetrahedron Letters*, 1139 (1973).

628. A. G. Anastassiou and S. W. Eachus, *Chem. Comm.*, 429 (1970).

629. C. W. Alexander, M. S. Hamdam, and W. R. Jackson, *J.C.S. Chem. Comm.*, 94 (1972).

630. H. Ferres, M. S. Hamdam, and W. R. Jackson, *J.C.S. Perkin II*, 936 (1973).

631. J. Elguero, R. Jacquier, and C. Marzin, *Bull. Soc. chim. France*, 4119 (1970).

632. J.-L. Aubagnac, J. Elguero, and R. Jacquier, *Bull. Soc. chim. France*, 3758 (1971).

633. J. Elguero and C. Marzin, *Bull. Soc. chim. France*, 3401 (1973).

634. A. Nuebauer, G. Litkei, and R. Bognár, *Tetrahedron*, **28**, 3241 (1972).

635. P. A. Boivin, P. E. Gagnon, L. Renaud, and W. A. Bridgeo, *Canad. J. Chem.*, **30**, 994 (1952).

636. P. E. Gagnon, J. L. Boivin, P. A. Boivin, and H. M. Craig, *Canad. J. Chem.*, **30**, 52 (1952).

637. P. E. Gagnon, J. L. Boivin, and A. Chisholm, *Canad. J. Chem.*, **30**, 904 (1952).

638. J. Bastús and J. Castells, *Proc. Chem. Soc.*, 216 (1962).

639. W. M. Jones, *J. Amer. Chem. Soc.*, **80**, 6687 (1958).

640. W. M. Jones, *J. Amer. Chem. Soc.*, **81**, 5153, 3776 (1959).

641. W. M. Jones, *J. Amer. Chem. Soc.*, **82**, 3136 (1960).

642. J. H. Rosenkranz and H. Schmid, *Helv. Chim. Acta*, **51**, 1628 (1968).

643. A. Hassner and M. J. Michelson, *J. Org. Chem.*, **27**, 3974 (1962).

644. J. Elguero, R. Jacquier, and D. Tizané, *Tetrahedron*, **27**, 123 (1971).

645. J. Elguero, R. Jacquier, and D. Tizané, *Bull. Soc. chim. France*, 1121 (1970).

646. C. Dittli, J. Elguero, and R. Jacquier, *Bull. Soc. Chim. France*, 4469 (1969).

647. C. Dittli, J. Elguero, and R. Jacquier, *Bull. Soc. chim. France*, 4474 (1969).

648. A. Rodgman and G. F. Wright, *J. Org. Chem.*, **18**, 465 (1953).

649. S. J. Cristol, A. L. Allred, and D. L. Wetzel, *J. Org. Chem.*, **27**, 4058 (1962); R. M. Moriarty, *J. Org. Chem.*, **28**, 2385 (1963).

650. T. Tsuchiya and V. Snieckus, *Canad. J. Chem.*, **53**, 519 (1975).

651. J. Elguero, C. Marzin, and D. Tizané, *Org. Magn. Resonance*, **1**, 249 (1969).

652. K. Berg-Nielsen, *Acta Chem. Scand.*, **27**, 1092 (1973).

653. D. H. Hunter and S. K. Sim, *J. Amer. Chem. Soc.*, **91**, 6202 (1969).

654. H. L. Snape, *J. Chem. Soc.*, 778 (1900).

655. D. H. Hunter and S. K. Sim, *Canad. J. Chem.*, **50**, 678 (1972).

656. B. Crookes, T. P. Seden, and R. W. Turner, *Chem. Comm.*, 342 (1968).

657. P. Karrer and A. Schlosser, *Helv. Chim. Acta*, **6**, 411 (1923).

658. P. Karrer, K. Escher, and R. Widmer, *Helv. Chim. Acta*, **9**, 301 (1926).

659. F. Schneider, *Annalen*, **529**, 1 (1937).

660. S. J. Kanewskaja, *J. prakt. Chem.*, **132**, 335 (1932).

661. W. J. Close, *J. Org. Chem.*, **15**, 1131 (1950).

662. A. H. M. Raeymaekers, L. F. C. Roevens, and P. A. J. Janssen, *Tetrahedron Letters*, 1467 (1967).

663. J. L. Wood and V. du Vigneaud, *J. Amer. Chem. Soc.*, **67**, 210 (1945).

664. R. Duschinsky and L. A. Dolan, *J. Amer. Chem. Soc.*, **67**, 2079 (1945); G. Swain, *J. Chem. Soc.*, 1552 (1948).

665. F. Lippich, *Ber.*, **41**, 2974 (1908).

666. H. D. Dakin, *Amer. Chem. J.*, **44**, 41 (1910).

667. H. Sobotka, M. F. Holzman, and J. Kahn, *J. Amer. Chem. Soc.*, **54**, 4697 (1932).

668. M. Bovarnick and H. T. Clarke, *J. Amer, Chem. Soc.*, **60**, 2426 (1938).

669. K. H. Dudley and D. L. Bius, *J. Heterocyclic Chem.*, **10**, 173 (1973).

670. W. J. Pope and J. B. Whitworth, *Chem. and Ind.*, **49**, 748 (1930); W. J. Pope and J. B. Whitworth, *Proc. Ray. Soc.*, **155A**, 1 (1936).

671. H. C. Brimelow, H. C. Carrington, C. H. Vasey, and W. S. Waring, *J. Chem. Soc.*, 2789 (1962).

672. E. J's-Gravenmade, G. D. Vogels, and C. Van Pelt, *Rec. Trav. chim.*, **88**, 929 (1969).

673. J. H. Bowie, E. Bullock, and A. W. Johnson, *J. Chem. Soc.*, 4260 (1963).

674. H. E. Carter, C. C. Sweeley, E. E. Daniels, J. E. McNary, C. P. Schaffner, C. A. West, E. E. van Tamelen, J. R. Dyer, and H. A. Whaley, *J. Amer. Chem. Soc.*, **83**, 4296 (1961).

675. D. G. Neilson, D. A. V. Peters, and L. H. Roach, *J. Chem. Soc.*, 2272 (1962).

676. J. P. Greenstein and M. Winitz, *Chemistry of the Amino Acids*, Vol. III, Wiley, New York, 1961, pp. 1971–1995.

677. A. Pawda, M. Dharan, J. Smolanoff, and S. I. Wetmore, Jr., *J. Amer. Chem.Soc.*, **95**, 1954 (1973).

678. C. Chapuis, A. Gauvreau, A. Klaebe, A. Lattes, J.-J. Périé, and J. Roussel, *Bull. Soc. chim. France*, 2676 (1973).

679. T. Yano, H. Kobayashi, and K. Ueno, *Bull. Chem. Soc. (Japan)*, **45**, 1249 (1972).

680. P. G. Hall and G. S. Horsfall, *J.C.S. Perkin II*, 1280 (1973).

681. K. Hofmann, *Imidazole and its Derivatives. Part I*, Interscience, New York, 1953.

682. P. Scheiner, *J. Amer. Chem. Soc.*, **90**, 988 (1968).

683. K. Alder and G. Stein, *Annalen*, **515**, 185 (1935).

684. R. Huisgen, L. Möbius, G. Müller, H. Stangl, G. Szeimies, and J. M. Vernon, *Chem. Ber.*, **98**, 3992 (1965).

685. P. Scheiner, J. H. Schomaker, S. Deming, W. J. Libbey, and G. P. Nowack, *J. Amer. Chem. Soc.*, **87**, 306 (1965).

686. J. E. Baldwin, J. A. Kapecki, M. G. Newton, and I. C. Paul, *Chem. Comm.*, 352 (1966).

687. R. S. McDaniel and A. C. Oehlschlager, *Canad. J. Chem.*, **48**, 345 (1970).

688. R. Huisgen and G. Szeimies, *Chem. Ber.*, **98**, 1153 (1965).

689. G. Bianchetti, R. Stradi and D. Pocar, *J.C.S. Perkin I*, 997 (1972).

690. J. F. Stephen and E. Marcus, *J. Heterocyclic Chem.*, **6**, 969 (1969).

691. W. A. Remers, G. J. Gibs, and M. J. Weiss, *J. Heterocyclic Chem.*, **4**, 344 (1967).

692. J. E. Baldwin, G. V. Kaiser, and J. A. Romersberger, *J. Amer. Chem. Soc.*, **86**, 4509 (1964).

693. J. E. Baldwin, G. V. Kaiser, and J. A. Romersberger, *J. Amer. Chem. Soc.*, **87**, 4114 (1965).

694. C. E. Olsen, *Acta Chem. Scand.*, **28B**, 425 (1974).

695. K. Kaas, *Acta Cryst.*, **29B**, 1458 (1973).

696. G. G. Lyle and M. J. Piazza, *J. Org. Chem.*, **33**, 2478 (1968).

3 NITROGEN HETEROCYCLES: SIX-MEMBERED RINGS

141

I. INTRODUCTION

The general order of presentation of the six-membered ring nitrogen hetero-cycles in this chapter begins with the monocyclic saturated, and partially saturated, compounds which are treated in order of increasing number of nitrogen atoms.   These are followed by the bicyclic five:six-membered ring systems with nitrogen atoms in the six-membered rings, the six:six-membered rings in order of increasing number of nitrogen atoms, and then reduced tricyclic systems containing at least one six-membered ring posessing a nitrogen heteroatom. Most of the photochemical reactions are grouped in a separate section. Similarly, the stereochemistry of the aromatic heterocycles which owe their asymmetry to a chiral substituent, to restricted rotation, or to molecular overcrowding is discussed in the final section.

II. PIPERIDINES

A.   Syntheses of Piperidines

Piperidines are more easily obtained from pyridines by reduction, but they have also been obtained by cyclization of open-chain amines. 2- and 3-Monosubstituted piperidines have only one chiral center. As the number of substituents is increased the relative stereochemistries of the substituents, that is cis or trans, become important and the number of possible geometrical isomers increases. These points are considered in the following methods of preparation. For further references on the chemistry of pyridine and its derivatives, see the monographs edited by Klingsberg and Abramovitch.[1]

1. Catalytic and Chemical Reduction of Pyridines

The reduction of monosubstituted pyridines to piperidines produces the same racemate irrespective of the method used. Polysubstituted pyridines, on the other hand, can yield the theoretical number of diastereoisomeric piperidines but the relative ratios of these is affected by the method of reduction. Generally, catalytic reduction favors a predominance, if not complete stereoselectivity, for the cis isomer. Catalytic reduction of 2,3-bismethoxycarbonyl-,[2] 2,6-dimethyl-,[3] and 4-alkyl-2-ethoxycarbonylpyridines[4] yield predominantly the *cis*-piperidines, whereas 4-hydroxy-2-methyl-,[5] 4-benzyloxy-2-carboxy-,[6] 2,3-tridecamethylene-,[7] 2-β-hydroxypropyl-3-methoxy-,[8] 3-ethyl-4-methoxycarbonylmethyl-,[9] 3-ethyl-4-3'-trichloro-2'-hydroxypropyl-,[10] and 2-acetonyl-3-alkyl- or alkoxypyridines[11] apparently yield exclusively the respective *cis*-piperidines. Tsuda and Kawazoe[12] studied the affect of the catalyst and position of methyl groups in dimethyl *N*-methylpyridinium iodides. Platinum oxide in water is slightly more or less selective than Raney nickel in ethanol, but the selectivity varies markedly with the position of substituents. *o*-Dimethyl derivatives give exclusively cis products, *m*-dimethyl derivatives (except 2,6) give predominantly cis products, and with *p*-dimethyl pyridinium methiodides the proportion of trans isomer can be quite high (cis to trans ratio for the 2,5-dimethyl isomer is 27:73 with Raney nickel). Another example of the latter case is 5-hydroxy-2-isopropylpyridine which yields about equal amounts of *cis*- and *trans*-5-hydroxy-2-isopropyl piperidines on reduction with Adam platinum oxide in acetic acid[13]—a system which usually favors a cis orientation. A Raney nickel catalyst in the presence of a base favors the formation of *trans*-piperidines particularly when one of the groups is an ethoxycarbonyl and can be epimerized by the basic medium, for example, with 4-alkyl-2-ethoxycarbonylpyridines.[4] Catalytic reduction of 3,5-dialkoxycarbonylpyridines with platinum oxide furnishes the *cis*-dicarboxylic esters because further reduction with LAH to the diol followed by bromination and cyclization with ammonia gives bispidine.[14] The latter is formed directly when 3,5-dicyanopyridine is reduced with Raney nickel in dioxane at 120 atm and 150°C.[15] Simultaneous reduction of the ring and the side chain occurs in 2-, 3-, and 4-benzoylpyridine on reduction with Adam platinum oxide in hydrochloric acid to 2-, 3-, and 4-piperidylphenyl carbinol. Two diasteromeric *racemic* hydrochlorides were separated in the case of the 2- isomer.[16] A chiral group can influence the reduction of a pyridine ring. *S*(+)-Picolylmethyl carbinol gives a 75:25 mixture of 2*S*,2'*S*(+)- and 2*R*,2'*S*(−)-2,2'-hydroxypropyl piperidines, and chiral picolylethyl carbinol behaves similarly.[17] The configuration of 2*S*,2'*S*(+) isomer was previously correlated with *S*(−)-pipecolinic acid.[18]

Chemical reduction of pyridines, in contrast with catalytic reduction, tends to produce a higher proportion of the trans isomer, although not necessarily more than the cis isomer. Reduction of 2,6-dimethylpyridine yields a 74:20 mixture of *cis-* and *trans-*2,6-dimethylpiperidines (compare with 90:10 in catalytic reduction).[3] Sodium and alcohol reduction of 2,3-tridecamethylene-,[7] 2-*n*-butyl-3-methyl-,[19] 2-*n*-butyl-3-methyl-1,2-dihydro-,[20] 2,6-dimethyl-,[21] and 2,6-diphenylpyridines[22] gives either largely the trans isomer or a ratio of trans to cis isomers greater than in the catalytic reduction. 2,4,6-Trimethylpyridine is an interesting example because reduction to a piperidine produces three chiral centers. Booth and co-workers[3] reduced it with sodium and ethanol and showed that the main product was the all-*cis*-trimethyl piperidine (**1**). A more detailed examination of this reduction confirmed that **1** was formed in larger amounts, but that *r*-2,*trans*-4,*cis*-6-trimethyl- (**2**), *r*-2,*cis*-4,*trans*-6-trimethylpiperidines (**3**), and *cis*-2,4,6-trimethyl-1,2,3,6-tetrahydropyridine (**4**) were also formed.

The proportions of **2**, **3**, and **4** can be increased at the expense of **1** by using sodium in *n*-butanol or even better by electrolytic reduction. 1,2,4,6-Tetramethylpyridinium sulfate is reduced by AlH_3 and the all-*cis*-piperidine is formed together with some of the *N*-methyl derivative of **4**.[23] Lithium aluminum hydride reduces 1,4,6-trimethyl-2-pyridone to a 15:22 mixture of *cis*-and *trans*-1,2,4-trimethylpiperidine, but the major product is 1,2,4-trimethyl-1,2,5,6-tetrahydropyridine.[24] Most of these isomers are obtained by distillation, glc or tlc and separations are described in the references cited. For the stereochemistry of reduction of the coenzyme nicotinamide adenine dinucleotide (NAD^+) see Section XIV.A of this chapter.

2. From Dehydropiperidines

Dehydropiperidines are converted into piperidines by catalytic or chemical reduction and by addition reactions across the double bond(s). Catalytic reduction of *N*-methyl 3- and 5-methyl-4-phenyl-1,2,5,6-tetrahydropyridines (10% Pd–C)[25] and 3-acetoxy-l-acetyl-2-ethoxycarbonyl-1,4,5,6-tetrahydropyridine (Raney nickel)[26] yields only the *cis*-disubstituted piperidines, whereas 1-benzoyl-4-ethoxycarbonylmethyl-3-ethyl-1,2,3,6-tetrahydropyridine (PtO$_2$) gives only *trans*-1-benzoyl-4-ethoxycarbonyl-methyl-3-ethylpiperidine.[27] The cis isomer of the last named compound is formed by a similar reduction of the double bond isomer 1-benzoyl-4-ethoxycarbonylmethylene(*exo*)-3-ethylpiperidine. These were interrelated with *cis*-1-benzoyl-4-cyanomethyl-3-ethylpiperidine from the alkaloid cinchonine and used to correct the previous stereochemistry of 4-ethoxy-carbonylmethyl-5-ethyl-2-oxopiperidine which was in error.[27] Catalytic reduction of 1,4-dihydro-5-β-indolylmethyl-1-methylnicotinamide produces a 20:25 mixture of *cis*- and *trans*-3-carbamoyl-5-β-indolylmethyl-1-methyl-piperidine.[28] Sodium borohydride reduces 2-cyanomethyl-3-methoxy-3,4,5,6-tetrahydropyridine to a 53:9 mixture of *cis*- and *trans*-2-cyano-methyl-3-methoxypiperidines but this may involve prior tautomerism of the double bond which can become exocyclic.[29]

The addition of cyanide ion to 3-benzyl-3-cyano-1-methyl-3,4,5,6-tetrahydropyridinium salts provides a mixtures of 3-benzyl-2,3-dicyano-1-methylpiperidine in which the two cyano groups are cis and trans to each other.[28] Hydroboration of 4-propyl- or 4-isopropyl-1-methyl-1,4,5,6-tetrahydropyridine followed by conventional oxidation produced a mixture consisting mainly (>73%) of *trans*-4-alkyl-1-methyl-3-hydroxypiperi-dines.[30] Similar hydroxylation of 6-alkyl-1-methyl-1,2,4,6-tetramethylpyri-dine (alkyl is Me,Et,*n*-Pr,*iso*-Pr) gave a mixture of the *cis*- (**5**) and *trans*-2-alkyl-4-hydroxy-1-methylpiperidine (**6**) and the *cis*-2-alkyl-5-hydroxy isomer

(**7**) (Equation 1). The cis isomer **5** is the most predominant and only small amounts of the 5-hydroxy compound (**7**) are formed. 2-Ethyl-1-methyl-1,2,5,6-tetrahydropyridine under identical conditions yields mainly *trans*-2-ethyl-3-hydroxy-1-methylpyridine but is contaminated with both the cis isomer and the 4-hydroxy isomers. The conditions of work-up of this reaction, if at room temperature or in boiling toluene with triethylamine, can produce different ratios of these isomers.[31] Hydroboration of 1-benzyl-(or 1-phenethyl)-4-methyl-1,2,5,6-tetrahydropyridine also gives mainly *trans*-1-benzyl(or 1-phenethyl)-4-methyl-3-hydroxypiperidine.[32] The regio-

and stereospecificity of hydroboration of N-substituted 3-(or 5)methyl-4-phenyl-1,2,5,6-tetrahydropyridine was established by analysis of the spectral data of the piperidinols formed.[32a]

Asymmetric hydroboration of 1-methyl-1,2,3,6-tetrahydropyridine with di-3-pinanylborane (derived from (+)-α-pinene) followed by oxidation provides a 70:21 mixture of 3-hydroxy- and 4-hydroxy-1-methylpiperidines. The 3-hydroxy derivative has a rotation, $[\alpha]_D^{27}$ + 1.56° (EtOH), and its absolute configuration is R because the $S(-)$ isomer is obtained from S-arginine by the conversion: $S(-)$-3-hydroxy-2-oxopiperidine $\rightarrow$ $S(-)$-3-hydroxypiperidine $\rightarrow$ $S(-)$-3-hydroxy-1-methylpiperidine $[\alpha]_D^{22}$ − 2.4° (EtOH).[33] Partial asymmetric reduction with a chiral hydroborating agent has also been observed by Grundon and co-workers[34] in the reduction of 2-alkyl(Me,Et,n-Pr)-3,4,5,6-tetrahydropyridine to chiral 2-alkylpiperidines. The reagents used are lithium bis(pinan-3-α-yl)phenyl(or methyl or n-butyl)borohydride and, when derived from (+)-α-pinene, enrichment of the R-2-alkylpiperidines is obtained. Conversely the S isomers are formed from the $(-)\alpha$-pinene reagent. The reverse chirality is observed when the reagent is bis- or tris(pinan-3-α-yl)borane. The overall enrichment of optical enantiomers is, however, at best 25%.

Pyridine 1-oxide and its 2-, 3-, and 4-alkyl and 4-phenyl derivatives react with alkyl thiols (e.g., t-BuSH) in the presence of acetic anhydride and triethylamine by addition of alkyl-S- and AcO- across the double bonds. In 4-alkylpyridine 1-oxides, the thiol groups attack positions 3 first and then 6, and the acetoxy group adds on to positions 1 and 2 (Equation 2). The alkylthio groups attack position 2, however, in 3-alkylpryidine 1-oxides.[35,36]

(2)

3. *From Piperidones*

Piperidinols are formed by reduction of piperidones or by addition of reagents across the C=O group. The orientation of the hydroxy group formed may be axial or equatorial and as in the reduction of cyclohexanone this is greatly influenced by both the reagent used and the nature and stereochemistry of other substituents in the ring. The most extensively studied compounds are 4-piperidones. Catalytic reduction of 1,3-dimethyl-4-piperidone hydrochloride, and the methochloride, gives predominantly cis-4-hydroxy-1,3-dimethylpiperidine (hydroxy group equatorial) with platinum in ethanol and water, whereas the free base, 1,3-dimethyl-4-piperidone, in dry dioxane or ethanol yields approximately equal amounts of cis- and trans-hydroxypiperidines.[37] Catalytic reduction of cis(meso)-3,5-diethoxycarbonylmethyl-1-methyl-4-piperidone forms a 2.5:1 mixture of the corresponding cis-cis- and trans-trans-piperidinols, but the latter is the main product on reduction with NaBH$_4$.[38] Balasubramanian and Padma[39] studied the effect of the medium on the catalytic reduction of methyl-phenyl-substituted 4-piperidones bearing three, four, and five substituents, and came to the conclusion that maximum yields of the 4-hydroxypiperidines with the OH-equatorial are obtained by reduction in acid media, and the presence of a methyl group α- to the carbonyl induces the formation of an axial OH. In neutral media only the axial OH form is obtained and in the case of 1,2,2-trimethyl-6-phenyl-4-piperidone the axial alcohol is formed irrespective of the medium. The presences of a phenyl group in the ring must have a strong influence on the stereochemistry of absorption on the catalyst. The nitrogen atom also must be important in this respect because the above results are different from those observed in the cyclohexanone series.

Chemical reduction of 4-piperidones generally produces two 4-hydroxypiperidines (Equation 3). The use of NaBH$_4$ and LAH gives a mixture

equatorial   axial   (3)

consisting predominantly, or almost entirely of the isomer with the hydroxyl group equatorial irrespective of other substituents in the ring.[40-43] The reverse is true in the Meerwein-Ponndorf reduction which produces a relatively higher proportion of the isomer with the axial hydroxy group.[40]

The stereoselectivity of NaBH$_4$ reduction of some 4-piperidones has been used to advantage by Witkop and Foltz[44] in converting 2S,4S(−)-2-carboxy-4-hydroxypiperidine (which occurs naturally in dates) into the 2S,4R(−) isomer by way of their 1-benzyloxycarbonyl derivatives. The 4-S-hydroxy group is thus oxidized to keto group which is then reduced back, but stereospecifically, to the epimeric 4-R-hydroxy derivative by NaBH$_4$. The 2S, 4R isomer forms a lactone readily and confirms the cis relationship of the carboxy and hydroxy groups.[45] The reduction of 2S-2-

carboxy-5-piperidone (from S-glutamic acid), however, yields exclusively the *trans*-hydroxy acid (OH equatorial), but its 1-benzyloxycarbonyl derivative is reduced to the allo-*cis*-hydroxy acid (OH axial) which can be dehydrated to a crystalline lactone.[46]

The reduction of 4-piperidone oximes is a useful method for preparing the 4-amino derivative, and as above an axial or an equatorial amino group can be formed depending on the reagent used. Reduction of 1-alkyl-4-hydroxyimino-3-phenylpiperidine and its 2-oxo derivative with LAH yields predominantly the cis isomer (NH_2 *ax*, Ph *eq*),[47-49] but sodium and ethanol produces mainly the trans isomer.[48] The effect of substituents α-to the hydroxyimino group is clearly exemplified in the reduction of 1-methyl-4-hydroxyimino-2,6-diphenylpiperidine and its 3,5-dimethyl derivative with LAH when a ratio of **8** to **9** of 70:30 or 20:80 is obtained if R is H or Me respectively (Equation 4).[50]

$$(4)$$

The stereospecificity of the addition of reagents other than hydrogen is discussed in the section on reactions of piperidones (Section II.D.2).

4. By Ring Closure

Several methods of cyclization of open-chain compounds into piperidines have been described and only when the asymmetry is at the carbon atoms involved in the cyclization are new chiral centers formed. The synthesis of $R(+)$-5-methyl-2-piperidone from $R(+)$-citronellal does not affect the asymmetric center (the $S(-)$ isomer is obtained by degradation of cevins).[51] Another example is the Dieckmann cyclization of R-2-benzyl-2-(ω-ethoxycarbonylpropionamido)propionic acid followed by hydrolysis and decarboxylation to $6S$-6-benzyl-6-methylpiperidine-2,5-dione which was finally reduced to $6S(-)6$-benzyl-6-methyl-2-piperidone (Equation 5). In other syntheses where the chiral substituent is involved, the stereochemistry can

$$(5)$$

be predicted from a knowledge of the mechanism of the reaction. The last mentioned piperidone is also formed in an interesting intramolecular cyclization of *R*-5-methyl-6-phenylhexanoylazide under the influence of uv light. There is obviously retention of configuration in the nitrene (singlet) insertion at the quaternary hydrogen atom (Equation 5).[52]

Intramolecular cyclization of 1,5-dicyano-1,5-dihydroxypentanes with alkylamines produces apparently only a single 2,6-dicyanopiperidine. When a nonbulky amine is used (e.g., ammonia, methylamine or ethylaamine) the most probable configuration of the cyano groups is cis.[53] McKenna and co-workers have studied the formation of quaternary piperidinium bromides from 1,5-dibromopentane with a phenyl or alkyl substituent in the chain and the secondary amines benzyl-, ethyl-, and isopropylmethylamines in an endeavor to note any stereospecificity in the product, that is, to see whether the *N*-methyl group in the piperidinium salt formed preferred to be axial or equatorial. The diastereomeric mixture of salts indicated that the more stable isomer predominated, that is, the one with the least steric interactions, but the stereoselectivity was low.[54] Ring closure of diethyl *meso*-1,4-diaminoadipate takes two courses. At low pH values it cyclizes to *trans*-3-amino-6-ethoxycarbonyl-2-piperidone, but at pH values above 10 it ring closes to 3,6-dioxo-2,5-diazabicyclo [2.2.2] octane which cleaves to *cis*-3-amino-6-ethoxycarbonyl-2-piperidone in methanolic HCl (Equation 6).[55]

In their concerted, but intensive, chemical studies on the synthesis of the quinazoline alkaloid febrifugine Baker and collaborators developed syntheses for the piperidine portion of the molecule. They prepared 2-

trans   cis

(6)

carboxymethyl-1-ethoxycarbonyl-3-methoxypiperidine by starting from racemic diethyl α-(5-benzyloxycarbonylamino-2-methoxypropionyl)malonic ester in several steps but obtained a mixture of cis and trans isomers.[56] Another synthesis from 2-benzamido-2-(2-tetrahydrofuranyl)propionic acid (m.p. 156°C) gave exclusively 2-carboxymethyl-1-benzoyl-3-hydroxy-piperidine, on successive treatment with hydrobromic acid and triethyla-mine, which must be the cis isomer because it cyclized readily to a lactone. Similar treatment of the isomeric 2-benzamido-2-(2-tetrahydrofuranyl) propionic acid (m.p. 117°C) gave the *trans*-piperidine which did not form a lactone easily.[57] Reductive cyclization of 3-phenyl-and 3-α-naphthyl-4-oxopentylnitrile with Raney nickel at 80°C produces mainly *cis*-2-methyl-3-phenyl- and *cis*-2-methyl-3-α-naphthylpiperidines respectively. Under similar conditions 4-benzoyl-4-phenylbutryonitrile gives *cis*-2,3-diphenyl-piperidine. Low temperature reduction of the first nitrile, on the other hand, provided *trans*-2-methyl-3-phenyl-6-piperidone.[58] A 1:1 mixture of *cis*-and *trans*-4,5-diaryl-2-piperidones is produced when a mixture of race-mates of ethyl 3,4-diaryl-4-cyanobutryrate is reduced with Raney nickel.[59] One has to be rather careful when using Raney nickel catalysts for preparing piperidines in which the substituents can epimerize by basic catalysts (e.g., aryl, ethoxycarbonyl) particularly if high temperatures are used.

1,3-Diphenylpropan-2-one undergoes an intramolecular Mannich reac-tion on treatment with methylamine and formaldehyde and *cis*-1-methyl-3,5-diphenyl-4-piperidone is the main product. However, the yield drops progressively as the ortho positions in the phenyl rings are substituted, and the reaction fails to yield a piperidine when all four positions are substi-tuted by methyl groups.[60] Further treatment of these and similar 4-piperi-dones with Mannich reagents provides bispidines.[60,61]

A different approach is used for the synthesis of *trans*-1-cyclohexyl-2,5-dimethyl-4-piperidone. It involves the addition of cyclohexylamine to the double bonds of the acyclic ketone 4-methylpenta-2,4-dien-3-one. Isomeri-zation of this piperidone into its cis isomer is said to involve a ring opened intermediate.[62]

5. *Miscellaneous*

Corey, Arnett, and Widiger[63] recently synthesized (+)-perhydrohistrion-icotoxin, the fully reduced form of the toxin histrionicotoxin (**10**) which Witkop and co-workers[64] isolated from the Columbian frog (this com-pound was formerly used as an arrow poison). The key step for making the piperidine ring is the Beckmann transformation of the cyclopentanone oxime **11** to the spiro-2-piperidone **12** which occurs with retention of con-figuration. The cis and trans isomers of the *d,l*-4,5-diethyl- and *d,l*-4,5-

diphenyl-2-piperidones are formed by a similar rearrangement of the corresponding 3,4-diethyl- and 3,4-diphenylcyclopentanone oximes, and they were reduced with sodium and butanol to the respective *cis*- and *trans*-3,4-diethyl- and 3,4-diphenylpiperidines.[65]

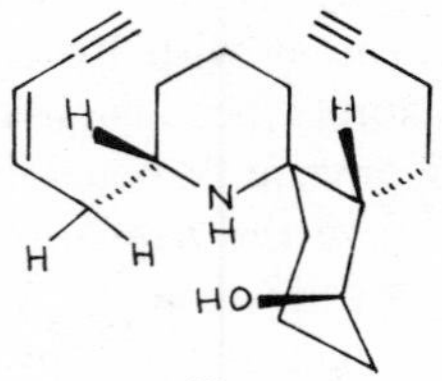

10

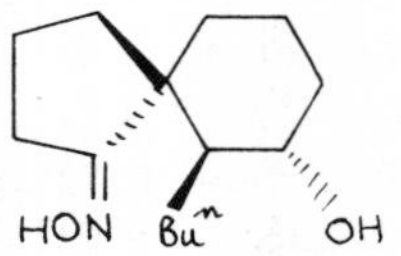

11 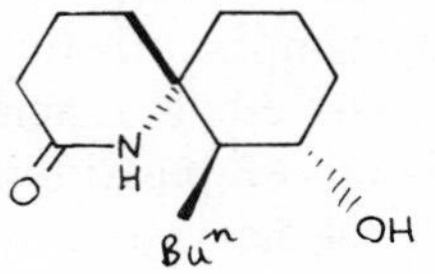   **12**

Reid and Bätz[66] found that 1:1 adducts are formed when phenylcyclobutendione reacts with enamines. These adducts are 1-hydroxy-2-oxo-7-phenyl-3-azabicyclo[4.1.0]hept-4-enes in which the phenyl group is endo or exo (**13**) depending on the substitution pattern in the enamines. A hydrogen atom on the enamine nitrogen atom favors the exo isomer and *N*-methyl and *N*-phenyl groups favor the endo isomer when R^2 and R^3 in **13**

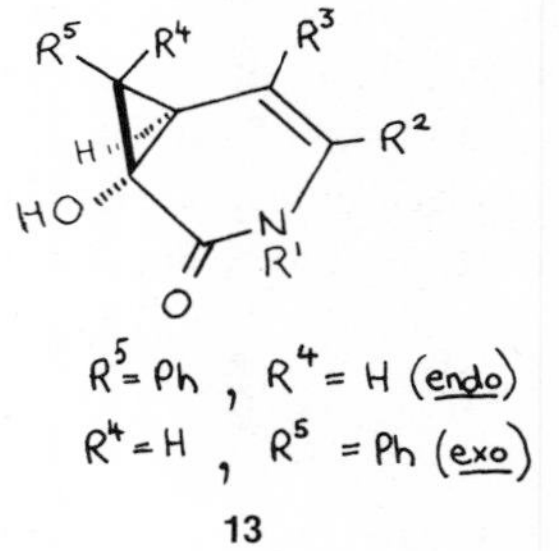

$R^5 = Ph$, $R^4 = H$ (endo)
$R^4 = H$, $R^5 = Ph$ (exo)

13

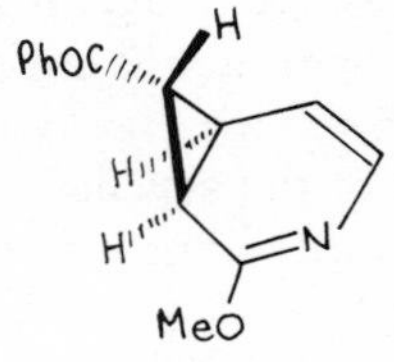

14

are Me and CO_2Et respectively. These are valuable intermediates for the preparation of variously substituted 2-pyridones into which they can be transformed. A similar ring system, 7-*exo*-benzoyl-2-methoxy-3-azabicyclo[4.1.0]hepta-2,4-diene (**14**), is formed by photochemical isomerization of 3-benzoyl-2-methoxy-3*H*-azepine.[67]

B. Configurations of Piperidines

The most useful physical tool to ascertain the relative configurations of substituents in piperidines has been ^{1}H nmr and more recently ^{13}C nmr. Most of the structures discussed in the previous and subsequent sections were derived mainly by this method and the relevant papers should be consulted for this information. Other physical methods such as ir, dipole moments, optical rotation, ord, and cd spectroscopy have been used, and the last three are for obvious reasons particularly necessary for characterizing and correlating chiral centers. The physical methods will be treated first followed by a discussion of the configuration of a number of chiral piperidines and related natural products. The number of natually occurring piperidines, however, is kept to a minimum because these are not the object of the present monograph.

For other than "finger printing" the compounds, ir spectroscopy is very useful in assessing the presence of α-hydrogen atoms antiperiplanar with the lone pair of electrons on the nitrogen atom and the relative configuration of hydroxyl substituents with respect to the nitrogen atom.[68] The former arrangement of the α-hydrogen atoms gives rise to Bohlmann bands in the region 2700–2800 cm^{-1}, that is, bands at slightly lower frequencies than the usual C–H stretching vibrations. These are described more fully in Chapter 4, Section VI.C. Sometimes these bands are not readily observed in rapidly inverting simple piperidines because the lone pair on the nitrogen atom is not always axial. However, if they do appear in the spectrum they strongly support an antiperiplanar arrangement of the α-hydrogen atom and the nitrogen lone pair, and can give valuable information[69] regarding the configuration. A caution was made about using Bohlmann bands in systems other than quinolizidines (e.g., in 4-piperidones) and whenever the piperidine ring is distorted from the normal "chair" conformation.[70]

Intramolecular hydrogen bonding is observed particularly in 3-hydroxy piperidines (i.e., 15) and is used to assess both the configuration and conformation of these molecules.[71−73] The configuration of ψ-conhydrine, 2-propyl-5-hydroxypiperidine (16), for example, must be trans because no

15

16

17

intramolecular hydrogen bonding is observed in the ir spectrum (compare 3-hydroxypiperidine).[74] Inter-piperidine hydrogen bonding is possible in some configurations of 2-hydroxy-α,β-dipiperidyl, for example, 2-hydroxy-tetrahydroanabasine (17).[75]

Lee and Kumler[76] measured the dipole moments of five- and six-membered cyclic imides. The dipole moments of 3-ethyl-3-phenyl(or 3-p-aminophenyl)-2,6-dioxopiperidine with the quasi equatorial aryl groups are consistent with theoretical values. Katritzky and his school have used dipole moment measurements to obtain the conformation of several piperidines (see Section C).

The chiroptical properties of N-nitrosoazetidine, pyrrolidine, and piperidine were examined in order to define a sector rule.[77] The results suggest that the signs of the originally proposed[78] sector rule must be reversed. The cd curves of piperidine nitroxides were measured and the spectra interpreted in terms of an octant rule similar to the one for ketones.[79] 4-Piperidones with a chiral substituent on the nitrogen atom, for example, PhCH(Me)-, gave cd curves which indicated that although the carbonyl chromophore is several σ bonds away from the chiral center it is significantly perturbed to exhibit a n-π* Cotton effect. The strength of the Cotton effect can give some information about the conformation of the molecule.[80] Cook and Djerassi[5] compared the chiroptical properties of optically active 1,2-dimethyl-3-piperidone with those of the analogues in which the hetero MeN< group was replaced by O, S, SO, and SO$_2$. The results clearly demonstrate that it is invalid to correlate rotatory effects with configuration in systems which only differ in the heteroatom. Cd studies of chiral 2-piperidones (δ-lactones) were reported recently.[80a]

The absolute configuration of (−)-3-methylpiperidine is R, and is established by comparison of cd curves of its N-nitroso- and N-methyl-mercaptothiocarbonyl derivatives with those of the respective (+)-camphidine derivatives which reveal opposite Cotton effects.[81] Similarly the configurations 4S(−)-4-ethoxycarbonylmethyl-2,2-dimethyl-, 4R,6S(−)-cis-4-ethoxycarbonylmethyl-2,2,6-trimethyl- and 4S,6S(+)-trans-4-ethoxycarbonylmethyl-2,2,6-trimethylpiperidines were established from comparisons of the cd data of their N-chloro derivatives. Rotatory contributions of α-methyl groups in the axial and equatorial positions were derived.[82] By using the D(−)- and L(+)-dimandelate salts El-Olemy and Schwarting separated the three isomers of the alkaloid anaferine, namely meso-, L(+)-, and D(−)-1,3-bis(2-piperidyl)-2-propanones (18). Comparison of the

18

ord curves of the hydrochlorides of the L(+) and D(−) enantiomers with those of (+)- and (−)-pipecolic acid gave the absolute configurations as *S,S* and *R,R* respectively.[83]

Sjöberg, Fredga, and Djerassi[84] introduced the dithiocarbamic ester group in order to obtain a Cotton effect in non-uv absorbing α-amino acids. Ripperger and Schreiber applied this method successfully to optically active piperidines and one example has been mentioned above. Ord and cd curves were obtained for methyl *N*-dithiocarboxylic esters of piperidines and pyrrolidines and used to assign the absolute configuration of α-substituents.[85] The *N*-dithiocarboxylate (in carbodithioate salts) group has a marked deshielding effect on the chemical shifts of the α-equatorial hydrogen atoms and is a valuable probe for observing these protons. The $-NCS_2^-$ group causes a downfield shift of $\alpha\text{-H}_{eq}$ of approximately 2.7 ppm.[86] When an α-alkyl group is in the equatorial position conformational ring inversion occurs on conversion into the sodium salt of the corresponding N-carbodithioic acid.[87]

Stereochemical assignments by ^{13}C nmr are coming more to the fore[88] and have been applied to 4-piperidones.[89] The large differences in chemical shifts generally observed for groups in different surroundings and the high accuracy with which these can be measured make it a very useful tool for configurational and conformational information. The nmr spectra of diastereomers are not usually very different, but the differences are increased considerably by lanthanide shift reagents. In order to make the differences even larger a chiral shift reagent can be used. In this way the optical purity of 2-aryl-2-piperidyl methanols have been determined.[90]

Optically active 5-hydroxy-2-*n*-propyl-(ψ-conhydrine),[13,74,91] 2-1'-hydroxypropyl-*,[74,92] 1-methyl-2-propyl-*,[93] 5-hydroxy-1,2-dimethyl-,[5] *cis*-4-amino-1-methyl-3-phenyl-,[48] *cis*-4-amino-3-methyl-1-phenyl-,[94] 1,*r*-2, *trans*-5 trimethyl-*trans*-4-hydroxy-*cis*-4-phenyl-*[95] *trans*-1,2,6-trimethyl(partial)-*,[21] 2-(α-hydroxy-α-phenyl)benzyl-*,[96] 1-benzyloxycarbonyl-2-carboxy-4-hydroxy-piperidines*[6] were prepared by optical resolution and the absolute configurations of those marked with an asterisk have been determined either by direct correlation with, or degradation to, products of known configuration. 2*S*,3*S*-1-Benzoyl-3-hydroxy-2-methyl-, 2*R*,4*S*-1-benzoyl-4-hydroxy-2-methyl-, and 2*R*-1-benzoyl-4-oxo-2-methylpiperidines were formed from racemic 1-benzoyl-2-methylpiperidine by microbial oxidation.[97] The enzymic resolution of *cis*-2-carboxy-4-hydroxypiperidine was achieved,[6] and the absolute configuration of several of the 3- and 4-hydroxy 2-carboxy-piperidines have been established.[44−46,98] The configurations of the following naturally occurring piperidines were also determined: *S*(−)-2-carboxy-1,1-dimethylpiperidine (betaine, homostachydrine)[99] and *S*-2-carboxyl-1,2,3,6-tetrahydropyridine (baikiain)[100] were correlated with *S*-aspartic acid; the all-*cis*-3*R*-hydroxy-2*R*-methyl-6*S*-11-oxododecylpiperidine (cassine, the papaya alkaloid),[101] its monomeric lactone[102] and its dimeric dilactone (capraine)[103]; and 2R,3S,6R-6(3-furyl)-2-(4-hydroxy-3-methyl-E-but-2-enyl)-3-methylpiperidine (alkaloid nuphamine).[104]

Racemization of isopelletierine (19) in ethanol occurs at room temperature and is much faster than in its N-methyl derivative. Since the asymmetric center is one carbon removed from the carbonyl group then the process is not occurring by ionization but by ring opening by way of an intermediate such as the enone 20.[105]

19   20   21

The racemic cis-3,5-cis-3',5'-tetramethyl-1,1'-spirobipiperidinium bromide (21) is synthesized from meso-2,4-dimethylglutaric anhydride and should be resolvable[106] (compare 1,1'-spirobipyrrolidinium salts in Chapter 3, Section III.C). Bis(α,α-dimethylglutarimido)mesitylene was obtained in two forms which were shown to be the cis and trans isomers by ir spectroscopy. The trans isomer must be a racemate.[107]

The first illustration of geometrical enantiomorphic isomerism was described by Lyle and Lyle,[108] they discovered that although cis-1-methyl-2,6-diphenyl-4-piperidone is meso and nonresolvable, on the other hand, its oxime (22) is not symmetric, and they separated it into its enantiomers by recrystallization of the (+)-10-camphorsulfonate salt. Lyle and Pelosi[109] further determined the absolute configuration of the (+) isomer of 22. A Beckmann rearrangement of this isomer gave the lactam 23 with a structure that follows from the known mechanism of the rearrangement in which the group anti to the oximino OH had migrated. Degradation of the lactam 23 gave the R(−)-diamine 24 which was prepared from R(−)-α-aminophenylacetic acid. Hence the absolute configuration of the (+)-oxime 22 is such that the hydroxy group is syn to the asymmetric carbon atom with the R configuration. This phenomenon also shows up in the nmr spectrum of syn- and anti-2,6-diphenyl-, 2,2-dimethyl-6-phenyl-, and 1,2,2-trimethyl-4-oximinopiperidines in which the anisotropy of the oximino group permits unambiguous assignment of the C–3, and C–5 protons.[110] Azines derived from cis-2,6-diphenyl-4-piperidone and hydrzaine exhibit similarly interesting ¹H nmr spectra.[111] Oximination of cis-1-alkyl-3,5-diphenyl-4-piperidones occurs with epimerization giving trans-1-alkyl-3,5-

diphenyl-4-oximinopiperidines. Apparently epimerization occurs prior to reaction with hydroxylamine because the *trans*-ketone reacts faster than the *cis*-ketone.[112]

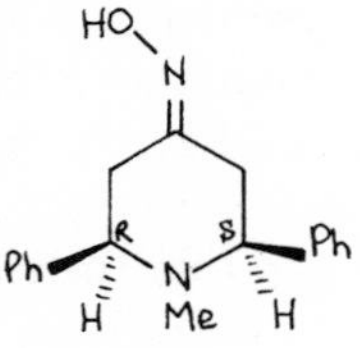

22 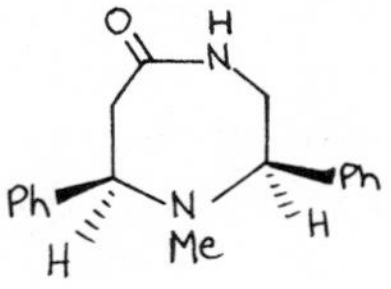   23 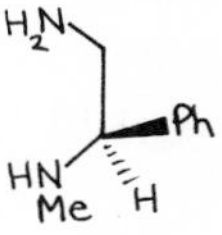   24

C. Conformation of Piperidines

Studies of the conformations of the piperidine system are particularly difficult because both nitrogen and ring inversion are rapid and the processes cannot be readily separated. The equilibria involved are shown in Scheme 1. The system is conformationally very flexible and ring inversion ($\Delta E^{\ddagger} > 42$ kJ mol^{-1}), is slower than nitrogen inversion ($\Delta E^{\ddagger} < 42$ kJ mol^{-1}), but it is still fast enough on the nmr time scale to produce time-averaged signals. A large volume of work has been published in this area and a detailed discussion cannot be given here. The essential features of the stereochemistry of the conformations and the equilibria, however, are described. Pyramidal nitrogen inversion was discussed by Lambert,[113] and the conformational analysis of organic compounds, including nitrogen heterocycles, was reviewed by Riddell and by McKenna.[114]

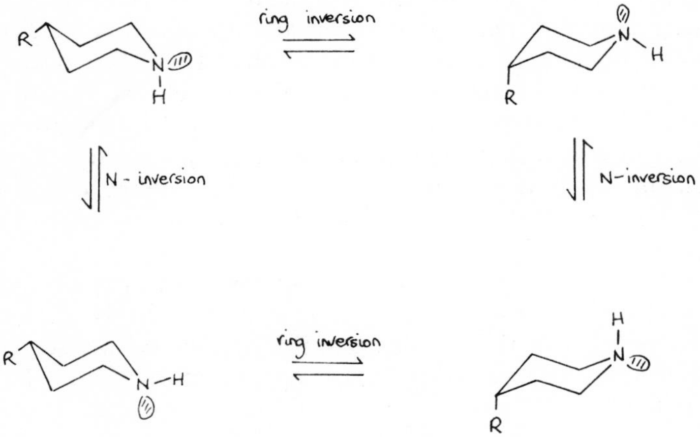

Scheme 1

One of the most important issues in the studies of the conformations of piperidines, and indeed of saturated nitrogen heterocycles generally, is the axial or equatorial orientation of the nitrogen lone pair of electrons. The unfortunate word "size" of the nitrogen lone pair was used in earlier work because the space requirements for the lone pair were compared with those of a hydrogen atom and a methyl group. The polarizable nature of the lone pair makes the word "size" meaningless. Aroney and Le Fèvre,[115] in 1958, measured the molecular polarizabilities of piperidine, N-methylpiperidine, and morpholine in benzene solution, and they concluded that the rings are in the chair form, and that the "volume" requirements were: lone pair $\approx$ methyl group $>$ hydrogen atom. This paper has initiated a considerable amount of excellent work in which several techniques were used and developed to arrive at the truth. In addition to the contributions of Le Fèvre and his school in this field, many names such as Lambert, Booth, Tsuda, and Eliel, among others, deserve mention. In particular, the work of Katritzky and his school stands out separately because of the depth and breadth of their approaches. Robb, Haines, and Csizmadia[115a] proposed a theoretical definition for the "size" of electron pairs and discussed the stereochemical consequences.

Further dipole moment results[116,117] led to the conclusion that the lone pair has a smaller steric requirement than the hydrogen atom. Lambert and co-workers[118] studied the nmr spectra at low temperatures and the ir spectra of piperidine, N-methyl- and N-t-butylpiperidines which were deuterated at C-3 and C-5 (to avoid complications from coupling). They concluded that there is a much higher population of the species with an axial lone pair in the 1-t-butyl derivative than in piperidine which has predominantly (but not completely) the lone pair in an equatorial orientation. These results were deduced from relative chemical shifts of α-equatorial and axial protons and Bohlmann bands in the ir spectra. They also concluded that the lone pair is the dominant factor (though not exclusively) in the relative shielding of the adjacent axial proton,[119] and that the axial-proton equatoral lone pair conformer would be favored not by attractive N–H· · · ·H forces but by repulsive lone pair H 1,3-axial interactions,[120] and that the proportion of axial NH conformers is reduced by the presence of an axial 3-methyl group.[121]

Suitably substituted piperidines were studied by Katritzky and collaborators and dipole moment results indicated that the N–H prefers the equatorial position in the gas phase and in noninteracting solvents, and that the energy difference between the NH equatorial and NH axial is approximately 1.68 kJ mol^{-1}.[122] Infrared studies in the gas phase[123] and in solution,[124,69] and microwave studies[125] are in agreement with this. In another approach the conformational equilibria in 1-t-butyl-3,5-dimethyl-4-

piperidones and 3,5-dideuterated derivatives were compared with those of the corresponding cyclohexanones and it was demonstrated that the steric requirement of the lone pair was smaller compared with that of a hydrogen atom.[126,127]

Booth[128] argued that if protonation of 3,5-dimethylpiperidine is faster than nitrogen inversion and N-hydrogen exchange with the medium is negligible, then it is possible to obtain an NH_{ax} to NH_{eq} ratio from the nmr spectra. The spectra of the piperidine in F_3CCO_2H and F_3CCO_2D gave this information from the vincinal J values between the $\overset{+}{>NHD}$ and the α-CH in the cations **25** and **25a**. He obtained a ratio of 54:46 for **25:26**. The

validity of this approach was disputed,[129] but further experiments by Booth and Little[130] confirmed and supported earlier claims. This method of kinetically controlled protonation as a measure of conformational equilibria in piperidines was refined by Robinson and co-workers[131] into a reliable method. It was performed by slowly diffusing the amine (vapor) through air onto the surface of an involatile acid. Under these conditions protonation is faster than N-inversion, is effectively irreversible, and is stereospecific with retention of configuration. The conclusion from these experiments is that the lone pair perfers an axial orientation but not by a large factor.

The orientation of N-alkyl substituents in piperidines, determined by dipole moment[132,133] and ir[124] measurements, agreed with the above-mentioned nmr data[118] that increase in the bulk of the alkyl group increases the proportion of axial N-lone pair. The effect of protonation on the changes in chemical shift of an N-methyl group varies according to whether an α-methyl group is present or not.[134] A long range effect was observed in 1-methylpiperidine methiodides in which 4-acetoxy,4-benzoyloxy, 4-chloro, and 4-hydroxy groups prefer the axial conformation due to transannular electrostatic interactions.[135] The preferred orientation of hydroxy and acetoxy groups on C-3 or C-4 of piperidines were compared by ^1H nmr spectroscopy with those in the corresponding piperidinium and N,N-dimethylpiperidinium ions. The proportion of the conformers with axial hydroxy or acetoxy groups were found to be larger in the cations than in the corresponding free bases.[135a] Intramolecular assistance was observed in the O-acylation of 4-hydroxypiperidines. The rates of esterification of piperidines with axial 4-OH groups by benzoyl chloride in tetrachloro-ethylene at 25°C were found to be four times as fast as their epimers with equatorial 4-OH group. A mechanism involving preliminary N-acylation followed by intramolecular transfer of the acyl group was postulated.[135b]

The equilibrium of 1,2,6-trimethylpiperidinium cation, which is toward the conformer in which all the methyl groups are equatorial, is discussed in terms of intermolecular interactions with solvent molecules and intra-molecular steric hindrance.[136] A computer program for strain energy minimization in six-membered rings gave conformational equilibria in 1-alkylpiperidines consistent with previous conclusions.[137]

The dipole moments, molar Kerr constants, and ^{1}H nmr data of several substituted 4-piperidones are explained in terms of "chair" conforma-tions,[138] but in the case of 2,2,6,6-tetramethyl-4-piperidone the ring is not rigid, and in 1,3,5-trimethyl-4-hydroxy-2,6-diphenylpiperidine there is a contribution of less than 16% from a "boat" conformer[139] (see below for ^{13}C nmr of "skew boat" conformers). The "chair" conformation for 1-methyl- and 1-acetyl-4-piperidone has been confirmed and there is no indication that the latter could be in the "boat" conformation to any appreciable extent.[140] Haller and co-workers[141,142] examined the nmr spectra of various substituted 4-piperidones and their oximes (and related pyrane and thiane derivatives) and evaluated the results in terms of $A^{1,3}$ (1,3-axial) interactions which were consistent with inverting "chair" con-formations. Aryl substituents in 6-aryl-2,2-dimethyl-4-piperidones and in the corresponding epimeric 4-hydroxy derivatives are equatorially dis-posed.[143] The conformational equilibria of many 1-alkyl 2,3-, 2,5-, and 3,5-dimethyl-4-piperidones in the presence of a base have been studied and the free energies were evaluated.[144]

N-Alkylpiperidine 1-oxides, like piperidinium salts, undergo ring in-version, and the evidence points out that the conformers with the axial oxygen atom are preferred,[145] but the ratio of axial to equatorial oxygen varies with the solvent. In 1-methylpiperidine 1-oxide the O_{ax} to O_{eq} ratio for CH_2Cl_2, $CDCl_3$, D_2O, and F_3CCO_2H is 3.1, 3.0, 1.6, and 6.2. The two diastereomeric 4-t-butyl-1-methylpiperidine 1-oxides were separated and their configurations were assigned.[146] Low barriers of nitrogen inversion were observed by esr spectroscopy for piperidine nitroxide and its 4-methyl, 4-isopropyl, 4-t-butyl, and 4,4-dimethyl derivatives.[147]

Intramolecular hydrogen bonding between an axial 3-hydroxy group and the piperidine nitrogen atom has been mentioned earlier (Section II.B) with respect to configuration. However, although the energy involved for hydrogen bonding is small, nevertheless it can affect the conformation. 3-Methyl-, 3-ethyl-, and 3-phenyl-3-hydroxypiperidines are predominantly in the hydrogen bonded conformation.[148] Such an interaction when the OH is at C–4 becomes less important because of the distance, and the ring would necessarily have to be in the "boat" form. The spectra of 1-methyl-4-hydroxypiperidine are explained in terms of two non–hydrogen bonded "chair" forms.[149] On the other hand, the 1- and 4-positions can be linked as in the lactone (26) but only when the 4-hydroxy and the 1-carboxymethyl

26   **27**

groups are cis to each other.[150] Weak hydrogen bonding is also possible between the 1- and a 4-substituent in amides such as 27.[151]

The phenyl group in 1-phenylpiperidine is shown in polarizability studies to be exclusively equatorial in benzene solution with the piperidine ring in the "chair" conformation[152] (compare the t-butyl group above). The benzyl group in 1-benzylpiperidines is particularly interesting and useful because the diastereotopic benzylic CH_2 protons are magnetically nonequivalent in the nmr spectrum if the piperidine carries an axial 3-alkyl substituent or an equatorial branched chain 3-alkyl group (e.q., t-butyl).[153] Also axial and equatorial N-benzyl groups on a quaternary nitrogen atom are readily distinguishable in this ring in the nmr,[154] and they are useful in studies of quaternization of piperidines (see Chapter 3, Section II.D.5). Alkyl groups in acetals and thioacetals derived from 4-piperidones are also magnetically nonequivalent and are used to interpret conformational properties. Thioacetals of 3-methyl-4-piperidones are much more readily perepared than the corresponding acetals.[155]

Variable temperature ¹H nmr (dnmr) studies in the piperidine series have not been as fruitful as in the aziridines (Chapter 2, Section I.A.3) because of rapid ring inversion which makes the effects present in the three-membered ring less marked in the six-membered ring. However, a few pertinent studies have been made. The coalescence temperature of 3,3,5,5-²H₄-1-methylpiperidine (252°K; i.e., $\triangle G^{\ddagger} = 49.8$ kJ mol⁻¹) is higher than that of 3,3,5,5-²H₄-piperidine(210°K; i.e., $\triangle G^{\ddagger} = 43.5$ kJ mol⁻¹)[118,156] and the effect is due mainly to ring inversion. A reexamination of these gave $\triangle G^{\ddagger}$ values of 32.5, 35.9, and 42.3 kJ mol⁻¹ for N-methylpiperidine, N-methyl-3,3,5,5-²H₄-4-piperidone, and N-chloropiperidine respectively.[157,158] The latter values are comparable with those obtained for N-trideuteromethyl-2-azabicyclo[2.2.2]octane (28,R = CD₃; $\triangle G^{\ddagger} = 35.1$ kJ mol⁻¹) and

N-chloro-2-azabicyclo[2.2.2]octane (**28**,R = Cl; $\Delta G^{\ddagger} = 44.4$ kJ mol^{-1}).[158] In the latter two compounds ring inversion is not possible and the effects must be due to nitrogen inversion. It is clear that the N-halogen atom raises the inversion barrier (compare with aziridines, Chapter 2, Section I.A.3). In N-chloropiperidine N-inversion is slowed down to the extent that it may be synchronous with ring inversion.[159] The fluorine atoms on nitrogen in 1-fluoro-2,6-dimethylpiperidine[160] and on C–4 in 4,4-difluoropiperidine[161] similarly slow down nitrogen inversion. Data from ^{19}F nmr of the latter

R = Me,Cl

28

gave a value of the free energy between the conformers with NH axial and NH equatorial of about 1.8 kJ mol^{-1}—which is of the same magnitude as obtained by other methods (see above).[161]

Conformational changes in N-acyl-,[162,163] N-sulfonyl-,[163] and N-nitroso-piperidines[163] have been observed by ^{1}H nmr spectroscopy and are attributed to restriction in the orientations of the anisotropic N-substituent. Alkyl-1-benzoylpiperidines reveal a coalescence of the ^{1}H nmr signals for similar reasons.[164] Restricted rotation of the N-substituent is also exemplified in 1-trichloromethylthio-, 1-phenylthio-, and 1-phenyl-2,2,6,6-tetra-methylpiperidines.[165] Chemical shifts induced by lanthanide reagents give valuable information about conformational equilibria in the N-nitroso derivatives of piperidine, 2-methyl, 2-ethyl, 2-propyl, and 4-methyl piperi-dines. The equilibrium proportions of these conformers were calculated from the relative induced chemical shift changes of the equatorial and axial populations of a particular proton with respect to a standard proton within the same molecule—but one which is common in all the molecules.[166]

^{13}C nmr spectroscopy is a useful tool for assigning preferred conforma-tions of piperidines. The conformations of cis-2-methyl-cis-5-methyl-r-4-phenyl-[δ], $trans$-2-methyl-cis-5-methyl-r-4-phenyl-[β], cis-2-methyl-$trans$-5-methyl-r-4-phenyl-[γ], and $trans$-2-methyl-$trans$-5-methyl-r-4-phenyl-[α] piperidines were deduced in this way.[167] Carbon 13 chemical shifts induced by protonation of piperidines were explained in terms of charge densities along the carbon skeleton, and the stereospecificity of the inductive effect was discussed.[168] The ^{1}H nmr (at 60, 100, and 220 MHz) and ^{13}C nmr spectra of α-, β-, and γ-1,2,5-trimethyl-4-hydroxy-4-phenylpiperidines indicate the relative configurations **29**, **30**, and **31** respectively. Whereas

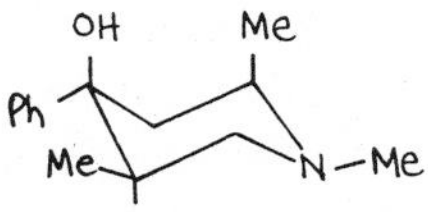

OH   Me   Ph   OH

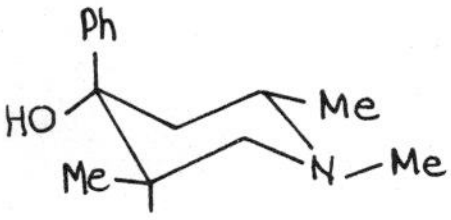

Ph   HO   Me   Me

Me   N—Me   Me   N—Me   Ph   N—Me   Me

α   β   γ

29   **30**   **31**

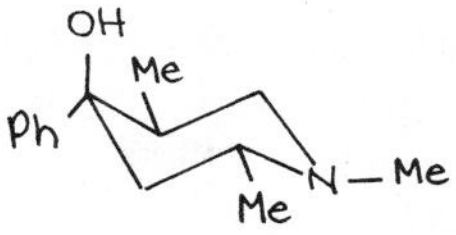

the β and γ isomers are in the chair forms shown in **30** and **31** the evidence for the α isomer points to a significant population in the "skew-boat" conformation **32** with a pseudoequatorial phenyl group.[169,170] These con-

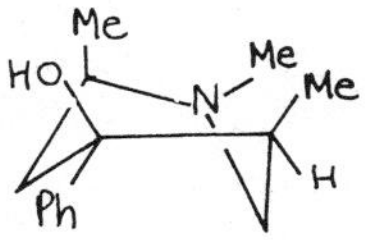

Me

HO   Me Me

N

Ph   H

32

formations, however, are altered on protonation or esterification of the alcohols. The stereochemistry of 1,2-dimethyl- and 1,3-dimethyl-4-hydroxy-4-phenylpiperidines was also sorted out by ^{13}C nmr[171] A recent interesting observation is the ^{13}C upfield shifts produced by the presence of a γ-nitrogen (oxygen and fluorine) atom in an antiperiplanar configuration.[172]

The pKa values of 3,3-dimethyl- and 3,3,5,5-tetramethylpiperidines were compared with those of the corresponding N-methyl derivatives in order observe the effects of 1,3-diaxial methyl interactions. The differences obtained were consistent with minor polar effects rather than steric hindrance of the solvated cations.[173] Studies of the epimerization of *cis*- and *trans*-4-*t*-butyl- and 4-methyl-2-ethoxycarbonylpiperidines gave ratios which were different from those observed in the corresponding cyclohexane esters. These were explained by dipolar repulsion effects between the ethoxycarbonyl group and the nitrogen lone pair. The pKa values of the ester hydrochlorides, that is, of the ammonium group, and of the carboxylic acids are dependent on the conformation of these molecules and hindrance to intramolecular hydrogen bonding.[4]

D.   Reactions of Piperidines

1.   Nitrogen Extrusion Reactions

The elimination of nitrogen from N-amino- and N-nitrosopiperidines has not been studied as extensively as in the aziridines. Overberger, Lombardino, and Kiskey[22] found that mercuric oxide caused the elimination of nitrogen from *N*-amino-2,6-diphenylpiperidine with formation of 1,2-diphenylcyclopentane mainly with retention of configuration. The *cis*-piperidine gives *cis*-diphenylcyclopentane in 65% yield (together with some diphenylcyclopentene) compared with the trans isomer (which is less stereospecific) which forms *trans*- (59%) and *cis*-1,2-diphenylcyclopentane (12%). This is most probably due to some isomerization of the *trans*- to the *cis*-piperidine prior to extrusion of nitrogen. They also found that the decomposition of *cis*- and *trans*-*N*-nitroso-2,6-diphenylpiperidines have stereospecificities similar to the above.[174] Yamada and co-workers examined this and related open chain systems, and they observed that in the latter case when optically active derivatives were used much racemization occurred[175]; this showed that the decomposition is not straightforward [compared with aziridines, Chapter 2, Section I.A.4.a; azetidines, Chapter 2, Section II.A.3; and pyrrolidines, Chapter 2, Section III.B.4.d].

2.   *Reduction and Addition Reactions of 4-Piperidones*

The reduction of several mono- and dimethyl 4-piperidones, their *N*-alkyl derivatives, and their hydrochlorides is achieved with various reagents, for example, $NaBH_4$, $LiBH_4$, Pt/H_2, $Al(isoPrO)_3$, and Li/NH_3–MeOH. The reductions are generally stereoselective and favor the equatorial piperidinol. This is not always the case, however, and where the reducing agents are Pt/H_2 or $Al(isoPrO)_3$ the axial piperidinol is favored.[41,176,177] The reactions of methyl and phenyl lithium on variously substituted 4-piperidones were also studied in detail and found generally to be stereoselective favoring the products with the axial hydroxyl group. There are several exceptions to this generality which depend on the stereochemistry of the piperidone and on the reagent.[178–181] One example is the addition of methyl lithium to *trans*-1-*t*-butyl-2,5-dimethyl-4-piperidone which gives a 20:80 mixture in favor of the equatorial alcohol, whereas a similar reaction with the cis isomer yields a 96:4 mixture in favor of the axial alcohol.[178] The epimeric 4-hydroxypiperidines are distinguished by their nmr spectra but some show differences in acid catalyzed etherification,[182] esterification with propionyl chloride,[183] acid catalyzed[184] and thionyl chloride mediated dehydration.[185] 1,3-Dimethyl-4-piperidone forms a cyanohydrin in which the hydroxy group is presumably equatorial.[186]

3.   *Epimerization of Piperidines and 4-Piperidones*

cis-2-Acetonyl-3-alkoxy(and 3-methyl)piperidines are readily epimerized to around a 30:70 mixture of cis and trans isomers under the influence of a base. The α-asymmetric center in these examples is one carbon removed from the active methylene group and the epimerization is best explained by ring opening of the enol followed by ring closure (Equation 7).[11] The rate

$$R = OMe, OEt, OPr^i, Me$$

at which the system comes to equilibrium varies with the solvent, temperature, and with the pH of the solution. cis- and trans-2,5-Dimethyl-4-piperidones epimerize at C–5 by way of the enol.[187] The base catalyzed cis-trans epimerization of 3,5-diphenyl-4-piperidones with varying alkyl groups on the nitrogen atom is in favor of the cis isomer and the equilibrium varies from 100:1 in the unsubstituted compound to 1:1 in the N-benzyl derivative. The ratio of cis and trans isomers is best determined from the ¹H nmr spectra of the alcohols obtained by reduction with LAH. No detectable epimerization occurs during these reductions.[42] The formation of an enol is clearly demonstrated in 2,6-dialkyl-3,5-bisalkoxycarbonyl-4-piperidones. The presence of an N-methyl group enhances the proportion of enol compared with the cyclohexane analogues, probably because the interaction of the N-methyl groups with the equatorial 2-6-dialkyl groups can be relieved by conversion into an axial N-Me in the enolic form.[188]

1,3,5-Trimethyl-4-piperidone epimerizes in the presence of t-butylamine, but a second slower reaction takes place whereby the NMe group is replaced by an N–Buᵗ group. This must involve ring opening with exchange of the basic groups. In the case of 1,2,6-trimethyl-4-piperidone no exchange with base occurs but the system epimerizes readily probably also by a ring opening and ring closure equilibrium (Equation 8). The 1,2,5-isomer behaves in the same way, and the equilibria favor the cis or equatorial isomer.[189] The methiodides of 1,3-di- and 1,2,5-trimethyl-4-piperidones undergo similar exchanges with isopropyl-, s-butyl-, and t-butyl-amines. The trans-t-butyl-2,5-dimethyl-4-piperidone has a skew-boat conformation.[190]

4. Rearrangements of Piperidines

Four different rearrangements of piperidines have been recorded. The first-essentially of the pinacol-pinacolone type, is the rearrangement of 1,2, dimethyl-4-hydroxypiperidin-4-yl-diphenyl carbinol to *trans*-4-benzoyl-*cis*-1,*r*-3-dimethyl-*cis*-4-phenylpiperidine by ZnCl₂–acetic anhydride.[186] The second is the rearrangement of the chloroformate esters of 3- and 4-hydroxypiperidines in the presence of a base. These give 5-chloropropyloxazolin-2-one and 6-chloroethyltetrahydro-1,3-oxazin-2-one respectively. It is a general reaction which has been observed with pyrrolidinols and azetidinols, and it involves intramolecular acylation of the nitrogen atom by the carbonyl chloride.[191] *cis*-1,2-Dimethylpiperidine oxide undergoes a retro-cycloaddition reaction on thermolysis, but the product recyclizes in a different manner into *cis*-2-methyl-3-oxa-2-azabicyclo[3.3.0]octane (Equation 9).[192]

$$\tag{9}$$

Wolff-Kishner reduction of 2-ethyl-1-methyl-3-piperidone proceeds smoothly to the corresponding piperidine, unlike the reduction of saturated hexahydro-1,2-dimethyl-3-azepinone in which ring cleavage occurs. It is not known if racemization takes place during the reduction.[193] A Clemensen reduction of optically active 2-ethyl-1,2-dimethyl-3-piperidone, however, causes a skeletal rearrangement and furnishes 2-1′-methylpropyl-1-methylpyrrolidine. The product is devoid of optical activity and suggests that the initial C–N bond cleavage must precede ring contraction (Equation 10).[194]

$$\tag{10}$$

5. *Quaternization of Piperidines*

The quaternization of piperidines has been intensively studied mainly by the respective schools of Katritzky and McKenna, and it was reviewed in 1970 by McKenna.[195] Consequently only a brief outline is warranted here.

The alkylation of piperidines is complicated because of the equilibria in Scheme 1, and every conformer is a target for the alkylating agent. The product mixture of alkylated piperidines is not, however, a measure of the conformational equilibria in the parent base. The Curtin-Hammett hypothesis holds true here; that is, the mixture of alkylated piperidines formed is determined by the free energy difference between the conformers in the transition state.

The direction of approach of the alkylating agent is the point which has attracted most attention. The composition of conformers is adequately assessed by ^{1}H nmr methods, although it is not entirely satisfactory in some benzyl salts,[196] and on occasions the use of X-ray analysis[197,198] was necessary. There is no evidence that isomerization occurs subsequent to quaternization.[199] The research has been extensive and many combinations of N-alkyl piperidines and of alkylating agents were studied but mainly with C-alkyl and C-phenyl substituted piperidines. The studies were also extended to decahydroquinolines, pyrrolidines, camphidines, tropanes,[200-203] and careful kinetic investigations in various solvents were made.[199,203-207] It was initially stated from the data available that the alkylating agent attacks the alkyl piperidine by an axial approach (33) with the entering alkyl group assuming an axial configuration with marked stereoselectivity.[202,207-209] Other work indicated that the alkylating agent attacked

axial attack

33

equatorial attack

34

equatorially,[204,207] and a controversy arose.[210,211] Further work showed that the alkylating agent approaches preferentially axially, provided that the N-substituent is not markedly bulkier than the incoming group.[212] The final outcome is that simple piperidines are methylated by an axial approach but that equatorial methylation occurs when the axial approach is sterically hindered.[213] The axial to equatorial ratio is nearly one in ethylation, and benzylation takes place by an equatorial attack (34).[204,205,213]

The direction of alkylation when α-methyl groups are present depends on the orientation of the α-methyl groups and the N-substituent.[203,214] Predominant equatorial methylation occurs in the tropane series[203,211] and in N-methyl 4-aza-5α-cholestane.[215] Methylation of N-alkyl(or aralkyl)-4-phenyl-4-formylpiperidines also proceeds by axial attack,[216] but the larger phenacyl group attacks preferentially equatorially.[217]

The amine boranes obtained by the reaction of borane with alkylpiperidines also can be formed by axial or equatorial attack. The ratio of axial to equatorial amine-BH$_3$ complex for various 2-alkyl(Me, Et, n-Pr, CH$_2$Ph) 1-methylpiperidines was determined by nmr spectroscopy at 33 and 77°C. The ratio varied with temperature, for example 1-methylpiperidine–BH$_3$ complex at 33°C is largely in favor of the axial-BH$_3$ diastereomer, but at 77°C there is a slight preference for the equatorial-BH$_3$ diastereomer.[218]

Dealkylation of quaternary alkylpiperidinium salts was studied by McKenna and collaborators.[219–221] The dealkylation was achieved with LAH and thiophenate ions. They concluded from rate and composition studies that the nucleophilic reaction more readily removes the alkyl group which is in the orientation that it had during its introduction in the original alkylation. Dealkylation of *trans*-1,2,5-trimethyl-4-piperidone methiodide, ethiodide, and isopropyliodide can be achieved with lithium in ammonia, but it is always the methyl group that is removed irrespective of the method of preparation of the quaternary iodide.[222]

The quadrupolar ^{14}N relaxation times for *cis*- and *trans*-2,6- and 2,5-disubstituted *N*-dimethylpiperidinium cations were found to be strongly dependent on the solvent, temperature, and stereochemical structure.[223]

6. *Fragmentation of 4-Chloropiperidines*

Grob and his school[224] studied in great detail the solvolysis of γ-chloro-amines. When the nitrogen lone pair α,β-σ bond and the C–Cl bond are aligned as in formula **35** or **35a** solvolysis of the halogen is accelerated by a synchronous mechanism which results in the fragmentation of the α,β bond (Equation 11). The degree of fragmentation (compared with substitution or elimination) is dictated by the ability with which the system can assume a conformation such as **35**. This is borne in the solvolysis of 4-chloropiperidine and its 1-methyl, 1-ethyl, 1-isopropyl, and *t*-butyl derivatives and in 4-chloro-2,2,6,6-tetramethylpiperidine in 80% ethanol. These solvolyze (with fragmentation) at rates 34–178 times faster than those of the corresponding cyclohexyl chlorides. 4-Chloro-1,2,2,6,6-pentamethylpiperidine, on the other hand, yields 87% of the 4-hydroxy derivative, but 4-chloro-

$$ \text{35a} \quad \text{or} \quad \text{35} \longrightarrow \quad (11) $$

1,4-dimethylpiperidine fragments (92%) at a rate that is 6.34 times that of the solvolysis of 1-chloro-1-methylcyclohexane.[225] Similar fragmentations of 4-tosyloxypiperidines further support the necessity for the stereoelectronic arrangement shown in formula **35** or **35a** for enhanced reactivity.[225a] Grob has reviewed the mechanism and stereochemistry of these and related fragmentations.[225b]

7.   *Miscellaneous*

Chemical reactions which do not affect chiral centers give products of known configuration, for example, the conversion of *cis*- and *trans*-2,3-bismethoxycarbonylpiperidine into *cis*- and *trans*-2,3-dimethylpiperidines respectively.[2]

Reduction of 3,4-epoxy-1-methyl-4-phenylpiperidine, with LAH or catalytically, is stereospecific and yields *cis*-3-hydroxy-1-methyl-4-phenyl-piperidine.[226]

The halogen atom in 3-chloropiperidine is displaced by nucleophiles at rates that are 10^4 times faster than in cyclohexyl chloride. The reaction follows first order kinetics, and the involvement of 1-azabicyclo[3.1.0]-hexane as intermediate is evidenced by the retention of configuration at C–3 in the displacement.[227] Intramolecular cyclization of 1-alkyl-3-bromopiperidine produces 1-alkyl-1-azoniabicyclo[3.1.0]hexanes which can be isolated as the perchlorate salts[228] (see Chapter 2, Section III. B.4.b and Chapter 4, Section VI.D).

(−)-3-Benzoyl-3-chloro-1-methylpiperidine, in methanolic sodium methoxide, cyclizes to (−)-2-methoxy-5-methyl-2-phenyl-1-oxa-5-azaspiro[2.5]-octane (**36**) without loss of optical activity. Decomposition of this epoxide in boiling aqueous hydrochloric acid or warm glacial acetic acid yields racemic 3-benzoyl-3-methoxy-1-methylpiperidine. The assistance of the *N*-methyl group in this reaction by involvement of an intermediate such as **37** was postulated and was confirmed by experiments using the [18]O labeled chloro ketone.[229]

The chemistry of 1,2,3,6-tetrahydropyridines (3-piperideines) was reviewed by Ferles and Pliml[229a] in 1970.

36   **37**

III.   REDUCED PYRIDAZINES

Most of the studies on reduced pyridazines have concentrated on the conformational properties of 1,2,3,6-tetrahydro- and hexahydropyridazines, on the extrusion of nitrogen from 3,4,5,6-tetrahydropyridazines, and on derivatives possessing a 3,6-methylene bridge. These and a few other properties are described under the following subheadings. Hexahydropyridazine will be called 1,2-diazane for simplicity. The chemistry of pyridazine and reduced pyridazines was reviewed in 1968 by Tišler and Stanovnik.[230]

A.   Syntheses of Reduced Pyridazines

Only a few simple hydropyridazines of stereochemical interest are worthy of mention here. The reduction of 1,4,5,6-tetrahydro-3,6-diphenylpyridazine with sodium amalgam in ethylene glycol at 130°C gives a 1:2 mixture of *cis*- and *trans*-3,6-diphenyl-1,2-diazane. These are oxidized by air to the corresponding 3,4,5,6-tetrahydro derivatives, that is, a $N = N$ double bond is formed.[231] The oxidation of an N–N single bond to a double bond in hydropyridazines can also be brought about chemically with manganese dioxide[232] or yellow mercuric oxide[233,234] under mild conditions. Dimethyl azodicarboxylate (DAD) has become a very useful reagent for introducing the N–N group to form 1,2-diaza heterocycles. It condenses readily with dienes to yield tetrahydropyridazines. Thus condensation of DAD with *trans-trans*-2,4-hexadiene gives a quantitative yield of 1,2-bismethoxycarbonyl-3,6-dimethyl-1,2,3,6-tetrahydropyridazine readily (Equation 12). The reaction with *cis-trans*-2,4 hexadiene is more sluggish but furnishes a 4:1 mixture of *trans*- and *cis*-1,2-bismethoxycarbonyl-3,6-dimethyl-1,2,3,6-tetrahydropyridazine. Hydrolysis and decarboxylation produce the respective *cis*- and *trans*-3,6-dimethyl-1,2,3,6-tetrahydropyridazines. Both these isomers react

$$(12)$$

with diazomethane in the presence of cuprous chloride, but whereas the trans isomer gives only one product: *trans*-2,5-dimethyl-3,4-diazabicyclo-[4.1.0]heptane (**38**), the cis isomer produces a 1:8 mixture of *cis*-2,5-dimethyl-3,4-diaza-*syn*-(**39**) and *anti*-bicyclo[4.1.0]heptanes (**40**).[234]

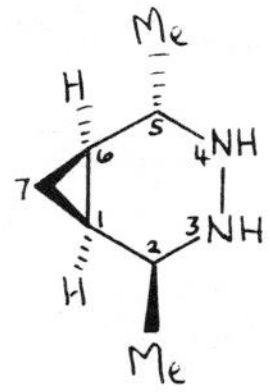

38 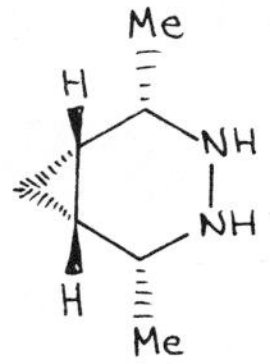   **39** 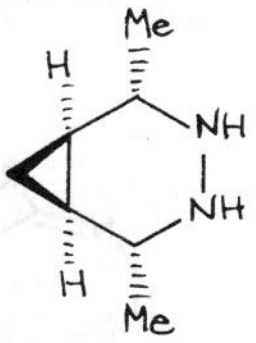   **40**

Similarly diethyl or dimethyl azodicarboxylate adds across the conjugated double bonds of cyclic systems with high cis stereospecificity. Examples of such dienes are cyclopentadiene,[235] 5-benzamidomethylcyclopentadiene,[236] fulvene and 6-acetoxyfulvene,[237] α-pyrone,[238] cycloheptatrienone and 2-substituted derivatives,[239] and cycloocta-1,3,5-triene[240] which yield the adducts **41** to **47**, respectively and are essentially 3,6-bridged 1,2,3,6-tetrahydropyridazines. 4-Methyl- and 4-phenyltriazolin-

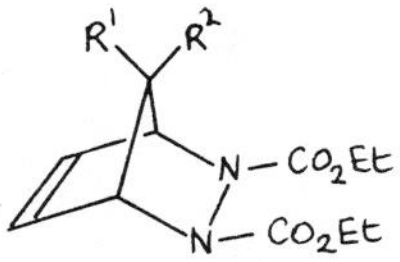

41 $R^1 = R^2 = H$ **45** $R = Me, Et, Ph$ **46** $R = H, OMe, OH$

42 $R^1 = CH_2NHCOPh$, $R^2 = H$

43 $R^1, R^2 = \,=CH_2$

44 $R^1, R^2 = \,=CHOAc$

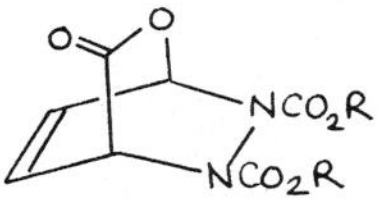

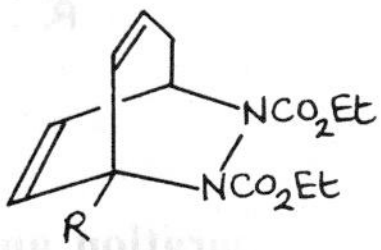

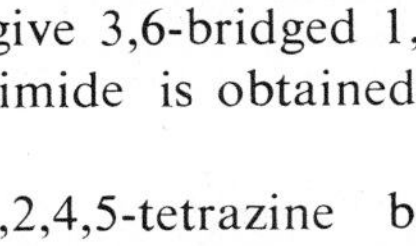

47

dione undergo Diels-Alder additions to conjugated double bonds in much the same way as above to give 3,6-bridged 1,2,3,6-tetrahydropyridazines, except that a 1,2-biscarboximide is obtained instead of a biscarboxylic ester.[241,242]

3,6-Bismethoxycarbonyl-1,2,4,5-tetrazine behaves as dienes toward cyclopropane and reacts with elimination of nitrogen and formation of 1,3-bismethoxycarbonyl-3,4-diazanorcara-1,3-diene (**48**) with cis stereospecificity. Variable temperature nmr studies of this bicyclo compound indicate that it undergoes inversion by degenerate valence isomerization (Scheme 2).[243]

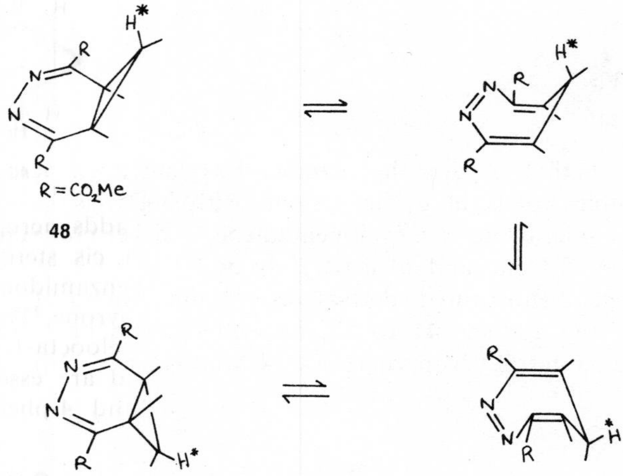

R = CO₂Me

48

Scheme 2

B.   Configuration and Conformation of Reduced Pyridazines

The relative configurations of some hydropyridazines are described in the previous section (III.A) and only a few more will be mentioned here. Bartlett and Porter[244] made *cis*- and *trans-d,l*-3,6-diethyl-3,6-dimethyl-3,4,5,6-tetrahydropyridazine (**49**) by oxidation of meso and *d,l*-3,6-diamino-3,6-dimethyloctane with iodine pentafluoride at −20°C. They were able to distinguish between them by resolving the *d,l*-amine into its enantiomers and oxidizing these to the optically active hydropyridazines.

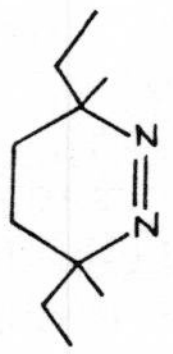

49

3*R*-3-Carboxy-1,2-diazane and its 5*S*-chloro and 5*S*-5-hydroxy derivatives were obtained from acid hydrolysis of the antibacterial nonamycin.[245]

Conformational analyses of 1,2,3,6-tetrahydropyridazines and 1,2-diazanes have been studied in great detail and will be considered in this order. Variable temperature ^{1}H nmr spectroscopic measurements of a large number of 1,2,3,6-tetrahydropyridazines are reported and clearly, in these molecules, ring inversion (**50** $\rightleftharpoons$ **51**) as well as nitrogen inversion is

taking place. Two distinct coalescence temperatures are observed with **50** ($R^1 = R^4 = Me, R^3 = R^2 = H$), **50** ($R^2 = R^3 = R^4 = H$, $R^1 = Me$), and **50** ($R^1 = CD_3, R^4 = Me, R^2 = R^3 = H$) indicating that both processes are occurring, but slowly, with the barriers of ring inversion smaller than those of nitrogen inversion, and with one substituent on the nitrogen atom axial and on the other equatorial.[246] Other examples showed that this is not necessarily always the case, that is, only one inversion barrier is sometimes observed and that experimental evidence suggests that the N-methyl groups could both be equatorial.[246] The energy barriers are of the order of 33–54 kJ mol^{-1}.[247] The spectra become more complicated when alkoxycarbonyl[248] or acetyl[249] groups are present on the nitrogen atoms because restricted rotation of these groups introduces other energy barriers. Nonetheless these have been examined in detail. The energy barriers for ring inversion are high, 75–78 kJ mol^{-1}, and are probably associated with the interaction between the N-acyl substituents.[248–252] Several 1,2-bisalkoxycarbonyl-1,2-dihydro-,1,2,3,6-tetrahydro-, and hexahydropyridazines were studied and hindered rotation of the alkoxycarbonyl groups was observed in all of them. Ring inversion occurred in the tetra and hexahydro derivatives but there was evidence that a "ring twisting" process was taking place in the 1,2-dihydro compounds.[253] The observation that the intensities of methyl signals in *cis*-1,2-bismethoxycarbonyl-1,2,3,6-tetrahydro-3,6-diphenylpyridazines varied with temperature was interpreted in terms of fast ring inversion but slow N–CO_2Me rotation.[254] This was refuted when it was found that although the methyl singlets had different heights their integrated areas were identical.[255]

Seventeen 1,2-bismethoxycarbonyl derivatives of **50** were examined in CDCl$_3$ and the equilibria **50** $\rightleftharpoons$ **51** and thermodynamic parameters were derived from ^{1}H nmr data. The conformer ratio was severely influenced by the nature and positions of substituents in the ring.[256] A large number of 1,2-disubstituted and trisubstituted 1,2,3,6-tetrahydropyridazines with a 3,6-methylene bridge were studied. These examples do not undergo ring inversion and only nitrogen inversion is observed. The ^{1}H nmr data are best explained by consecutive trans-trans inversion processes at the two nitrogen atoms (i.e., **52** ⇌ **53**).[257] There is also ^{13}C nmr evidence that trans ⇌ trans rather than cis ⇌ cis nitrogen inversion occurs in the reduced derivatives of **52**[258] and thermodynamic parameters of many substituted deriva-

tives of **52** have been tabulated. The esters **52** ($R^3 = CO_2R$) also display coalescences due to restricted rotation.[257,259] In addition to nitrogen inversion and restricted rotation, the saturated derivatives of **52** and **53** show yet another energy demanding conformational change which is called "bridge flipping."[260] It is an equilibrium between two "twisted" conformations, **54** ⇌ **55**, in which the degree of flexibility increases from $n=1$ to $n=4$.[261]

Lehn and Anderson examined several 1,2-diazanes by ^1H dnmr spectroscopy and found that the ring inversion energy barriers were only slightly lower than those of the corresponding 4,5-dehydro derivatives.[246,247] The dipole moment data for 1,2-dimethyl-1,2-diazane as measured by Katritzky and co-workers was explained in terms of four conformers in the equilibria **56** to **59** with about equal population of axial-axial, axial-equatorial, and equatorial-equatorial conformers.[262] This did not agree with earlier ^1H nmr results,[247] but a reexamination of this data showed that it could be explained in terms of about an equal population of three conformers: **56**, **57**, and **59**.[262,263] Conformers **56** and **58** are equivalent when $R^1 = R^2 =$ H, and when $R^1 = R^2 =$ Me ^1H nmr data indicates a 64:36 ratio for **57:58**, that is, there is an absence of conformers **56** and **59** when 1,3-syn-axial methyl interactions are strong. When the 4-substituents are Me- and p-$NO_2C_6H_4$- the proportions for **57** ($R^1 =$ Me, $R^2 = p$-$NO_2C_6H_4$), **57** ($R^1 = p$-$NO_2C_6H_4$-, $R^2 =$ Me), **58** ($R^1 =$ Me, $R^2 = p$-$NO_2C_6H_4$), and **58** ($R^1 = p$-$NO_2C_6H_4$) are 26, 38, 15, and 21% respectively.[263] Another physical method which is used to give some information about the orientation of nitrogen lone pairs is photoelectron spectroscopy. Nelsen and co-workers[264–266] applied this method to 1,2-diazanes and related systems.

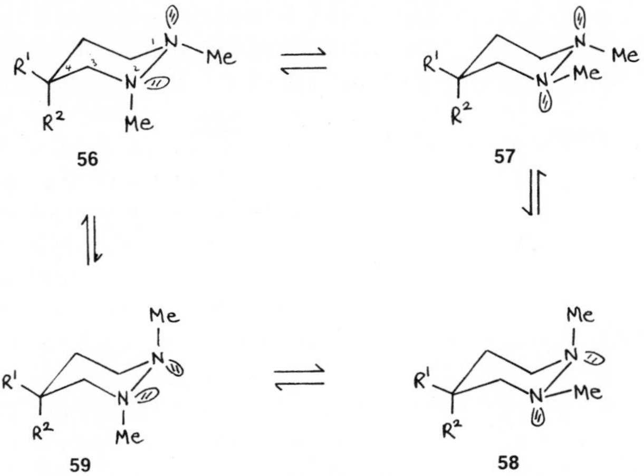

Scheme 3

They found that *cis-* and *trans*-1,2,3,6-tetramethyl-1,2-diazanes exist mainly in the 1-methyl-axial,2-methyl-equatorial conformer and that 1,2-dimethyl-1,2-diazane is mainly in the diequatorial conformer **57** ($R^1 = R^2 =$ H), but that more work was necessary to confirm this. Katritzky and co-workers[267] studied the equilibria and kinetic processes of this system further and showed that the conformational changes can be explained in terms of "passing" and "nonpassing" inversion processes which are different in the magnitude of the steric interactions in the transition state. Koch and Zollinger[268] concluded from ¹H nmr data that in *trans*-4,5-dihydroxy(or acetoxy)-1,2-diazanes and related compounds the predominant conformation is curiously that in which all substituents are axial.

Esr measurements of *cis-* and *trans*-1,2,3,6-tetramethyl-1,2-diazane support a structure in which the hydrazine portion of the molecule is not planar.[269]

C. Reactions of Reduced Pyridazines

The reactions of stereochemical interest of reduced pyridazines are dominated by those in which nitrogen is extruded from 3,4,5,6-tetrahydro derivatives; usually with formation of a cyclic hydrocarbon. Thermolysis of *meso(cis)-* and *d,l(trans)*-3,6-diethyl-3,6-dimethyl-3,4-5,6-tetrahydropyridazine (**49**) caused the elimination of a nitrogen molecule and formation of the respective cyclobutanes with high, but not complete, retention of cis and trans configurations. A large amount of 1-isopentene was produced, but Bartlett and Porter[244] were unable to detect any differences in the radical pair of singlet and triplet states by using sensitizers. Similarly thermolysis of *cis-* and *trans-*3,6-diphenyl-3,4,5,6-tetrahydropyridazines gave higher proportions of *cis-* and *trans-*1,2-diphenylbutanes, but they were each contaminated with substantial amounts of the *trans-* and *cis-*diphenylbutanes respectively.[231,270] Thermal and photochemical decomposition of the three diazabicycloheptanes **38**, **39**, and **40** gave the same products, namely, *trans-cis*, *cis-cis*, and *trans-trans* penta-2,5-dienes respectively. These results violate the pattern that the stereochemistry of photochemical reactons is opposite to that of thermal reactions.[232,234] However, in this case an acyclic product is formed so it is possible that two steps are involved in each reaction making the overall product of reaction the same irrespective of whether it is induced thermally or photochemically, that is, a net crossover.

Thermolysis and photolysis of 3,4,5,6-tetrahydropyridazines with a $(CH_2)_n$ bridge across C–3 and C–6 have attracted much attention[271–275] and has been studied to a large extent by Allred and co-workers.[276–279] Nitrogen is lost and a C–C bond is formed in the process (Equation 13). Examples such as **60** [R^1 = OMe or OBz, $R^2 = R^3 = R^4$ = H, n = 1 (*exo*)] **60** [R^2 = OMe or OBz, $R^1 = R^3 = R^4$ = H, n = 1 (*endo*)],[276–278] **60** [$R^1 = R^3$ = D, $R^2 = R^4$ = H, n = 1 (*exo*)], and **60** ($R^2 = R^4$ = D, $R^1 = R^3$ = H, n = 1 (*endo*)][280]

60   **61**   **62**   (13)

yield predominantly the cyclopropanes with inversion of configuration; that is, they give predominantly **61** rather than **62**. These occur with net crossover because the products from photolysis and thermolysis appear to be the same. The rates of thermolysis are greatly enhanced by the presence of a fused alicyclic ring; for example, the relative rates of **63**, **64**, and **65** are

63   **64**   **65**

1.0, 6.7 × 10⁴, and 1.1 × 10¹⁷ respectively.[281,282] Similarly the relative rates for **63**, **66**, **67**, and **68** are 1, 8.8 × 10², 6.7 × 10⁴, and 1.1 × 10¹⁷

66   **67**   **68**

respectively.[240] The rate enhancements of 10^{17} are about the largest observed in cyclopropanes and suggest a transition state such as **69**. This is evidenced by the products formed, that is, the reaction **70** → **71**,[283] and the formation of 1,4-dihydrobenzene and N-phenyl- or N-phenylsulfonyl-1,4-dihydropyridine from **60** (R^1 and $R^3 = -CH_2-$, $R^2 = R^4 = H$, $n = 1$)[284] and **60** (R^1 and $R^3 =$ -NPh- or -NSO$_2$Ph-, $R^2 = R^4 = H, n = 1$) respectively.[235,285] The latter reaction was developed into an unambiguous synthesis of N-substituted 1,4-dihydropyridines.[285a] A most interesting extention of these reactions is the formation of *trans*-tricyclo[3.1.0.0^{2,4}]-

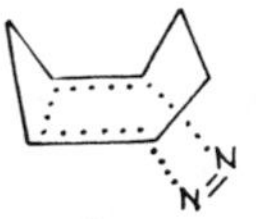

69

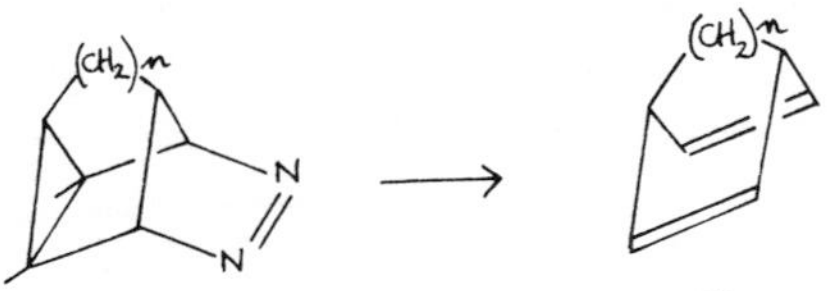

70    $n = 0,1,2,3$

71

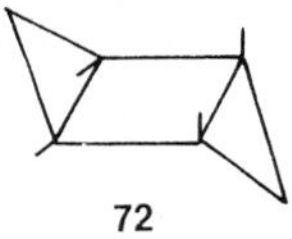

72

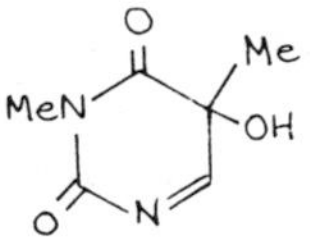

73

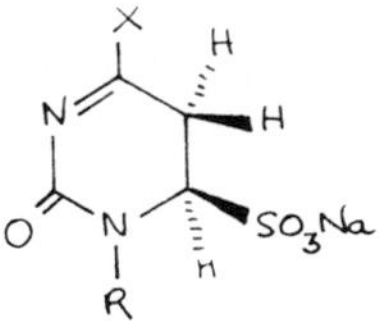

74   X = OH, NH$_2$

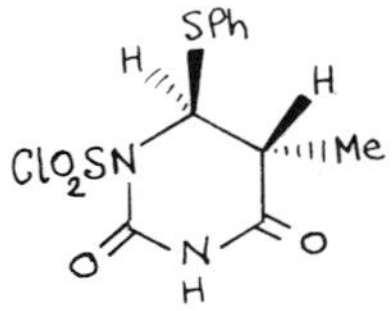

75

hexane (**72**), in 10–15% yield, by photolysis of **60** (R^1 and $R^3 = -N = N = CH_2-$, $R^2 = R^4 = H, n = 1$).[233]

IV. REDUCED PYRIMIDINES

The pyrimidine ring is widely distributed in nature and a vast number of derivatives are chiral in one form or another. The chiral portion of the molecule is not generally in the pyrimidine nucleus, for example, nucleosides (see Section XIV.D), but the chiral properties of the molecule as a whole are affected by its presence. Only the stereochemistry of the reduced pyrimidine ring and some derivatives will be discussed here because this monograph is not concerned with natural products. Fleeting reference, however, is made to a few relevant examples which are derived from natural sources. The general chemistry of pyrimidines has been admirably reviewed by Brown.[286]

A.   Synthesis of Reduced Pyrimidines

The addition of hypobromite or hypochlorite to the C–5, C–6 double bond of uracil or thymine is essentially trans,[287,288,289] and treatment of the product with base, for example, Ba(OH)$_2$, Ag$_2$O, or NaOH,[287] produces the cis-5,6-glycol.[290] cis-5,6-Dihydroxy 5,6-dihydro-thymine, thymidine, and 1,3-dimethylthymine are isomerized to the trans-5,6-glycols by boiling water.[291] This isomerization could occur by direct nucleophilic displacement of OH by water with inversion of configuration, by dehydration to the 1,6-dehydro derivatives and rehydration, or by ring-chain tautomerism (i.e., by way of ring opening). The first two are the most probable, and Hahn and Wang[292] isolated a 1,6-dehydro derivative 2,3,4,5-tetrahydro-5-hydroxy-5-methyl-2,4-dioxopyrimidine (73) together with cis- and trans-5,6-dihydroxy 5,6-dihydro-thymine by ^{137}Cs-γ-irradiation of thymine. Hydration of 73 would give mainly the trans-glycol. The cis-thymine glycol is also prepared from thymine by hydroxylation with permanganate.[293] Specific syntheses of chiral thymine glycols,[293a] and the isomerization of cis-(+)-thymine glycol into its trans-(−) isomer which probably involves ring-chain tautomerism,[293b] have been reported. ^{60}Co-γ-Irradiation of uracil gives a mixture of cis- and trans-5,6-dihydro-5,6-dihydroxyuracil among other products.[294] For further details about similar 5,6-dihydro-5,6-dihydroxy pyrimidines, see Brown's monographs.[286] The addition of sodium bisulfite to uracil, cytosine, and, cytidine-5-phosphate to give adducts such as 74 is most probably trans.[295–297]

Chlorosulfonyl isocyanate adds across the double bond of a mixture of cis- and trans-1-phenylthioprop-l-enes to give a high yield of trans-1-chlorosulfonyl - 5 - methyl - 2,4 - dioxo - 6 - phenylthiohexahydropyrimidine (75).[298]

Kunieda and Witkop[299] extended the study of the reaction of dimethyloxosulfonium methylide (Me$_2$$\overset{-}{S}$=$\overset{+}{C}H_2$) with double bonds in heterocycles to

1,3-disubstituted uracils, thymines, and nucleosides. All the adducts had the general structure 76.

76   77   78

Barbituric acids and thiobarbituric acids with a variety of chiral side chains at C–5 were synthesized from optically active malonic esters or cyanoacetic esters by conventional methods.[300–302] The absolute configurations in some of the side chains are known because those of their precursors are known.[300–302] C–5 in barbituric acid is not an asymmetric center, but it becomes one if N–1 or N–3 is substituted. Three such optically active examples are described, namely, 5-cyclohex-1'-en-1'-yl-5-methyl-, 5-ethyl-5-phenyl-, and 5-methyl-5-phenyl-1-methylbarbiturates[303] (see Section IV.B). Amide carbonyl groups are not readily reduced by NaBH$_4$. 5,5-Dibenzyl-1,3-dimethylbarbiturate, however, is unusual because it is rapidly reduced by NaBH$_4$ (1 hr, 25°C) to the 4-hydroxy derivative which is further reduced to   trans-5,5-dibenzyl-4,6-dihydroxy-1,3-dimethylhexahydro-2-oxopyrimidine (77) stereospecifically. Mineral acids or alkalis epimerize the diol into its cis isomer. The pyrimidine ring of the diol is completely degraded on exhaustive borohydride reduction to 2-benzyl-3-phenylpropanol.[304]

The thiourea derived from R-4-methylsulfinylbut-3-enyl isothiocyanate (sulforaphene, from radish seeds) cyclizes slowly in ethanolic ammonia mainly to 1'R,4S-4-methylsulfinylmethyl-2-thiohexahydropyrimidine (78) in 86% optical purity. This is basically a stereoselective addition of a thioureido-NH across an olefinic double bond. The R absolute configuration of the product was deduced by reduction to the 4-methylthiomethyl derivative and the synthesis of this from R-1,3-diamino-2-methylthiopropane obtained from R-β-methionine of known absolute configuration.[305]

79

The stereochemistry of the reversible equilibrium between dihydro-orotic acid and orotic acid by dihydroorotate dehydrogenase from *Zymobacterium oroticum* was resolved by Blattmann and Rétey.[306] The absolute configuration of the biologically active dihydroorotic acid is S because it can be prepared from natural S-aspartic acid. The addition and removal of H–5 and H–6 by the enzyme occurs with trans stereospecificity making the *pro-S*-5-hydrogen atom (79) the one that is labilized by the enzyme–NAD$^+$ complex.

A naturally occurring uracil with an optically active side chain, from a bacillus, which was recently synthesized is S(+)-5-(4',5'-dihydroxypentyl)uracil.[307]

B.   Configuration of Reduced Pyrimidines

The configurations of several reduced pyrimidines have been discussed in the previous section (IV.A), but a few more examples are described here. The ord and cd curves of a number of barbiturates which owe their optical activity to a chiral center on the C–5 side chain such as S-5-alkyl-5(2′-pentyl)- and S-5-alkyl-5(3′-hydroxy-1′-methylpentyl)barbiturates were examined.[300,308] The nature of the 5-alkyl group, the replacement of the 2-oxo group by a thio-oxo group, and the solvent used have a specific effect on the chiroptical properties, indicating that some absorptions in the pyrimidine ring are active. Knabe and co-workers[303] have determined the absolute configurations of the three optically active 5-cyclohex-1′-en-1′-yl-5-methyl-, 5-ethyl-5-phenyl-, and 5-methyl-5-phenyl-1-methylbarbiturates (see Section IV.A of this chapter), in which C–5 is a chiral center, by degradation with alkali to chiral malonamic acid and conversion into chiral ethanolamines of known absolute configuration. Several isomeric pairs of 5,5-dialkylbarbiturates, which differ only by the cis-trans arrangement of a C–5 olefinic or carbocyclic side chain, have been separated, for example, 5-crotylalkyl-,[309] 5-1′-chloro-3′-propenyl-,[310] 5-2′-methylcyclopentyl-,[311] and 5-5-benzylidine-,[312] but not 5-vinyl barbiturates.[313]

The configurations of capreomycidine (80)[314–316] and viomycidine hydrohalide (81),[317,318] obtained from the degradation of polypetide antibiotics from bacteria, were determined. An X-ray study of the hydrobromide of the latter confirmed its identity and configuration.[319] An X-ray

study of the d,l-bisulfite adduct of 2-aminopyrimidine (82) showed that it was a 3,4-adduct.[320]

C.   Conformation of 1,3-Diazanes (Hexahydropyrimidines)

1,3-Diazane is a conformationally flexible molecule capable of ring inversion and of pyramidal inversion[113] at two nitrogen atoms. The most interesting feature about 1,3-diazanes is the relative orientation of the lone pair of electrons on the nitrogen atoms. For simplicity only nitrogen inversion is depicted in Scheme 4. Ring inversion in **83** and **84** produces the mirror images of **85** and **86** respectively. The mirror image of **86**, however, is identical with **86**, but the mirror image of **85** is not identical with **85**. In

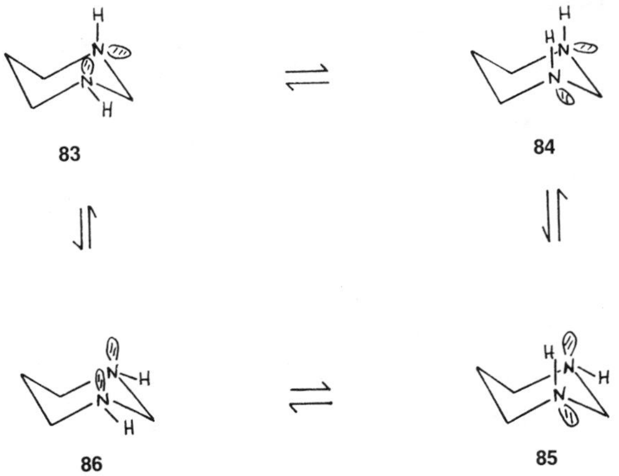

83   ⇌   **84**

86   ⇌   **85**

Scheme 4

unsymmetrically substituted 1,3-diazanes the system becomes more complicated.

Ring inversion has been observed by variable temperature ¹H nmr spectroscopy in N-alkyl derivatives, and thermodynamic activation parameters were determined.[321,322] The free energy change is of the order of 48 kJ mol⁻¹ and comparable with those of other six-membered saturated heterocycles.[323] The conformation **86**, in which the unshared electron pairs on the nitrogen atoms are syn axial, is less favored than the others because of the proximity of the lone pairs of electrons. This destabilizing phenomenon has been called the "rabbit-ear effect" by Eliel and co-workers[324] and is only another example of the "anomeric effect"[325] (which has been called the "Edward-Lemieux effect")[326] which is normally observed in carbohydrate chemistry. Several alkyl and deuterated alkyl 1,3-diazanes were examined and all showed this effect[327,324] which was believed to be of substantial magnitude. ¹H nmr examination of 1-methyl-1,3-diazane revealed that the N-H showed a strong preference for being axial[327] but in 1,3-dimethyl- and 1,3,5-trimethyl-1,3-diazanes the N-alkyl groups were oriented partly diequatorial and partly axial-equatorial[328] suggesting that the rabbit-ear effect was not all that strong. Katritzky and co-workers[329] measured the dipole moments of N-methyl- and N-ethyl-1,3-diazanes and showed that the conformational free energy difference between axial and equatorial N–Me and N–Et groups are slightly smaller than in the corresponding N-alkylpiperidines, and deduced that the repulsion between two axial nitrogen lone pair of electrons was about 1.26 kJ mol^{-1} (0.3 kcal mol^{-1}). The free energy difference for the N-isopropyl group was surprisingly higher. Eliel and co-workers[330] studied the 1,3-dialkyl-1,3-diazanes further and while agreeing with previous data estimated that the repulsion of the lone pairs was approximately 4.2–6.3 kJ mol^{-1} (1.0–1.5 kcal mol^{-1}) which was higher than previously stated. They also showed that the equilibrium in 1,2,3-trimethyl-1,3-diazane lies towards the N-methyl-axial N-methyl-equatorial conformer, making the forces between three vicinal equatorial groups override the rabbit-ear effect. Katritzky and co-workers[331] measured the first overtone N–H frequency in the ir spectra and the dipole moments of 1,3-diazane, 1,3-thiazane, 1,3-oxazane, and 1,4-diazane and found the results compatible with each other. In all cases the presence of a heteroatom β- to an NH group causes a predominance of the N–H axial conformer. Two possible explanations are: (a) 1,3-syn axial attraction of a lone pair of electrons and a hydrogen atom and (b) dipolar repulsion forces between two 1,3-syn axial lone pair of electrons. MO calculations of interactions in 1,3-diazane by Chen and Jesaitis[332] revealed that there is true physical meaning to the "rabbit-ear effect," and they concluded that lone pair interactions include more than a dipole-dipole electrostatic portion; the angle between the lone pairs is quite different from the zero degree they depicted, and that solvation and/or aggregation is important.

Katritzky and co-workers[136] developed a computer program for strain energy minimization in six-membered rings and used it to calculate conformational equilibria. These are in close agreement with those observed previously in 1,3-diazanes,[329] namely, the percentages of alkyl groups diequatorial, at equilibrium, of 1,3-dimethyl-, 1,3-diethyl-, 1,3-diisopropyl-, 1,3-di-t-butyl-, 1-t-butyl-3-methyl-, 1-t-butyl-3-ethyl-, and 1-t-butyl-3-isopropyldiazanes are 55,59, about 100, 100, 69, 75, and about 100. They also showed that J_{gem} values and chemical shifts of the methylene protons between two heteroatoms of a large number of N-alkyl 1,3-diazanes, 1,3-oxazanes, and 1,3,5-triazanes could not give an accurate indication of the orientation of the lone pair on the nitrogen atom.[333] Although, however, there is some theoretical justification for a direct correlation of this J_{gem} value and chemical shift with the orientation of the lone pair,[334] this is clearly not borne out in practice.

The dipole moment of 1,3-dinitro-1,3-diazane is similar to the calculated value for the structure **87** when θ is 101°30′.[335] The dipole moments of 5-nitro-1,3-diazanes (**88**, R = H,Me,Et,n-Pr) are found to be high for the alkyl derivatives and are interpreted in terms of the preponderance of the conformer with the nitro group equatorial (89) when R = H, but axial (88) when R is an alkyl group.[336] The ratio of nitro group axial to equatorial in 1,3-diethyl-5-methyl-5-nitro-1,3-diazane is 6:1,[337] and this is consistent with previous results. Monoquaternization of 1,3-dialkyl-5-methyl-5-nitro-1,3-diazane with methyl or acyl halides gives a mixture of isomers, which could be separated by tlc and assigned by nmr spectroscopy, and whose ratios are different from the conformer ratios in the starting material. The ratios of the quaternary salts formed are consistent with steric effects and with the preference of the nitro group for axial orientation when it is gemminal to a methyl group.[338]

The changes in the ^{1}H nmr spectra with temperature of N-methyl, N-ethyl-, and N-isopropyl-N',N'-bishexahydropyrimidinylmethanes is consistent with the presence of two diasteroisomeric sets of conformtions, racemic and meso, which interconvert by ring inversion.[339]

Ir[290] and ^{1}H nmr[288,289,292] evidence strongly supports that the preferred conformations for 5,6-dihydrouracils and many of its 5,6-disubstituted derivatives are between two inverting "half-chairs" with the HN–CO–NH–CO bonds almost in one plane (90 ⇌ 91). The conformer ratios for several

5- and 6-bromo, chloro, methyl, and hydroxy uracils have been determined from J values and chemical shifts of protons at C–5 and/or C–6.[288–290]

D.   Reactions of Reduced Pyrimidines

Very few reactions of hydropyrimidines are of stereochemical interest other than the degradation of N-methylbarbituric acids to ascertain the configurations of the asymmetric centers (see Section IV.B of this chapter). Kondo and Witkop[340] discovered that catalytic reduction of thymidine[341] in the presence of rhodium on alumina is stereospecific and yields ($-$)-dihydrothymidine (92). Acid hydrolysis of the nucleoside gave pure $S(-)$-5,6-dihydrothymine which ring opens to $S(+)$-β-ureidoisobutyric acid. This is related to the enantiomer of the $R(-)$-β-aminobutyric acid obtained from *in vivo* metabolism of thymidine. Reduction of thymidine with NaBH$_4$ in methanol gives a mixture of epimeric dihydrothymidine methanol adducts which yield, on mild acid hydrolysis, racemic *cis*-4-methoxy-5-methyl-2-oxohexahydropyrimidine (93). Photolytic reduction of thymidine or photolysis of $S(-)$-dihydrothymidine[342] causes cleavage of the pyrimidine ring.

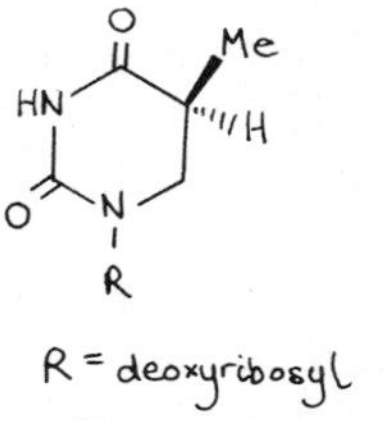

R = deoxyribosyl

92

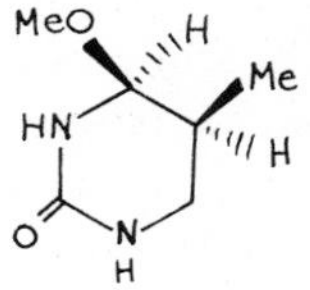

93

V.   REDUCED PYRAZINES

Many of the examples under this heading are hexahydro-derivatives, that is, piperazines. A large part of the stereochemistry of mono-, di-, tri-, and tetrasubstituted piperazines was elucidated prior to 1935. More recently the interest has been on diketopiperazines and conformational problems of the piperazine ring. The general chemistry of pyrazines was reviewed by Pratt in 1957[343] and updated by Cheeseman and Werstiuk in 1972.[344]

A.   Synthesis of Reduced Pyrazines

Reduction of polysubstituted pyrazines with sodium and alcohol or catalytically[345-348] yields a mixture of stereoisomers from which the individual bases have been separated, and in many cases characterized as cis or trans, that is, *d,l* or meso (see the next section). The ratios of products sometimes varied with the conditions of reduction, for example, the ratio of isomeric piperazines in the catalytic reduction of 2,3,5,6-tetramethyl-pyrazine varied with temperature and pressure.[347] Reduction of 2,3-dihydro derivatives of known relative stereochemistry, that is, *trans-* and

cis-2,3-dihydro-2,3,5,6-tetraphenylpyrazine from the condensation of d,l-
and meso 1,2-diaminostilbene with benzil, gave a mixture of piperazines
in which the relative configuration at one set of adjacent carbon atoms was
known.[349]

The best way of making piperazines of known configuration is by direct
synthesis rather than by reduction, as in the preparation of (+)- and (−)-2-
methyl-1,4-diphenylpiperazine by condensation of (−)- and (+)-N,N'-
diphenyl-1,2-propylenediamine and ethylene bromide (note inversion of
rotation occurs on cyclization). These were also obtained by resolution of
the d-camphor-β-sulfonate salt of the racemate.[350]

Recently Cignarella and Gallo[350a] prepared cis- and trans-4-benzyl-2,6-
dimethyl-3-oxopiperazines by the condensation of 2-amino-1-benzylamino-
propane and ethyl α-bromopropionate. These were separated and converted
into cis- and trans-2,6-dimethylpiperazines by reduction and hydrogenolysis.
The isomers were identified by their ¹H nmr spectra.

Diketopiperazines are formed when α-amino ester salts are basified or
when α-amino acids are heated. They are sometimes referred to as anhydrides. α-Amino acids, other than glycine, yield the optically active cis-3,6-
disubstituted 2,5-diketopiperazines with the same configurations at the
chiral centers which correspond to those of the α-amino acids. Dipeptides

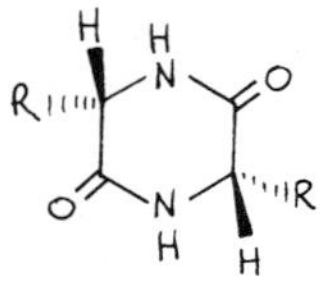

94

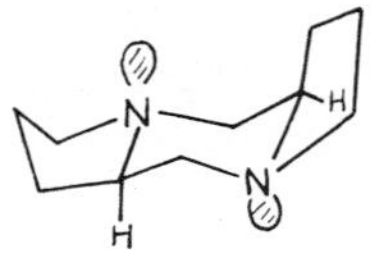

95

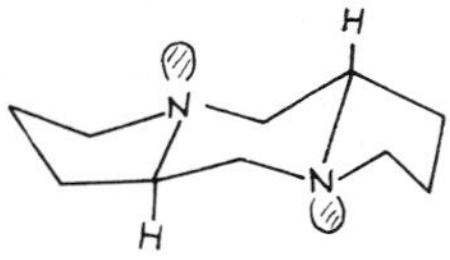

96

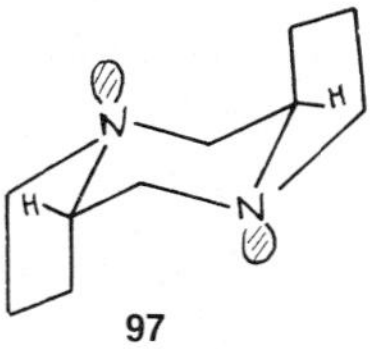

97

also cyclize to diketopiperazines, and in this way "mixed" 2,5-diketopipera-
zines can be obtained (i.e., **94** with different R groups). Fischer and co-
workers[351] clearly showed that cis-3,6-disubstituted diketopiperazines are
optically active but that the trans isomers are meso because of internal
compensation of optical rotation. They showed that D-alanyl-D-alanine
and L-alanyl-L-alanine cyclized into optically active diketopiperazines
whereas L-alanyl-D-alanine and D-alanyl-L-alanine both gave the same

optically inactive diketopiperazine. Proline and its derivatives yield tricyclic compounds. 5-Oxo-S-proline gave the optically active tricyclic diketopiperazine with acetic anhydride, and this can be reduced to the optically active 1H,5H-dipyrrolo[1,2-a:1'2'-d]piperazine (**95**). S-Proline, on the other hand, racemizes under these conditions and provides the meso mixture of **96** and **97**. The last two are only conformational isomers of the R,S compound, interconvertible by nitrogen inversion.[352] Dipeptides of glutamic acid cyclize at 150°C to diketopiperazines (e.g., **98**), and their carboxyethyl side chain can be cyclized into the ring nitrogen atom to form bicyclic

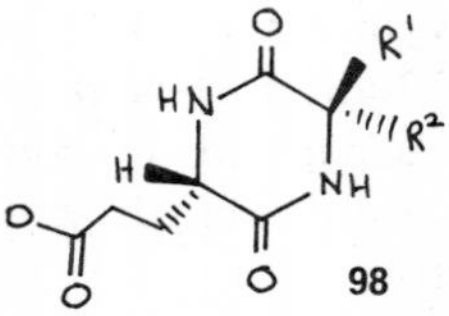

R' = H, R² = Me, CH₂Ph
and   R' = CH₂Ph , R² = H

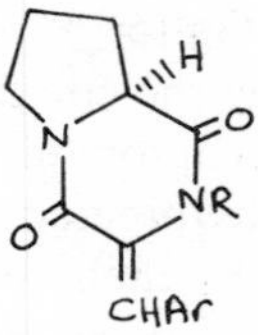

98

99
R = Ac

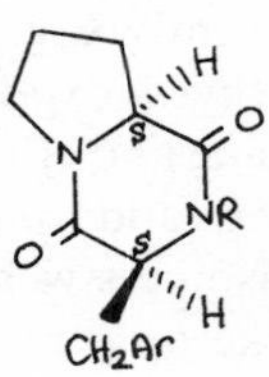

100

triones without epimerization of the asymmetric centers if trifluoroacetic anhydride, but not acetic anhydride was used.[253] Catalytic reduction of the arylidene derivatives of the diketopiperazine from glycyl-S-proline (**99**) (derived from the arylidene azlactone and S-proline) proceeds with high stereospecificity. The product is the S,S-diketopiperazine **100**, which on hydrolysis furnishes the S-arylamino acid, for example, S-phenylalanine, S-dopa, and N-methyl-S-phenylalanine. No asymmetric induction, however, occurs with alkylidene derivatives or with arylidene derivatives of the ketopiperazine from glycyl-S-leucine.[354] N-Pyruvyl-S-proline methylamide cyclizes rapidly with kinetically controlled stereospecificity to the hydroxy diketopiperazine **101** in acidic or basic aqueous solutions. This isomerizes

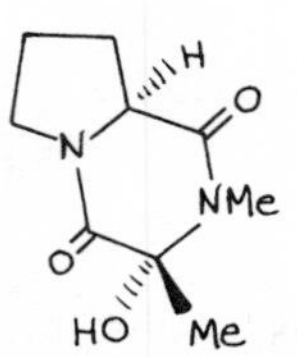

101

to the thermodynamically more stable epimer in which the stereochemistry at the hydroxyl-bearing carbon atom is inverted. The OH group can be displaced by a thiol group with optical induction.[355] For further information on diketopiperazines see Pratt[343] and Greenstein and Winitz,[356] and Section V.C.

B. Configuration of Reduced Pyrazines

The optical resolution of 2-methyl-1,4-diphenylpiperazine has already been mentioned in the previous section (V.A.). 2,3-Diethyl-[357] and diphenyl-piperazines[348] exist in cis and trans forms. These are separable and only one isomer, obviously the trans form, can be resolved into optically active enantiomers. 2,5-Dimethylpiperazine was also separated into *cis* and *trans* forms[358,359] and only one form was resolved by way of the *N*-methylene camphor derivative. The optically active form in this case is the cis isomer, the trans isomer being meso [compare with diketopiperazines in the previous section (V.A.)].[359] 2,3,5-Trimethylpiperazine should theoretically exist in four racemic forms, that is, eight stereoisomers; only two racemates have been isolated, but have not yet been resolved, into their optical antipodes.[360] 2,3,5,6-Tetramethylpiperazine is a very interesting stereochemical case and was studied in detail by Kipping.[361] It has four carbon atoms which in turn can be chiral and should have 2^4 (i.e., 16) stereoisomers. However, the symmetry of this molecule is such that these are narrowed down to the five geometrical isomers shown in the flat formulas **102–106**, and the

corresponding "chair" formulas **102a–106a**. Ring inversion of the "chair" formulas **102a–106a** will give the other "chair" conformers **102b–106b** respectively. Taking these isomers in turn, **102** is meso (plane of symmetry), **103** is racemic, **104** is meso (plane of symmetry), **105** is meso (rotating axis of symmetry, i.e., after rotating its axis through 180° it becomes identical with its mirror image), and **106** is racemic. Kipping succeeded in isolating all five tetramethylpiperazines, which he named α, β, γ, δ, and ε, and attempted their optical resolution, but he could only deduce the structure of one of these with certainty. He resolved this isomer by way of the the d-α-camphor-π-sulfonate salt which gave the laevo antipode $[\alpha]_{546}$ − 135°, and by way of the d-methylene camphor derivative he obtained the dextro antipode $[\alpha]_{546}$ + 135.5°. These could have either structure **103** or **106**. He demonstrated that it must be **103** because it gave two different 1,4 derivatives when benzenesulfonyl and toluene-p-sulfonyl substituents were introduced in turn. The racemate **106** should give only one geometrical isomer of 1(or 4)-benzenesulfonyl-4(or 1)-toluene-p-sulfonyl derivative.[361,362,363] The five isomers **102–106** were examined by ¹H nmr spectroscopy and their stereochemistry was assigned.[364] Hayashi[349] isolated five geometrical isomers of 2,3,5,6-tetraphenylpiperazine by reduction of tetraphenylpyrazine and trans-2,3-dihydro-2,3,5,6-tetraphenylpyrazine.

Hanby and Rydon[365] demonstrated geometrical isomerism in diquaternary salts of piperazine. They isolated the cis (**107**, only one conformer is drawn) and the trans isomer (**108**)of 1,4-bis [β-hydroxy(acetoxy and chloro)-ethyl]-1,4-dimethylpiperazinium salts, and concluded that the cis-diiodides of the chloroethyl salts were sensitive to light in contrast with the trans-diiodides. The structure of the trans-diiodide was confirmed by an X-ray study. A similar example is the oxidation of 1,4-diphenylpiperazine to a

107   108

1:10 mixture of cis- and trans-1,4-diphenylpiperazine 1,4-dioxides which have axial-equatorial and axial-axial oxygen atoms respectively. These formed complexes with several inorganic substances (e.g., hydrogen peroxide, perchlorates, ferricyanides) and the stereochemistry was deduced from the expected instability of the hydrogen peroxide complex.[366] The ratio of N-oxides formed is consistent with the known preferential axial N-oxidation, (compare the N-oxidation of 4-t-butyl-1-methylpiperidine)[146] (see the next section [II.C]).

An X-ray study of 2,6-bis(bromomethyl)-1,4-diphenylpiperazine revealed that although the piperazine ring is in the "chair" form, this form is slightly distorted with a C–N–C bond angle of 120°.[367] Karle[368] determined the structure of the diketopiperazine from S-prolyl-S-leucine by single crystal analysis. The diketopiperazine ring has a folded conformation with a dihedral angle of 143° between the two planar peptide units. The molecular structure of the antiviral compound bisdethio(methylthio)acetylaranotin and its absolute configuration was determined by X-ray analyses.[369] Ord

109

and cd spectral and X-ray studies of related compounds have also been made.[370]

C. Conformation of Piperazines

Piperazine is a conformationally flexible molecule in which ring inversion and pyramidal atomic inversion at each of the nitrogen atoms occurs.[113] Most of the physical evidence points out to a preference for a chair form, or almost chair form if substituted, for piperazine. Early measurements of electric moments of N,N'-dichloro-, N,N'-dichloro-2,6-dimethyl-, N,N'-dimethyl-, 2,5-dimethyl-N,N'-diformyl-, N,N'-biscyanomethyl,-[371] N,N'-dinitroso-, and N,N'-dinitropiperazines[372] support a flexible chair conformation rather than a crown or boat conformation. For a particular chair conformation there are three possibile invertomers, **110–112**, but the last conformer in which the lone pair of electrons on the nitrogen atom are

110   **111**   **112**

both equatorial is generally excluded, or of such a low population as to be neglected, because of the known preference of substituents for the equatorial orientation. Thus conformer **112** would readily undergo ring inversion to the preferred conformer **110**. This is also justified by the evidence from physical data (see below).

Polarizability studies by Aroney and Le Fèvre[152] supported the chair conformation for 1-phenylpiperazine with the phenyl group equatorial, but with piperazine and its 1,4-dinitroso and 1,4-dimethyl derivatives both chair and boat forms are present with the N–H and N–Me bonds sometimes axial and sometimes equatorial—unlike the N-phenyl groups which are always equatorial. Allinger and co-workers concluded,[117] from dipole moment measurements, that the methyl groups in N,N'-dimethylpiperazine are equatorial to the extent of about 95%. From the N–H frequency of the first overtone in the ir spectrum and from dipole moment measurements, Katritzky and co-workers[331] showed that the N–H bonds in piperazine prefer the equatorial orientation to about the same extent as in piperidine. The N–H bond in 1-methyl- and 1-*t*-butylpiperazine also shows a preference for the equatorial conformation resembling the piperidine analogues. Improved and accurate measurements of the dipole moments of 1-alkyl-4-*t*-butylpiperazines concede that the conformers with the *N*-alkyl group equatorial and lone pair axial predominate.[373] Horikoshi and collaborators[373a] determined the ratio of conformers in *N,N'*-dimethylpiperazine by dielectromeric titration in dioxane at 30°C. The ratio of 0.880:0.116:0.004 for **110** (R = Me), **111** (R = Me), and **112** (R = Me) which they obtained is consistent with the above and with data by Allinger and co-workers.[373b]

^{1}H nmr spectra at low temperature of piperazines reveal that the free energy for ring inversion is comparable with that of other hexahydroazanes, that is, $\triangle G^{\ddagger} = 37–53$ kJ mol^{-1} (cyclohexane $\triangle G^{\ddagger} = 42$ kJ mol^{-1}).[323,374,375] ^{1}H nmr measurements of *cis*-1-phenyl-3,5-dimethylpiperazine indicate that the predominant conformer in which all substituents are equatorial (**113**) is strongly favored. Unlike the trans isomer, the spectrum of the cis isomer is unaltered on cooling.[376]

113

A study to determine the requirements for obtaining magnetically nonequivalent benzylic CH_2 protons in *N*-benzylpiperazines was made by Lyle and Thomas.[377] *N*-Benzylic CH_2 protons are nonequivalent if the conformer is chiral and if the benzyl group has a vicinal equatorial alkyl substituent, for example *cis*-1-benzyl-2,3- or -2,5-dimethylpiperazines, but not in *cis*-1-benzyl- and 1,4-dibenzyl-3,5-dimethylpiperazines.

Mono- and diprotonation of 1,4-dialkylpiperazines were studied by ^{1}H nmr spectroscopy in water and deuterium oxide. The effects of pH and pKa values were discussed in terms of conformations and conformational changes of mono- and dications.[378–381] The interconversion of *cis*- and *trans*-1,4-dimethylpiperazine dihydrochloride was examined and the free energy changes support the contention that the effective "size" of the N–H protons increases with respect to uncharged NH protons due to increased hydration around the charge.[382]

Thermodynamic and geometric parameters for the most probable conformations of 3,6-disubstituted 2,5-dioxopiperazines were evaluated.[383]

D.   Reactions of Reduced Pyrazines

5,6-Diphenyl-2,3-dihydropyrazine undergoes a Diels-Alder reaction with diethyl fumarate to form a 1:1 adduct. This is not the expected 1,4-adduct; it turned out to be a 2,5-adduct, 7-*endo*-ethoxycrabonyl-8-*exo*-ethoxy-carbonyl-1,6-diphenyl-2,5-diazabicyclo[2.2.2]oct-5-ene[384,385]   **(114).**   *N*-

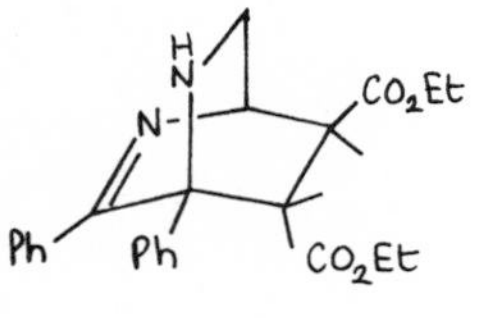

114 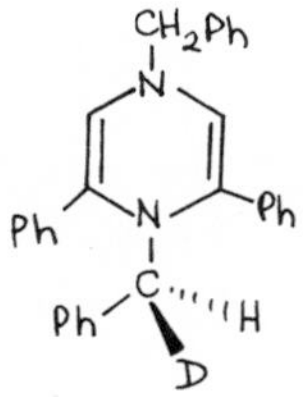   **115** 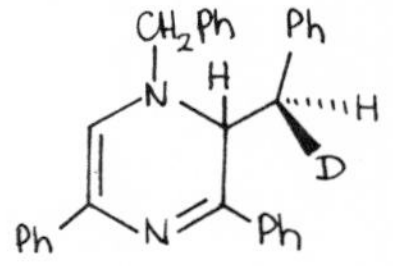   **116**

Methyl maleimide adds in a similar fashion to give the endo adduct.[385] The structures of the products suggest that the overall nonpolar 1,3-double bonds might have migrated to 2,4- double bonds, which would be polar enough for satisfactory addition. The alternative explanation, a re-arrangement of 1,4- to a 2,5-adduct, is less likely.

1,4-Dibenzyl-1,4-dihydro-2,5-diphenylpyrazine, prepared from *N*-benzyl-diphenacylamine hydrobromide and benzylamine, rearranges thermally by a [1,3]-sigmatropic shift of the 1-benzyl group to 1,2-dibenzyl-1,2-dihydro-2,6-diphenylpyrazine in high yield.[386] The rearrangement occurs with very little cross over (i.e., intramolecular) and is highly stereospecific. By using the chiral monodeuterobenzyl derivative **115** it was shown that the rearrangement gave the 1,2-dihydropyrazine **116**, demonstrating that inversion occurred at the benzylic carbon atom.[386,387] Another rearrangement ob-served in the dihydropyrazines is the photochemical conversion of 2,3-dihydro-2,3,5,5,6,6-hexamethylpyrazine 1,4-dioxide into a mixture of the *trans*- (**117**, major) and *cis*-dioxaziridines (**118**, minor). This conversion takes place when the wavelength of light is below 300 nm. Only the mono-oxaziridine is formed by irradiation at longer wave lengths.[388]

Steric inhibition to quaternization was noted in 1,2,2,4,5,5-hexamethylpi-perazine with ethylene, trimethylene, and *p*-xylylene dibromides. Whereas

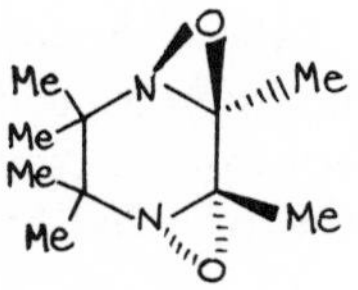

117 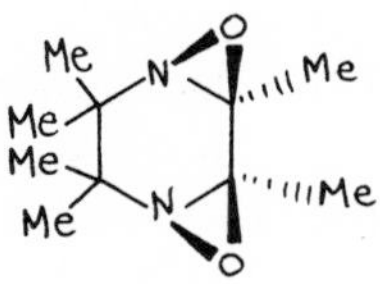   **118** 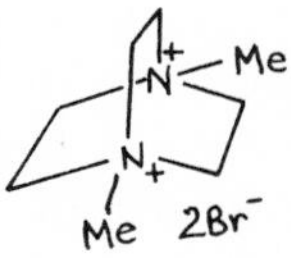   **119**

1,4-dimethylpiperazine and ethylene dibromide yield 1,4-dimethyl-1,4-diazoniabicyclo[2.2.2]octane dibromide (119), the hexamethylpiperazine and ethylene, trimethylene, or *p*-xylylene dibromide only give its dihydrobromide (and presumably the corresponding olefin), but only under the influence of uv light does the latter reaction with ethylene dibromide furnish the quaternary salt, ethylene bis-(1,2,2,4,5,5-hexamethyl-1-piperazinium)-dibromide (120).[389]

120

The stereochemistry of 2,5-dioxopiperazines (diketopiperazines, cyclo-dipeptides) derived from α-amino acids is still attracting attention. Brockmann and Musso[390] made *cis*-(*d,l*)- and *trans*(meso)-3,6-bishydroxymethyl-2,5-dioxopiperazine (serine anhydride) among others, and Lassen and Stammer[391] studied their hydrolysis at various concentrations of hydrochloric acids and compared these with that of 2,5-dioxopiperazine. The cis-trans isomerization of several cyclodipeptides, in ethanolic sodium ethoxide at 30–75°C and in aqueous solutions at 250°C, were examined in detail by Eguchi and Kakuta,[392] and they obtained the thermodynamic constants for the equilibria. Their compounds could be divided into three groups: those in which cis $\simeq$ trans ratios (i.e., cyclo-ala.ala., cyclo-leu.leu, and cyclo-phe.phe); those with trans > cis ratios (i.e., cyclo-pro.ala, cyclo-pro.leu, cyclo-pro.phe, and cyclo-pro.val); and in cyclo-pro.pro and cyclo-hyp.hyp (hyp = 4-hydroxy-*S*-proline) only the cis isomers were found at equilibrium (i.e., cis ≫ trans). The last example was looked at more carefully and cyclo-*S*-hyp-*S*-hyp, cyclo-*S*-hyp.*R*-α-hyp, and cylo-*R*-α-hyp.*R*-α-hyp (α-hyp is *allo*-4-hydroxyproline) were examined. It was found that the equilibria largely favored cyclo-*S*-hyp.*S*-hyp and provided a very good method for converting *R-allo*-4-hydroxyproline into *S*-4-hydroxyproline.[393] Similar results were obtained by Vicar and Bláha.[394] Schmidt and co-workers[395] used the ability of some of these dioxopiperazines to ionize in the presence of a base and to reprotonate stereospecifically, to synthetic use. Noting that cyclo-*S*-pro.*S*-pro does not lose its configuration in the process, they treated it with a base followed by ethyl bromide or ethylthiol and isolated the ethyl (121, R = Et) and ethylthio derivatives (121, R = SEt) respectively with retention of configuration. Repetition of this sequence in order to prepare the disubstituted derivatives, however,

121   122   123

gave disubstituted racemic products. Oxidation of the ethylthio compound
121 (R = SEt) with perbenzoic acid gave the corresponding sulfone **121**
(R = SO$_2$Et), but oxidation with hydrogen peroxide in acetic acid caused
displacement of the SEt group and the hydroperoxide **121** (R = −OOH)
was formed. The latter was reduced to the alcohol **121** (R = OH) which
reacted with ethylthiol in the presence of ZnCl$_2$ to form the original starting
material (**121**, R = SEt). All the reactions proceeded with retention of
configurations, but attempts to involve the second asymmetric carbon atom
always led to the thermodynamically more stable R,S or a mixture of S,S
and R,R isomers because the products were optically inactive.[396] An
exception to this is the formation of a di- or polysulfide bridge between
the two chiral centers which must necessarily be cis (i.e., **122**), and the
S configuration is retained. These compounds (**122**) are called epidi (or
poly) thio-S-prolyl-S-proline anhydrides and are found in some naturally
occurring antiviral antibiotics (e.g., aranotin, Chapter 3, Section V.B, and
formula   **109**).[397,398] $3S,6S$-3-Benzyl-6-hydroxymethyl-1,4-dimethyl-3,6-
epitetrathio-2,5-dioxopiperazine was isolated from cultures of a fungitoxic
Hyalodendron species.[398a]

The decarboxylation of *cis*- or *trans*-3-carboxy-3-(3-indolyl)methyl-1,6-
dimethyl-2,5-dioxopiperazines (**123**, R^1 = R^2 = R^3 = H, R^4 = CO$_2$H; R^1 = Cl,
R^2 = R^3 = H, R^4 = CO$_2$H; R^1 = H, R^2 = R^3 = OMe, and R^4 = CO$_2$H) is
highly stereospecific in that both isomers produce the same 6:1 ratio of
cis to trans decarboxylated derivatives (**123**, R^4 = H). This may be due to
stereospecific ketonization of the enol formed after decarboxylation, and is
affected by the N-methyl group. If this group is absent the cis to trans ratio
alters to 3:2.[399]

VI.   1,3,5-TRIAZANES

Stereochemical interest in 1,3,5-triazanes has been mainly in the conformational properties of the ring and of substituents on the nitrogen atoms.
The ring is flexible and the free energy of inversion of 1,3,5-trimethyl,
triethyl, triisopropyl, and tri-*t*-butyl ($\triangle$G‡ in cyclohexane: 49.8, 47.7, 44.8,
and 41.8 kJ mol^{-1} respectively)[400] are of the same order as those of 1,3-

diazanes, 1,4-diazanes (piperazines), morpholines, and related systems.[322,323,401]

Carter, McIvor, and Miller[402] examined the condensation product of acetaldehyde and methylamine and found that the ^{1}H nmr is complex, but may be explained by an equilibrium between a monomeric azomethine and two 1,2,3,4,5,6-hexamethyl-1,3,5-triazanes—probably **124** and **125**. Katritzky and co-workers[333] showed that J_{gem} values and chemical shifts

124   **125**   **126**

of the methylene protons between two nitrogen atoms cannot be used to give reliable information regarding the orientation of the nitrogen lone pair in diaza- and triazaazanes, and thus ^{1}H nmr could not give information about the conformations from these compounds. They measured the electric dipole moments of 1,3,5-trialkyl-1,3,5-triazanes and interpreted their results in terms of equilibrating conformations. The tri-t-butyl compound was considered as an equilibrium between the triequatorial conformer and the diequatorial-mono-axial conformer **126** which consisted of 85% of the latter. This is the first example of a freely inverting N-t-butyl group which prefers an axial orientation.[403] By using the dipole moments of mixed 1,3,5-trialkyl-1,3,5-triazanes, Katritzky and co-workers[404] calculated the conformational free energies for axial N-alkyl groups in this system. The low values obtained for this system compared with the N-alkyl-1,3-diazanes and piperidines were explained by decreased syn-1,3 interactions of the axial N-alkyl group in the axial conformation and the unfavorable interactions of the N-alkyl groups in the triequatorial conformers. The dipole moments of 1,3,5-trinitroso- and 1,3-dinitro-5-nitroso-1,3-5-triazanes were found to be lower than those calculated for "chair" conformations with all the N-substituents equatorial.[335]

Bushweller, Lourandos, and Brunelle[405] studied the ^{1}H nmr spectra of 1,3,5-trimethyl-1,3,5-triazane and found one coalescence temperature at −59°C due to ring inversion but rapid nitrogen inversion, and a second one at −100°C due to slow nitrogen inversion. The spectrum at −144°C

127

was explained by the presence of the diequatorial-monoaxial conformer **127** almost exclusively. This is the first unambiguous measurement of the barrier for nitrogen inversion in a six-membered ring where nitrogen inversion is free from all rate-retarding factors, other than the energy required to rehybridize the nitrogen atom to a polar transition state and the strain associated with the six-membered ring. This free energy is 30.1 kJ mol^{-1} (7.2 ± 0.1 kcal mol^{-1}) at $-122.5°C$.

An X-ray crystallographic study of acetaldehyde ammonia trihydrate by Lund[406] revealed that it has 2,4,6-trimethyl-1,3,5-triazine rings stacked between water molecules. These molecules are arranged with their oxygens forming hydrogen bonded six-membered rings (with respect to the oxygen atoms only) in a chair conformation packed, and hydrogen bonded, with the triazane in a chair conformation having all the methyl groups equatorially disposed. All that can be said about the orientation of the nitrogen lone pairs is that they are axial and involved in hydrogen bonding. The crystals can be dehydrated without destroying the triazane ring.

The piperidine trimer from the trimerization of 3,4,5,6-tetrahydropyridine[407] or from the reaction of N-chloropiperidine with 3,4,5,6-tetrahydropiperidine[408] is a 1,3,5-triazane derivative. These were prepared by Schöpf and co-workers who showed the existence of α and β forms which were in equilibrium with an open form. The α form probably has the all-cis configuration **128** and the β form has the cis-cis-trans configuration **129**. At pH 9–10 the α form isomerizes to isotripiperidine **130**.[409,410] Armarego and Sharma[411] obtained 36% yield of a 1-oxoisoindole trimer by boiling 2,3-

128   **129**

130

dihydro-3-hydroxyisoindol-1-one in thionyl chloride. Its physical data supported the structure: *cis, cis, trans*-triazino[2,1-*a*;4,3-*a'*;6,5-*a'*]triisoindole-6(10*b*H),12(16*b*H), 18(6*b*H)-trione (**131**).

131

VII.   1,2,4,5-TETRAAZANES

1,2,4,5-Tetraazane is the last of the monocyclic six-membered nitrogen heterocycles to be discussed, and as the other members of the system, it is conformationally mobile but has only two carbon atoms that can bear substituents. Hence 3,6-disubstituted 1,2,4,5-tetraazanes should exist as cis and trans geometrical isomers which are symmetrical but different. Only a small number of these were prepared by Kauffmann and co-workers[412] from aliphatic aldehydes by decomposition of their styrene-deviatives with sodium hydrazide, or from aldehyde and hydrazine at 5°C. These are crystalline compounds with fairly sharp melting points and are probably the trans isomers. They behave very much like dimers by dissociating into monomers on heating above their melting points or when in aqueous solution.

N-Methyl and ethyl 1,2,4,5-tetraazanes are stable enough for physical measurements, and Anderson and Roberts[413] first examined the ^{1}H nmr spectra of the 1,2,4,5-tetramethyl derivative. At $-87°C$ the methylene groups appeared as AB quartets and the free energy of activation for ring inversion (48.9 kJ mol^{-1}) was of the same order as in other azanes (compare with Chapter 3, Sections III. B., IV.C, and V.C). They interpreted their results, after excluding the 1,3-diaxial isomer **132** (R = Me), in terms of equilibrating conformers with the 1- and 4-methyl groups axial and the 3- and 6-methyl groups equatorial, that is, **133** (R = Me) and **134** (R = Me). Other authors[414,415] also discarded conformer **132** and concluded that the tetramethyl compound is predominantly in the conformation with two axial and two equatorial methyl groups, but they could not agree between conformer **133** or **134**. Detailed analyses of the variable temperature ^{1}H

132   **133**   **134**   **135**   **136**

nmr, ir spectroscopic, and dipole moment data were made, by Katritzky and co-workers,[416] of the tetramethyl and tetraethyl derivatives together with similar but condensed systems, and they considered all the conformers **132–136**. They concluded that conformer **132** can be excluded, that the methyl derivative is a rapidly equilibrating mixture of the monoaxial **136** (R = Me, 70%) and the noncentrosymmetric (**133**, R = Me, 30%) conformer, and the tetraethyl derivative is a mixture of conformers **136** (R = Et; 33%), **133** (R = Et; 2%), and **134** (R = Et; 65%).

137   **138**

X-Ray, ¹H nmr, and dipole moment data of 1,4-dimethyl-1,2,4,5-tetraazane were explained by Ansell, Erickson, and Moore[417] in terms of the equilibrium **137** ⇌ **138** in which there was slow ring inversion but rapid nitrogen inversion. This is consistent with previous results and the inversion barrier is comparable with the one mentioned above. More recent X-ray studies support the centrosymmetric structure **137** in the crystals.[417a] Hammerum[418] gave evidence that 2,5-diacyl(or thioacyl) groups in 1,4-dialkyl-1,2,4,5-tetraazanes increase the nitrogen inversion barrier, that is, increase the coalescence temperature. N-acyl groups generally lower N-inversion barriers but in this case the acyl group causes hindered inversion in the adjacent N-alkyl nitrogen atom by obstruction with the alkyl group and by its electronegative properties.

Photoelectron spectroscopy can give empirical estimates of the dihedral angles between lone pairs. An estimate for 1,2,4,5-tetraazane suggests that it is in the noncentrosymmetric conformer **133** (R = H).[256]

Reduced 3,6-bridged 1,2,4,5-tetraazanes have been prepared by Diels-Alder reactions. The addition of excess of azodicarboxylic esters to α-pyrone yields 5,6,7,8-tetra(alkyoxycarbonyl)-5,6,7,8-tetraazabicyclo[2.2.2]-oct-2-ene (**139**). Attempts to make tetraazabarallene from this adduct failed.[238] This is not surprising because adducts such as **140**, obtained from 4,4-dimethyl-4*H*-pyrazoles and 4-phenyl-1,2,4-triazolin-3,5-dione, readily lose a nitrogen molecule.[419,420]

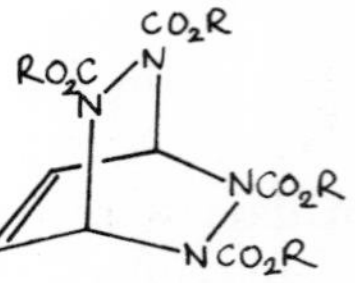

139

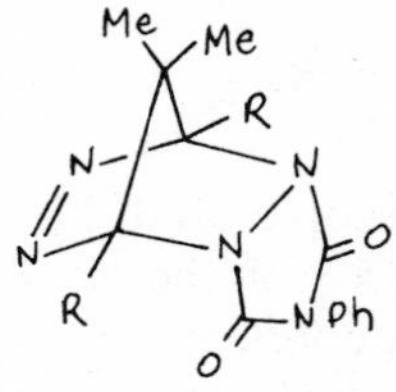

140

VIII. REDUCED FUSED FIVE:SIX-MEMBERED RINGS

Most of the stereochemical interest in the reduced fused five:six-membered rings with nitrogen atoms in the six-membered ring is in the cis-trans relationship at bridgehead carbon atoms. Reduced 1- and 2-pyrindines are discussed in this section, together with reduced purines which are best included under this heading.

A. Perhydro-1-pyrindines

Prelog and Szpilfogel[421] first obtained perhydro-1-pyrindine by sodium and alcohol reduction of 6,7-dihydro-5*H*-1-pyrindine (2,3-trimethylene-pyridine) and tentatively assigned the trans structure **141**. The cis isomer **142** was later prepared by catalytic reduction of 2-cyanoethylcyclopenta-none which must involve the reduction of the nitrile to an amine followed by cyclization, dehydration, and further reduction.[422] 3- and 4-Methyl

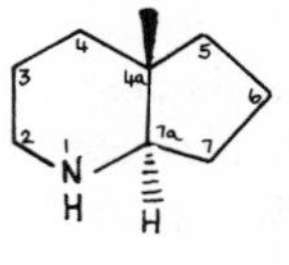

141

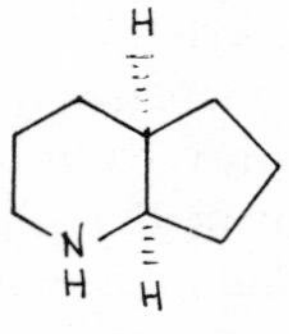

142

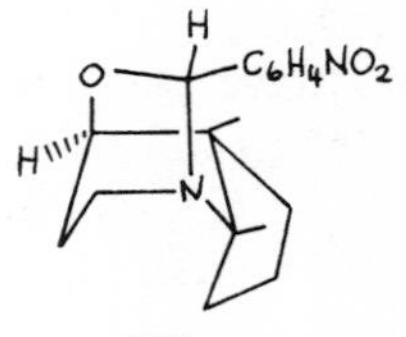

143

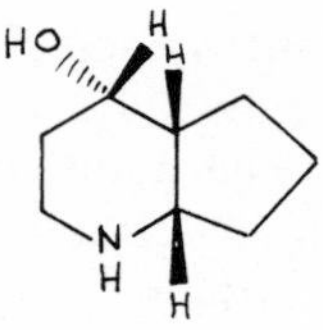

144

derivatives were also prepared by this method and presumably the methyl groups are trans to the bridgehead hydrogen atoms. 4a-Ethoxycarbonyl-[423] and 4a,6-dimethylperhydro-1-pyrindines[424] were similarly formed using a nickel catalyst, but these may well be mixtures of cis and trans isomers. Both isomers **141** and **142** were obtained by separate reductions of 3,4,4a,5, 6,7-hexahydro-2H-1-pyrindine which must be an intermediate in the reduction previously described.[425] Sodium and alcohol reduction gave the trans isomer **141**, and catalytic reduction with Adam catalyst in acetic acid gave the cis isomer **142**. Mistryukov[426] proved the cis configuration of the above by preparing perhydro-cis-4-oxo-1-pyrindine,[427] and removing the oxygen by N-benzoylating, making the 4,4-dithioketal derivative with ethane-1,2-dithiol, and desulfurizing with Raney nickel followed by hydrolysis to authentic **142**. He reduced the 4-oxo group, with sodium borohydride, to an 8:1 mixture of perhydro-4-hydroxy-1-pyrindines. The predominant isomer only gives a 1,3-oxazine (**143**) readily by reaction with p-nitrobenzaldehyde, thus establishing the configuration of the alcohol as **144**. The cis base **142** was also obtained by catalytic reduction of 1-benzoyl-2,3,4,5,6,7-hexahydro-1-pyrindine followed by hydrolysis.

A fungal alkaloid isolated from *Rhizoctonia leguminicola* was shown to be a *trans*-trihydroxyperhydro-1-pyrindine with the relative configuration depicted in **145**. The configuration at C–5 is not yet known.[428]

145

B. Perhydro-2-pyrindines

Prelog and Metzler[429] described the preparation of *cis*- (**146**) and *trans*-perhydro-2-pyrindine (**147**) from 3,4-trimethylenepyridine by catalytic reduction (PtO$_2$/AcOH) and by reduction with sodium and alcohol respectively. Catalytic reduction (PtO$_2$/AcOH) of 2,6-dichloro-3-ethyl-4,5-trimethylenepyridine[430] and 1,3,3-trimethyl-3,4,5,6-tetrahydro-5H-2-pyrin-

146   **147**   **148**

dine[431] gave presumably the *cis,cis*-4-ethyl- and mainly the *cis, cis*-1,3,3-trimethylperhydro-2-pyrindines respectvely. Ayerst and Schofield[432] obtained authentic *cis*-perhydro-2-pyrindine from LAH reduction of *cis*-1,3-dioxoperhydro-2-pyrindine which was in turn prepared from the known *cis*-2-carboxycyclopentylacetic acid. Authentic *trans*-perhydro-2-pyrindine was formed in low yield when the mixture of amidic acids from the known *trans*-2-carboxycyclopentylacetic acid was reduced with LAH. This, however, proved to be different from the product described by Prelog and Metzler,[429] which was most probably a mixture of monounsaturated hydro-2-pyrindines. These results were confirmed by Jewers and McKenna.[433]

5-Chloro-1-azatwistane (Section X.C) hydrolyzes to a mixture of *endo*- and *exo*-6-formyl-*cis*-perhydro-2-pyrindines by ring cleavage.[434] These were synthesized by an application of the ingeneous and novel oxidation, developed by Taylor, McKillop, and co-workers,[435] of cyclohexene to formylcyclopentane by thallium trinitrate. *N*-Acetyl-*cis*-1,2,3,4,4a,5,8,8a-octahydroisoquinoline gives a 9:1 mixture of *endo*- and *exo*-6-formyl *cis*-perhydro-2-pyrinidines by this method.[434]

Reduction of (2'-methyl-5',6'-dihydro-3',4'-pyrido)ferrocene and its 1'-methyl derivative with $NaBH_4$ is stereospecific and furnishes the 1',2,'5',6'-tetrahydro-2'-*endo*-methyl derivatives **148** (R = H) and **148** (R = Me). There is obviously considerable steric interference at the side next to the iron atom, and the 2'-*exo*-methyl derivative related to **148** (R = Me) is formed when (5',6'-dihydro-3',4'-pyrido)ferrocine methiodide is treated with methyl magnesium iodide. Like the hydride ion in the reduction, the methyl group also attacks from the side opposite to the iron atom.[436]

cis-2-Methylperhydro-2-pyrindine is readily obtained by direct methylation but on oxidation it yields two *N*-oxides. These have not been identified but must be diastereomeric having the *N*-oxygen atom, with respect to the piperidine ring, axial in one isomer and equatorial in the other.[432]

N-Methyl-*cis*-perhydro-2-pyrindine forms a methiodide which undergoes a β-Hofmann fragmentation giving *N,N*-dimethyl-*cis*-2-vinylcyclopentylmethylamine (**151**). The anti stereochemistry of the fragmentation is consistent with the cleavage of the more probable conformation **149** rather

| 149 | 150 | 151 |

than **150** in order to produce the olefin **151**[432,433] (see Chapter 2, Section III.E.4). The four isomers, α-, β-, γ-, and δ-skytanthine (**152–155**), at least three of which occur naturally in *skythanthus* oil, were prepared from the corresponding nepetalinic acids, and their methiodides were subjected to the Hofmann elimination reaction. The olefinic amines produced in each

152	153	154	155

case were related to the stereochemistry of the respective quaternary base and to an anti elimination mechanism as in **149**.[437] Studies of related naturally occurring 2-pyrindines have been made,[438] and an X-ray crystallographic analysis of 5-hydroxyskytanthine (as its methiodide) and tecomanine (as its methoperchlorate), both alkaloids from *Tecoma stans*, revealed the absolute configurations **156** and **157** respectively.[439]

156	157

C.   Reduced Purines

The stereochemistry of reduced purines becomes interesting when the 4,5-bond is reduced. Very few examples of these have been prepared[440] but their configurations were not known.[441] Tafel reduced uric acid in sulfuric acid electrolytically. At low current densities he obtained 5,6-dihydro-4-ureidouracil because the initially formed 1,6-dihydro compound was rapidly hydrolyzed in the acid medium. At high current densities, on the other hand, "purone," perhydro-2,8-dioxopurine (**158**, $R^1 = R^2 = R^3 = R^4 = H$) was also formed.[442] *N*-Methylated uric acids behaved similarly, that is, formed the respective *N*-methylperhydro-2,8-dioxopurines.[443] The

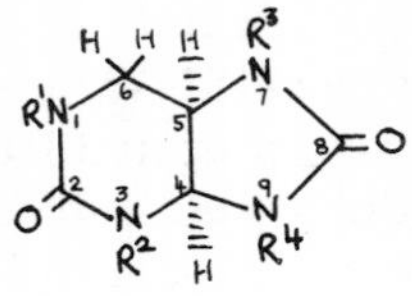

158 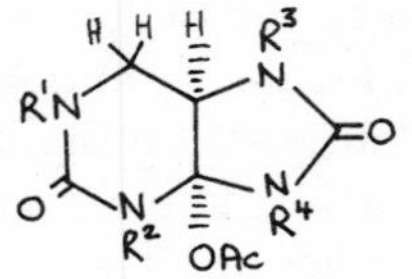   159 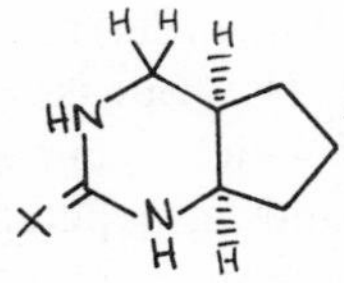   160

stereochemistry at C–4 and C–5 in these compounds is most probably cis, but this has not been proved unequivocably (see below).

Armarego and Reece[444,445] reduced 2,8-dioxopurine catalytically (PtO₂/HCl) and, as in the electrolytic reduction, obtained 5,6-dihydro-5-ureido-uracil. The ureido compound was cyclized by boiling in acetic anhydride to cis-5-acetoxyperhydro-2,8-dioxopurine (159, $R^1 = R^2 = R^3 = R^4 = H$). The cis configuration of the purine was deduced from the analogy of the ¹H nmr pattern of signals for H–5 and H–6,6 with the corresponding protons of cis-3,4,5,6-tetrahydro-5,6-trimethylene-2-oxopyrimidine (160, $X = O$) and its 2-amino derivative (160, $X = NH$). The magnitudes of the coupling constants are also consistent with the cis configuration. They repeated Tafel's electrolytic reduction of uric acid and of purin-2,8(1H,7H)-dione at high current density and isolated purone (158, $R^1 = R^2 = R^3 = R^4 = H$); they showed from the signal patterns of the ¹H nmr spectra that the stereochemistry at the bridgehead hydrogen atoms was cis. 1,3,7,9-Tetramethylpurin-2,8(1H,7H)-dione iodide; 1,3,9-trimethylpurin-2,8(1H,7H)-dione iodide; and 1,3,7-trimethyl- and 1,7,9-trimethylpurin-2,8(1H,7H)-dione are reduced by NaBH₄ at different rates, but all give cis-perhydropurines, that is, 1,3,7,9-tetramethyl-(158, $R^1 = R^2 = R^3 = R^4 = Me$); 1,3,9-(158, $R^1 = R^2 = R^4 = Me$, $R^3 = H$); 1,3,7-(158, $R^1 = R^2 = R^3 = Me$, $R^4 = H$); and 1,7,9-trimethyl-cis-perhydro-2,8-dioxopurines (158, $R^1 = R^3 = R^4 = Me$, $R^2 = H$). 2,8-Diamino-1,7,9-trimethylpurine diiodide is similarly reduced by NaBH₄ to cis-2,8-diamino-1,7,9-trimethyl-1,4,5,6-tetrahydropurine (161). The unmethylated analogue, cis-2,8-diamino-1,4,5,6-tetrahydropurine, is

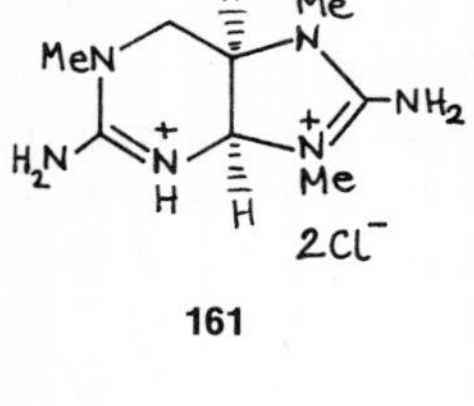

161 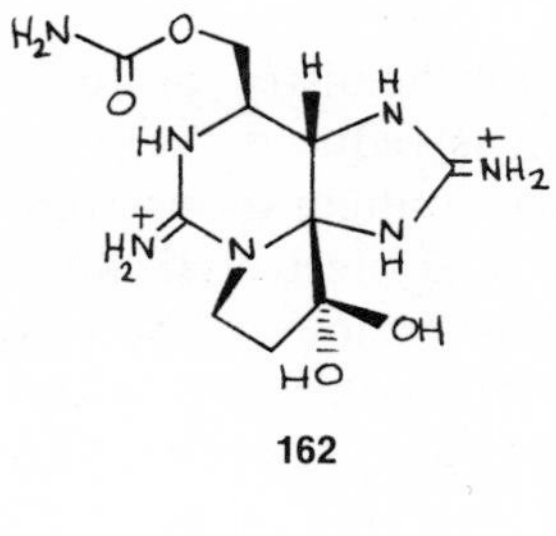   162

formed in the electrolytic reduction of 2,8-diaminopurine at high current densities.[445]

Rapoport and co-workers[446] studied the chemistry of saxitoxin, a neurotoxin from the Alaskan butter clam and from cultures of *Gonyaulax catanella*, and deduced a reduced 2,8-diaminopurine structure for it. It defied X-ray crystallographic analysis because it had not given a satisfactory crystalline derivative. Schantz and collaborators[447] also examined the structure intensively and finally succeeded in obtaining a crystalline *p*-bromobenzenesulfonate. Single crystal X-ray analysis gave the structure and absolute configuration (162) for saxitoxin, which differs from that of Rapoport only in a few details.

IX. REDUCED QUINOLINES

The stereochemical aspects of 1,2,3,4-tetrahydroquinolines and of decahydro(perhydro)quinolines are discussed in this order. Very few other reduced quinolines with three-dimensional properties of interest are known, and they have been included in Sections IX.A.2 and A.3. The stereochemistry of unsaturated chiral quinolines is described in Sections XIII.D and XIV.C of this chapter.

A.   1,2,3,4-Tetrahydroquinolines (Including a Few Dihydroquinolines)

1. *Syntheses of 1,2,3,4-Tetrahydroquinolines*

Thomas[448] reduced 2,4-dimethylquinoline with sodium and alcohol to an oily mixture of *cis*- and *trans*-1,2,3,4-tetrahydro-2,4-dimethylquinolines. By careful recrystallization of the α-bromocamphor-π-sulfonate salts all four optical antipodes were isolated. While confirming these results, Plant and Rosser[449] also reduced 3,4-dimethylquinoline by a variety of methods but obtained only one 1,2,3,4-tetrahydro-3,4-dimethylquinoline. This is probably the trans isomer. Reduction of 3,4-dimethyl-2-quinolone with sodium amalgam and alcohol produced a 6:1 mixture of two 1,2,3,4-tetrahydro-3,4-dimethyl-2-oxoquinolines. The major component is most probably the trans isomer. Further reduction of the mixture provided the same (*trans*)-tetrahydro-3,4-dimethylquinoline as above. Witkop and co-workers[450] reduced kynureniç acid, 2-carboxy-4-quinolone, photochemically or with

$NaBH_4$ to d,l-2-carboxy-1,2,3,4-tetrahydro-4-oxoquinoline, kynurenine yellow. Its N-acetyl methyl ester was stereospecifically reduced to cis-1-acetyl-2-carboxy-1,2,3,4-tetrahydro-4-hydroxyquinoline because treatment with acetic anhydride or dicyclohexyl carbodiimide furnished the lactone (163).

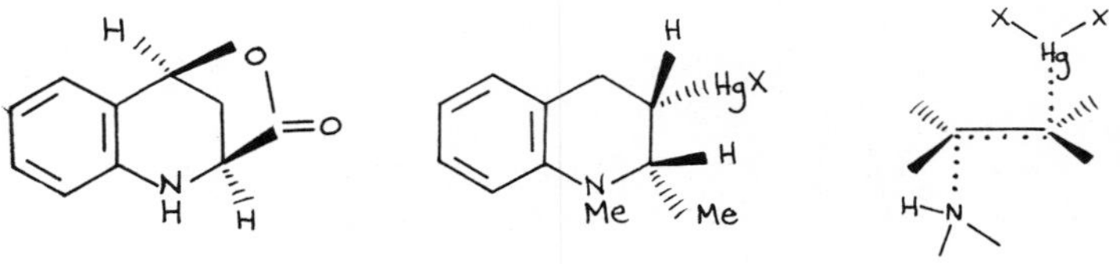

163   164   165

The intramolecular aminomercuration of olefinic amines has been used to prepare tetrahydroquinolines, for example, the cyclization of $trans$-o-but-2-enyl-N-methylaniline to the quinoline **164**. The stereochemistry of the aminomercuration reaction is trans (i.e., **165**) and the mercury compound **164** or similarly cyclized compounds could be helpfully used for preparing derivatives of known stereochemistry.[451] Some N-benzoyl-N-acrylamides of o-substituted anilines undergo intramolecular cyclization, on irradiation, with a 1,5-migration of the ortho substituent (Equation 14). Migration takes place when R^2 is CO_2Me or $CONH_2$, but when R^2 is

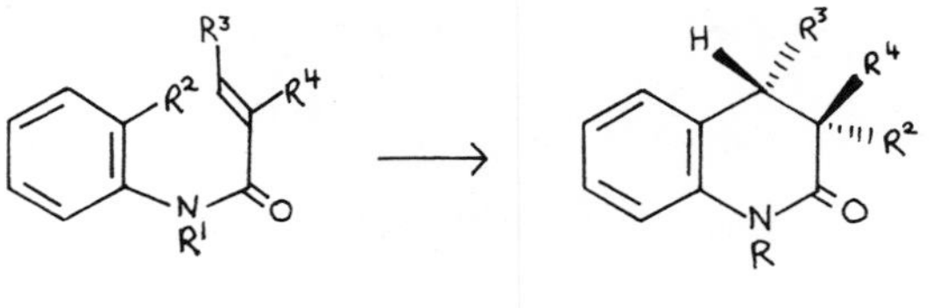

(14)

CO_2H decarboxylation occurs, and when $R^2 = OMe$ cyclization takes place at the unsubstituted ortho position with reduction of the 2-oxo function.[452] Another intramolecular cyclization which is brought about by polyphosphoric acid is the formation of 1-(1,2,3,4-tetrahydro-2-oxoquinol-4-yl)-1,2,3,4-tetrahydroisoquinoline (m.p. 210–211°C) from N-phenethylindol-3-ylacetamide. This reduced quinolylisoquinoline together with its diastereoisomer (m.p. 184–185°C) are obtained by catalytic reduction of 1-(quinol-2-on-4-yl)-3,4-dihydroisoquinoline, but the stereochemistry of these two isomers has not been established.[453]

In a novel quinoline synthesis *N*-methyl anilines react with *N*-methyl- and *N*-phenylmaleimides in the presence of benzoylperoxide by a [2 + 2]π cycloaddition to give the *cis*-3,4-dicarboximide derivatives of 1,2,3,4-tetra-hydroquinoline (**166**) in fair yields. It is obviously a free radical reaction

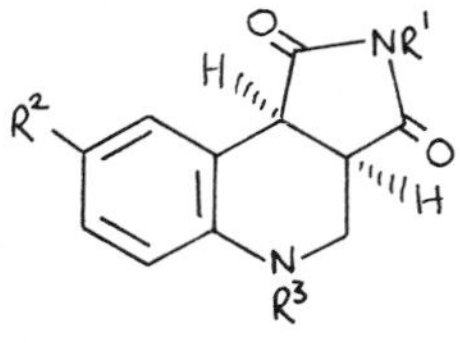

166

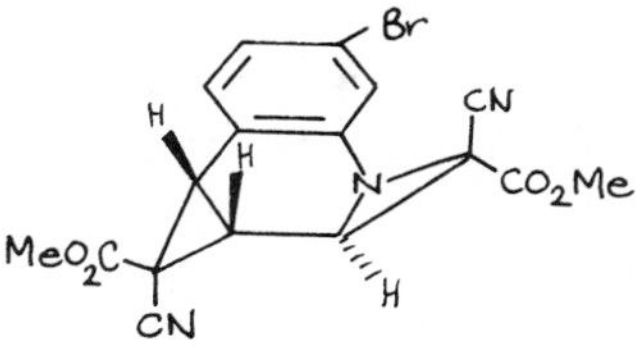

167

and the stereochemistry of the carbonyl groups is by definition cis.[454] The same results are achieved by irradiation.[454a] These products should be a good source of *cis*-3,4-disubstituted tetrahydroquinolines. 1,2,3,4-Tetra-hydroquinolines can also be obtained by addition reactions from quinolines. Methyl α-bromocyanoacetate adds across the 1,2 and 3,4 double bonds of quinoline to produce the bis adduct 1,2-dicyano-1,2-bismethoxycarbonyl-1a,2b,2,7b-tetrahydro-1*H*-azirino[1,2-*a*]cyclopropa[*c*]quinoline. An X-ray analysis of the di-adduct **167** from 7-bromoquinoline revealed that the two three-membered rings are oriented trans to each other.[455] Dimethyl acetylenedicarboxylate adds across the enamine double bond of 1,4-dihydro-1-methylquinoline and furnishes the cis adduct (**168**).[456]

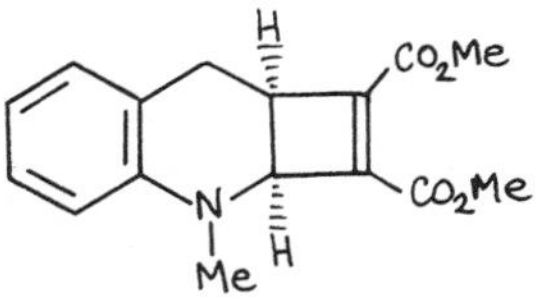

168

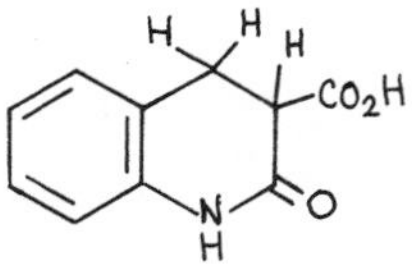

169

2. *Configuration and Conformation of 1,2,3,4-Tetrahydroquinolines*

In attempting to resolve 3-carboxy-1,2,3,4-tetrahydro-2-oxoquinoline (**169**) into its optical antipodes, Leuchs[457] found that one quinoline salt crystal-lized out of solution in over 95% yield. Decomposition of the salt gave the (+)-acid indicating that a second order asymmetric transformation (kinetic resolution) had taken place.[458] This was due to the instability of the chiral center, caused by enolization, which sets up the equilibrium:

$$d\text{-acid-}d\text{-base} \rightleftharpoons l\text{-acid-}d\text{-base}$$

The salt from the d-acid and d-base is obviously less soluble, it crystallizes out and disturbs the equilibrium until all the acid is converted into it. The free (+)-acid racemizes slowly in acetic acid.

The α-amino acid isolated from *Ithomid* and *Heliconian* butterflies is S-2-carboxy-1,2,3,4-tetrahydro-8-hydroxyquinoline. Its absolute configuration was assigned from the change observed in optical rotation from methanol to methanolic hydrochloric acid solutions[459] (see Clough-Lutz-Jirgenson rule, Chapter 2, Section III.B.2). (−)-4-Carboxy-1,2,3,4-tetrahydro-2,2-dimethylquinoline (**170**), obtained by optical resolution of the brucine salt was assigned the S configuration because of the similarity of its ord curve with that of the known $S(-)$-1-carboxy-1,2,3,4-tetrahydronaphthalene. The chiral acid was reduced to the alcohol, converted to the iodide, and cyclized to 2,3-benzo-6,6-dimethyl-1-azabicyclo[2.2.1]hept-2-ene (**171**, $n = 1$) which has the absolute configuration $1R,4S$.[460] The chemical shift of the C–8 protons, in the ^{1}H nmr spectra in $CDCl_3$, of the diastereo-

170   171   172

isomeric salts of 2-N-isopropylaminomethyl-6-methyl-7-nitro-1,2,3,4-tetrahydroquinoline and α-methoxy-α-trifluoromethylphenylacetate are slightly different and can be used to evaluate the purity of the base. This effect is lost if the polarity of the solvent is increased (e.g., in CD_3OD) and can be explained if a salt-pair is assumed in which the 8-proton is affected slightly differently in the two diastereoisomers by the phenyl group of the acid as in formula **172**.[461]

Heller's dimer,[462] prepared by reduction of 2-methylquinoline with zinc and acid, was shown to have the structure **173** (R = H).[463] 6-Bromo-2-methylquinoline gave a similar dimer which was also obtained by direct bromination of the former dimer. An X-ray analysis of the dibromo derivative confirmed the head-to-head-trans structure (**173**, R = Br).[464]

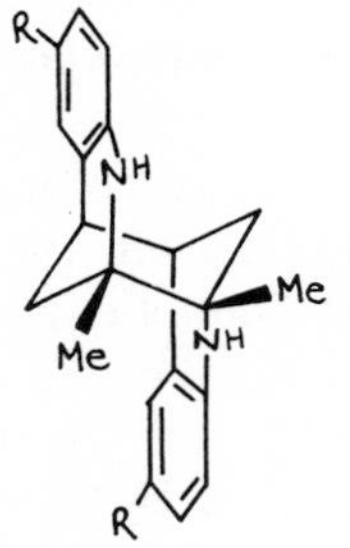

173

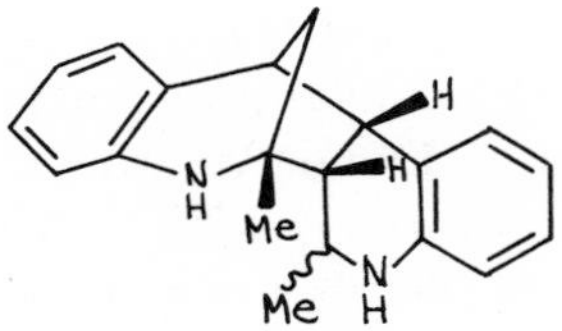

174

Electrolytic reduction of 2-methylquinoline, on the other hand, gave the *endo*-quinolinoquinoline **174**, and its structure was deduced from the ¹H nmr evidence.[465] It differs from the former C–2,C–4:C–2,C–4 dimer in being a C–2,C–4:C–3,C–4 dimer.

Booth[466] measured the chemical shifts and coupling constants of twenty 1,2,3,4-tetrahydroquinolines, their methiodides and N-acyl derivatives, by ¹H nmr spectroscopy and obtained valuable information regarding the conformation of these molecules. He concluded that the tetrahydroquinoline is a flexible half-chair structure which is rapidly inverting between the two conformers **175** and **176**. The nitrogen atom in all but the quaternary salts is also undergoing pyramidal inversion. 1,2,3,4-Tetrahydroquinoline is most probably equally populated with conformers **175** and **176**, but this is not the case if substituents are present in the reduced ring. A methyl

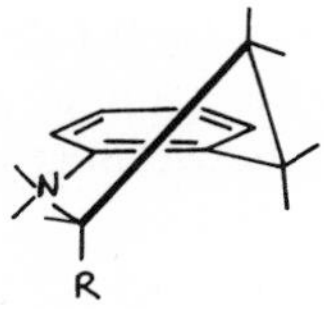

175

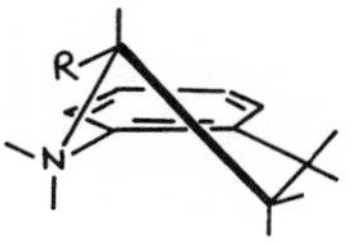

176

group at C–2 would have a strong tendency for the equatorial orientation and therefore would favor the conformer which will keep it in this conformation. The spectra suggest that 1-benzoyl-1,2,3,4-tetrahydro-2-methyl-(and 4-methyl)quinoline and 1,2,3,4-tetrahydro-1,1-dimethyl-2-phenylquinolinium iodide have rigid half-chair conformations. Also, whereas the *gem*-dimethyl groups in 1,2,3,4-tetrahydro-2,2,4-trimethylquinoline are non-equivalent, in their 1-methyl and 1-methyl methiodide derivatives they become equivalent. The *gem*-dimethyl groups in 1,2,3,4-tetrahydro-1,2,2-trimethylquinoline, on the other hand, are magnetically equivalent. These last results could not be readily explained.

The last example in this subsection is a dihydroquinoline but is best included at this point. The bright orange product from the reaction of 2-(3,3,3-trichloro-2-hydroxypropyl)quinoline and alcoholic sodium hydroxide, obtained by Einhorn,[467] was shown, by Woodward and Kornfeld,[468] to be 3-acetyl-2-carboxy-1,2-dihydroquinoline (**177**). It is a relatively stable compound and one optical enantiomer was obtained by recrystallization of its brucine salt. The latter evidence excluded the 1,4-dihydroquinoline (**178**) as a possible alternative.

177

178

3.   Reactions of Di- and Tetrahydroquinolines

Acid catalyzed disproportionation of 3,4-disubstituted 1,2-dihydroquinolines into 1,2,3,4-tetrahydroquinolines proceeds with some stereospecificity (Equation 15). Tilak and co-workers[469] showed that 1,2-dihydro-7-methoxy-1,3,4-trimethyl-, 1,2-dihydro-7-methoxy-1,3-dimethyl-4-phenyl-, and 1,2-dihydro-7-methoxy-1-methyl-3,4-tetramethylenequinolines disproportionated to yield about 65:35 mixtures of *cis*- and *trans*-1,2,3,4-tetrahydroquinolines in each case. However, when the 3,4-trimethylenequinoline **179** (R^1,R^2 = –[CH$_2$]$_3$–) was subjected to similar treatment the cis product was obtained exclusively. There is evidence from mass spectrometric measurements for the presence of a stable carbonium ion such as **180**,[470] and the reaction may well proceed by this intermediate. A similar study of 2,3-disubstituted 1,4-dihydroquinolines was made and showed that the *cis*-tetrahydroquino-

line predominated in 2-phenyl-3-methyl- and 2,3-trimethylene-1,4-dihydro-
1-methylquinolines. The trans isomer was the main product, however,
when the two 2- and 3-positions were linked by a penta- and hexamethylene
bridge.[471]

The reality of Booth's "half-chair" conformations for 1,2,3,4-tetrahydro-
quinolines was demonstrated in the acid catalyzed hydrolysis of *cis*-(**181**)
and *trans*-4-arylamine-2-methyl-1,2,3,4-tetrahydroquinolines (**182**). The
rates of hydrolysis of the isomers with the pseudo-axial 4-arylamino group
182 were higher than those with the pseudo-equatorial 4-arylamino group
181 which form, in each case, the kinetically controlled pseudo-axial
4-hydroxy derivatives. These were slowly converted into an equilibrium
mixture of *cis*- and *trans*-1,2,3,4-tetrahydro-4-hydroxy-2-methylquinoline.[472]

B.   Decahydroquinolines

1.   *Syntheses of Decahydroquinolines*

a.   *By Catalytic Reduction of Quinolines or Partially Reduced Quinolines.*
Complete reduction of quinoline or 1,2,3,4-tetrahydroquinoline yields deca-

hydroquinolines. Hückel and Stepf were first to separate these into the cis and trans forms.[473] Several reducing methods have been used[473] but catalytic reduction is the most convenient. Higher temperatures are required with a nickel catalyst for complete reduction,[474] and for smaller scale preparations colloidal platinum in acetic acid is most satisfactory and gives a higher yield of cis isomer.[475,476] It is not known what the cis-trans ratios are for all the different preparations, but it appears that the trans isomer in some cases is formed in much larger proportion. It is thus much easier to obtain the trans isomer in pure form because it can be separated by freezing it out of the mixture[474,475] leaving the cis enriched isomer. This is then purified by recrystallization of both the *N*-benzoyl derivative and the hydrochloride. More recently a commercial mixture was separated by fractional distillation with an efficient spinning band column but the smaller amount of high boiling *cis*-decahydroquinoline required further purification by way of the above mentioned derivatives.[477] Practically pure *cis*-decahydroquinoline is prepared by catalytic reduction of quinoline first with Raney nickel, followed by hydrogenation in concentrated hydrochloric acid using platinum black as catalyst.[478] Catalytic reduction of 2,3,8-trimethylquinoline gives the decahydroquinoline, but this is probably a mixture of geometrical isomers.[479] Even when it is possible to say what proportions of *cis*- and *trans*-decahydroquinolines are formed from substituted quinolines (e.g., by glc) it is not always easy to separate the isomers.[480]

Reduction of 2-cyanoethylcyclohexanones yields decahydroquinolines which most probably consist mainly, if not entirely, of the cis isomer (Equation 16).[424,481] The octahydroquinolines **183** are undoubtedly inter-

$$R = H, Ph, Me \qquad\qquad \mathbf{183} \tag{16}$$

mediates. Henshall and Parnell[425] prepared the 4a-methyl-2,3,4,4a,5,6,7,8-octahydroquinoline **183** (R = Me) by cyclizing 3-(2,2-ethylenedioxy-1-methylcyclohexyl)propylamine after hydrolysis. Catalytic reduction of it gave *cis*-4a-methyldecahydroquinoline, but reduction with sodium and alcohol gave the trans isomer. Catalytic reduction of such octahydroquinolines always seems to give a higher proportion of cis isomer if the medium is strongly acidic, but when it is alkaline, as in the hydrogenation of 1,2,3,4,-4a,5,6,7-octahydro-2,7-dioxo-4a-(2-phenoxyethyl)quinoline, the trans isomer is produced in much larger amounts.[482] *trans*-Decahydroquinoline is also derived from sodium and alcohol reduction of 5,6,7,8-tetrahydroquinoline.[421]

b. *By Cyclization Reactions.* cis- and *trans*-Decahydroquinolines were unequivocally synthesized by cyclization of *cis*- and *trans*-2-(3-bromopropyl)-cyclohexylamines, and this confirmed the assigmnents of all the earlier work.[483] Cyclization of 1,2-disubstituted cyclohexanes of known relative configuration are useful for making quinolines of known stereochemistry.

184   185   186

However, a check must always be made in order to ensure that inversion has not occurred. The ring closure of *trans*-1-N-ethoxycarbonylethyl-*N*-methyl-2-methoxycarbonylcyclohexane produces *trans*-3-ethoxycarbonyl-1-methyl-4-oxodecahydroquinoline, whereas the cis isomer also forms the *trans*-decahydroquinoline. Obviously the ethoxycarbonyl group was inverted in order to give the more stable *trans*-decahydroquinoline. This compound can also be obtained by heating 1-ethoxycarbonylmethylcarbonylcyclohex-1-ene with *N*-methylamine.[484] Several cyclizations to decahydroquinolines somewhat similar to the above have been reported, but the stereochemistry was not always established.[485,486]

Dry distillation, with soda lime, of *cis*- or *trans*-1-amido-2-carboxyethylcyclohexanes provided the respective *cis*- or *trans*-2-benzyl-, 2-p-tolyl-, and 2-phenyl-3,4,4a,5,6,7,8,8a-octahydroquinolines without affecting the asymmetric centers. Catalytic reduction of these octahydroquinolines apparently afforded only one (two are possible) 2-substituted decahydroquinoline in each case.[487] Sokolov and co-workers[488] obtained a mixture of 2-methyl-4-oxodecahydroquinoline (184) from 1(but-3-enonyl)-cyclohex-1-ene and ammonia, and devised methods for converting it into four geometrical isomers. 6-Cyano-2,3,4,4a,5,6,7,8-octahydro-2-oxo-6-phenylnaphthalene (185) undergoes a hydroxyl ion induced intramolecular cyclization to the 3,8a-ethano-2,7-dioxo-cis-decahydroquinoline (186).[489]

Methanolic HCl causes the intramolecular cyclization of the ethylenedioxy ketal of 4-(2-aminomethylvinyl)-4-ethylcyclohex-2-en-1-one into the ketal of cis-4a-ethyl-7-oxo-1,2,4a,5,6,7,8,8a-octahydroquinoline.[489a] Oppolzer and Fröstl[489b] used a new stereoselective approach by converting the open chain N-acyl-N-(pent-4-en-1-yl)-N-buta-1,3-dienylamine into predominantly N-acyl-cis-1,2,3,4,4a,5,6,8a-octahydroquinolines in good yields by heating at 190–215°C. It should be possible to reduce these octahydroquinolines to decahydroquinolines without upsetting the stereochemistry.

   c.   By Beckmann and Schmidt Rearrangements, and Annelations. Hydrindan-1-one oximes undergo a Beckmann rearrangement generally with complete retention of configuration at the bridgehead carbon atom and with the formation of 2-oxodecahydroquinolines (Equation 17). When a methyl group is present at the bridgehead carbon atom α to the oximino group the yields are lower due to fragmentation of the five-membered ring, but the steric course of the reaction is unchanged.[490] Also, the presence of a double bond in the six-membered ring, for example, in 2,3,3a,4,7,7a-hexahydroindan-1-one, does not alter the specificity of the reaction.[491] However, a Schmidt reaction (HN_3–H_2SO_4) on the hydrindan-1-ones pro-

$$\text{(17)}$$

ceeds in the other direction also affording the 1-oxodecahydroisoquinolines with retention of configuration.[490] Earlier results with cis-trans mixtures of hydrindanones demonstrated that both decahydroquinolines and decahydroisoquinolines are formed in comparable amounts.[492]

A reaction which may be general for the synthesis of angularly substituted hydroquinolines is an application of the methyl vinyl annelation method. Methyl vinyl ketone condenses with cyclic enamines derived from hydropyridines to yield 4a-substituted 7-oxodecahydroquinolines (Equation 18).[493] The quinoline formed apparently has the cis configuration.[494]

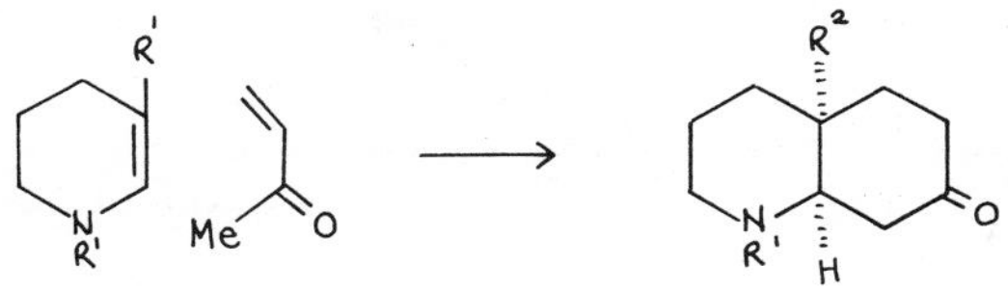

(18)

2. Configuration of Decahydroquinolines

The cis-trans configuration of decahydroquinoline have been firmly established by synthesis,[483] but each of these is still a racemate. In 1915 Mascarelli and Nigrisoli[476] resolved the trans form into its enantiomers by recrystallization of its *d*-bromocamphorsulfonate salt. Popvici and co-workers[495] later found that although this method is good for preparing the (+) antipode it is not satisfactory for obtaining the (−) antipode. They found that the latter is obtained in better yields from recrystallization of the (+)-tartrate salt. A detailed study of the biological hydroxylation of (±)-*trans*-1-benzoyldecahydroquinoline and its optical forms was made by Johnson and co-workers.[496] The microorganism *Sporotrichum sulphurescens* oxidized the racemic trans isomer to the three hydroxy derivatives **187**, **188**, and **189** which are the 4aS,5S,8aR, the almost racemic, and the

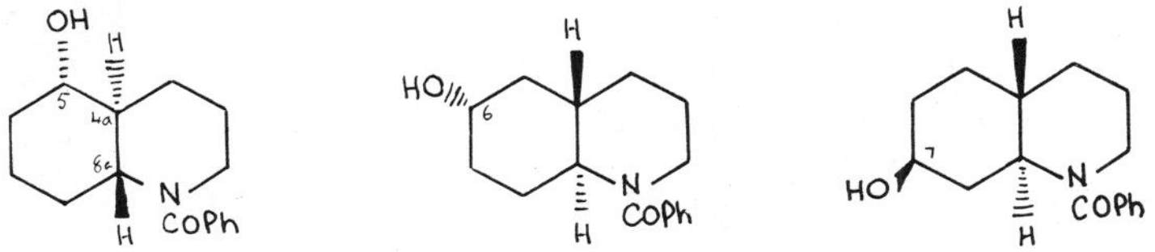

187

188

189

4aS,7S,8aS isomers respectively in 80–90% total yield. *trans*-4aR,8aS(+)-1-Benzoyldecahydroquinoline was converted into a 65:35 mixture of 4aS,6S,8aS-(+)- **188** and 4aS,7S,8aS-(+)- **189**, whereas the 4aS,8aR(−)-benzoyldecahydroquinoline was hydroxylated to an 87:13 mixture of 4aS,5S,8aR-(−)- **187** and 4aR,6R,8aR-(−)- enantiomer of **188**. The absolute configurations were established by oxidizing the alcohols to the respective ketones and applying the octant rule to the ord and cd data.[496,497] Complete reduction and hydrolysis gave the absolute configuration of *trans*-(−)-4aR,8aS-[α]$_D$ − 4.6° and *trans*-(+)-4aS,8aR-decahydroquinolines[α]$_D$ + 4.8°.[496] From this data and the known rigidity of the decahydroquinoline molecule, Johnson and co-workers were able to define a coordinate system for the contour of the active site of the hydroxylating enzyme.[498]

Ripperger and Schreiber[499,500] examined the ord and cd spectra of
N-chloro- and N-nitroso-*trans*-(−)-decahydroquinolines, compared them
with quadrant rules that they previously deduced for N-halo and N-nitroso
saturated heterocycles, and confirmed the above absolute configurations.
Roberts and Thomson[501] prepared the nitroxide of *trans*-(−)-2,2,8a-tri-
methyl-4-oxodecahydroquinoline (190) and related isomers and measured
their ord and cd spectra. The data suggested that the signs of the back

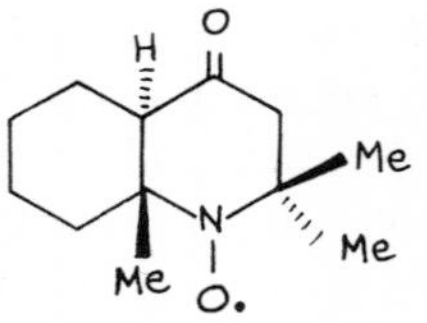

190

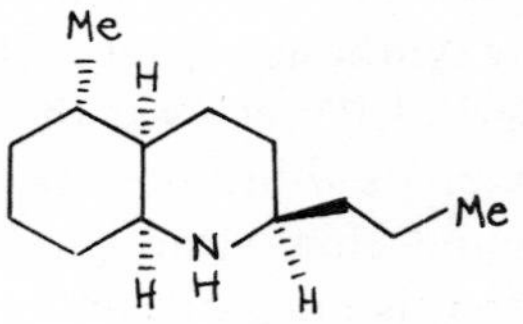

191

octants of the nitroxide chromophore are the same as those for the carbonyl octant rule, and they deduced the absolute configuration.

The relative, but not absolute, configuration of *d,l-trans*-4a-*trans*-8a-
trans-5-methyl-*r*-2-propyldecahydroquinoline (191) (*d,l*-pumiliotoxin C
from the skin of the colored Panamian frog *Dendrobates pumilio*) was
elucidated by a multistep synthesis starting from *cis*-1,2,3,4,4a,5,8,8a-
octahydro-2-oxoquinoline.[491]

3.   Conformation of Decahydroquinolines

Armarego[477] published the ^{1}H nmr spectra of *cis*- and *trans*-decahydro-
quinolines and observed that the band envelope of the protons, as in
cis- and *trans*-decalins,[502] was very narrow for the cis isomer and broad
for the trans isomer. The broad band envelope is explained by the conformationally rigid trans structure with large differences in chemical shifts
(∼0.5 ppm) between the axial and equatorial protons. The narrow band
envelope is partly due to the more flexible cis structure, which is an equilibrium between the two extreme chair-chair conformers 192 and 193,
because the axial protons of one conformer become the equatorial protons

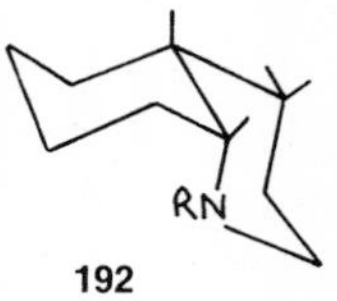

192

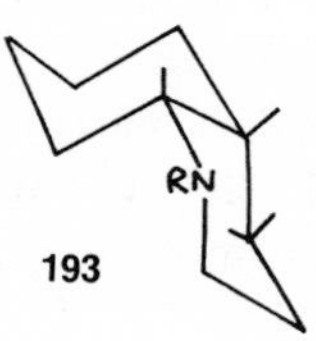

193

of the other. Further data, however, showed that in cis-decahydroquinoline the conformer **192** is predominant and the narrow band envelope is most probably caused by alteration of the shielding and deshielding of the equatorial and axial protons respectively because of the cis configuration of the system.[503] Booth and Bostock[504] examined the ^{1}H nmr spectra in more detail, assigned some of the proton chemical shifts, and showed that the cis isomer is almost exclusively in the conformation **192**. Booth and Griffiths studied the ^{1}H, ^{13}C, and ^{19}F nmr spectra of several 1-substituted cis-decahydroquinolines. Conformer **192** was the predominant one when R = H and Me, but conformer **193** was preferred when R = CD_2CH_3 and CD_2CF_3. Similarly N-benzoyl, N-phenylsulfonyl, N-phenylcarbamoyl, and N-nitroso groups cause this switch of preferred conformation to **193**.[478] In a very comprehensive paper on piperidine, alkylpiperidines, and decahydroquinolines, they assigned the ^{13}C nmr chemical shifts of all the carbon atoms of trans-decahydroquinoline and of the carbon atoms of the two conformers of cis-decahydroquinoline and some of its derivatives. From these values they deduced that for R = H the ratio of conformers **192** and **193** at −74°C was 93.5 : 6.5, which confirmed previous conclusions.[506]

Nitrogen inversion occurs in the decahydroquinolines and the cis isomer is particularly interesting because of 1,4-syn axial nonbonded interactions across the two rings shown in Scheme 5. From Scheme 5 it can be seen why the conformers **192a** and **192b** are preferred in the unsubstituted compound. Of these two, the conformer **192a** with the lone pair in the "inside" orientation is much more favored in the unsubstituted compound. However, if a large group is present on the nitrogen atom the alternative twin-chair conformations **193a** and **193b** are preferred in which the repulsive interactions of the N-substituent with the C–8 methylene protons are relieved. Such groups as ethyl,[505,506] nitroso, phenylsulfonyl, and phenyl-amido[478] cause this alteration in predominant conformation. In trans-decahydroquinoline the lone pair can assume an axial or equatorial orientation. Whereas little is known about the preferred orientation in the unsubstituted compound, Eliel and Vierhapper showed, by ^{13}C nmr spectroscopy, that an equatorial N-methyl group is strongly preferred.[507] They[508] also examined the conformational equilibria in N-methyldecahydroquinoline and derivatives in which the substituents forced the N-methyl group into preferred axial or equatorial conformations by nonbonded steric interaction. By using ^{13}C nmr spectroscopy they evaluated the conformational energies for the equilibria N–Me(ax) $\rightleftharpoons$ N–Me(eq) and obtained $\Delta G°$ values between 7.8 and 10.3 kJ mol^{-1}. A similar study of N-methylpiperidine gave slightly lower $\Delta G°$ values which were less reliable

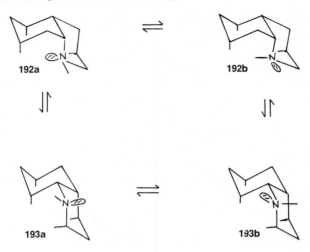

Scheme 5

because of the lack of a suitable model for the conformer with an axial
N-methyl group. Katritzky and co-workers[508a] took advantage of the equilibria in Scheme 5 in order to obtain some idea of the relative steric requirements of the nitrogen "lone pair" and a hydrogen atom. They studied
the equilibria between *cis,cis*- and *trans,cis*-1-methoxycarbonyldecalins and
cis- and *trans*-4-methoxycarbonyl-1-methyl-*cis*-decahydroquinolines in the
presence of methoxide ions. They used conductometric analysis and isolated the *cis,cis*-quinoline derivative by preferential reaction with methyl
iodide. They concluded that in non–hydrogen bonding solvents the con-

former with the "lone pair" "inside" the molecule **194** was more stable
than the other conformer **195**, but in hydrogen bonding solvents they were
of comparable stability. This demonstrates that steric interactions are in
the order: H-solvated lone pair > H–H > H-free lone pair.

   The lanthanide shift reagent tris(dipivaloylmethanato)europium was used
successfully by Booth and Griffiths[509] to obtain the complete assignment

of all the induced proton signals in *trans*-decahydroquinoline. The equilibrium binding constant at 31°C is 12 liters mol⁻¹. Analysis of the coupling constants of the protons in *cis*- and *trans*-4,4-disubstituted 2-methyldecahydroquinolines and the effect of solvent on these has been reported.[510]

4. Reactions of Decahydroquinolines

a. *Dehydrogenation and Quaternization.* Prior to the availability of spectroscopic methods, the best way of differentiating between *cis*- and *trans*-decahydroquinolines was by catalytic dehydrogenation. It is not surprising from the present knowledge of the reactions at a catalyst surface,[511] that the cis isomers are dehydrogenated faster than the trans isomers.[512]

McKenna and co-workers[195,208,215] examined the alkylation of *N*-alkyldecahydroquinolines and concluded that alkyl groups prefer to attack the molecule axially (compare the quaternization of piperidines, Section II.D.5). The relative reactivity of *cis*- and *trans*-decahydroquinolines towards alkylating agents, for example, ethyl, 2-dimethylaminoethyl, and 3-chloropropyl halides was demonstrated by competitive reactions. Two molecular equivalents of a mixture of *cis*- and *trans*-decahydroquinolines were reacted with one molecular equivalent of the halide. The product was a mixture of 1-alkyl-*cis*-decahydroquinoline and *trans*-decahydroquinoline halide inferring that the cis base reacted faster.[513]

In a study of the von Braun reaction of 1-methyl-*trans*-decahydroquinoline with cyanogen bromide at low temperatures, Fodor and co-workers[514–516] obtained a 9:1 mixture of Me axial and Me equatorial 1-cyano-1-methyl-*trans*-decahydroquinolinium bromides. The structure of the latter was confirmed by an X-ray analysis (contrast with the strong preference of an equatorial *N*-methyl group in 1-methyl-*trans*-decahydroquinoline, last section [3]). Quaternization with methyl iodoacetate also produced an 8:2 mixture of Me axial and Me equatorial 1-methoxycarbonylmethyl-1-methyl-*trans*-decahydroquinolinium iodides giving some indication of the direction of attack of the alkylating agent.[516]

b. *Reactions of 4-Oxodecahydroquinolines.* All four racemates of 2-methyl-4-oxodecahydroquinoline were obtained from a mixture prepared by cyclizing 1-but-2-enoylcyclohex-1-ene (Section IX.B.1.b). The isomer **(196)** and the cis and trans mixture of the epimeric 2-methyl-4-oxo compounds were separated by crystallization. Treatment of **196** with hydrochloric acid then sodium bicarbonate gave a 3.5:5 mixture of the cis **197** and trans **196** isomers. The former was transformed into the latter on storage or in boiling water to the extent of 30%, demonstrating the lability of the system. The isomers with the epimeric 2-methyl group were separated

196   197

by way of their N-benzoyl derivatives.[488] In contrast 1-benzoyl-4-oxo-
trans-decahydroquinoline is isomerized on acid hydrolysis to 4-oxo-*cis*-
decahydroquinoline hydrochloride.[517] Replacement of N–H by N–Me, on
the other hand, does not alter the relative stability of the trans-fused
system.[518] The stability of the fused system is therefore strongly affected
by the nature and position of the substituents.

Reduction of the oxo group in 4-oxodecahydroquinolines was investigated by Mistryukov[41] and by Sokolov and co-workers.[519,520] The 4-hydroxy
compounds formed with axial or equatorial OH groups were consistent
with electrostatic and steric effects in the 4-oxo compounds. 4-Oxo-*trans*-
decahydroquinoline, its 1-benzoyl and its 1-methyl methiodide derivatives,
and 1-benzoyl-4-oxo-*cis*-decahydroquinoline gave the 4-OH-equatorial and
4-OH axial isomers in the ratios 86:14, 81:19, 53:47, and approximately
98:2 respectively on reduction with $NaBH_4$.[41] In contrast, reduction with
aluminum isopropoxide or catalytically (e.g., of 1-benzoyl-4-oxo-*trans*-
decahydroquinoline) yields entirely the axial alcohol.[517] Attempts to differentiate between the four 4-hydroxydecahydroquinolines by the ability
of their O-acetates to undergo $O \rightarrow N$ migration of the acetyl group was
not wholly satisfactory.[521] Five isomeric 4-hydroxy-2-methyldecahydro-
quinolines were obtained by reduction of the 4-oxo compounds and also
their 1-methyl derivatives by further methylation with formaldehyde and
formic acid.[519,520] Many of these were identified by ir spectra since intramolecular hydrogen bonding between the OH group and the ring nitrogen
atom in some of the isomers was evident.[522] Acid hydrolysis of 1-benzoyl-
4-hydroxydecahydroquinolines, in contrast to 1-benzoyl-4-oxo-*trans*-deca-
hydroquinoline (see above), does not result in epimerization of any chiral
center.[523]

The addition of Grignard or related reagents to 2-methyl- and 1,2-
dimethyl-4-oxo-*trans*-decahydroquinolines is not always highly stereoselective and depends on whether the 2-methyl group is axial or equatorial.
When the 2-methyl group is equatorial *n*-butyl magnesium bromide yields
traces of the alcohol with the *n*-butyl group axial (**198**, $R^2 = H$ or Me,
$R^2 = n$-Bu), and approximately 82% of the epimeric alcohol with the butyl
group equatorial (**199**, $R^1 = H$ or Me, $R^2 = n$-Bu). But-1-yn-3-enyl magne-

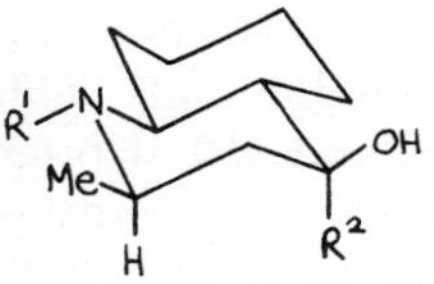

198 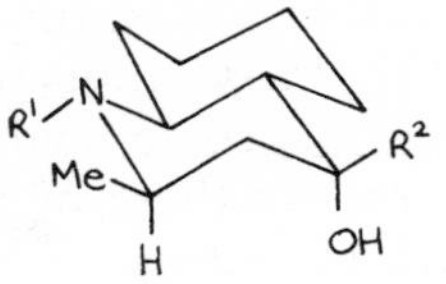   **199** 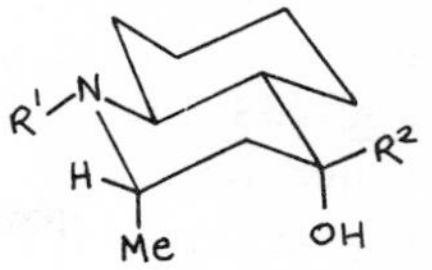   **200**

sium bromide, on the other hand, furnishes about equal amounts of epimeric alcohols. The 4-oxodecahydroquinoline with an axial 2-methyl group, however, gives only the epimer in which the hydroxyl group is axial (**200**, R^1 = H,Me, R^2 = n-Bu, or –C≡C–CH = CH$_2$).[524] The reactivity of 1-methyl-4-oxodecahydroquinolines with phenyl lithium is similar and some partial cis-trans isomerization may occur.[518]

   c. *Fragmentation of Decahydroquinolines.*   Grob and co-workers[525,526] extended their investigations on the fragmentation (by solvolysis) of 4-chloropiperidines (Equation 11) to the decahydroquinoline series. Here again they found that the orientation of the N–C–C–C–X system (compare with formula **35**) had a great influence on the rates and mode of reaction. They examined the solvolysis, in 80% ethanol, of the 1-methyl-tosyloxy-decahydroquinolines, **201–207** (only preferred conformations for the reac-

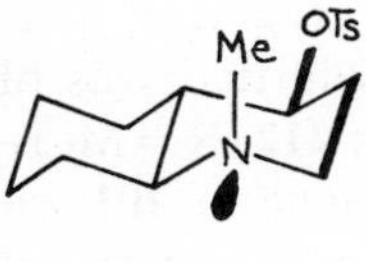

201 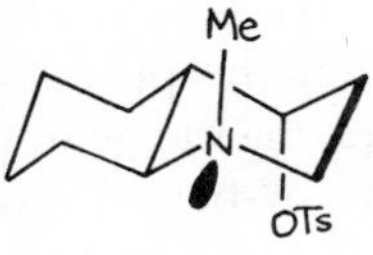   **202**    **203** 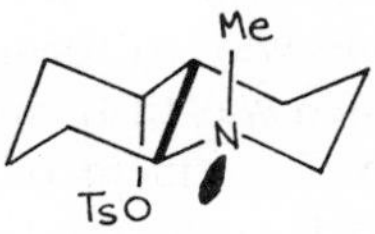   **204**

205    **206** 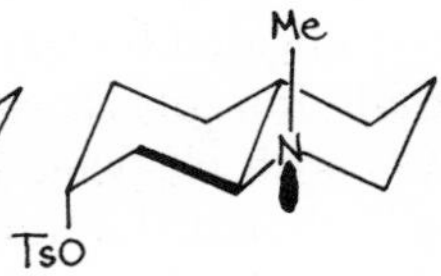   **207** 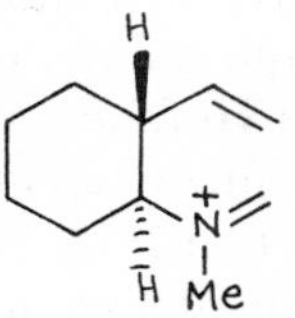   **208**

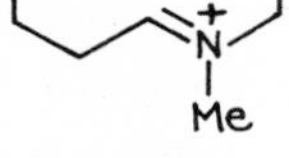

209

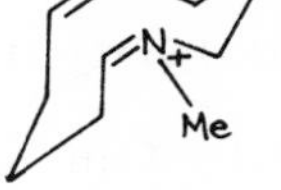

210

tion are shown), and compared the rates with those of the corresponding
α- and β-decahydronaphthyltoluene-p-sulfonates. The quinolines **201, 203,**
and **205** fragmented to the olefinic compounds **208, 209,** and **210** at rates
which were 4870, 46, and 56 times those of the respective *trans*-α-tosyl-
oxydecalins with quantitative fragmentation. The compounds **204, 206,**
and **207** solvolyzed by substitution and elimination (i.e., S_N1 and E_1) only
at rates which were 2–8 times slower than those of the corresponding
decalins. The retardation is caused by the inductive effect of the nitrogen
atom. Elimination as well as fragmentation occurred with the isomer **202,**
indicating that a conformational change to the skew-boat structure **211**
must take place before fragmentation. Theoretical justification for the

Grob fragmentation was provided by Hoffmann and co-workers.[527]
Marshall and Babler[528] found that α-methylsulfonyloxy-*trans*-decalin fragmented along the bond between the two rings to give cyclodeca-1,5-diene.
They argued that Grob's compound, **201,** should also fragment in the
same way to form **212** (R = Me), and that the product **208** that was obtained was the result of a subsequent Cope rearrangement: **212** (R = Me) →
208. Treatment of the *O*-methanesulfonyl-*N*-benzyl analogue of **201** with
$NaBH_4$ gave mixtures consisting predominantly of **213** together with substantial amounts of **214.** The reaction thus either proceeds by 4a,8a C–C
bond fragmentation followed by a Cope rearrangement or by both this
mechanism and the one indicated by Grob, that is, 2,3 C–C fragmentation.

The Hofmann degradation of *cis*- and *trans*-1,1-dimethyldecahydro-
quinolinium salts occurs with cleavage of the N–C_2 bond and formation
of *cis*- and *trans*-1-allyl-2-dimethylaminocyclohexane. The equatorial H–2
is eliminated in the trans and cis isomers (in either "chair-chair" conformation) if the mechanism involves the coplanar $B^- \, ^\cdot H–C–C–NMe_2$ system[529] (see Section VIII.B. formula **149**; Chapter 2, Section III.E.4).

X.   REDUCED ISOQUINOLINES

Considerable interest in the chemistry and stereochemistry of di- and
tetrahydroisoquinolines, reduced in the heterocyclic ring, arose from the

early discoveries of the presence of this nucleus in a number of alkaloids (e.g., the aporphine, cularine, morphine, emetine). Only a brief discussion of the stereochemistry of 1,2,3,4-tetrahydroisoquinolines is warranted here, and for further reading excellent texts on isoquinoline alkaloids are available.[530] Much of the stereochemistry of the decahydroisoquinolines was studied for purely chemical interest and for the possible discovery of new drugs because of the known physiological properties of the reduced isoquinoline nucleus.

A. 1,2,3,4-Tetrahydroisoquinolines (Including a Few Dihydroisoquinolines)

1. Syntheses of 1,2,3,4-Tetrahydroisoquinolines

The Pictet-Spengler[531] synthesis, which involves the intramolecular cyclization of 2-phenylethylamines and aldehydes in the presence of acid, has been commonly used for preparing chiral tetrahydroisoquinolines. S-3,4-Dihydroxyphenylalanine (dopa) and formaldehyde yield optically active S-3-carboxy-1,2,3,4-tetrahydro-6,7-dihydroxyisoquinoline.[532,533] In the same reaction using acetaldehyde two 1-methyl derivatives are formed, but there is stereoselectivity and the major product is the optically active cis isomer **215** (Equation 19).[532,534] The structure was confirmed by an X-ray

$$ (19) $$

analysis of the ethyl ester hydrochloride.[532] These compounds occur naturally in velvet beans (*Mucuna* species).[533,534] 1,2,3,4-Tetrahydro-6-methoxy-2-methyl-1-phenylisoquinoline and its epimer are similarly prepared from R(+)-2-amino-1-(3-hydroxyphenyl)ethanol and benzaldehyde followed by methylation of the phenolic oxygen (CH_2N_2) and the ring nitrogen atom ($HCHO–HCO_2H$). The absolute configurations were deduced by ord spectral comparisons.[535] In another application several *erythro*- and *threo*-methyl-3-amino-2,3-diarylpropionates and formaldehyde produced *trans*- and *cis*-3-aryl-4-methoxycarbonyl-1,2,3,4-tetrahydroisoquinolines respectively.[536] If the acid used in this reaction is formic acid then the N-methyl products are formed.[537]

A Bischler-Napieralski[538] cyclization ($POCl_3$–P_2O_5) of (+)-N-formyl-3-phenyl-2-propylamine gave (+)-3,4-dihydro-3-methylisoquinoline which was reduced with palladium black to 1,2,3,4-tetrahydro-3-methylisoquinoline. A mixture of the optically active cis- and trans-1,3-dimethyl derivatives was similarly prepared by starting with the N-acetyl propylamine.[539]

Uskoković and co-workers[540] used a different approach for preparing 1,2,3,4-tetrahydro-4-hydroxyisoquinolines. 2-Benzyl-4-hydroxy-6,7-dimethoxyquinolinium chloride reacted with methyl magnesium iodide to form the adduct 2-benzyl-1,2,3,4-tetrahydro-6,7-dimethoxy-1-methyl-4-oxoisoquinolinium iodide which on reduction with Pd–C/H_2 in acetic acid, followed by formaldehyde and Raney nickel reduction, gave a mixture of racemic cis- and trans-4-hydroxy-1,2,3,4-tetrahydro-6,7-dimethoxy-1,2-dimethylisoquinolines. The relative configurations were determined by ^{1}H nmr and ir spectroscopy.

Dimeric reduced isoquinolines can be produced either by chemical or by photochemical reduction. When isoquinoline is treated with zinc dust in acetic anhydride a 1:1 mixture of d,l(trans)- and meso(cis)-N,N'-diacetyl 1,1',2,2'-tetrahydro-1,1'-biisoquinolyl is formed. Further catalytic reduction of these gives the octahydro derivatives which can be deacetylated by hydrolysis with hydrobromic acid without epimerization. Catalytic reduction of 1,1'-biisoquinolyl dihydrochloride with platinum in ethanol produces exclusively cis(meso)-1,1',2,2',3,3',4,4'-octahydro-1,1'-biisoquinolyl. A methylene bridge between the nitrogen atoms of the isomeric octahydro-1,1'-biisoquinolyls is introduced by means of formaldehyde, for example, in 216 (R = H) and 217 (R = H).[541] A mixture of the methyl 216 (R = Me)

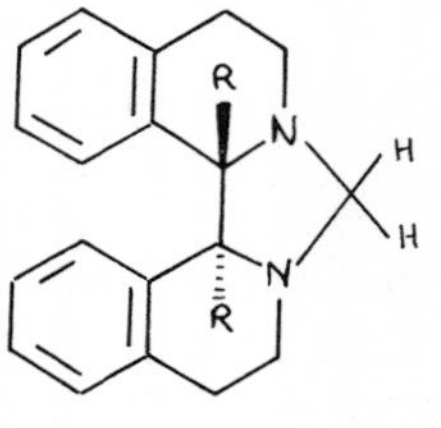

216

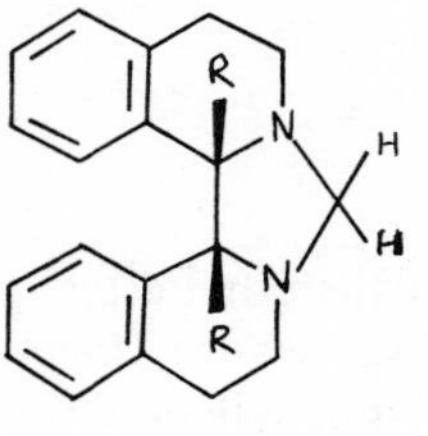

217

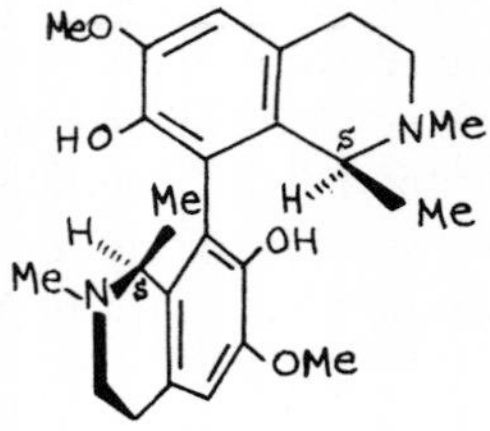

218

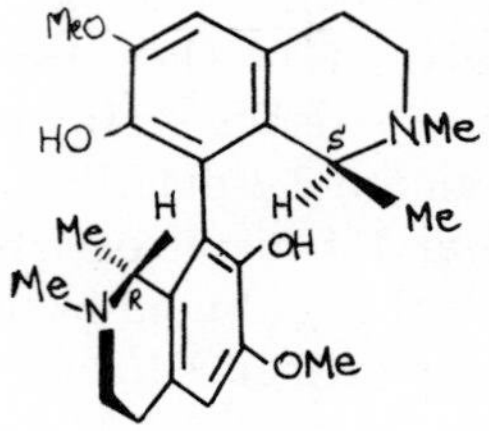

219

and **217** (R = Me) is formed when 3,4-dihydro-1-methylisoquinoline is photolyzed in methanol, with benzophenone as sensitizer, and can be distinguished by ^{1}H nmr spectroscopy. 3,4-Dihydro-6,7-dimethoxyisoquinoline behaves similarly and the methylene bridge is removed by reduction with aluminum amalgam.[542] Bobbitt and co-workers[543] prepared biisoquinolyls by phenolic coupling of two molecules of 1,2,3,4-tetrahydro-7-hydroxy-6-methoxy-1-methylisoquinoline. In this case the isoquinoline rings are linked by the C–8,C–8' positions and three isomers are formed depending on the conditions. Oxidation of the racemic isoquinoline (i.e., R,S) in the presence of platinum gave only a racemic mixture of 1,1',2,2',

R = Me or Et

220

3,3',4,4'-octahydro-7,7'-dihydroxy-6,6'-dimethoxy-1,1',2,2'-tetramethyl-8,8'-biisoquinolyls in which the configurations at C–2 and C–2' are S,S **(218)** and R,R. Electrolytic oxidation is less stereoselective and gave a mixture of three isomers: **218**, the R,S-pair **219**, and the enantiomer of **218**. Two rotamers are possible in the S,S and R,R isomers because of restricted rotation about the pivot 8,8-bond. The rotamers of the R,S isomer are identical.

The 1,2- and 3,4- double bonds in 2-methylisoquinolinium salts are reactive towards olefins and towards nucleophiles. Bradsher and collaborators[544] found that methyl (and ethyl) vinyl ether adds onto 2-methylisoquinolinium salts to form the 1,4-adduct **220**. The addition is highly stereospecific and the structure **220** was confirmed by an X-ray analysis. Cyclopentadiene adds in a similar fashion to 2-ethyl-3-methyl- and 2,3-dimethylisoquinolinium salts.[544a] The adducts have the cyclopentene ring over the benzene ring of the isoquinoline as would be predicted from theoretical considerations of the repulsive forces involved. The structure of the 2-ethyl adduct was also confirmed by an X-ray study.

Two molecules of isoquinoline methiodide condense with one molecule of an active methylene compound (e.g., nitromethane, malonic ester) in the presence of alkoxide to give the 1,2,3,4-tetrahydroisoquinoline dimer interlocked with the active methylene compound. Apparently the active methylene group adds across the 1,2- double bond of the quaternary salt which then reacts with another molecule of the quaternary salt at C–1 (i.e., a C–4,C–1' bond) followed by the formation of a C–3,C–4' bond and attack of the active methylene carbon on C–3' to produce 10-substituted 8,17-dimethyl-8,17-diazadibenzo[*c.j*]tetracyclo[7.3.1.0^{2,9}.0^{7,16}]tridecanes (Equation 20). The structures of the tridecanes were derived from

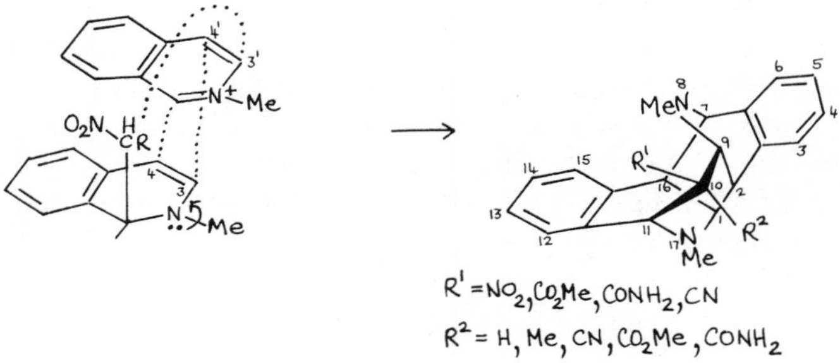

$$R^1 = NO_2, CO_2Me, CONH_2, CN$$
$$R^2 = H, Me, CN, CO_2Me, CONH_2$$

(20)

^{1}H nmr spectra of protio and deutero substituted derivatives prepared from the respective deuterated active methylene compounds and deuterated isoquinolinium salt.[545] A similar dimerization is observed when 1-cyano-1,2-dihydro-2-methylisoquinoline is treated with 2-equivalents of NaH. Elimination of CN$^-$ occurs and the rings are linked together as in the above (i.e., C–4,C–1′ and C–3,C–4′) except that C–2 is directly attached to C–1′, not by way of an active methylene group (i.e., **220a**). The structure **220a** was deduced from ^{1}H nmr data.[546]

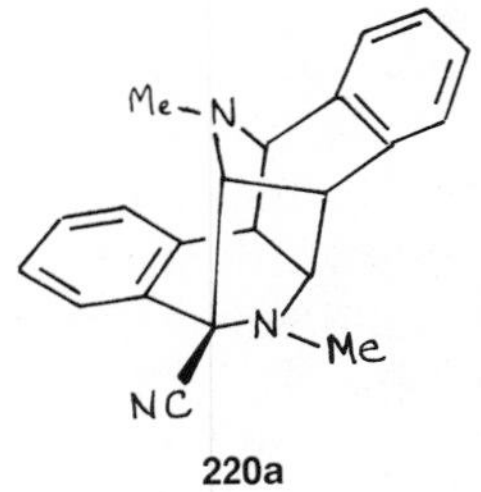

220a

2. Configuration and Conformation of 1,2,3,4-Tetrahydroisoquinolines

Several 1-benzyl-1,2,3,4-tetrahydroisoquinolines were resolved into their optically active enantiomers and the absolute configurations were determined in connection with studies on the structure of benzyl isoquinoline alkaloids. Ord, optical rotations, nmr, and X-ray methods were used for complete structure elucidation.[530,539,547–552] Like the 1,2,3,4-tetrahydroquinoline, the 1,2,3,4-tetrahydroisoquinoline also exists in interconverting half-chair conformations (compare with **175** and **176**, Section IX.A.2 of this Chapter). Haimova and co-workers[536,537] reduced the *cis-* and *trans-*3-aryl-4-methoxycarbonyl-1,2,3,4-tetrahydroisoquinolines (see Section X.A.1) to the corresponding 4-hydroxymethyl derivatives and showed by ¹H nmr spectroscopy that the substituents at C–3 and C–4 have pseudo-axial and pseudo-equatorial orientations, and that the more stable conformations for the cis isomers are those in which the hydroxymethyl groups are axial and hydrogen bonded (intramolecular) with the nitrogen atom **221.**

| 221 | 222 | 223 |

The *N*-nitroso group in 1,2,3,4-tetrahydro-*N*-nitrosoisoquinoline caused large chemical shift differences in the ¹H nmr spectra between the α axial and α equatorial hydrogen atoms. The data was interpreted in terms of the two conformers **222** and **223** in ratios that varied with the solvent. In benzene solution the ratio of **222** to **223** was 1 : 2.5 and in carbon tetrachloride it was 1.5 : 1.[553] The isopropyl methyl groups in *N*-substituted 1-isopropyl-1,2,3,4-tetrahydro-6,7-dimethoxyisoquinoline are magnetically nonequivalent. This is caused by both anisotropic effects, restricted rotation, and by the presence of a chiral center at C–1. The chemical shift differences between the methyl groups and the *J* values between the isopropyl hydrogen atom and the methyl groups varies with the substituents on the nitrogen atom. The largest chemical shift difference (0.73 ppm) yet observed for isopropyl methyl groups is found in this system.[554]

3. Reactions of 1,2,3,4-Tetrahydroisoquinolines

General discussions on the degradation of optically active 1,2,3,4-tetrahydroisoquinolines will be found in texts on the chemistry of isoquinoline alkaloids,[530] and will not be treated here. The stereospecificity of the enzyme α-chymotrypsin, which catalyzes the hydrolysis of esters and amides derived from *N*-acetyl derivatives of naturally occurring L-(*S*)amino acids, has been thoroughly discussed in several monographs.[555,556] The specificity of the enzyme towards 1,2,3,4-tetrahydro-3-methoxycarbonyl-2-oxoisoquinoline is anomalous in that it hydrolyzes the *R* isomer and not the *S* isomer. This system has been investigated in much detail and considerable information regarding the active site of the enzyme has accumulated, but some ambiguities are still unsolved.

During the cyclization of optically active *N*-dimethoxyethyl-*N*-methyl-α-benzyl-3,4-dimethoxybenzylamine, Knabe, Dyke, and co-workers[557] observed that a rearrangement occurred and optically active 3-benzyl-3,4-dihydro-6,7-dimethoxy-2-methylisoquinolinium perchlorate was formed. Migration of the benzyl group probably took place after the initial cyclization because optically active 1,2-dihydro-6,7-dimethoxy-2-methyl-1-(1-phenylethyl)isoquinoline rearranged to optically active 1,2-dihydro-6,7-dimethoxy-2-methyl-3-(1-phenylethyl)isoquinolinium salt (224). This was converted into optically inactive 1-cyano-1,2,3,4-tetrahydro-6,7-dimethoxy-2-methyl-1-(1-phenylethyl)isoquinoline (225) (Equation 21).[557a] The zero rotation of 225 is due to fortuitous compensation of the three chiral centers because the perchlorate derived from it is optically active. The relative stereochemistry of the three chiral centers in the last named compound is

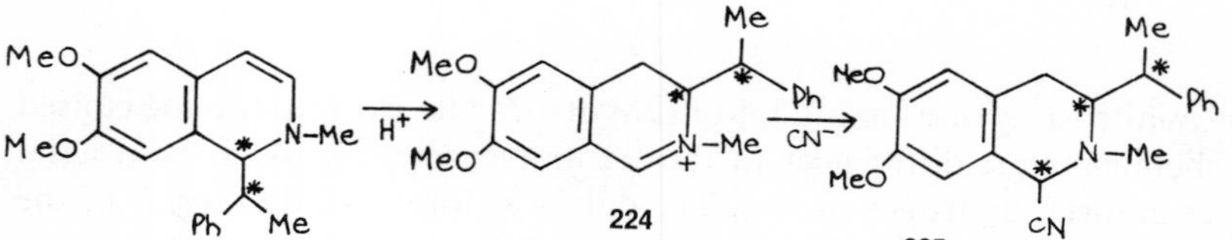

(21)

224

225

not known. Dyke and co-workers[557b] demonstrated that the cyclization-rearrangement of the R-allylaminoaldehyde, 226, into 3S-3-allyl-1,2,3,4-tetrahydro-6,7-dimethoxy-2-methylisoquinoline (Equation 22) was completely stereospecific. The C–1 → C–3 migration of the benzyl group,

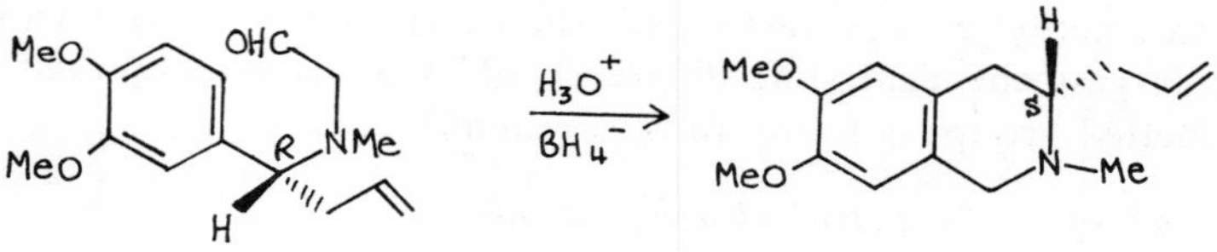

(22)

226

studied by Knabe, Dyke, and co-workers, was shown to be concerted and intramolecular. It is affected by the size of the substituent on the nitrogen atom, and 1-aryl and 1-alkyl groups do not migrate. The yield of product drops in dilution. The data so far is consistent with a mechanism requiring a bimolecular transition state in which the 1-benzyl group of one molecule is transferred to C–3 of the second molecule.[557] It is possible that at higher temperatures the rearrangement proceeds by a free radical mechanism. The acid catalyzed rearrangement of 3-substituted 1-allyl-2-methyl-1,2-dihydroisoquinoline to 3-substituted 3-allyl-2-methyl-3,4-dihydroisoquinolinium cation by way of an aza Cope rearrangement (intramolecular suprafacial sigmatropic [3.3]shift) also produced unusual products, such as

2,3-benzo-7-chloro-9-methyl-9-azabicyclo[3.3.1]non-2-ene (the tetrahydroisoquinoline with a 3-carbon bridge between C–1 and C–3), which are similar to the presumed intermediates in the rearrangement.[557c]

B. Decahydroisoquinolines

1. Syntheses of Decahydroisoquinolines

a. *By Catalytic Reduction of Isoquinolines and Partially Reduced Isoquinolines.* Witkop[558] reduced isoquinoline in acetic acid containing sulfuric acid with platinum oxide and isolated a mixture containing 70–80% of *cis*- and at least 10% of *trans*-decahydroisoquinoline, from which the cis isomer was obtained in high purity. Pure *trans*-decahydroisoquinoline is best prepared by reduction of 1,2,3,4-tetrahydroisoquinoline first by lithium in ethylene diamine followed by catalytic reduction. The proportion of cis isomer is relatively small in this case, and the method is just as satisfactory for the preparation of *trans*-N-methyldecahydroisoquinoline.[559] From the catalytic reduction of 4-hydroxy,[560] 5-hydroxy,[561] 7-hydroxy-2-methyl,[562,563] and 4-ethoxycarbonyl[564] 1,2,3,4-tetrahydroisoquinolines the respective *cis*-decahydroisoquinolines were isolated. Catalytic reduction of 7-hydroxy-8-methyl[565] and 5-chloro-8-hydroxy-2-methyl[566] 1,2,3,4-tetrahydroisoquinolines gave mixtures from which the respective *cis*- and *trans*-decahydroisoquinolines were separated. Apparently the use of a Raney nickel catalyst favors the formation of the trans isomer,[567,568] whereas reductions with platinum oxide in acidic medium favor the cis isomer. This is in agreement with the contention of Auwers and Skita[569] which predicts that reductions in neutral or basic media give high proportions of trans isomer and in acidic media they give the cis isomer. The presence and position of substituents in the benzene ring of isoquinoline have a marked effect on the stereochemistry of reduction. Catalytic reduction of 2-methyl-5-nitroisoquinolinium tosylate and of 5-hydroxyisoquinoline with PtO₂ in acid medium furnishes a 2:1 mixture of the corresponding cis- and trans-fused decahydroisoquinolines.[570] Under similar conditions 5-bromo-2-methyl-8-nitroisoquinolinium tosylate produced an approximately 1:2.2 mixture of cis- and trans-fused 8-amino-2-methyldecahydroisoquinolines.[571]

Reduction of 2-acetyl-,[572] 2-benzoyl-,[573] and 2-methyl-1,2,3,4,6,7,8,8a-octahydro-6-oxoisoquinolines[574,575] gave cis or a mixture of cis and trans isomers of the respective decahydroisoquinolines. Augustine[576] examined in detail the reduction of 2-benzoyl-1,2,3,4,6,7,8,8a-octahydro-6-oxoisoquinoline under·a variety of conditions and concluded that in all cases a mixture of *cis*- and *trans*-2-benzoyl-6-oxodecahydroisoquinoline was formed, but the ratio varied from 25:75 with 10% Pd–C in ethanol to 85:15 with 30% Pd–C in ethanolic-aqueous HCl. This is consistent with the Auwers-Skita rule and demonstrates the need for detailed experimentation for the attainment of optimum yields. Reduction of 2-methyl-1,2,3,4-5,6,7,8-octahydro-7-oxoisoquinoline (double bond common to both rings) gives only cis-2-methyl-7-oxodecahydroisoquinoline with Pd–C in neutral or acidic medium.[574] 2-Methyl-1,2,3,4,4a,5,6,7-octahydro-7-oxoisoquinoline with PtO$_2$–AcOH furnished cis-2-methyl-7-oxodecahydroisoquinoline,[577] but Pd–C gave 9:1 and 6:3 ratios of cis and trans isomers in acidic and neutral media respectively.[574] The influence of the catalyst, and the medium, is important because 2-benzoyl-1,2,3,4,6,7,8,8a-octahydro-6-oxoisoquinoline is reduced by Rh–Al$_2$O$_3$ in about 10% ethanolic HCl medium to cis-2-benzoyl-6-oxodecahydroisoquinoline, whereas Pd–C in ethanol gives the trans isomer.[578]

b. By Cyclization of 1,2-Disubstituted Cyclohexanes.   Authentic trans-decahydroisoquinoline was synthesized from trans-4-methoxycarbonyl-5-methoxycarbonylmethylcyclohex-1-ene, by way of the diacid, cyclic anhydride and imide which was prepared from dimethyl fumarate and butadiene (Equation 23).[579] It was identical with Witkop's[558] compound which was

differentiated from the cis isomer by being less readily dehydrogenated with palladium black. Abramovitch and Muchowski[580] obtained a mixture of cis- and trans-1-cyano-2-bis(ethoxycarbonyl)methylcyclohexane, from a Michael condensation of diethyl malonate and 1-cyanocyclohex-1-ene, which was cyclized to a 68:32 mixture of cis- and trans-4-ethoxycarbonyl-3-oxodecahydroisoquinolines. These were separated by chromatography and hydrolyzed, decarboxylated, and reduced to cis- and trans-decahydro-isoquinolines which were identical with Witkop's compounds. A similar series of reactions starting from 4-t-butyl-1-cyanocyclohex-1-ene gave cis- and trans-6-t-butyldecahydroisoquinoline.[581–583] Kimoto and co-workers[584–586] synthesized cis- and trans-4-hydroxydecahydroisoquinolines, starting from cis- and trans-hexahydrophthalic anhydrides, by conventional methods without epimerizing the asymmetric centers. They prepared the 7-hydroxy-2-methyldecahydroisoquinolines from the lactone 227, which was converted into the key intermediate 228, and by epimerization into the trans isomer 229. The structure of the latter was confirmed by conver-

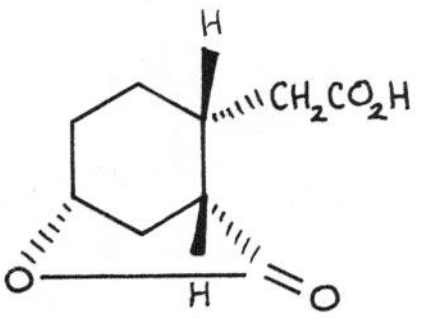

227 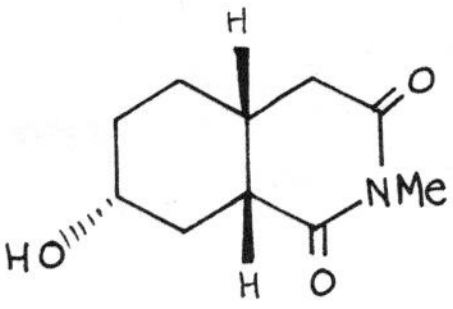   228 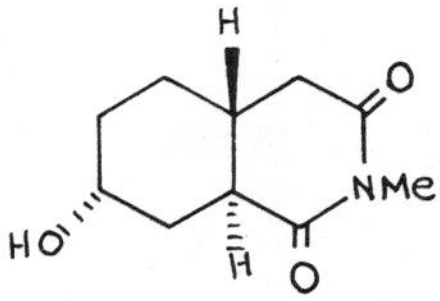   229

sion into *trans*-2-methyldecahydroisoquinoline.[587,588] Itoh[589] prepared all four isomers of 4-hydroxymethyl-2-methyldecahydroisoquinoline from reduction and cyclization of the cyclohexylidene diester **230**, which provided about equal amounts of the epimeric esters (**231**), followed by conventional

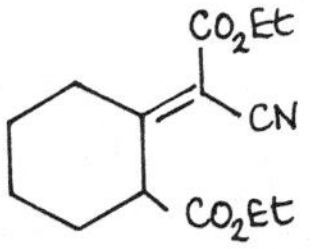

230 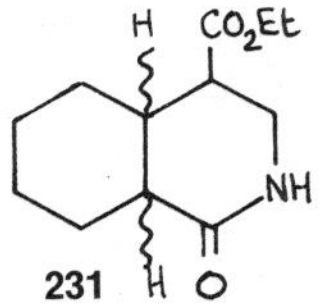   231

chemical elaboration of each of these. The 4*a*-phenyldecahydroisoquinoline which Boekelheide and Schilling[590] prepared by cyclization of 1-aminomethyl-2-bromoethyl-2-phenylcyclohexane is most probably the cis isomer.

c. *By Cyclization of 1-Aminoethylcyclohex-1-enes.* Grewe and coworkers[591-593] cyclized 1-methylaminoethylcyclohex-1-ene with acetaldehyde at pH 6 and obtained 4*a*-hydroxy-1,2-dimethyldecahydroisoquinoline. Arylpyruvaldehydes in methanol also underwent cyclization but a cyclic acetal resulted which was given the structure **232**. Grob and Wohl[594]

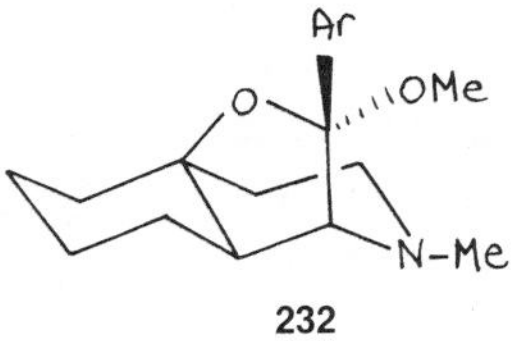

232

reexamined Grewe's synthesis and showed that it was stereospecific leading to cis-fused 4*a*-substituted (depending on the nucleophile present) decahydroisoquinolines. This is a trans addition of the methyleneimine intermediate and the nucleophile to give the cis-fused system (Equation 24). *cis*-4*a*-Hydroxy-2-methyldecahydroisoquinoline obtained in this way was the isomer of the product formed from addition of HOBr to 2-methyl-1,2,3,4,5,6,7,8-octahydroisoquinoline (double bond common to both rings), that is, *trans*-8*a*-bromo-4*a*-hydroxy-2-methyldecahydroisoquinoline, fol-

$$Y^- = OH, OMe, OAc, Cl \qquad (24)$$

lowed by debromination. The *trans*-bromohydrin was converted to *cis*-4a,8a-epoxy-2-methyldecahydroisoquinoline by alkali then reduced to *trans*- and *cis*-8a-hydroxy-2-methyldecahydroisoquinolines by LAH and Raney Ni–H_2 respectively. If the reaction in Equation 24 is performed in the presence of chloride ions the *cis*-4a-chloro- and *cis*-4a-chloro-2-methyl decahydroisoquinolines are formed. This is a kinetically controlled reaction because when the *cis*-hydroxy compound **233** (Y = OH) or its trans isomer are treated with hydrogen chloride in acetic acid only the thermodynamically more stable *trans*-4a-chloro-2-methyldecahydroisoquinoline is formed. The structures of the chloro compounds were confirmed by reduction to the known *cis*- and *trans*-2-methyldecahydroisoquinolines.[594,595] Schneider[596,597] observed similar cyclizations with the hydroxymethylcyclohexenes (**234**, R = CH_2Ph, and $CH_2CH=CH_2$) and proposed the trans structure for

234

the 4a-hydroxydecahydroisoquinolines that resulted. If these postulated structures are true, then the vicinal hydroxymethyl group may well have upset the stereochemistry of the nucleophilic attack on the intermediate carbonium ion in the transition state.

    d.   *Miscellaneous.*   N-Methyl-N-(2'-oxocyclohexyl)methyl 2-(p-methoxybenzoyl)ethylamine cyclizes under Mannich conditions into a mixture of two stereoisomers of 4a-hydroxy-2-methyldecahydroisoquinoline. The trans-fused structure **235** was assigned to one of the isomers mainly from ir data, but the second isomer could be either one of the three other possibilities.[598]

    The pyrrolidine enamine of 1-methyl-4-piperidone undergoes the annelation reaction with methyl vinyl ketone, as does cyclohexanone, and pro-

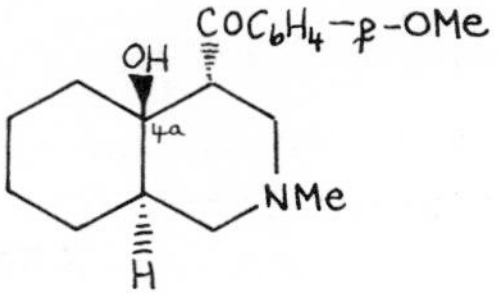

235

duces 2-methyloctahydroisoquinoline which is hydrogenated with Pd–C to cis-2-methyl-6-oxodecahydroisoquinoline.[599] A similar condensation with 1-acetyl-2-arylmethyl-3-ethoxycarbonyl-4-piperidone forms 2-arylmethyl-1-ethoxycarbonyl-6-hydroxy-6-methyl-3-azabicyclo[3.3.1]nonan-9-one. The latter rearranges in the presence of alkoxide into the 8a-ethoxycarbonyl-4a-hydroxydecahydroisoquinoline in which the stereochemistry is most probably trans (Equation 24a).[600,601]

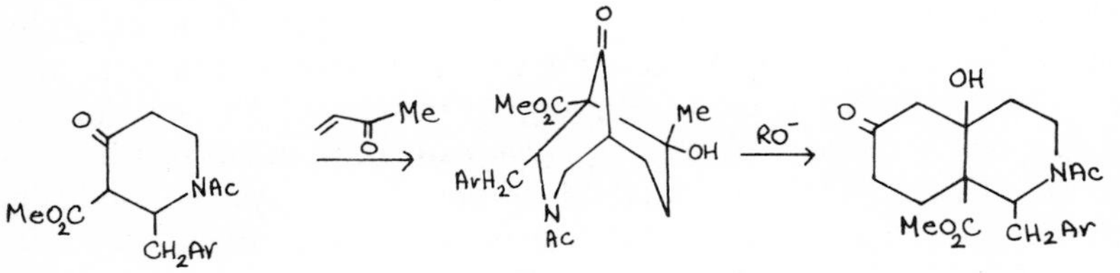

(24a)

Lithium diphenyl copper adds across the double bond of 1,2,3,4,6,7,8,8a-octahydro-2-methyl-6-oxoisoquinoline in a regio- and stereospecific manner. The structure of the product, cis-2-methyl-6-oxo-4a-phenyldecahydro-isoquinoline, was confirmed by an X-ray analysis.[602]

Epoxidation of 1,2,3,4,5,6,7,8-octahydro-2-methylisoquinoline yields a mixture of two 4a,8a-epoxides. Performic acid, however, gives almost exclusively the isomer **236** in which the epoxy group is cis to the benzyl group. The authors postulated a transition state (**237**) involving π-bonding with the benzene ring.[603]

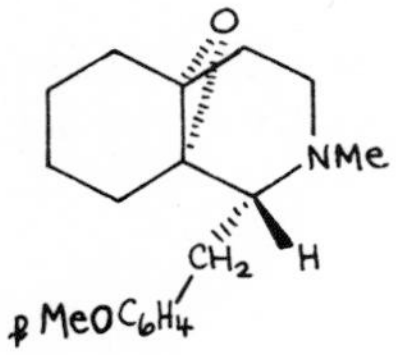

236

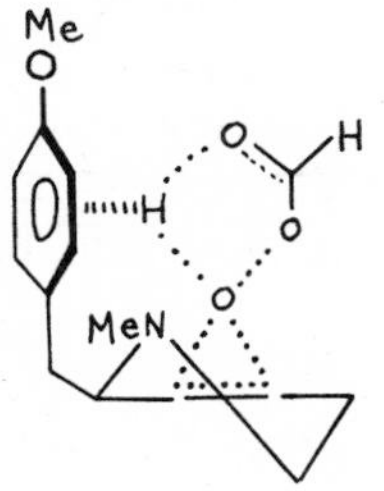

237

2.   Configuration and Conformation of Decahydroisoquinolines

The configurations of several decahydroisoquinolines have been mentioned in the previous section and most of them were assigned by synthesis and spectroscopic studies, and a few by X-ray analyses.

The optical resolution of *trans*-2-methyldecahydroisoquinoline was achieved by Witkop[568] by recrystallization of the (+)-bitartrate salt. The (+)-base was identical with the 2-methyldecahydroisoquinoline obtained from the degradation of Yohimbine and established the constitution of the isoquinoline portion of the alkaloid. *cis*- and *trans*-2-Benzoyl-6-oxodeca-hydroisoquinolines were resolved into their enantiomers. The absolute configurations of the *trans*-4aS,8aS and 4aR,8aR isomers were derived from evaluation of the ketone regions of the ord and cd curves which had positive and negative Cotton effects respectively. The absolute configuration of the *trans*-4aS,8aS and *cis*-4aS,8aR isomers were confirmed by their preparation (by sequences which did not affect the chiral centers) from naturally occurring chinchonine by way of the *t*-butyl ester of meroquinene of known absolute configuration. The carbonyl group of the *cis*-4aS,8aR enantiomer is stereospecifically reduced by *Sporotrichum exile* (QM-1250) to the alcohol, and this affords a very good method for obtaining unreacted, optically pure *cis*-4aR,8aS-1-benzoyl-6-oxodecahydroisoquinoline directly and its enantiomer by oxidation of the biologically prepared alcohol.[578]

The ^{1}H nmr spectra of the decahydroisoquinolines resemble those of decahydroquinolines; that is, the carbocyclic protons of the cis isomer display a narrow band envelope and those of the trans isomer show a broad band envelope.[477] The spectrum of *trans*-decahydroisoquinoline remains unchanged on cooling to $-60°C$, but that of the cis isomer broadens and splits at $-30°C$.[594] The latter is consistent with equilibria similar to those in *cis*-decahydroquinoline (Scheme 5). ^{1}H nmr and ir spectra of cis and trans axial and equatorial 5-, 6-, 7-, and 8-monohydroxy-2-methyl-decahydroisoquinolines were measured.[604] Several of these exhibit intra-molecular hydrogen bonding in the ir (e.g., **238**), and this is useful in assigning configuration and conformation.[566] The orientation axial ($\nu_{max} = 1000$ cm^{-1}) and equatorial ($\nu_{max} = 1040$ cm^{-1}) can be deduced from ir spectra and from the $W_{1/2}$ value (width-at-half-height) of the CH—OH proton.[605,584] The ir spectra of all four 4-hydroxymethyl-2-methyldecahy-droisoquinolines give valuable information about their structures, namely,

238

239

240

239, 240, 241 ⇌ **242**, and **243** ⇌ **244**.[589] The p*Ka* values of the 5-, 6-, 7-, and 8-hydroxy-,[605] and 4-hydroxymethyl-2-methyldecahydroisoquino-lines[589] were recorded but did not give very valuable information regarding the stereochemistry. The p*Ka* values for the 4-hydroxy derivatives were, however, more useful.[584]

Although *trans*-decahydroisoquinoline is a rigid stable chair-chair structure, by judicious substitution Abramovitch and Struble[582,583] succeeded in transforming this into a twist-boat conformation **(245)** by inserting a 6-*t*-butyl group, an 8*a*-ethoxycarbonyl group, and a 3-oxo group.

241

242

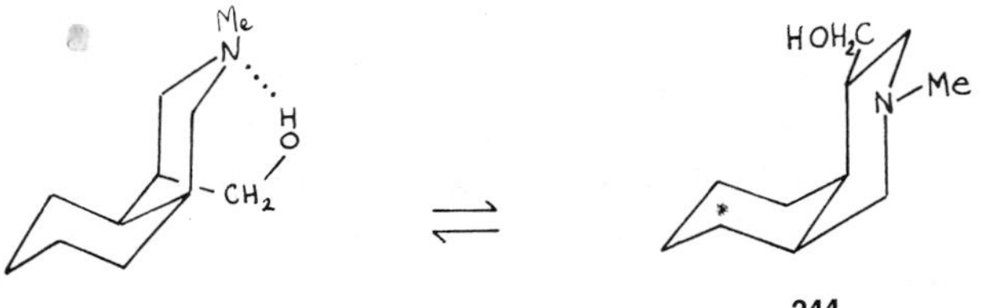

243

244

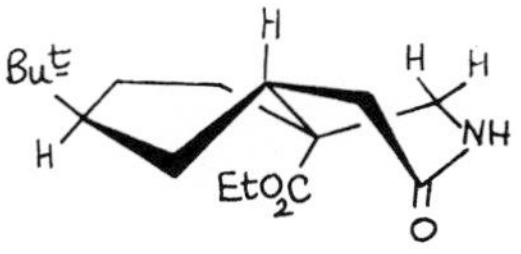

245

3. Reactions of Decahydroisoquinolines

The heterolytic fragmentation observed in 4-chloropiperidines (Section II.D.6) and tosyloxydecahydroquinolines (Section IX.B.4.c) was shown to occur also in 4*a*-chloro-2-methyldecahydroisoquinolines. Grob and co-workers[606] demonstrated that in the presence of a base (or potassium cyanide) fragmentation of the cis isomer occurs as in Equation 25. The

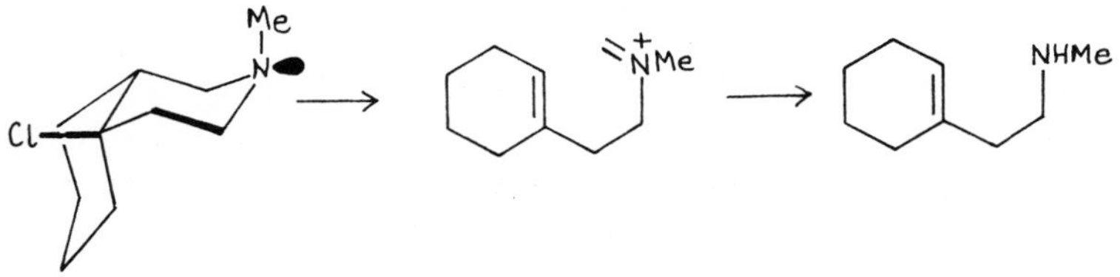

(25)

trans isomer yields a fragmented product (e.g., **246** with KCN) together with octahydroisoquinolines from an elimination reaction. The mechanisms are obviously different for the two isomers. The fragmentation of the trans isomer probably goes through the carbonium ion **247**. In 80%

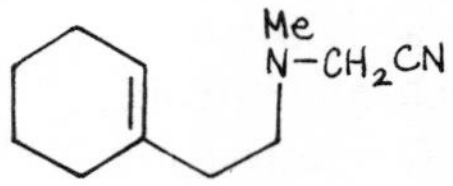

246 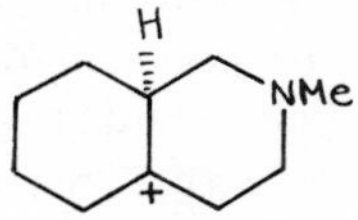   **247**

ethanolic triethylamine, the first order rates constants are 9.80×10^{-2} and 1.26×10^{-4} sec^{-1} for the cis and trans isomers respectively. When compared with the respective *cis*- and *trans*-4a-chlorodecalins, *cis*-4a-chloro-2-methyldecahydroisoquinoline is 132 times faster, but the *trans*-isoquinoline is only 0.22 times faster.

The reaction of *cis*- or *trans*-4a-chloro-2-methyldecahydroisoquinoline with hydrochloric acid is different from the above and 5-, 6-, and 7-chloro-decahydroisoquinolines are formed. The reaction involves elimination of HCl and multiple isomerization of the double bond, followed by addition of HCl, to yield the thermodynamically more stable chloro derivatives.[585] *trans*-4a-Bromo-2-methyldecahydroisoquinoline behaves similarly. The isomerization and effect of temperature are depicted in Scheme 6. Grob and

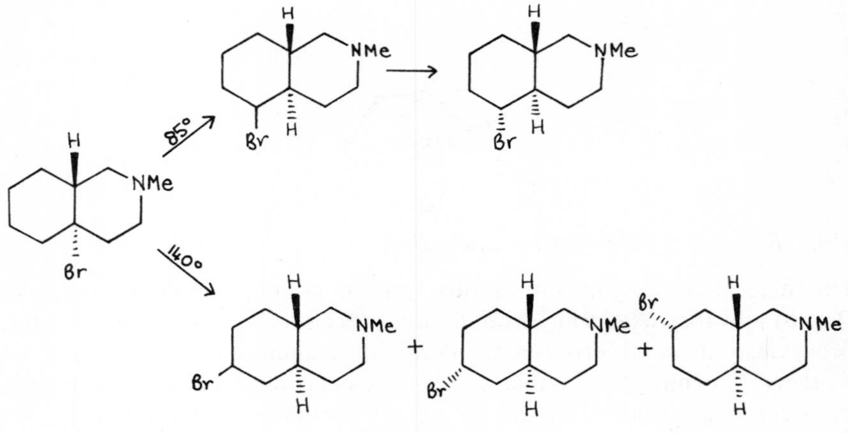

Scheme 6

Wohl[607] studied this system in detail and found that the epimeric bromo compounds are equilibrated to a mixture of 74% equatorial and 26% axial bromine orientation with LiBr in methyl ethyl ketone. *cis*-4a-Hydroxy-2-methyldecahydroisoquinoline in boiling 60% perchloric acid yields a mixture of 5-, 6-, and 7-hydroxy-*trans*-2-methyldecahydroisoquinolines by elimination of water, and multiple double-bond migration followed by hydration.[608]

Hofmann degradation of *cis*- and *trans*-2-methyldecahydroisoquinoline methiodides gives *cis*- and *trans*-1-*N*,*N*-dimethylaminomethyl-2-vinylcyclohexane **(248)** respectively,[609] which is consistent with the mechanism discussed previously (Section IX.B.4.c and Chapter 2, Section III.E.4).

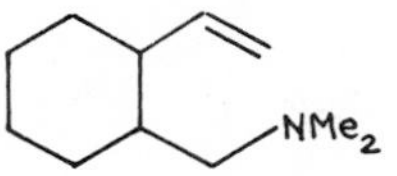

248

The stability of cis and trans isomers of 1,3-dioxodecahydroisoquinoline and its 4a- and 8a-methyl derivatives were revealed by equilibrating each of them with an equal amount of 5% Pd–C at about 300°C under pressure in the absence of solvent. The cis:trans ratios, evaluated by ir spectroscopy, were 32:68, 71:29, and 60:40 respectively and demonstrated the influence of the bridgehead methyl group on the equilibria.[610]

C.   Azatwistanes

1-Azatwistanes, 1-azatricyclo[4.4.0.^{1,6}0^{3,8}]decanes **(249, R = lone pair)**, are compounds with a bridgehead nitrogen atom but are included after the heading of reduced isoquinolines because they are readily formed by the intramolecular cyclization of *cis*-decahydroisoquinolines. *cis*-1-Alkyl-6-(equatorial)methanesulfonyloxy(or *p*-toluenesulfonyloxy)decahydroisoquinolines in boiling methyl cyanide or toluene undergo intramolecular quaternization to yield *N*-alkyl 1-azatwistane salts **(249)**.[559,573] If the alkyl group is benzyl then the parent compound can be obtained by hydrogenolysis of the salt. A methyl substituent at C–4a does not interfere with the reaction.[611] In a

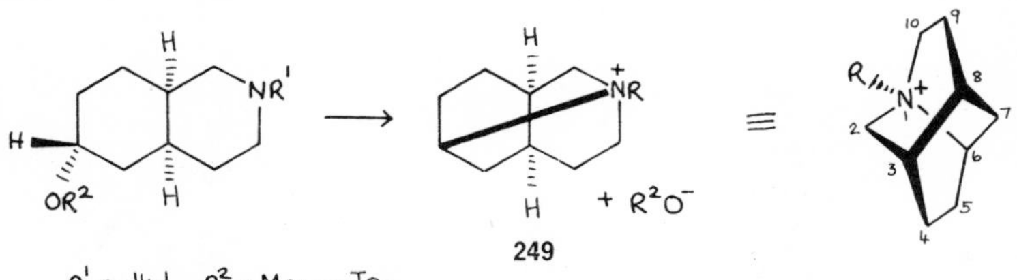

249

R¹ = alkyl, R² = Mes or Tos

second synthesis a 4:3 mixture of *cis*- **(250)** and *trans*-(with respect to the nitrogen atom)5-chloro-1-azatwistanes **(251)** is formed when *cis*-2-chloro-1,2,3,4,4a,5,8,8a-octahydroisoquinoline is photolyzed in 80% trifluoracetic acid, and can be separated by way of their *p*-toluenesulfonate salts. Reduction of these with sodium in isopropanol yields 1-azatwistane together with the product of Cl–C–C–N fragmentation *cis*-1,2,3,4,4a,5,8,8a-octahydroisoquinoline (Equation 26). The structure of the hydrochloride salt of *trans*-5-chloro-1-azatwistane (**251**) was confirmed by an X-ray analysis.[612]

The chloro-1-azatwistanes solvolyze in an interesting manner. In aqueous dioxane the cis isomer yields the α-carbinolamine 1-azaisotwistane **252** ($R^1 = H, R^2 = OH$ or $R^1 = OH, R^2 = H$) which is formed by a Wagner-Meerwein rearrangement. The isotwistane then hydrolyzes into a mixture of *endo*- and *exo*-6-formyl *cis*-perhydro-2-pyrindines (Chapter 3, Section VIII.B).[434] The general reaction with various nucleophiles (e.g., OH⁻, MeO⁻, *t*-BuO⁻, and PhCH₂O⁻) is shown in Equation 27.[613] The product from methanolysis of the azatwistane **250** was identified unequivocally as 10-methoxy-6-azatricyclo[4.3.1.0³,⁸]decane (**252**, $R^1 = H$, $R^2 = OMe$) by an

X-ray analysis of its hydrochloride.[614] Solvolysis of the chloroazatwistane **250** produces the twistane carbonium ion **253** which is in equilibrium with the rearranged isotwistane carbonium ions **253a** and **253b**. The latter requires a 1,3-hydride shift as well as a Wagner-Meerwein rearrangement. The carbinolamine derived from the carbonium ion **253a**, that is, **252**

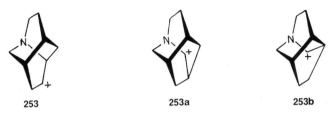

| 253 | 253a | 253b |

($R^1 = H, R^2 = OH$ and $R^1 = OH, R^2 = H$), ring opens to the aldehydes in Equation 27. The existence of the carbonium ion **253b** is demonstrated by ring opening of the α-carbinolamine derived from it and conversion into the all-*cis*-5-formyl-2-pyrindine (Equation 28).[615]

$$(253) \xrightarrow{OH^-} \quad \longrightarrow \quad \equiv \qquad (28)$$

The hydrocarbon twistane, tricyclo[4.4.0.0^{3,6}]decane, has been obtained in optically active forms by Tichý,[616,617] who determined its absolute configuration. The optically active forms of 1-azatwistane have not yet been prepared, but a related compound, $(-)$-4-azatricyclo[4.4.0^{1,6},0^{3,8}]decan-5-one (**255**), was synthesized by Tichý and collaborators[618] from $2S,5S(+)$-*endo*-bicyclo[2.2.2]octan-2,5-dicarboxylic acid (**254**), of known absolute configuration, by way of the $(+)$-amino acid (Equation 29). The absolute configuration (*P*-helicity) **255** follows from the synthesis. An interesting feature about this compound is the relatively nonpolar amide carbonyl group which has a high ir stretching frequency ($\nu_{max} = 1710$ cm-1 in CCl$_4$)

$$\qquad \rightleftharpoons \qquad \longrightarrow \qquad (29)$$

| 254 | | 255 |

because of the cage structure. In an unrelated synthesis 6-N-chloro-N-methylaminomethylbicyclo[2.2.2]oct-2-ene cyclizes intramolecularly in methanol to 2-methoxy-4-methyl-4-azatwistane.[618a]

Oxa-, thia-, and oxathiatwistanes are described in Part II, Chapter 2, Section XI.D; Part II, Chapter 3, Section XIII; and Part II, Chapter 4, Sections II.E and IV.D respectively.

XI. REDUCED FUSED SIX : SIX-MEMBERED RINGS WITH MORE THAN ONE NITROGEN ATOM

A. Reduced Naphthyridines

cis- and trans-Decahydronaphthyridines are capable of existence as were the quinolines and isoquinolines. Armarego[477] prepared trans-decahydro-1,5-, 1,6-, 1,7-, and 1,8-naphthyridines by reduction of 1,5-, 1,6-, 1,7-, and 1,2,3,4-tetrahydro-1,8-naphthyridines with sodium in boiling ethanol. Catalytic reduction of 1,5-naphthyridine with PtO_2 in $AcOH–H_2SO_4$ gave an approximately 2:1 mixture of cis- (256) and trans-decahydro-1,5-naphthyridines (257) which were separated by chromatography. These results are in agreement with the Auwers-Skita rule.[569] The ^1H nmr spectra of cis- and trans-decahydro-1,5-naphthyridines were used to distinguish between them because, as in the quinolines (Section IX.B.3 of this chapter) and isoquinolines (Section X.B.2) the proton signal envelopes were narrower in the cis than in the trans isomers. trans-Decahydro-1,6-, 1,7-, and 1,8-naphthyridines all had broad band envelopes for the proton signals. trans-Decahydro-1,5-naphthyridine is a meso compound and the cis isomer is racemic. The

256   257

reverse is true for decahydro-1,8-naphthyridine, and the cis and trans isomers of decahydro-1,6- and 1,7-naphthyridines are racemic. None of these have been obtained in optically active forms.

Catalytic reduction of 2,6-dihydroxy-1,5- and 1,7-naphthyridines in acetic acid (PtO_2) give high yields of cis-2,6-dioxodecahydro-1,5- and 1,7-naphthyridines respectively. The former is also formed in 95% yield when racemic diethyl 4,5-diaminosuberate is kept at pH 7.5, and the meso-diamino ester similarly gives trans-2,6-dioxodecahydro-1,5-naphthyridine. Reduction of these dioxonaphthyridines with LAH produced cis- and trans-decahydro-1,5-naphthyridines identical with those described earlier.[619]

Takata[620] prepared decahydro-1,8-naphthyridine and its 3- and 3,6-dimethyl derivatives by cyclization of the corresponding 3-(3-aminopropyl)-2-piperidones to the amidines **258** followed by reduction with sodium and amyl alcohol. Zondler and Pfleiderer,[621] on the other hand, converted cyanoethyl cyclic aldimines, such as **259,** into the respective aminopropyl compounds which cyclized into decahydro-1,8-naphthyridines (Equation 30). The unsubstituted base is obviously trans because it is identical with

258   (30)   **259**

the above. The trans structure is probably valid also for the 3-methyl and 3,6-dimethyl derivatives but not necessarily for those which produce naphthyridines with bridgehead substituents. The intramolecular cyclization of 1,1,1-tris(2-cyanoethyl)acetone produces a 4a,8a-tetramethyleneperhydro-1,8-naphthyridine (**260**) which must be cis.

260

4-Cyanopyridine methiodide dimerizes in 0.8*N*-NaOH in methanol at 0°C to the cis-fused tricyclic octahydro-2,7-naphthyridine (**262,** R = H) (*endo*-4,9-dimethyl-7,11-dicyano-4,9-diazatricyclo[6.2.2.0^{2,7}]dodeca-5,11-diene). At still lower temperatures the intermediate head-to-head [2 + 2]-dipyridine dimer **261** is formed in appreciable amounts. The naphthyridine can be monodeuterated specifically with DCl, and the monodeutero derivative **262** (R = D) isomerizes by a sigmatropic shift or by way of the intermediate **261**[622,623] (Equation 31). 2-Cyanopyridine methiodide undergoes an identical series of reactions except that the 6,8-dicyanotricyclododecadiene is obtained instead of the 7,11-dicyano derivative (**262**).[624]

B. Reduced Cinnolines

The stereochemistry of reduced cinnolines has attracted very little attention and only one reference is made here.

In a synthesis of azamorphan analogues, Kametani and collaborators[625] have cyclized 2-carboxymethyl-2-*m*-methoxyphenylcyclohexanone with methylhydrazine to 4*a*-(*m*-methoxyphenyl)-2-methyl-3-oxo-2,3,4,4*a*,5,6,7,8-octahydrocinnoline. This was subsequently reduced with LAH or catalytically to 4*a*-(*m*-methoxyphenyl)-2-methyldecahydrocinnoline which was presumed to be trans-fused. The latter gave an azamorphan by a Pictet-Spengler reaction (Equation 32).

C. Reduced Phthalazines

Only a few of the hydrophthalazines that have been prepared have some three-dimensional properties of interest.

cis-4a,5,6,7,8,8a-Hexahydrophthalazines are formed by a stereoselective synthesis when a cycloaddition reaction is involved. The reaction of choice is the cycloaddition of 3,6-disubstituted 1,2,4,5-tetraazines and a cyclohexene. Extrusion of a nitrogen molecule follows the addition, and the required 1,4-disubstituted cis-4a,5,6,7,8,8a-hexahydrophthalazines are formed.[626] Steric factors affect the ease of reaction. Wilson and Warrener[627] have successfully used this with cyclohexa-1,3-dienes related to norbornene. The cis-hydrophthalazines obtained undergo thermal fragmentation by a retro (i.e., $\pi 4s + \pi 2s$) Diels-Alder reaction and form a useful new route to fulvenes or furans and pyridazines (Equation 33).

$$R^2 = CF_3, CO_2Me$$
$$X = O$$

(33)

In a more conventional synthesis cis- or trans-1-acetyl-2-ethoxycarbonyl-[628] or 1,2-di(2,4-dinitrophenoxycarbonyl)[629] cyclohexanes are treated with hydrazine or pyrazoline respectively and produce the hexahydrophthalazines apparently without epimerization.

Conformational changes in the reduced pyridazine ring of hydrophthalazines are similar to those observed in the 1,2-diazanes (hexahydropyridazines—see Section III.B). Nakamura[630] noted ¹H nmr spectral changes in 2,3,5,10-tetrahydro-1H-pyrazolo[1,2-b]phthalazines when the temperature was varied. He measured the free energies involved and concluded that an equilibrium between two trans-fused N-inverting (synchronously) conformers, 263 ⇌ 264, was established.

263   264

Nakamura and Kamiya[631] found that 1-hydroxy-2,3,5,10-tetrahydro-1H-pyrazolo[1,2-b]phthalazine-5,10-diones undergo ring-chain tautomerism. When this system carries a methyl or a phenyl group in the 5-position the tautomerism is stereospecific and in favor of the isomer with the 1-OH and 3-Me or 3-Ph groups in the cis configuration (Equation 34).

(34)

Price, Sutherland and Williamson[249] examined *trans*-4a,8a-dibromo-2,3-bisethoxycarbonyldecahydrophthalazine in their study of ring and nitrogen inversion in N,N'-diacylhexahydropyridazines. They measured the free energy for the barrier of rotation of the N-ethoxycarbonyl groups without interference of the ^{1}H nmr spectra from the effects of ring inversion.

D.   Reduced Quinazolines

Active interest in the stereochemistry of highly reduced quinazolines began recently, although the first record of a perhydroquinazoline was made by Mannich and Hieronimus in 1942.[632] They claimed that 3-benzyl-2-oxo-1,2,3,4,5,6,7,8-octahydroquinazoline disproportionated, on boiling in 20% hydrochloric acid, into 3-benzyl-2-oxo-2,3,5,6,7,8-hexahydroquinazoline and 3-benzyl-2-oxodecahydroquinazoline, but the stereochemistry of the latter was not assigned.

In 1964–1965, Tsuda,[633] Goto,[634] Woodward,[635] and their respective collaborators independently determined the structure of the potent non-protein neurotoxin "tetrodotoxin" which is present in certain varieties of the Japanese puffer fishes. Several X-ray studies of derivatives were reported, but Woodward and Gougoutas[636] obtained the complete structure by single crystal X-ray analysis which gave the absolute configuration 265.

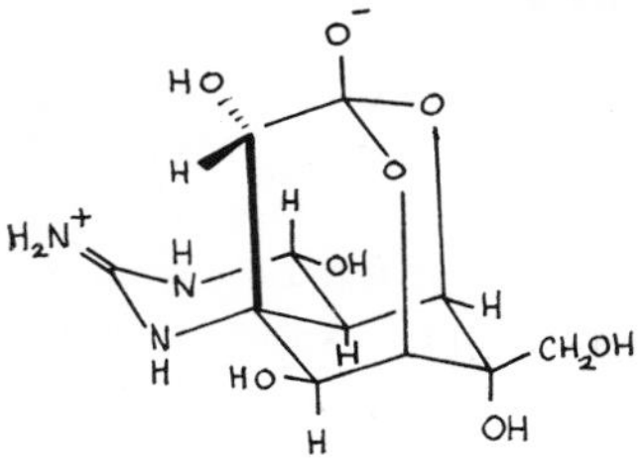

265

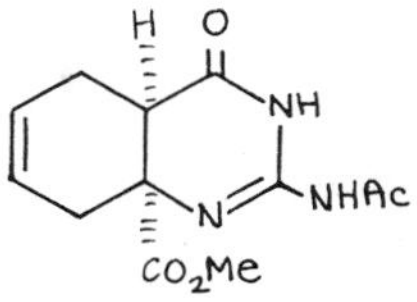

266

It is a 2-iminohydroquinazoline and has stimulated much work towards its synthesis and the synthesis of simpler analogues for biological evaluation. The total synthesis of tetrodotoxin was achieved in 1972 by Kishi and his collaborators[637] in Japan. In this synthesis the heterocyclic ring was formed in the last stages. In a different approach Keana and co-workers[638] prepared 2-acetamido-8a-methoxycarbonyl-4-oxo-3,4,4a,5,8,8a-hexahydroquinazo-line (266) and the 2,4-dioxo derivative, by a Diels-Alder reaction between 2-acetamido-6-methoxycarbonyl-4-oxopyrimidine and butadiene. The stereochemistry at the ring junction is undoubtedly cis. In yet another approach by Armarego,[639] which was intended for the preparation of simpler analogues of tetrodotoxin, 2-oxo-1,2,3,4,5,6,7,8-octahydroquinazoline was fused with nitroacetic acid. The product cis-8a-nitromethyl-2-oxodecahy-droquinazoline (267, R = NO₂) was formed in quantitative yield. The addition of the elements of nitromethane was completely stereospecific and was demonstrated by conversion into cis-8a-methyl-2-oxodecahydroquinazoline (267, R = Me), and by an unambiguous synthesis of the latter from cis-1-amino-trans-1-methyl-r-2-ethoxycarbonylcyclohexane which was obtained

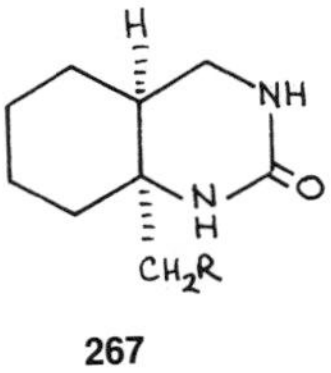

267

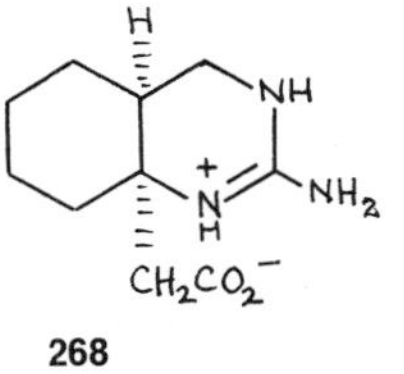

268

by chemical elaboration of the product from a Diels-Alder reaction without altering the relative stereochemistry of the asymmetric centers. The cis-nitro compound 267 (R = NO₂) was converted into the amino acid 268 by reactions which did not affect the stereochemistry at C–4a and C–8a.[640]

Bauer and Nambury[641] prepared 3-benzenesulfonyloxy- and 3-benzoyl-*trans*-2,4-dioxoperhydroquinazolines when the disodium salt of *trans*-1,2-bis-*N*-hydroxyamidocyclohexane was treated with benzenesulfonyl chloride and when its dibenzoyl derivative was treated with a base, respectively. The Lossen rearrangement had occurred with retention of configuration. Armarego and Kobayashi[642] prepared *cis*- and *trans*-hexahydroanthranilic acids, and from these they obtained a variety of hexahydro- and octahydroquinazolines in which the stereochemistry of the asymmetric carbon atoms was unaltered. They converted the anthranilic acids into the respective *cis*- and *trans*-1-amino-2-aminomethylcyclohexanes which reacted with formaldehyde to form *cis*- and *trans*-decahydroquinazolines respectively,[643] and with *S*-methylisothiouronium sulfate the diamines gave the corresponding *cis*- and *trans*-2-amino-3,4,4*a*,5,6,7,8,8*a*-octahydroquinazolines.[644]

Catalytic reduction of 2-amino- or 2-amino-4-chloro-5,6,7,8-tetrahydro-[642] and 1,2,3,4,5,6,7,8-octahydro-2-oxoquinazolines[645] with platinum oxide in acetic acid produces exclusively the respective *cis*-2-amino-3,4,4*a*,-5,6,7,8,8*a*-octahydro- and *cis*-2-oxodecahydroquinazolines in high yields. Zigeuner and co-workers[646] also obtained the spirodecahydroquinazolines **269** by catalytic reduction of the corresponding 1,2,3,4,5,6,7,8-octahydro-2-oxo derivatives. The stereochemistry of the latter compounds is undoubtedly cis at the ring junction but has not been established. They suc-

R=H,Me,Et

269   **270**   **271**

ceeded in forming the adducts (e.g., **270**, X = O,S; R = H,Me,CH$_2$Ph) by heating xylenols and 1,2,3,4,5,6,7,8-octahydro-2-oxo(or thio)quinazolines in concentrated HCl for long periods. The stereochemistry of the adducts also is not known.[647] The condensation product $(C_{11}H_{12}N_2O)_x$ from the reaction of urea and but-3-en-2-one was shown by a detailed study to be a *trans*-2-oxodecahydroquinazoline with the structure **271**.[648]

In a biosynthetic type of synthesis *o*-aminobenzaldehyde and *S*-5-methyl-3,4,5,6-tetrahydropyridine were condensed together in a stereospecific manner and gave 5*aR*,8*S*-8-methyl-5,5*a*,6,7,8,9-hexahydropyrido[2,1-*b*]quinazolinium salt **(272)** which was isolated as the picrate. The structure of this product was confirmed by an X-ray analysis.[649] Armarego and Kobayashi[643,644] prepared *trans-4aR,8aS(−)-, 4aS,8aR(+)-, cis-4aS,8aS-*

272

(−)-, and 4aR,8aR(+)-decahydroquinazolines by unambiguous syntheses, and from the chiral intermediates *cis-* and *trans-*1-amino-2-aminomethyl-cyclohexanes the four optical enantiomers of 2-amino-3,4,4a,5,6,7,8,8a-octahydroquinazolines of known absolute configuration were synthesized. All had plain ord curves. *trans*-4aR,8aS-4-Oxo-3,4,4a,5,6,7,8,8a-octahydro-quinazoline **(273)**, on the other hand, exhibited a (+)-ve Cotton effect with a maximum at 264 nm.[644]

273   **274**   **275**

The reduced ring of 1,2,3,4-tetrahydroquinazolines, like those of 1,2,3,4-tetrahydroquinolines and isoquinolines (Sections IX.A.2 and X.A.2 respectively), should be in the inverting half-chair conformation. There is no experimental verification of these conformations, but the ^{1}H nmr spectra of the 1,3-trimethylene- **(274, R = H)**, 2-methyl-1,3-trimethylene- **(274, R = Me)**, and 2-phenyl-1,3-trimethylene-1,2,3,4-tetrahydroquinazolines **(274, R = Ph)**, and of 1,3-dimethylene-2-phenyl-1,2,3,4-tetrahydroquinazoline **(275)**, which have more rigid structures, were measured.[650] The orientation of the 2-substituents are as shown in the structures **274** and **275**.

The rigid chair-chair conformation of *trans*-decahydroquinazoline **(276)** is evident from the signal patterns of the two C-4 protons in the ^{1}H nmr spectra and from the absence of any changes in spectra on heating or cooling. The chemical shifts and coupling constants of *trans*-decahydroquinazoline (276), its 1- and 3-methyl and its 1,3-dimethyl derivatives fell clearly into a pattern from which it was deduced that the lone pair of electrons on the nitrogen atoms were respectively 1,3-diequatorial, 1-axial-3-equatorial, 1-equatorial-3-axial, and 1,3-diaxial.[651] These deductions were not consistent with observations in 1,3-diazanes (hexahydropyrimidines) (see Section IV.C), and the dipole moments of these methyl-*trans*-decahydroquinazolines were reexamined. The data showed that the apparently clear-cut ¹H nmr results were fortuitous, and a cautionary note was made regarding the assignments of the orientation of the nitrogen lone pair of electrons from the chemical shifts, and J_{gem} values of methylene protons on carbon atoms adjacent to nitrogen atoms.[652]

cis-Decahydroquinazoline exists in two possible extreme conformations, 277 and 278 (neglecting nitrogen inversion). The ¹H nmr spectra of the unsubstituted[643] compound, all the 3-substituted[640] and 8a-substituted derivatives[639, 640, 653] have signal patterns for the C–4 protons which are consistent only with the conformation 277. The other conformation, 278, must be present in small amounts because the proton signals of *cis*-decahydroquinazoline coalesce at about 160°C.[643] Moreover, Armarego and Reece[645] demonstrated that by inserting a 1-benzyl and 1-methyl group in *cis*-decahydroquinazoline only the conformation 278 is evident in the ¹H nmr spectrum. The presence of 2-imino, 2-oxo, or 2-thiono groups do not alter the spectral patterns of the parent compounds, that is, the conformations are unaltered.[645] Katritzky and collaborators,[654] however, found that if a second oxygen atom is inserted at C–4, as in *trans*- and *cis*-2,4-dioxo-decahydroquinazolines 279–281, the dihydrouracil rings attain a half-chair conformation and make it difficult to distinguish between the two cis conformers 280 and 281 by ¹H nmr spectroscopy.

For further reading on reduced quinazolines see Armarego,[655] and for unsaturated quinazolines see Section XIV.B of this chapter. The stereochemistry of the Tröger base and related compounds is described in Chapter 4, Section VIII.

E.   Reduced Quinoxalines

1,2,3,4-Tetrahydroquinoxalines are generally prepared from o-phenylene-diamines with a two carbon unit. d,l-1,2,3,4-Tetrahydro-2-methoxycar-

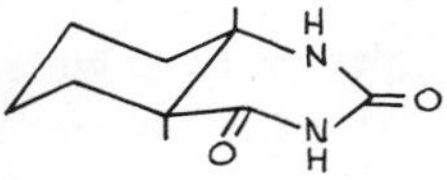

279 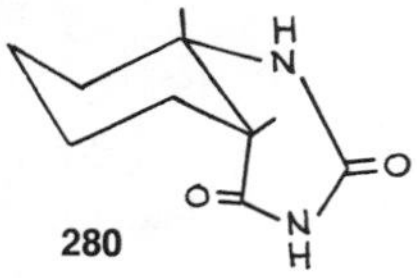   **280** 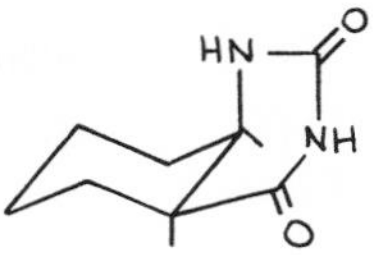   **281**

bonyl-1,4-ditosylquinoxaline is obtained from N,N'-ditosyl-o-phenylenedi-amine and methyl 2,3-dibromopropionate.[656] Catalytic reduction of 2,3-dialkylquinoxalines furnishes the cis-2,3-disubstituted 1,2,3,4-tetrahy-droquinoxalines,[657,658] and reduction with sodium and alcohol gives the trans isomer.[659] Interestingly, LAH reduces 2,3-dimethylquinoxaline (and its 5- and 6-aza derivatives) stereospecifically to cis-1,2,3,4-tetrahydro-2,3-dimethylquinoxalines. In order to account for this specificity the transition state structure **282** is postulated.[660] Haddadin, Alkaysi, and Saheb[659] found that NaBH$_4$ also reduced 2,3-dimethyl-, 2,3-tri-, tetra-, penta-, and hexamethylenequinoxaline 1,4-dioxides to yield predominantly, if not ex-clusively the cis-disubstituted tetrahydroquinoxalines. The mechanism of this reduction does not proceed by way of the mono-N-oxide because an authentic mono-N-oxide produced a 1:1 mixture of cis and trans isomers.

(+)- and (−)-1,2,3,4-Tetrahydro-2-methylquinoxalines were isolated by Schultz and co-workers[661] from an optical resolution of the dibenzoyl-d-

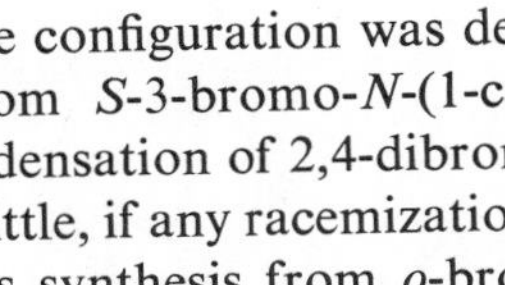

282

tartrate salts. The absolute configuration was deduced from an unambigu-ous synthesis starting from S-3-bromo-N-(1-carboxyethyl)-2-nitroaniline (**283**) prepared by the condensation of 2,4-dibromonitrobenzene and S-ala-nine (Equation 35). Very little, if any racemization occurred in this sequence but attempts to effect this synthesis from o-bromonitrobenzene were un-

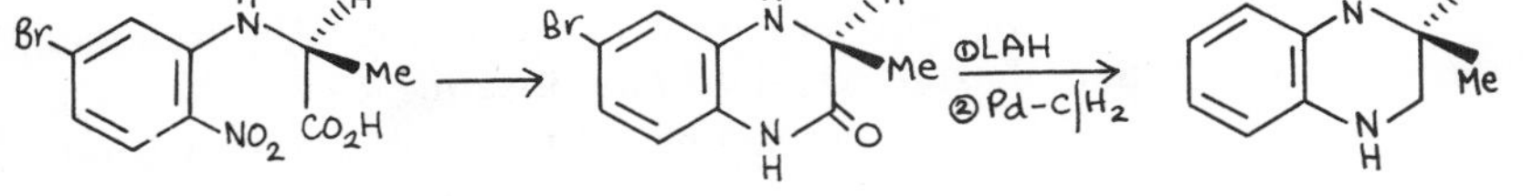

(35)

283

successful. The optical rotation of the enantiomers varied considerably with the solvent used, for example, the S enantiomer had $[\alpha]_D^{24} + 60.2°$ (THF), $-6.1°$ (CHCl$_3$), $+ 35.8°$ (95% EtOH), and also varied with temperature.[661] Tosylation of 2-methyl-, 2-amido, 2-carboxy-, 2-ethoxycarbonyl-, and 2-hydroxymethyl-1,2,3,4-tetrahydroquinoxalines takes place specifically on N-1 as evidenced by ^{1}H nmr comparisons with 1,2,3,4-tetrahydro-1-tosylquinolines and quinaldines. The absolute configuration of the R enantiomers of the above optically active amido, carboxy, ethoxycarbonyl, and hydroxymethyl compounds were deduced by conversion into 2S-1,2,3,4-tetrahydro-2-methylquinoxaline.[662]

2,3-Disubstituted 1,2,3,4-tetrahydroquinoxalines, and their 6- and 7-aza derivatives, exist in two inverting "half-chair" conformations which are in rapid equilibrium even at $-87°$C.[658] It is possible sometimes to distinguish between cis and trans isomers of 1,2,3,4-tetrahydroquinoxalines from the chemical shifts and coupling constants of the methine protons in the ^{1}H nmr spectra,[657,658] but the spectra are not necessarily clear in all cases. The spectra of the N,N'-dibenzoyl and N,N'-diformyl derivatives are more useful for making this distinction because the signals are apparently clear in all cases.[657]

cis- and *trans*-Decahydroquinoxalines are known. Mousseron and Combes[663] synthesized a decahydroquinoxaline by heating *cis*-1-chloro-2,2'-chloroethylaminocyclohexane (obtained from cyclohexene oxide and ethanolamine followed by reaction with PCl$_5$) with aqueous ammonia at 130°C for 5 hr in a bomb. This is the trans isomer (m.p. 150–151°C[664]) and was synthesized by reduction of the product from *trans*-1,2-diamino-

(36)

cyclohexane and glyoxal.[665] It has also been prepared by catalytic reduction of quinoxaline with Pd–C at 180°C and 50 atm. Similarly *trans*-2-methyl-decahydroquinoxaline was prepared but the 2,3-dimethyl and 2,3-diphenyl derivatives required lower temperatures in the hydrogenation.[666] Broadbent and co-workers[665] found that *cis*-decahydroquinoxaline can be obtained also by catalytic reduction of quinoxaline with rhodium-on-aluminum oxide at 100°C and 360 atm or with Raney nickel (W6); or by reduction of

1,2,3,4-tetrahydroquinoxaline with PtO_2 in acidic medium.[667,668] Authentic *cis*- and *trans*-decahydroquinoxalines were prepared from *cis*- and *trans*-1,2-diaminocyclohexanes by reaction with chloroacetic acid or diethyl oxalate to form the respective 2-oxo- or 2,3-dioxodecahydroquinoxalines followed by reduction with LAH (Equation 36).[668] Reduction of a mixture of cyclohexane-1,2-dione and ethylene diamine with Pt–H_2 in water, and a mixture of *trans*-1,2-diaminocyclohexane and glyoxal with Pt–H_2 in ethanol, gave respectively *cis*- and *trans*-decahydroquinoxaline, albeit in low yields.[669]

Brill and Schultz investigated the catalytic reduction of 2-oxo-1,2,5,6,7,8-hexahydroquinoxaline with a variety of catalyst and solvent combinations and found that the ratios of *cis*- and *trans*-2-oxodecahydroquinoxalines altered slightly with the conditions used. In most cases the cis isomer predominated or was the exclusive product. Reduction with sodium and amyl alcohol produced 58% of the trans isomer only, and electrolytic reduction in aqueous H_2SO_4 gave 8% and 13% of *trans*-decahydroquinoxaline and its 2-oxo derivative respectively.[669]

cis-Decahydroquinoxaline is a symmetrical molecule and therefore meso. The trans isomer is racemic and was resolved into its enantiomers by recrystallization of the dibenzoyl-*d*-tartrate salt and then the dibenzoyl-*l*-tartrate salt.[668] The absolute configuration was established by Gracian and Schultz[670] from 1*R*,2*R*-1,2-diaminocyclohexane (obtained by a Schmidt reaction from 1*R*,2*R*-dicarboxycyclohexane of known absolute configuration) by a synthesis, as shown in Equation 36, which gave *trans*-4*aR*,8*aR*-(+)-decahydroquinoxaline.

The 1,4-dialkyl-4*a*,8*a*-dicyanodecahydroquinoxalines, formed by the reaction of cyclohexane-1,2-dione and *N,N'*-dialkylethylene diamines and HCN, are the cis isomers because the C–2 and C–3 methylene protons appear as one rather sharp singlet in the ¹H nmr spectrum. This is consistent with the presence of the two rapidly inverting cis conformations **284** and **285**.[671]

For further information about reduced quinoxalines, see the review by Cheeseman.[672]

284   **285**

F.   Reduced Tetraazanaphthalenes

1.   Reduced Pteridines

Reduced pteridines are of considerable biological importance because many of them are essential cofactors in fundamental metabolic processes. Unfortunately very little is known about their stereochemistry which could be very important in understanding the finer points about their steric relationship with the respective enzymes and substrates.

Several pteridines are known to undergo reversible water (or alcohol) addition across the 3,4- double bond[673,674] and the structure of one such adduct, 2-amino-4-ethoxy-3,4-dihydropteridine hydrobromide, has been confirmed by an X-ray analysis.[675] An asymmetric center is created at C–4 when the hydrate is formed but normally in aqueous solution the racemate is obtained. In the presence of an enzyme, however, the hydration process is dissymmetric and an asymmetric center is formed at C–4. Evans and Wolfenden[676] have provided experimental verification of this by studying the hydroxy-deamination of 4-aminopteridine with the enzyme adenosine deaminase which has hydratase activity. The overall equilibrium is shown in Equation 37. The intermediacy of the chiral hydrate **286** is evidenced by an initial rapid rise followed by slow decay of optical rotation.

Albert and Serjeant[677] obtained a pteridine dimer from the reaction of 4,5-diaminopyrimidine and ethyl pyruvate in one of their attempts to prepare 6-(or 7)methyl-7-(or 6)oxo-7,8-(or 5,6)dihydropteridine. The dimer, 5,6,7,8-tetrahydro-6-oxo-7-(5,6,7,8-tetrahydro-7-oxopteridin-6-ylmethyl)-7-

286

methylpteridine (**287**) has a chiral center and was resolved into its enantiomers, $[\alpha]_D^{20} \pm 180°$, by chromatography on a cellulose column. The pteridine dimer, drosopterin (**288**), which Schlobach and Pfleiderer[678] isolated from the eye pigment of *Drosophila melanogaster*, is also chiral and detailed studies of its chiroptical properties have been described. The chirality in this case is due to restricted movement of the two pterin rings. Archer and Mosher[658] reduced 2,4-diacetamido-6,7-dimethylpteridine catalytically to a tetrahydro derivative which was apparently the cis isomer.

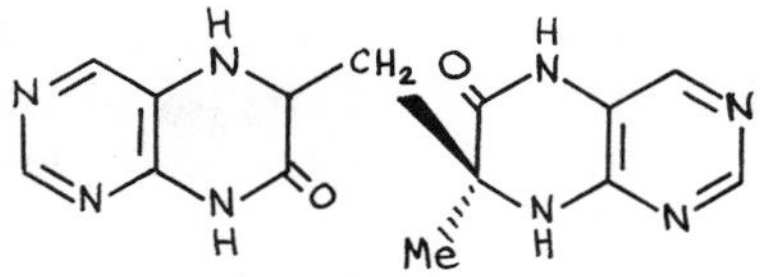

287

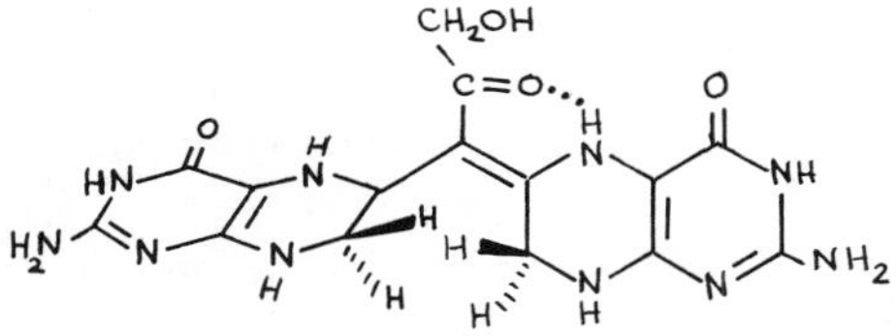

288

Folic acid has an asymmetric center in the S-glutamic acid side chain but when it reduced to 5,6,7,8-tetrahydrofolic acid (**289**, $R^1 = R^2 = H$); a second chiral center at C–6 is introduced. Catalytic reduction with PtO_2–H_2 in acetic acid leads to an equal mixture of diastereoisomers $[\alpha]_D^{27} + 14.9°$, l,S and d,S. Only one of these diastereoisomers is biologically active and the biologically synthesized tetrahydrofolic acid has $[\alpha]_D^{27} - 16.9°$; that is, it is the l,S form.[679] Catalytic reduction of folic acid with the soluble optically active rhodium catalyst [(py$_2$)(−)- or (+)-(N-1-phenylethyl form-amide)RhCl(BH$_4$)]$^+$Cl$^-$ is stereoselective and yields a predominance of one diastereoisomer. The stereospecificity of the reaction was tested biologically, but the results showed that the stereospecificity was not very high.[680] The more stable N^5,N^{10}-methylenetetrahydrofolate, $[\alpha]_D^{23} - 36°$[681] (**289**, $R^1,R^2 = -CH_2-$), was resolved into its optically active forms by chromatog-raphy on a TEAE-cellulose column. The faster moving d,S isomer, $[\alpha]_D^{23.5} + 165°$, was biologically active and the slower moving l,S isomer, $[\alpha]_D^{23.5} - 82°$, was inactive.[682,683] The cd curves of the diastereoisomers have almost

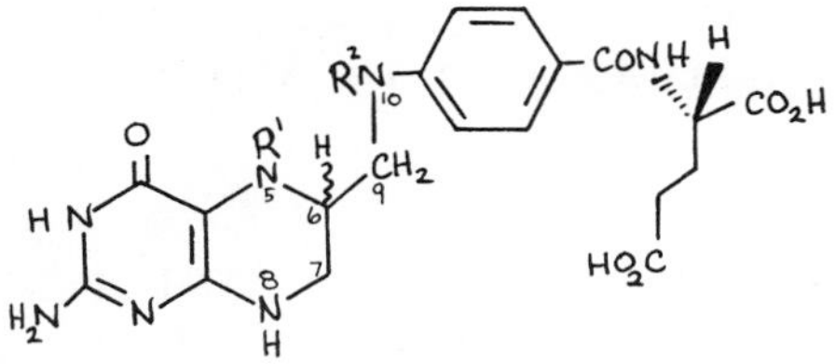

289

exactly opposite elipticities.[684] There is therefore a change in the sign of rotation in converting **289** ($R^1 = R^2 = H$) to **289** ($R^1, R^2 = -CH_2-$). An important observation made by Benkovic and co-workers[685] is the reduction of N^5, N^{10}-methenyltetrahydrofolate to N^5, N^{10}-methylenetetrahydrofolate with $NaBH_4$, which they found to be highly stereoselective, and they showed that the hydride inserted by the reagent is the one that is removed by the enzyme 5,10-methylenetetrahydrofolate dehydrogenase (from baker's yeast).

Cosulich, Smith, and Broquist[686] took advantage of the differential solubilities of the calcium salts in water of the diastereoisomers of 5-formyltetrahydrofolic acid (folinic acid) in order to obtain the biologically active *l,S* isomer ($[\alpha]_D^{23} - 15.1°$).

The chemistry and biochemistry of pterins, folates, and reduced folates are admirably discussed in a monograph by Blakley,[687] and the mechanism of action of folic acid cofactors is reviewed by Benkovic and Bullard.[688]

2. *Reduced 1,4,5,8-Tetraazanaphthalenes*

Decahydro-1,4,5,8-tetraazanaphthalene is formed from ethylene diamine and excess of glyoxal or 1,2-dichloro-1,2-diethoxyethane. The stereochemistry is probably trans at the ring junction.[689] When a 2:3 molecular ratio of ethylene diamine and glyoxal are mixed, a 20% yield of a compound with the molecular formula $C_{10}H_{14}N_4O_2$ is formed. The chemistry and spectra lead to one of two possibilities which was settled by an X-ray analysis. The structure of the compound is shown in formula **290**.[690]

290

XII. REDUCED SIX-MEMBERED RINGS CONTAINING NITROGEN ATOMS AND TWO OTHER FUSED RINGS

A. Reduced Acridines

9,10-Dihydroacridine (acridan) is not planar (**291**) and the molecule can invert rapidly. The inversion is clearly demonstrated in the ^1H nmr spectra of *N*-acetyl-, *N*-chloroacetyl-, and *N*-iodoacetylacridans at low temperatures. The singlet signal of the C–9 methylene protons at room temperature becomes an AB quartet at −75°C, and the free energy changes for the three derivatives are 48.9, 54.8, and 54.8 kJ mol⁻¹ respectively. At lower temperatures, furthermore, rotation of the N-acyl group is hindered, and the methylene protons of the N-iodoacetyl group also appear as a quartet.[691]

291

The ¹H nmr spectra of a series of alkylacridans with 9-alkyl and 9,9-dialkyl groups, ranging from Me to t-Bu, are consistent with a preferred conformation in which the heterocyclic ring is in the boat form and the more bulky 9-substituents are in the pseudo-axial orientation.[691a]

cis- and trans-1,2,3,4,4a,9,9a,10-Octahydroacridines were prepared by Perkin and Sedgwick in 1924 by reduction of 1,2,3,4-tetrahydroacridine with tin and HCl.[692] They assigned the cis and trans structures to the higher (m.p. 82°) and lower melting (m.p. 72°) bases respectively and resolved both of them into their optically active forms by recrystallization of their d- and l-camphorsulfonate salts.[692a] Cartwright and Plant[692b] later applied the generalization that the trans isomers are higher melting and revised the above assignments. Booth[693] examined the ¹H nmr spectra of both octahydroacridines and confirmed beyond doubt the latter assignments. He also measured the spectra of the cis-N-methyl- and N-acetyloctahydro-acridines and was able to assign specific conformations for the cis-10-methyl compound (similar to 292) and its methiodide as 293, but in the case of the N-acetyl derivative a mixture of these conformations was revealed. Booth and co-workers[694] observed that the methiodides of cis- and trans-10-methyl-1,2,3,4,4a,9,9a,10-octahydroacridines are resistant to Hofmann elimination and that this is consistent with the absence of trans colinearity of the four centers in the structures 293 and 294 involved in such a reaction (see Chapter 3, Sections VIII.B, IX.B.4.c, X.B., and XII.A; Chapter 2, Section III.E.4). The alternative cis conformer, (292), in which

292

293

294

the four centers are adequately placed for facile elimination, is not observed in the ^{1}H nmr spectrum.

Four of the five possible geometrical isomers of perhydroacridine are known. Adkins and Coonradt[695] obtained a perhydroacridine (α) in a pure state by catalytic hydrogenation (Ni) of 1,2,3,4-tetrahydroacridine. Masamune and co-workers[696] obtained the α and β isomers, by catalytic reduction (Pt-AcOH) of *trans*-1,2,3,4,4a,9,9a,10-octahydroacridine, and a mixture of β and γ isomers, by reduction of *cis*-octahydroacridine, which are separable. The *trans*- and *cis*-N-methyloctahydroacridines, however, gave a mixture of the α and γ, and γ and a small amount of a new isomer (δ) respectively on similar hydrogenation. The α, β, and γ isomers were identified by ir studies[697] as the trans-trans (295), trans-anti-cis (296), and trans-syn-cis (297) isomers. The ir evidence was based essentially on evaluation of the Bohlmann bands derived from the C–H stretching frequencies of C–H hydrogen atoms in antiperiplanar configuration relative to the nitrogen lone pair of electrons (see Section II.B). These were confirmed by

295   α 296 β

γ

297

^{1}H nmr studies of the *N*-alkyl derivatives.[698, 699] A mixture of α-, β-, and γ-perhydroacridines was also formed from 2-(cyclohexan-2-onyl) cyclohexanone and an amine. The γ isomer was separated by way of its *N*-nitroso derivative.[699] The δ isomer, which is obtained in very small amounts, is probably a mixture of *cis*-syn-*cis*- and *cis*-anti-*cis*-perhydroacridines.

The course of alkylation of the *N*-alkyl derivatives of α- and β-perhydroacridines was investigated and in each case the two isomeric quaternary salts were formed when the alkyl groups were different.[700] The relative proportions of the isomeric quaternary salts are consistent with the steric interactions involved in the alkylation reactions. Hofmann degradation of the methiodides of the *N*-methyl derivatives of α- (295) and γ-perhydroacridines (297) produced the respective *N*-methyl compounds (i.e., demethylation). In the β isomer, however, elimination did occur and *trans*-1-dimethylamino-2-(cyclohex-1-en-3-ylmethyl)cyclohexane was obtained. Inspection of formula **296** will reveal that only in this case is there an anti-periplanar H–C–C–NMe$_2$ system which is set up for facile elimination in agreement with the product obtained.[696]

Reference should be made to the monographs by Albert and by Acheson for further information on reduced acridines.[701]

B.   Reduced Phenanthridines

cis- and *trans*-1,2,3,4,4a,5,6,10b-Octahydrophenanthridines (**298**) were prepared by Pictet-Spengler or Bischler-Napieralsky cyclization of the appropriate 2-arylcyclohexylamines (**299**) without epimerization, followed by stereoselective catalytic reduction of the C,N double bond in the latter

cis or *trans*	*cis* or *trans*, R=OMe	
	R' = H, R^{3}CO–	
298	**299**	**300**

synthesis. Their conformations were derived from their ^{1}H nmr spectra.[702,703] Irradiation of *N*-benzoylenamines of cyclohexanones (**300**) in methanolic solution were stereospecific and produced only the *trans*-octahydrophenanthridinones.[704]

Reductive cyclization of *trans*-1-carboxy-2-(*o*-nitrophenyl)cyclohex-4-ene followed by reduction of the double bond and the amide carbonyl furnished *trans*-5,6,6a,7,8,9,10,10a-octahydrophenanthridine (**301**). The cyclohexene was obtained by a Diels-Alder reaction between *o*-nitro-*trans*-cinnamic acid and butadiene, but if *o*-nitro-*cis*-cinnamic acid was used in the first place then the final product was the isomeric *cis*-5,6,6a,7,8,9,10,10a-octahydrophenanthridine.[705] The cis isomer was also prepared from *o*-nitro-phenylbutadiene and acrylic acid by essentially the same synthesis.[706] Photolysis of *N,N*-dimethylaniline and cyclohexene yields *cis*-*N*-methyl-5,6,6a,7,8,9,10,10a-octahydrophenanthridine.[454a] The ^{1}H nmr spectra could

301

302   303   304

not be used to decide readily which conformations are preferred, but it appears that the *N*-benzoyl derivatives of both isomers have somewhat distorted conformations.[707]

Masamune and co-workers[696] reduced catalytically the *trans*-octahydrophenanthridine (**301**) with Pt in acetic acid to the *trans-anti-trans*-perhydrophenanthridine (α isomer, **302**). Similar reduction of the cis isomer of compound **301** gave β-perhydrophenanthridine (trans-syn-cis **303**), which was also prepared by reduction (Pt) of trans **298** (R = H) in ethanolic solution. β-Perhydrophenanthridine, together with γ-perhydrophenanthridine (cis-syn-cis **304**), was formed when the *cis*-octahydrophenanthridine (**298**, R = H) was reduced (Pt) in ethanol. The structures were assigned from ir studies and provide further evidence that a hydrogen atom on a nitrogen atom is more space demanding than a nitrogen lone pair of electrons.[697] No Hofmann elimination was observed in any of the *N*-methyl methiodides of these perhydrophenanthridines.[696]

C. Reduced Benzoquinolines and Benzoisoquinolines

One of the best examples of hindered nitrogen inversion by obstruction is the *N*-tosyl derivative of 1,2,3,4-tetrahydro-4-oxobenzo[*h*]quinoline (**305**).

305   306

The hindrance to inversion is revealed by the changes in the ^{1}H nmr spectral patterns of the hydrogen atoms drawn in formula **305**.[708]

Horii and co-workers[709–713] prepared several cis and trans derivatives of 1,2,3,4,4a,9,10,10a-octahydrobenzo[*f*]quinolines (**306**) and deduced the preferred conformations by ^{1}H nmr spectroscopy. Reductive cyclization of

1-(2-cyanoethyl)-2-tetralone afforded a 20:15 mixture of *cis*- and *trans*-benzo[*f*]quinolines (**306**, $R^1 = R^2 = R^3 = H$), and the isomeric 2-(2-cyanoethyl)-1-tetralone gave a 35:46 mixture of the corresponding *cis*- and *trans*-1,2,3,4,4a,5,6,10b-octahydrobenzo[*h*]quinolines.[714] The magnetic non-equivalence of the benzylic CH_2 signals in *cis*- and *trans*-1-benzyl-1,2,3,4,-4a,5,10,10a-octahydrobenzo[*g*]quinolines was used as a stereochemical probe to differentiate between the isomers.[715]

Oppolzer[716] extended his studies on thermal intramolecular cycloaddition reactions to the synthesis of octahydrobenzo[*f*]- and octahydrobenzo[*h*]isoquinolines (Chapter 2, Section III.B.1.a). The reaction in Equation 38 proceeded with cis stereospecificity when $X = O$, $Y = H_2$, or $X = H_2$ and $Y = O$, but with *trans* stereospecificity when $X = Y = H_2$. Similarly in the reaction in Equation 39 the product was the *cis*-benzo[*f*]isoquinoline when

$X = O$ and the trans isomer when $X = H_2$. The intermediate *o*-quinomethine is formed by thermal opening of the cyclobutene ring to give the endo (**307**) or the exo (**308**) transition states to account for cis and trans stereo-

specificities respectively. The cyclizations are probably under steric control because of the possible equilibration of these transition states.

D.   Reduced Phenazines

In 1938 Le Fèvre, Turner, and collaborators[717] suggested a folded molecule for 9,10-dihydro-9,10-dimethylphenazine. The dipole moment was difficult to interpret, but this may be because of nitrogen inversion as well as folding, that is, there are three possible orientations of the two methyl groups (ax-eq, ax-ax, and eq-eq) for each folded conformation.

Reduction of 1,2,3,4-tetrahydrophenazine with sodium amalgam in ethanol gives one isomer, *trans*-1,2,3,4,4a,5,10,10a-octahydrophenazine, but catalytic reduction using Raney nickel yields a mixture of cis and trans isomers.[718] Miyano and Nakao[719] heated *o*-phenylene diamine with cyclohexane-1,2-diol in the presence of sodium at 110°C and obtained a poor yield of *cis*- and *trans*-octahydrophenazines. Irrespective of whether *cis*- or *trans*- or a mixture of *cis*- and *trans*-cyclohexane diols were used, the *cis*-octahydrophenazine predominated. The cis isomer was more soluble in ether than the trans isomer and afforded easy separation of isomers. Bromination of *cis*-5,10-diacetyl-1,2,3,4,4a,5,10,10a-octahydrophenazine gave the 7,8-dibromo and not the 6,8-dibromo derivative. Catalytic reduction of 6,8-dibromo-1,2,3,4-tetrahydrophenazine with Pd–C in acetic acid and acetic anhydride produced the 5-acetyl-*cis*-octahydrophenazine.[720]

Godchot and Mousseron[721] separated α-, β-, and γ-perhydrophenazines (m.p. 135, 95, and 62°C respectively) from the reduction of 1,2,3,4,5,6,7,8-octahydrophenazine. Four perhydrophenazines are theoretically possible. Clemo and McIlwain[722] reduced 1,2,3,4-tetrahydrophenazine, 1,2,3,4,5,6,-7,8-octahydro-, and *cis*- and *trans*-1,2,3,4,4a,5,10,10a-octahydrophenazines and obtained α- or α,β- and γ-perhydrophenazines. These results and dehydrogenation experiments suggested that the α isomer was probably the *cis-trans*-perhydrophenazine (**309**) but attempts to resolve any of these into their optically active forms were unsuccessful.

309

310   **311**   **312**

E. Miscellaneous

Stereoselective cyclization occurred in the Pictet-Spengler synthesis of 1-substituted 3-methoxycarbonyl-1,2,3,4-tetrahydrocarbolines from (±)-tryptophan methyl ester and various aldehydes (Equation 40). The ratio of 1,3-*cis*- and 1,3-*trans*-substituted hydrocarbolines varied with the aldehyde used, for example, when $R = CO_2Et$ the *cis*-carboline was formed exclusively, and when $R = o$-nitrophenyl the cis to trans ratio was $1:3$.[723] Yamada and co-workers[724] carried out the above reaction with *S*-tryptophan and acetaldehyde. Asymmetric induction occurred and the tetrahydro-β-carboline formed was converted into $S(-)$-tetrahydroharman of known absolute configuration.[725] The tetrahydroharman formed, however, was only 50% optically enriched.

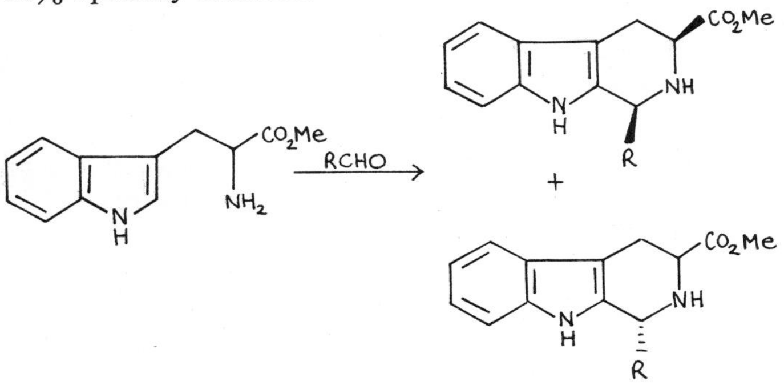

The reduction of linear systems, such as compounds **310** with $NaBH_4$, invariably yield the cis-fused products (i.e., **311**);[726] whereas similar reductions of the C,N double bond in the angular compounds **312** produce the trans or cis and trans products.[727]

The ^{1}H nmr spectra of 1-methyl-2,3-dihydro-1*H*-benzo[*d.e*]isoquinoline showed that although there is some strain because of the two *peri* methylene groups the molecule is conformationally mobile (Scheme 7). For inversion

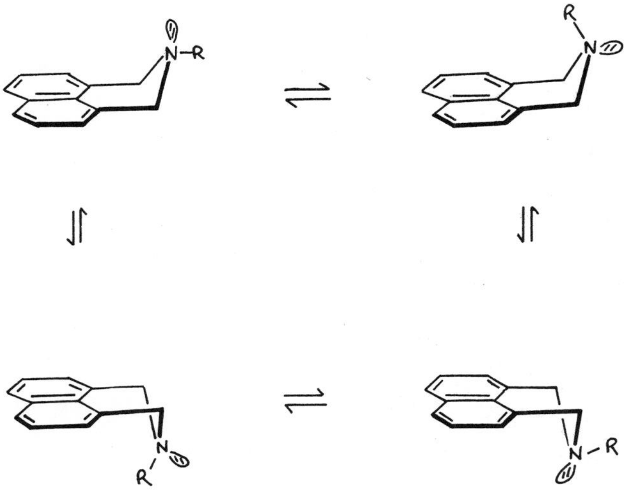

Scheme 7

to occur the peri methylene groups would need to be pushed apart and have to distort the aromatic system. This reveals itself in a measurable nitrogen inversion ($\triangle G^{\ddagger}$ = 42.7 kJ mol^{-1} in CD$_3$OD) and is a new demonstration of the peri effect.[728]

XIII.   PHOTOCHEMICAL REACTIONS

The effect of ultraviolet light on nitrogen heterocyclic compounds produces interesting intermediates and dimeric products. A few of these are described in this section because of their stereochemical structures. Moriarty[728a] has recently discussed the three-dimensional aspects of cyclobutane and of dimers of heterocycles which possess a cyclobutane structure.

A.   Pyridines

Irradiation of pyridine in aqueous solution converts it into Dewar pyridine (313) which is not planar. The bicyclic compound decomposes back to pyridine with a half life of about 2 min. The presence of this structure is confirmed by adding NaBH$_4$ whereby 5-azabicyclo[2.2.0]hex-2-ene (314, R = H) is formed.[729] 5-Aza-6-oxobicyclo[2.2.0]hex-2-ene (315, R^1 = R^2 = R^3 = H), together with its dimer, is formed on irradiation of 2-pyridone. Better yields were obtained when 1,3,4-trisubstituted derivatives are used, for example, bicycle 315 (R^1 = Me, R^2 = CN, R^3 = OMe) is formed in 89%

yield. These are generally more stable at room temperature and give good ^{1}H nmr spectra, but they decompose with first order kinetics on heating.[730] When 2,3,4,5,6-penta(perfluoroethyl)pyridine is irradiated with uv light (>200 nm) another valence isomer, penta(perfluoroethyl)-1-azabicyclo-[2.2.0]hexa-2,5-diene (316), is formed as a stable colorless liquid. On

further irradiation it undergoes a second isomerization to the 1-azapris-
mane, 2,3,4,5,6-penta(perfluoroethyl)-1-azatetracyclo[2.2.0.^{2,6}0.^{3,5}]hexane
(317). The ^{19}F nmr spectra of these are consistent with the structures
drawn.[731]

Photolysis of 1,2-dihydro-1-methoxycarbonylpyridine, prepared from
pyridine, NaBH$_4$, and ClCO$_2$Me, gave N-methoxycarbonyl-5-azabicyclo-
[2.2.0]hex-2-ene (314, R = CO$_2$Me). 1,4-Dihydro-1-methoxycarbonylpyri-
dine, also formed in the above reduction, undergoes a [2 + 2] cycloaddition
reaction with maleic anhydride.[732] 1,4-Dihydro-3,5-diethoxycarbonylpyri-
dine and its 1-methyl derivative dimerize on photolysis to the head-to-tail
anti (318) and head-to-tail syn (319) dimers with structures which were
established by ^{1}H nmr spectroscopy and X-ray analysis. Further irradiation
of the syn isomer produces the cage structure 320.[733] In 1960, Taylor and
Paudler[734] produced dimers of 2-pyridone and 1-methyl-2-pyridone by uv
irradiation and formulated them as 3,4-dimers from chemical and dipole
moment measurements. Taylor, Paudler, and Kuntz[734a] had thought that
they produced valence tautomers of pyridine by irradiating 2-amino- and

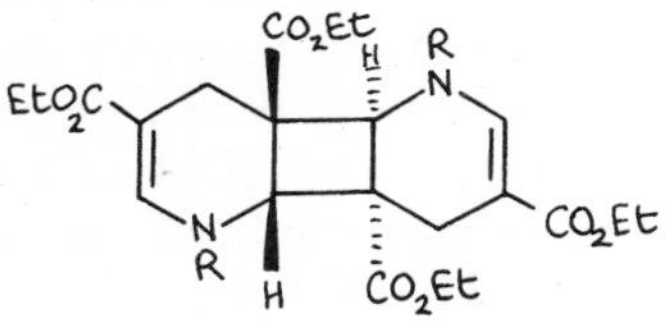

318

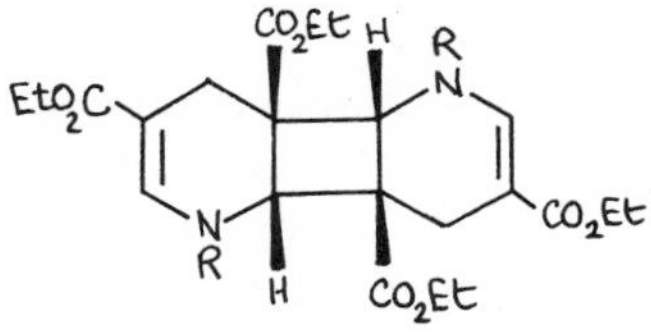

319

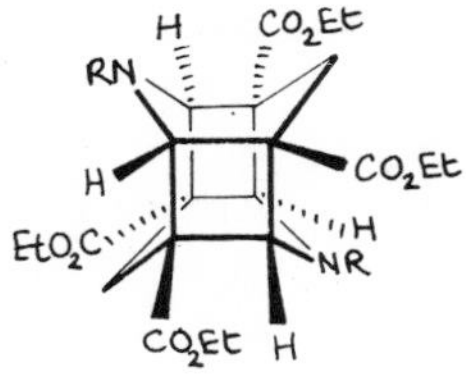

320

2-amino-5-chloro-pyridines, but later[735] they showed that the products
were the 1,4-anti-trans dimers 321 (R = H and R = Cl). The diamino dimer,
after reduction, was hydrolyzed to the dioxo dimer 322 which was identical
with the reduction product of the 2-pyridone dimer obtained previously.
The original 2-pyridone 3,4- dimer was therefore revised to the 1,4-anti-
trans structure 323 similar to that of 2-aminopyridine (i.e., 321). Irradiation
of 2-amino-, 2-amino-3-, 4-, 5-, and 6-methylpyridines, and a corresponding series of 2-pyridones also gave similar photodimers which were interrelated chemically. The configuration of the dimers was firmly established

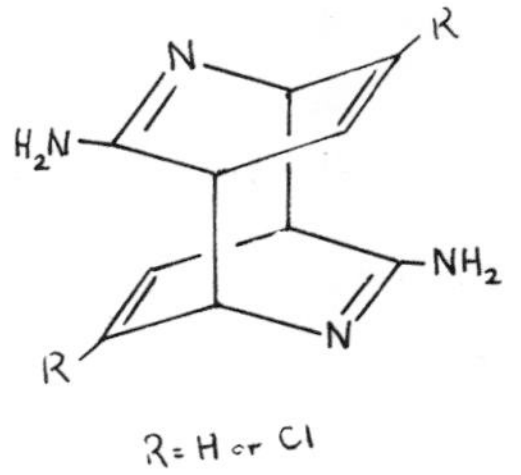

R = H or Cl

321

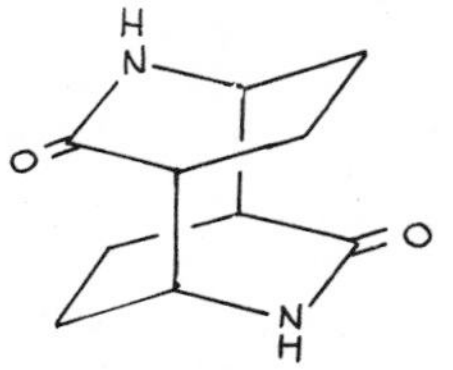

322

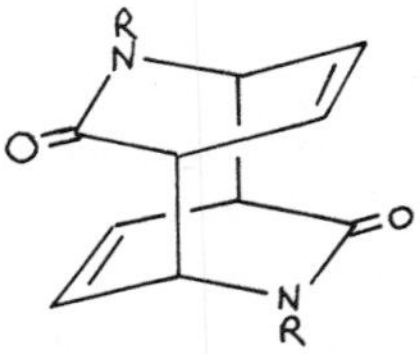

323

from detailed ¹H nmr studies and from their physical and chemical properties.[736] Almost simultaneously, Paquette and co-workers[737,738] and deMayo, Stothers, and collaborators[739] arrived at the same conclusions for the 2-pyridone photodimers. The latter workers confirmed the structure by dipole moment measurements, and Paquette and co-workers concluded that the ¹H nmr spectra of the 1-methyl-2-pyridone photodimer, 3,7-dimethyl-3,7-diazatricyclo[4.2.2.2]dodeca-9,11-diene-4,8-dione are best explained by the two interconverting structures 324 and 325.

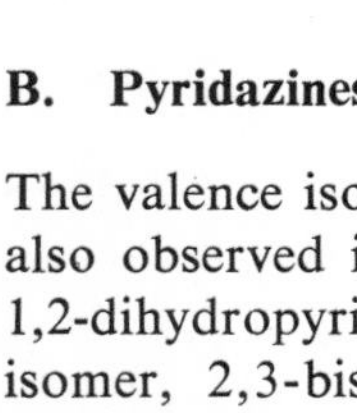

324

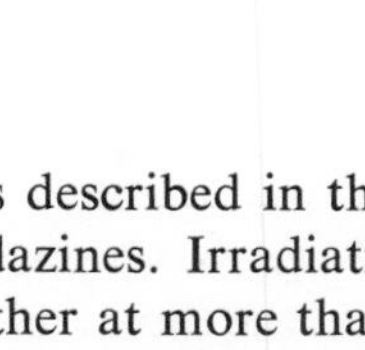

325

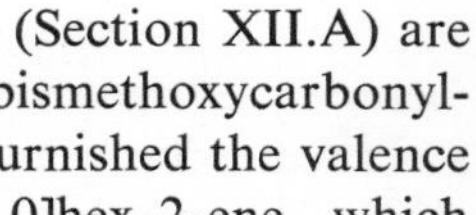

326

B.   Pyridazines

The valence isomerizations described in the pyridines (Section XII.A) are also observed in the pyridazines. Irradiation of 1,2-bismethoxycarbonyl-1,2-dihydropyridazine in ether at more than 285 nm furnished the valence isomer, 2,3-bismethoxycarbonyl-5,6-diazabicyclo[2.2.0]hex-2-ene, which was reduced with diimide to the bicyclohexane 326.[740] The syn and anti dimers of pyridazines 327 and 328 were prepared by a [2 + 2]π cyclo-addition reaction, not in a photochemical reaction, but by treatment of 1-methylpyridazinium methosulfate, or its 3-methoxy or 3-methyl derivatives, with potassium cyanide.[741,742]

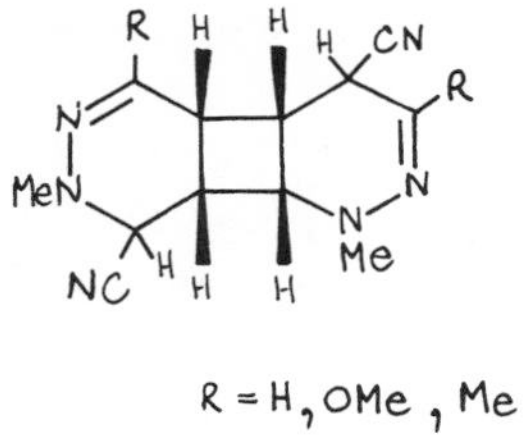

R = H, OMe , Me

327 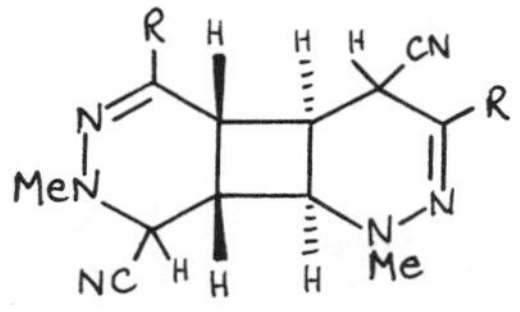   328

C. Pyrimidines

Considerable interest has been demonstrated in the study of the effect of uv radiation on pyrimidines because of the damaging properties of uv light on nucleotides and nucleic acids. The pyrimidines that were extensively investigated were consequently those commonly found in ribonucleic and deoxyribonucleic acids, that is, uracils and thymines. The essential reactions and the nature of the products only are detailed here but not exhaustively.

Elad, Rosenthal, and Sasson[743] examined the photolysis of 1,3-dimethyluracil in a variety of solvents, and in frozen aqueous solutions, and obtained all four possible cyclobutane dimers, **329–332**. The relative proportions of these varied with the solvent used, and a reaction by way of triplet

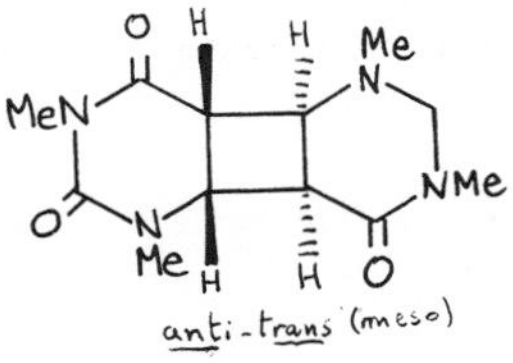

anti-trans (meso)

329

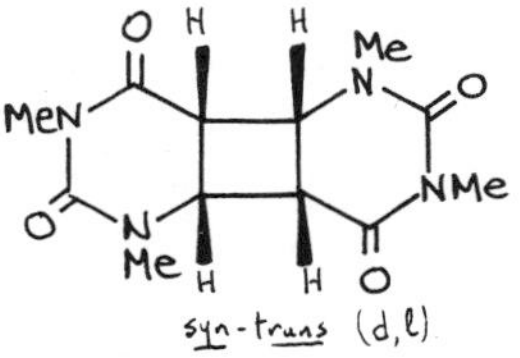

syn-trans (d,l)

330

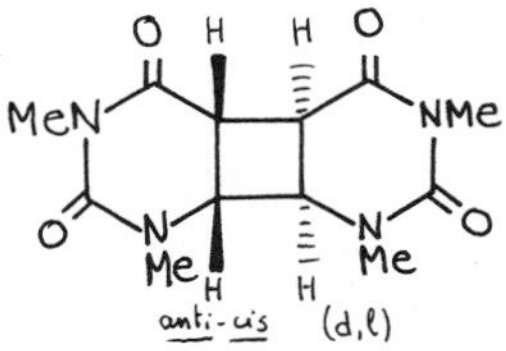

anti-cis (d,l)

331

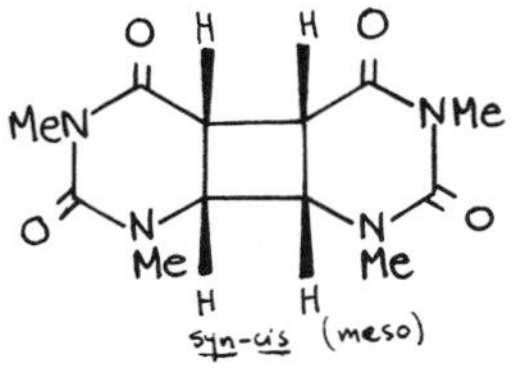

syn-cis (meso)

332

precursors was proposed. The structure of the dimer **329** was confirmed by an X-ray study.[744] Nnadi and Wang[745] found that there was a dramatic increase in the yield of 1,3-dimethyluracil and uridine dimers when water was replaced by deuterium oxide as solvent, at various concentrations, but that the proportion of the pyrimidine hydrate **333** was still high under the conditions used. The mechanism of the dimerization of uracil, 6-methyluracil and 6-methyluridine was explained in terms of lactam-lactim tautomers,[746] steric effects,[747] and metastable dimeric species.[748] Very little dimerization of 1,3-dimethyluracil occurs in the presence of photosensitizers

333   **334** $R = Bu^t, OAc$ **335** $R = Bu^t, OAc$

and the syn photodimers are monomerized faster than the anti dimers by irradiation or by aluminum chloride.[749] The $[2 + 2]\pi$ cycloaddition of 1,3-dimethyluracil and *t*-butyl vinyl ether or vinyl acetate is regioselective when the solution is irradiated in the presence of a photosensitizer. The cis isomers **334** and **335** are formed predominantly. When ketene diethyl acetal (1,1-diethoxyethylene) is used, on the other hand, the cis adduct (with similar regiospecificity) was accompanied with a little of the trans isomer. The trans adduct is readily isomerized to the cis adduct in the presence of alumina.[750] Photolysis of uracil in the presence of vinylene carbonate and acetone was similarly stereoselective and gave a 5:2 mixture of the *syn*- and *anti*-[2 + 2] adducts **336** and **337**.[751]

336   **337**

Thymines respond to uv radiation in the same way as uracils, and four cyclobutane dimers similar to **329–332** (except that they contain two bridgehead methyl groups) are produced. The analogous four *N*-methylcyclobutane dimers derived from 1,3-dimethylthymidine have been separated.[752] The relative proportions of these dimers varied considerably with the solvent used,[753] and similar dimerizations of 1-carboxymethylthymine (and uracil)[754] and methyl orotate (and orotic acid)[755] proceeded without decarboxylation. The structures of these dimers were derived by chemical hydrolysis,[756] reduction,[757] and other reactions[758] in which the cyclobutane ring remained intact. [13]C nmr[759] and X-ray[760] analysis have been used for structure elucidation. Photolysis of 4-(2-hydroxypyrimidin-6-yl)thymine gave a pyrimidine tetramer (**338**),[761] and a thymine trimer (**339**)[762,763] was described. Both these structures were confirmed by X-ray studies.

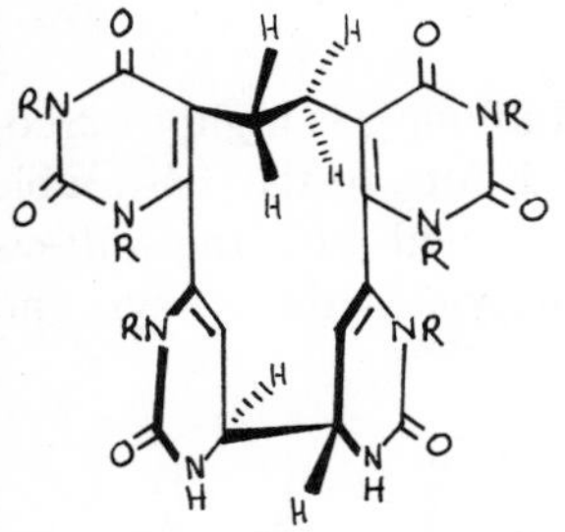

338

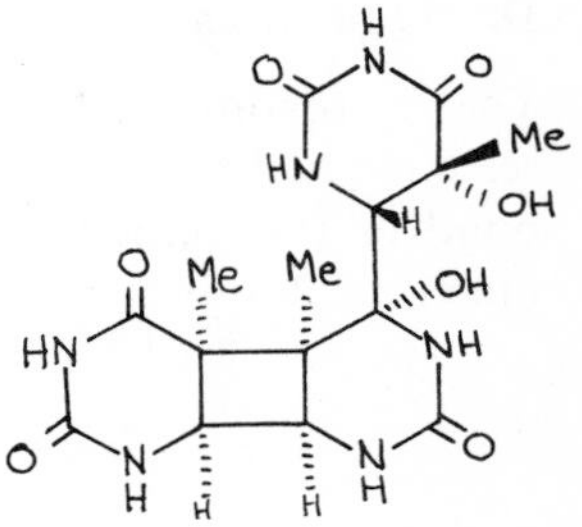

339

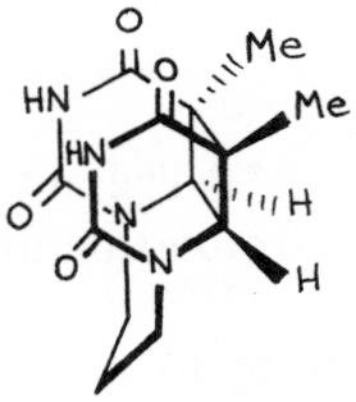

340

Leonard and co-workers[764,765] studied the photodimerization of systems in which the nitrogen atoms of two thymine or two thymidine rings are linked together (N_1–N_1 and N_1–N_3) by a carbon chain. They obtained intramolecular photodimers such as **340** with structures which were established by X-ray analyses. These studies were extended to "abbreviated" dinucleotides, for example, 5'-deoxy-5'-(1-thyminyl)thymidine, and photoadducts like **341** and **342** were formed in comparable amounts.[766] Frank

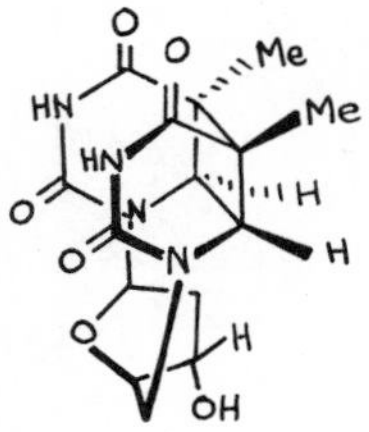

341

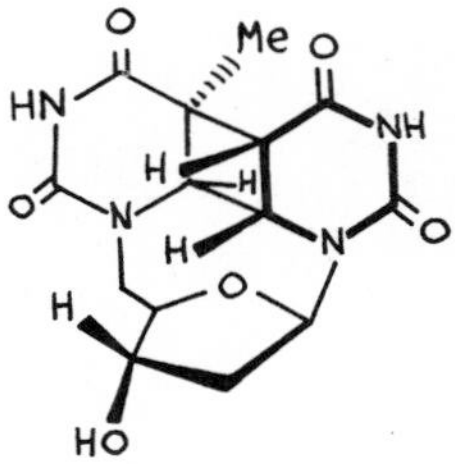

342

and Paul found that in the crystals of 1,1′-trimethylene bisthymine the two rings are placed such that both inter- and intramolecular photodimerization can take place and have discussed the possibilities of both reactions.[767] There is evidence that such molecules stack in aqueous solution and the results are therefore not surprising.[767a]

D. Quinolines

The photoaddition of olefins to carbostyrils (2-quinolones) is highly stereo-selective and regioselective. These are [2 + 2]π additions to the 3,4-double bond of the quinoline and are sometimes accompanied with the anti-cis dimer of 2-quinolone (Equation 41). Cyclopentene yields the cis-syn and

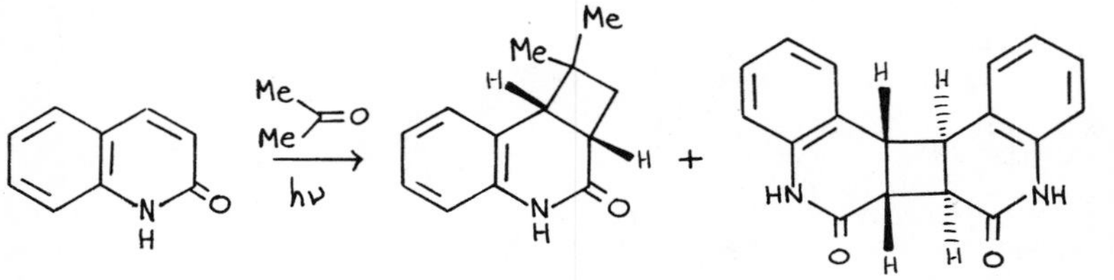

$$(41)$$

cis-anti adducts in a 1:3 ratio.[768,769] The anti-cis dimer of 2-quinolone had been previously prepared by uv irradiation and its structure was confirmed by dipole moment measurements.[734]

The photolysis of 5,6,7,8-tetrahydro-2-quinolone produces the 1,4- di-mer,[770] 1,3,4,7,8,9,10,11-octahydro-2H,5H-4a,11:10a,5-bis(iminomethano)-benzo[a,e]cyclooctane-13,16-dione similar to the 2-pyridone dimer (Section XIII.A), and an X-ray analysis confirmed the structure (343).[771] In the

343

344

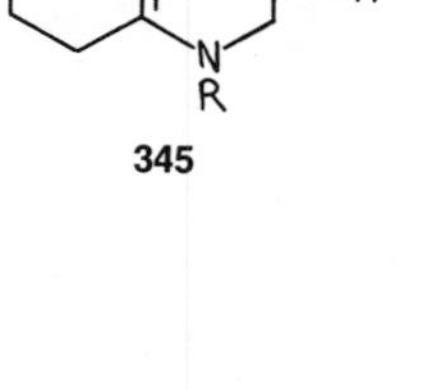

345

presence of diphenyl acetylene (tolan), at 350 nm, the pentacyclic adduct **344** is formed which breaks up into **345** with uv light of shorter wavelength (254 nm).[772,773]

Photocycloaddition of methyl α-methylacrylate and 1-benzyl-1,2,3,4,5,6,-7,8-octahydro-2,6-dioxoquinoline occurs across the bridgehead carbons, and a 1:1 mixture of the regiospecific adducts **346** ($R^1 = Me$, $R^2 = CO_2Me$) and **346** ($R^1 = CO_2Me$, $R^2 = Me$) is obtained.[774] A 1,3-addition of the ele-

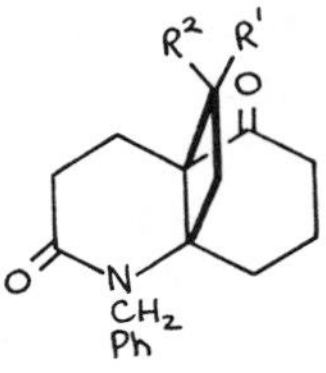

346

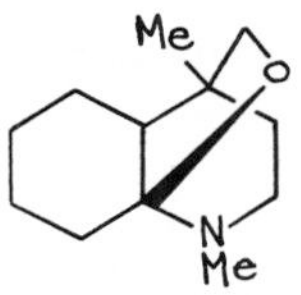

347

ments of methanol was observed when 1,4-dimethyl-2,3,5,6,7,8-hexahydro-quinolinium perchlorate in methanol was photolyzed in the presence of chloride ion in the form of HCl or LiCl. No reaction occurred in the absence of Cl ions. The product, 7,10-dimethyl-10-aza-11-oxatricyclo-[5.3.2.0^{1,6}]dodecane (**347**) probably has the two rings of decahydroquinoline fused in the trans configuration.[775]

E. Isoquinolines

Isocarbostyrils (1-isoquinolones) react with olefins under the influence of uv light in a regio- and stereospecific manner, as the carbostyrils, to form the 3,4-adducts, for example, **348** together with the anti-cis dimer of the isocarbostyril[776] (compare with 2-quinolones Section XIII.D, Equation 41). 3-Isoquinolone and its 1-methyl derivative yield the photodimer with the 1,4-anti configuration **349**.[777]

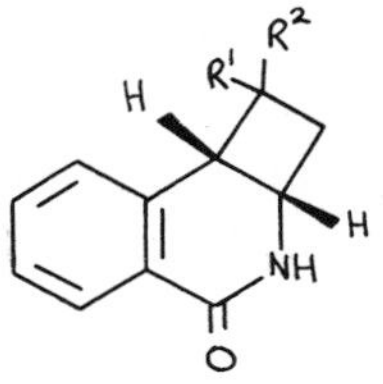

348

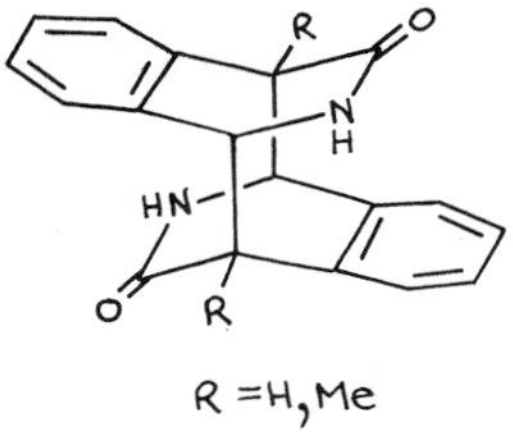

R = H, Me

349

XIV.   AROMATIC COMPOUNDS

This section contains a number of six-membered aromatic nitrogen heterocycles in which the stereochemical properties such as steric hindrance, asymmetry in the side chain, restricted rotation, intramolecular overcrowding, and cis-trans isomerism of substituents are described.

A.   Pyridines

Brown and Kanner[778] measured the ionization constants of a number of alkyl pyridines in order to determine the effect of alkyl groups on the basicity of pyridine. Alkyl groups were found to increase the basic strength, but the estimated pKa value of 2,6-di-t-butylpyridine was 1.4 units higher than the measured value. It was even a weaker base than pyridine by 0.8 pKa units. This was explained by steric crowding around the nitrogen atom and obstructing attack by a hydrated proton. Essery and Schofield[779] observed similar effects in substituted 4-aminopyridines and found that the presence of 3,5-dialkyl substituents caused a base weakening effect in 4-dimethylaminopyridines. The explanation put forward for this effect is that the 3,5-alkyl groups caused steric inhibition of resonance in the cations **350** and **351** which is the important factor that makes 4-aminopyridine such a strong base (pKa 9.29).

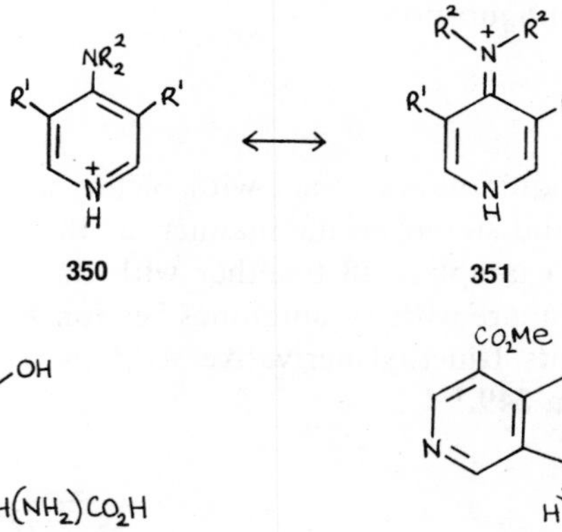

350   **351**

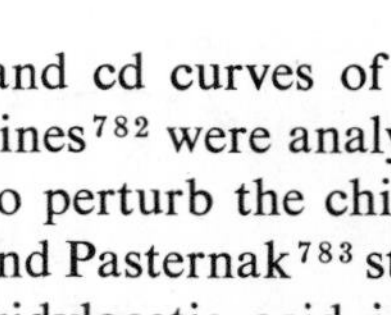

352 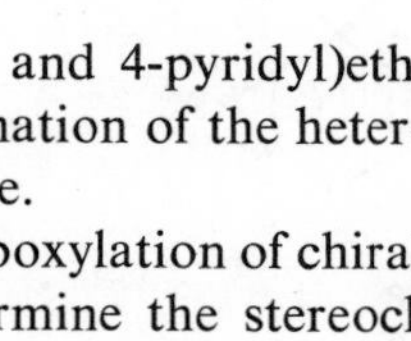   **353**

The ord and cd curves of $S(-)$-1-(2-, 3-, and 4-pyridyl)ethanols[780,781] and ethylamines[782] were analyzed, and protonation of the heterocyclic ring was found to perturb the chiral chromophore.

Doering and Pasternak[783] studied the decarboxylation of chiral 2-(1-ethyl-1-methyl)pyridylacetic acid in order to determine the stereochemistry of the reaction. The acid decarboxylated in neutral medium although it was stable in acid and in alkaline media. The 2-(2-butyl)pyridine formed was racemic, and they showed that racemization occurred during the decarboxylation reaction because optically active 2-(2-butyl)pyridine was stable under the conditions of reaction.

The structure **352** was established for (−)-leucaemine,[784] and it was found to racemize on long boiling in water.[785] The pyrindine alkaloid, cis-5,6-dihydro-6-hydroxy-4-methoxycarbonyl-7-methyl-7H-2-pyrindine (**353**) was isolated from a Jasminum species,[786] and trans-5,6-dihydro-5,7 dimethyl-7H-1-pyrindine was present in California petroleum.[787]

The syn and anti isomers of 2-benzoylpyridine phenylhydrazone have been isolated and identified by their uv spectra.[788] Two isomeric forms, cis and trans, of 2-(4-dimethylaminophenylazo)pyridine were separated and the trans structure was assigned to the more stable red higher melting compound.[789]

The biological reduction of nicotinamide adenine dinucleotide (NAD+) (**354**, R = H) is discussed in this section because the stereochemistry of reduction of the pyridine ring, which is essentially the insertion of a hydride ion on to C–4 (with concomitant loss of positive charge on the nitrogen atom), is influenced by the chirality of the rest of the molecule and the enzymes used. The dehydrogenase enzymes that require NAD+ as coenzyme are of two types: class A, for example, lactate dehydrogenase; and class B, for example, yeast glucose 6-phosphate dehydrogenase. The difference be-

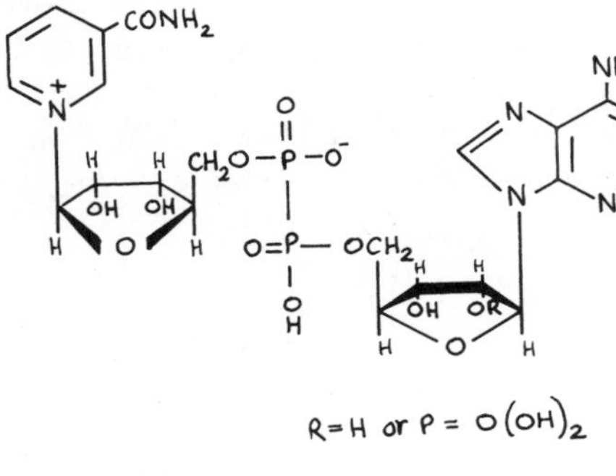

R = H or P = O(OH)$_2$

354

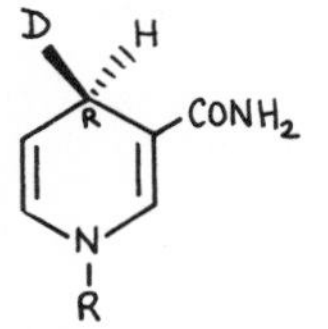

355

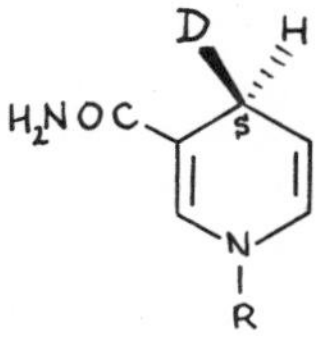

356

tween these two enzymes is in the stereospecificities of the hydride transfer on to C–4 of the nicotinamide ring.[790,791] The enzymes of class A transfer the hydride, deuteride, or tritide ion to the A face (arbitrary) (e.g., **355**), and the enzymes of class B to the B face (e.g., **356**). Cornforth and co-workers[792] established the absolute configuration of the reduction by converting the deutero NAD²H (**355**) from the class A enzyme into the methanol adduct followed by oxidation to chiral monodeuterosuccinic acid which was identical with the previously determined R isomer (Equation 42). The arbitrarily chosen configuration was, coincidentally, the same as

$$\textbf{355} \longrightarrow \qquad \xrightarrow{[o]} \qquad \qquad (42)$$

the one that was determined experimentally. Coenzyme II, (NADP⁺, **354**, $R = -P = O(OH)_2$) also operates with two classes of enzymes.[790]

B. Quinazolines

The quinazoline alkaloid peganin (Vasicine) **357**, from *Peganum harmala*, is generally isolated in a racemic form.[655] Späth and co-workers[793] obtained the alkaloid with a laevo rotation by using mild isolation procedures. They resolved the *d,l* base by way of the *d*-tartrate salt and found that it racemized slowly in 5% aqueous HCl at 100°C, and on repeated sublimation.[794] The absolute configuration of peganine is not known.

357

358   **359**   **360**

Armarego and Smith[795] demonstrated the effect of peri interactions in quinazolines. Quinazoline is known to form the hydrated cation **358** ($R^1 = R^2 = H$) in aqueous solution. 4-Methylquinazoline cation is only slightly hydrated ($\sim 15\%$); however, the proportion of hydrated cation can be increased by increasing the size of the 4-substituent from methyl to ethyl to isopropyl. This is caused by increased steric interaction with the peri C–5 hydrogen atom because the increased electron releasing properties of the 4-alkyl substituents should decrease nucleophilic attack at C–4 and decrease hydration. The cation of 4,5-dimethylquinazoline is almost completely hydrated (**358**, $R^1 = R^2 = Me$) and shows that the methyl groups are crowding each other. This interaction is relieved by covalent hydration. Related interactions in 3,4-dihydro-4-imino-3,5-dimethylquinazoline (**359**) force the system to undergo a Dimroth rearrangement to 5-methyl-4-methylaminoquinazoline (**360**) much faster than in 3,4-dihydro-4-imino-2,3-dimethylquinazoline.[796]

2-Morpholino-4-methylthiocarbamoyl, thioureido, and thioformamido quinazolines undergo prototropic tautomerism. The tautomers are distinguishable by ir and nmr spectroscopy and the thiocarbamate tautomers **361** and **362** have been isolated.[797] The ir spectra of 4-cyanamino-2-morpholinoquinazoline reveal a hydrogen bonded cyanoimino form (**363**) in the solid state.[798]

361   362   363

C. Diheterocycles with Restricted Rotation and Molecular Overcrowding

Fielding and Le Fèvre[799] determined the dipole moment of 2,2'-bipyridyl and interpreted their results in terms of a planar conformation with the nitrogen atoms trans to each other. Uv data, however, suggest that the interplanar angle between the aromatic rings in the dipyridyl is larger than in diphenyl because there is less conjugation in the former. Uv and ¹H nmr spectra of diquaternary salts of 4,4'-bipyridyl possessing aryl groups α to one of the nitrogen atoms were studied. The data indicate that the aryl groups are prevented from becoming coplanar with the heterocyclic ring because of interference by the quaternizing alkyl groups.[800]

Spotswood and co-workers prepared[801,802] several methyl substituted 2,2′-bipyridyls with an alkane bridge between the quaternized nitrogen atoms (364, $n = 2,3,4$) and examined their spectral properties. The calculated

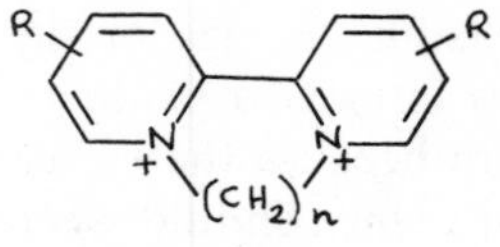

364

interplanar angles are consistent with the data. When the methyl groups in (364, R = Me) are in the 3,3′-positions the compounds are shown to be in rigid twisted conformations. The ¹H nmr proton signals in the unsubstituted bipyridyls (364, R = H) were analyzed as ABMX systems and the differences in the chemical shifts of the corresponding protons, that is, 3 and 3′, 5 and 5′, were related to the probable conformations of the salts. The spectral data provide evidence for a mobile and a frozen conformation.

The uv spectra of 4,4′-biquinazolinyls are very similar to those of quinazoline indicating noncoplanarity, whereas those of the 2,2′-biquinazolinyls have long wavelength absorption bands suggesting extensive conjugation and coplanarity of the heterocyclic rings.[803]

Early attempts to resolve phenylpyridines were unsuccessful,[804] but later the N-methyl salt of 3,3′-bismethoxycarbonyl-2,2′-bipyridyl (365) was resolved by way of the d-camphorsulfonate salt and was found to racemize slowly.[805] 4,7-Diamino-1,10-dimethyl-5,6-diazaphenanthrene was also obtained in optically active forms but was less stable.[806] Similarly, optically active 1,1′-, 4,4′-, 5,5′-, 8,8′-biisoquinolyls,[807] and 4,4′- and 5,5′-biquinolyls[807a] were prepared, and they racemized readily in acid solution. The effect of solvent on the racemization of 4,4′-biquinolyl (366) was studied in some detail by Crawford and Ingle.[808] The half life did not vary much

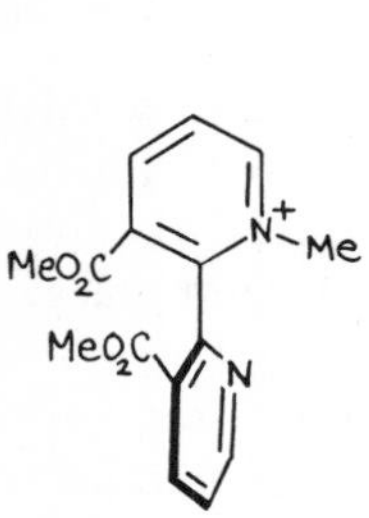

365

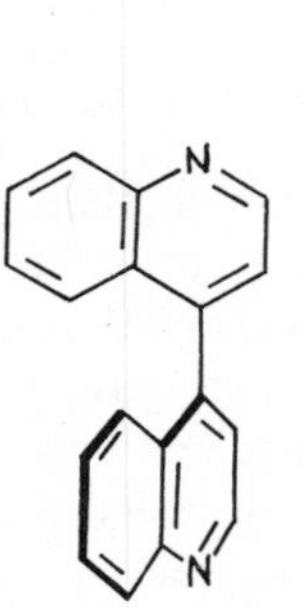

366

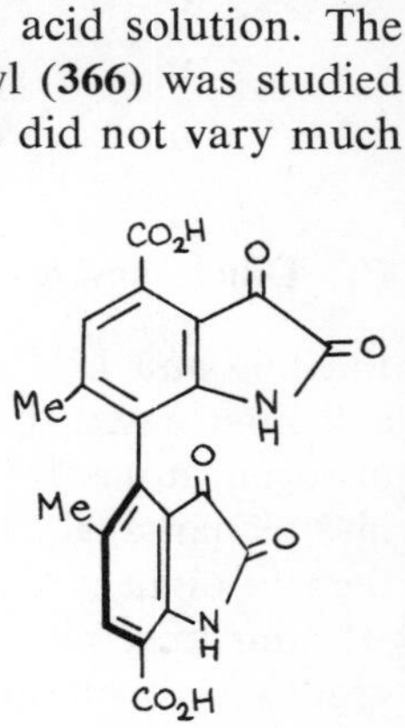

367

from one solvent to another but it was increased considerably in aqueous acid as the acid concentration was increased, that is, at pH 0.68, $t_{1/2} = 118$ min and at pH 3.0, $t_{1/2} = 44$ min. The effect was explained in terms of changes in the molecular dimensions of 4,4'-biquinolyl on protonation. (−)-4,4'-Dicarboxy-3,3'-dimethyl-6,6'-dinitrobiphenyl was reduced to the 6,6'-diamine and converted into 4,4'-dicarboxy-6,6'-dimethyl-2,2',3,3'-tetra-oxo-7,7'-biindolyl (**367**) without racemization.[809]

Fields and co-workers[810] obtained 9-(2-pyridyl)anthracene from the reaction of a 4a-azoniaanthracene salt and benzyne by way of the intermediate azoniatriptycene adduct (Equation 43). They applied this synthesis to the preparation of 13,14-bis(2-pyridyl)pentaphene (**368**) which can exist in three *d,l* pairs, but only one is obtained which may be a mixture of the three isomers. In the mono *N*-methyl derivative of **368** four *d,l* pairs are possible but only two are observed by ^{1}H nmr spectroscopy. It is possible that the pyridine rings are not locked and that at room temperature the twisted pentaphene skeleton is undergoing inversion accompanied by rotation of both pyridine rings so that only two of the four *d,l* pairs are observed. The phenanthrene derivatives **369** were also prepared and their uv spectra revealed the interactions between the two pyridyl groups in the form of intramolecular charge transfer effects.[811,812]

When the azonia adduct **370**, formed from azoniaanthracene and ketene diethyl acetal, was hydrolyzed it gave the ketone which rearranged to 2-acetoxy-1-(2-pyridyl)naphthalene on heating with acetic anhydride. A

(43)

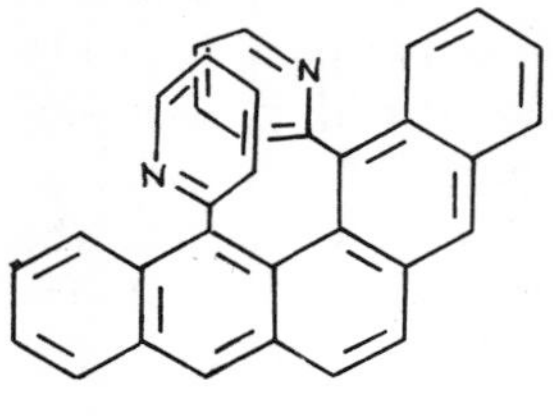

368

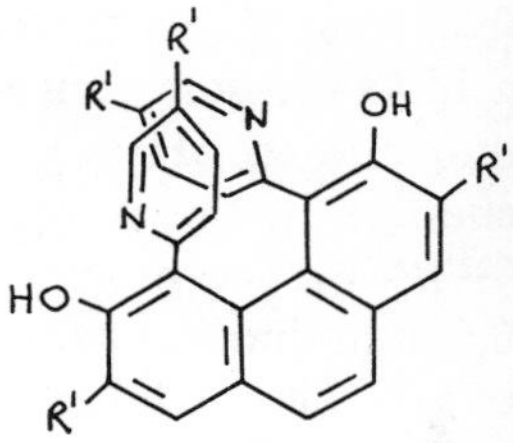

369

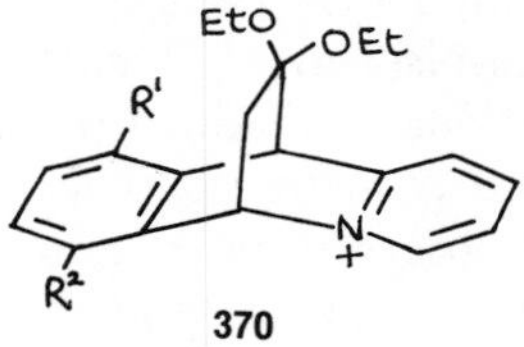

370

similar series of reactions by way of the intermediate **370** ($R^1 = Bu^t$, $R^2 =$ OAc) furnished 2,5-diacetoxy-8-*t*-butyl-1-(2-pyridyl)naphthalene (**371**) in which there was considerable interaction between the peri *t*-butyl and pyridyl groups. This strain is relieved when the 2-acetoxy group is hydro-

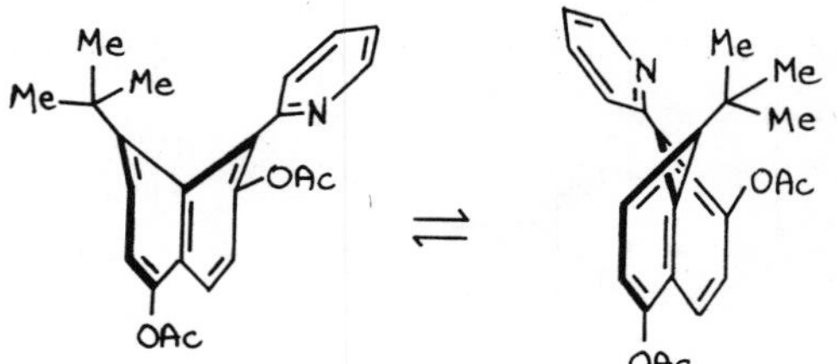

371

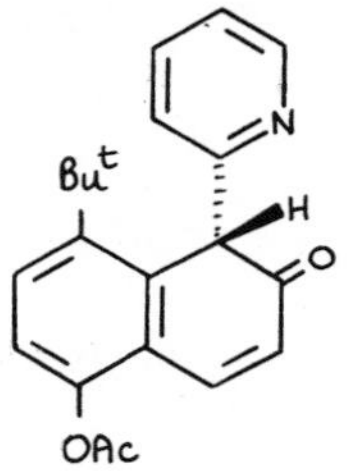

372

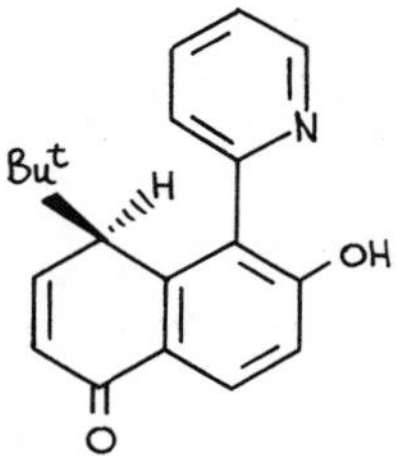

373

lyzed because the 2-hydroxy derivative obtained enolizes to the ketone
372. If both acetoxy groups are hydrolyzed then the ketone **373** is formed.[813]

The heterohelicene, 4,7,16-triacetyl-3-phenyl-3,4-dihydropyrazino[3,4-c:
6,5-c']dicarbazole (**374**), prepared from cyclohexane-1,4-dione bisphenyl-

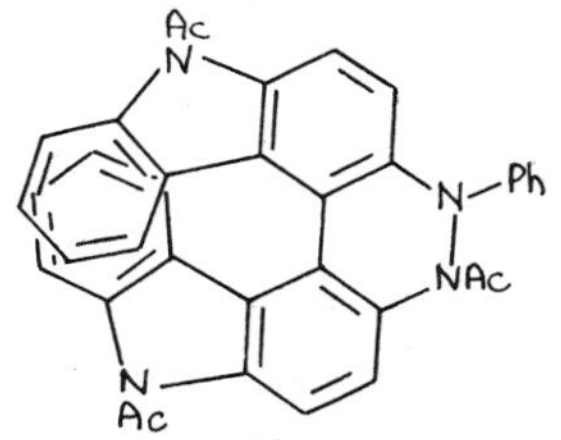

374

hydrazone, was partially resolved on an acetyl cellulose column but the
optical rotations observed were very weak for this type of molecule.[814]

D. Miscellaneous

The rate of nucleophilic substitution of the halogen atom in 6-chloro-
phenanthridine is enhanced when methyl groups are present in the 1,10-
positions. The enhancement is explained by the relief of strain in the
transition state (**375**) of the 1,10-dimethyl compound.[815]

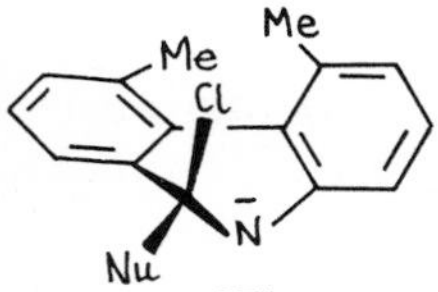

375

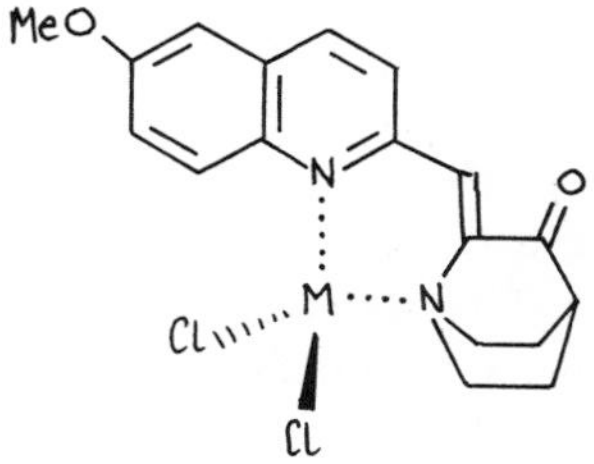

376

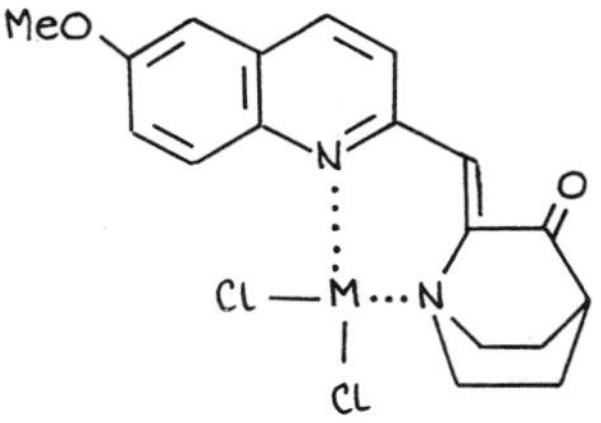

377

The cobalt, nickel, and copper complexes of *trans*-2-(6-methoxy-2-quino-lyl)methylene-3-quinuclidinones were prepared. Cobalt formed tetrahedral complexes (**376,** M = Co) and copper formed square planar complexes (**377,** M = Cu), but the complexes of nickel were particularly interesting. Nickel gave the two types of complexes, the less stable tetrahedral complexes (**376,** M = Ni, and analogue with H replacing OMe) and the more stable square planar complexes (**377,** M = Ni, and analogue with H in place of OMe) which were interconvertible in hot solvents such as methylene chloride, ethanol, and *n*-butanol.[816]

Hindered internal rotation of the *N*-dimethylamino group in 6-dimethyl-aminopurine hydrochloride in aqueous solution was studied by ^{1}H nmr spectroscopy and the rates of rotations; activation parameters and spin lattice relaxation data were evaluated.[817] *N*-2'-aminoaryl 6-purinoylamides (e.g., **378**) were isolated in two stable conformations. Both forms are

378

stabilized by intramolecular hydrogen bonding between the NH_2 group and N–1 or N–7. Models suggest a folded structure for one isomer in which the aryl ring may be close enough to the purine residue as to interfere with the π system.[818]

The stereochemistry of purine and pyrimidine nucleosides and nucleotides, polynucleotides and nucleic acids is beyond the scope of this monograph, but several texts are available on these subjects.[819,820]

XV.   REFERENCES

1. E. Klingsberg, Ed., *Pyridine and its Derivatives*, Interscience, New York, Part 1, 1960, Part 2, 1961, Part 3, 1962, Part 4, 1964; R. A. Abramovitch, Ed., *Pyridine and its Derivatives*, Vol. 14, supplement 2 and 3, 1974, supplement 4, 1975, Interscience, New York.

2. A. Šilhánková, D. Doskočilová, and M. Ferles, *Coll. Czech. Chem. Comm.*, **34,** 1976 (1969).

3. H. Booth, J. H. Little, and J. Feeney, *Tetrahedron*, **24,** 279 (1968).

4. D. E. Caddy and J. H. P. Utley, *J.C.S. Perkin II*, 1258 (1973).

5. M. M. Cook and C. Djerassi, *J. Amer. Chem. Soc.*, **95**, 3678 (1973).

6. G. Jollés, G. Poiget, J. Robert, B. T. Terlain, and J.-P. Thomas, *Bull. Soc. chim. France*, 2252 (1965).

7. V. Prelog and U. Geyer, *Helv. Chim. Acta*, **28**, 1677 (1945).

8. B. R. Baker and F. J. McEvoy, *J. Org. Chem.*, **20**, 136 (1955).

9. M. Uskoković, C. Reese, H. L. Lee, G. Grethe, and J. Gutzwiller, *J. Amer. Chem. Soc.*, **93**, 5902 (1971).

10. G. Grethe, H. L. Lee, T. Mitt, and M. Uskoković, *J. Amer. Chem. Soc.*, **93**, 5904 (1971).

11. D. F. Barringer, Jr., G. Berkelhammer, S. D. Carter, L. Goldman, and A. E. Lanzilotti, *J. Org. Chem.*, **38**, 1933 (1973).

12. M. Tsuda and Y. Kawazoe, *Chem. and Pharm. Bull.* (*Japan*), **18**, 2499 (1970).

13. L. Marion and W. F. Cockburn, *J. Amer. Chem. Soc.*, **71**, 3402 (1949).

14. F. Galinovsky and H. Langer, *Monatsh.*, **86**, 449 (1955).

15. F. Bohlmann, N. Ottawa, and R. Keller, *Annalen*, **587**, 162 (1954).

16. K. E. Crook and S. M. McElvain, *J. Amer. Chem. Soc.*, **52**, 4006 (1930).

17. G. Fodor and G. A. Cooke, *Tetrahedron*, **1**, supplement 8, 113 (1966); G. Fodor and E. Bauerschmidt, *J. Heterocyclic Chem.*, **5**, 205 (1968).

18. H. C. Beyerman, L. Maat, A. van Veen, A. Zweistra, and W. von Philipsborn, *Rec. Trav. chim.*, **84**, 1376 (1965).

19. N. J. Leonard and B. L. Ryder, *J. Org. Chem.*, **18**, 598 (1953).

20. J. Huet, *Bull. Soc. chim. France*, 973 (1964).

21. R. K. Hill and J. W. Morgan, *J. Org. Chem.*, **31**, 3451 (1966); R. K. Hill, T. H. Chan, J. A. Joule, *Tetrahedron*, 147 (1965).

22. C. G. Overberger, J. G. Lombardino, and R. G. Kiskey, *J. Amer. Chem. Soc.*, **79**, 6430 (1957).

23. A. Šilhánková, D. Doskoćilová, and M. Ferles, *Coll. Czech. Chem. Comm.*, **34**, 1985 (1969).

24. M. Holík, A. Tesařová, and M. Ferles, *Coll. Czech. Chem. Comm.*, **32**, 1730 (1967).

25. K. M. J. McErlane and A. F. Casy, *Canad. J. Chem.*, **50**, 2149 (1972).

26. H. Plieninger and S. Leonhäuser, *Chem. Ber.*, **92**, 1579 (1959).

27. R. J. Sundberg and F. O. Holcombe, Jr., *J. Org. Chem.*, **34**, 3273 (1969).

28. F. Troxler, *Helv. Chem. Acta*, **56**, 374 (1973).

29. Z-i. Horii, K. Morikawa, and I. Ninomiya, *Chem. and Pharm. Bull.* (*Japan*), **19**, 2230 (1969).

30. M. Ferles, P. Štern, P. Trška, and F. Vyšata, *Coll. Czech. Chem. Comm.*, **38**, 1206 (1973).

31. P. Štern, P. Trška, and M. Ferles, *Coll. Czech. Chem. Comm.*, **39**, 2267 (1974).

32. M. A. Iorio, P. Ciuffa, and G. Damia, *Tetradhedron*, **26**, 5519 (1970).

32a. M. A. Iorio, G. N. Barrios, E. Menichini, and A. Mazzeo-Farina, *Tetrahedron*, **31**, 1959 (1975).

33. R. E. Lyle and C. K. Spicer, *Tetrahedron Letters*, 1133 (1970).

34. J. F. Archer, D. R. Boyd, W. R. Jackson, M. F. Grundon, and W. A. Khan, *J. Chem. Soc.* (*C*), 2560 (1917).

35. B. A. Mikrut, F. M. Hershenson, K. F. King, L. Bauer, and R. S. Egan, *J. Org. Chem.*, **36**, 3749 (1971).

36. R. S. Egan, F. M. Hershenson, and L. Bauer, *J. Org. Chem.*, **34**, 665 (1969).

37. E. A. Mistryukov, *Izvest. Akad. Nauk S.S.S.R.*, *Ser. khim.*, 1793 (1965).

38. K. Tsuda and K. Sakia, *Chem. and Pharm. Bull. (Japan)*, **7**, 199 (1959).

39. M. Balasubramanian and N. Padma, *Tetrahedron*, **24**, 5395 (1968).

40. M. Balasubramanian and N. Padma, *Tetrahedron*, **19**, 2135 (1963).

41. E. A. Mistryukov, *Izvest. Akad. Nauk S.S.S.R.*, *Ser. khim.*, 1788 (1965).

42. G. Settimj, M. T. Carani, G. Gatta, and S. Chiavarelli, *Gazzetta*, **100**, 703 (1970).

43. D. Taub, C. H. Kuo, and N. L. Wendler, *J. Chem. Soc. (C)*, 1558 (1967).

44. B. Witkop and C. M. Foltz, *J. Amer. Chem. Soc.*, **79**, 192 (1957).

45. J. W. Clark-Lewis and P. I. Mortimer, *J. Chem. Soc.*, 4268 (1961).

46. H. C. Beyerman and P. Boekee, *Rec. Trav. chim.*, **78**, 648 (1959).

47. K. Hohenlohe-Oehringen, *Monatsh.*, **96**, 262 (1965).

48. K. Hohenlohe-Oehringen and H. Bretschneider, *Monatsh.*, **96**, 246 (1965).

49. K. Hohenlohe-Oehringen, *Monatsh.*, **94**, 1222 (1963).

50. W. Ziraikus and R. Haller, *Arch. Pharm.*, **305**, 493 (1972).

51. O. Jeger, V. Prelog, E. Sundt, and R. B. Woodward, *Helv. Chim. Acta*, **37**, 2302 (1954).

52. S-i. Yamada and S. Terashima, *Chem. Comm.*, 511 (1969).

53. H. E. Johnson and D. G. Crosby, *J. Org. Chem.*, **27**, 1298 (1962).

54. D. R. Brown, J. McKenna, and J. M. McKenna, *J. Chem. Soc. (B)*, 567 (1969).

55. H. Newman, *J. Heterocyclic Chem.*, **11**, 449 (1974).

56. B. R. Baker, R. E. Schaub, F. J. McEvoy, and J. H. Williams, *J. Org. Chem.*, **17**, 132 (1952).

57. B. R. Baker, F. J. McEvoy, R. E. Schaub, J. P. Joseph, and J. H. Williams, *J. Org. Chem.*, **18**, 153 (1953).

58. M. Julia, O. Siffert, and J. Bagot, *Bull. Soc. chim. France*, 1007 (1968).

59. E. Gellert, R. Rudzats, R. E. Summons, and B. R. Worth, *Austral. J. Chem.*, **24**, 843 (1971).

60. D. Misiti, G. Settimj, P. Mantovani, and S. Chiavarelli, *Gazzetta*, **100**, 495 (1970).

61. R. Haller and H. Unholzer, *Arch. Pharm.*, **305**, 855 (1972).

62. V. F. Kucherov and N. I. Shvetsov, *Izvest. Akad. Nauk S.S.S.R.*, *Ser. khim.*, 263 (1961).

63. E. J. Corey, J. F. Arnett, and C. N. Widiger, *J. Amer. Chem. Soc.*, **97**, 430 (1975).

64. T. Tokuyama, K. Uenoyama, G. Brown, J. W. Daly, and B. Witkop, *Helv. Chim. Acta*, **56**, 2597 (1974).

65. C. F. Koelsch and C. H. Stratton, *J. Amer. Chem. Soc.*, **66**, 1881 (1944); C. F. Koelsch and R. F. Raffauf, *J. Amer. Chem. Soc.*, **66**, 1857 (1944).

66. W. Reid and F. Bätz, *Angew. Chem. Internat. Edn.*, **10**, 735 (1971), *Annalen*, **762**, 1 (1972).

67. M. Ogata, H. Matsmuoto, and H. Kanō, *Tetrahedron*, **25**, 5217 (1969).

68. S. Wolfe, H. B. Schlegel, M.-H. Whangbo, and F. Bernardi, *Canad. J. Chem.*, **52**, 3787 (1974).

69. P. J. Krueger and J. Jan, *Canad. J. Chem.*, **48**, 3236 (1970).

70. C.-Y. Chen and R. J. W. Le Fèvre, *Tetrahedron Letters*, 1611 (1965).

71. R. E. Lyle, and D. H. McMahon, W. E. Krueger, and C. K. Spicer, *J. Org. Chem.*, **31**, 4164 (1966).

72. H. S. Aaron and C. P. Ferfuson, *Tetrahedron*, **30**, 803 (1974).

73. G. Hite, E. E. Smissman, and R. West, *J. Amer. Chem. Soc.*, **82**, 1207 (1960).

74. J. Sicher and M. Tichý, *Coll. Czech. Chem. Comm.*, **23**, 2081 (1958).

75. C. Schöpf, F. Braun, H. Koop, G. Werner, H. Bressler, K. Neisuis, and E. Schmadel, *Annalen*, **658**, 156 (1962).

76. C. M. Lee and W. D. Kumler, *J. Amer. Chem. Soc.*, **83**, 4586 (1961).

77. W. Gaffield, L. Keefer, and W. Lijinsky, *Tetrahedron Letters*, 779 (1972).

78. G. Snatzke, H. Ripperger, C. Horstmann, and K. Schreiber, *Tetrahedron*, **22**, 3103 (1966).

79. R. Ramasseul, A. Rassat, and P. Rey, *Tetrahedron*, **30**, 265 (1974).

80. Y. Kashman and S. Cherkez, *Tetrahedron*, **28**, 1211 (1972).

80a. H. Meguro, T. Konno, and K. Tuzimura, *Tetrahedron Letters*, 1309 (1975).

81. H. Ripperger and K. Schreiber, *Tetrahedron*, **23**, 1841 (1967).

82. H. Ripperger and H. Pracejus, *Tetrahedron*, **24**, 99 (1968).

83. M. M. El-Olemy and A. E. Schwarting, *J. Org. Chem.*, **34**, 1352 (1969).

84. B. Sjöberg, A. Fredga, and C. Djerassi, *J. Amer. Chem. Soc.*, **81**, 5002 (1959).

85. H. Ripperger and K. Schreiber, *Tetrahedron*, **21**, 407 (1965).

86. H. Booth and A. H. Bostock, *Chem. Comm.*, 637 (1967).

87. T. P. Forrest and S. Ray, *Chem. Comm.*, 1537 (1970).

88. J. B. Stothers, *Carbon 13 N.m.r. Spectroscopy*, Academic, New York, 1972, p. 464; N. K. Wilson and J. B. Stothers, *Topics Stereochem.*, **8**, 1 (1974).

89. A. J. Jones and M. M. A. Hassan ,*J. Org. Chem.*, **37**, 2332 (1972).

90. F. I. Carroll and J. T. Blackwell, *J. Medicin. Chem.*, **17**, 210 (1974).

91. R. K. Hill, *J. Amer. Chem. Soc.*, **80**, 1611 (1958).

92. R.K. Hill, *J. Amer. Chem. Soc.*, **80**, 1609 (1958).

93. R. Lukes and M. Smetacková, *Coll. Czech. Chem. Comm.*, **6**, 231 (1934).

94. W. F. M. Van Bever, C. J. E. Niemegeers, and P. A. J. Janssen, *J. Medicin. Chem.*, **17**, 1047 (1974).

95. D. Fires and P. S. Portoghese, *J. Medicin. Chem.*, **17**, 990 (1974).

96. P. S. Portoghese, T. L. Pazdernik, W. L. Kuhn, G. Hite, and A. Shafi'ee, *J. Medicin. Chem.*, **11**, 12 (1968).

97. R. A. Johnson, H. C. Murray, L. M. Reinke, and G. S. Fonken, *J. Org. Chem.*, **34**, 2279 (1969).

98. J. W. Clark-Lewis and P. I. Mortimer, *J. Chem. Soc.*, 189 (1961).

99. A. V. Robertson and L. Marion, *Canad. J. Chem.*, **37**, 829 (1959).

100. F. E. King, T. J. King, and A. J. Warwick, *J. Chem. Soc.*, 3590 (1950); N. A. Dobson and R. A. Raphael, *J. Chem. Soc.*, 3643 (1958).

101. W. Y. Rice, Jr. and J. L. Coke, *J. Org. Chem.*, **31**, 1010 (1966); R. J. Highet, *J. Org. Chem.*, **29**, 471 (1964).

102. H. Rapoport, H. D. Baldridge, Jr., and E. J. Volcheck, Jr., *J. Amer. Chem. Soc.*, **75,** 5290 (1953).

103. J. L. Coke and W. Y. Rice, Jr., *J. Org. Chem.*, **30,** 3420 (1965).

104. Y. Arata and T. Ohashi, *Chem. and Pharm. Bull. (Japan)*, **13,** 1365 (1965).

105. F. Galinovsky, G. Bianchetti, and O. Vogl, *Monatsh.*, **84,** 1221 (1953).

106. C. R. Noller and C. E. Pannell, *J. Amer. Chem. Soc.*, **77,** 1862 (1955).

107. R. Adams and D. C. Blomstron, *J. Amer. Chem. Soc.*, **75,** 2375 (1953).

108. R. E. Lyle and G. G. Lyle, *J. Org. Chem.*, **24,** 1679 (1959).

109. G. G. Lyle and E. T. Pelosi, *J. Amer. Chem. Soc.*, **88,** 5276 (1966).

110. R. Haller and W. Ziriakus, *Arch. Pharm.*, **305,** 741 (1972).

111. W. Ziriakus and R. Haller, *Arch. Pharm.*, **305,** 814 (1972).

112. R. Andrisano, A. S. Angeloni, and G. Gottarelli, *Tetrahedron*, **30,** 3827 (1974).

113. J. B. Lambert, *Topics Stereochem.*, **6,** 19 (1971).

114. F. G. Riddell, *Quart. Rev.*, **21,** 364 (1967); J. McKenna, *"Conformational Analysis of Organic Compounds,"* Royal Institute Chemical Lectures, No. 1 1966.

115. M. Aroney and R. J. W. Le Fèvre, *J. Chem. Soc.*, 3002 (1958).

115a. M. A. Robb, W. J. Haines, and I. G. Csizmadia, *J. Amer. Chem. Soc.*, **95,** 42 (1973).

116. R. J. Bishop, L. E. Sutton, D. Dineen, R. A .Y. Jones, and A. R. Katritzky, *Proc. Chem. Soc.*, 257 (1964).

117. N. L. Allinger, J. G. D. Carpenter, and F. M. Karkowski, *Tetrahedron Letters*, 3345 (1964); *J. Amer. Chem. Soc.*, **87,** 1232 (1965).

118. J. B. Lambert, R. G. Keske, R. E. Carhart, and A. P. Jovanovich, *J. Amer. Chem. Soc.*, **89,** 3761 (1967).

119. J. B. Lambert and R. G. Keske, *Tetrahedron Letters*, 2023 (1969).

120. J. B. Lambert, S. D. Bailey, and B. F. Michel, *Tetrahedron Letters*, 691 (1970).

121. J. B. Lambert, D. S. Bailey, and B. F. Michel, *J. Amer. Chem. Soc.*, **94,** 3812 (1972).

122. R. A. Y. Jones, A. R. Katritzky, A. C. Richards, R. J. Wyatt, R. J. Bishop, and L. E. Sutton, *J. Chem. Soc. (B)*, 127 (1970).

123. R. W. Baldock and A. R. Katritzky, *J. Chem. Soc. (B)*, 1470 (1968).

124. M. Tsuda and Y. Kawazoe, *Chem. and Pharm. Bull. (Japan)*, **16,** 702 (1968).

125. C. C. Costain, P. J. Buckley, and J. E. Parkin, *Quart. Rev.*, **21,** 368 (1967).

126. P. J. Brignell, A. R. Katritzky, and P. L. Russell, *Chem. Comm.*, 723 (1966).

127. P. J. Brignell, A. R. Katritzky and P. L. Russell, *J. Chem. Soc. (B)*, 1459 (1968).

128. H. Booth, *Chem. Comm.*, 802 (1968).

129. J. McKenna and J. M. McKenna, *J. Chem. Soc. (B)*, 644 (1969).

130. H. Booth and J. H. Little, *J.C.S. Perkin II*, 1846 (1972), *Tetrahedron*, **23,** 291 (1967); B. Bianchin and J. J. Delpuech, *Tetrahedron*, **30,** 2859 (1974).

131. P. J. Crowley, M. J. T. Robinson, and M. G. Ward, *J.C.S. Chem. Comm.*, 825 (1974).

132. R. J. Bishop, L. E. Sutton, D. Dineen, R. A. Y. Jones, A. R. Katritzky, and R. J. Wyatt, *J. Chem. Soc. (B)*, 493 (1967).

133. R. A. Y. Jones, A. R. Katritzky, A. C. Richards, and R. J. Wyatt, *J. Chem. Soc. (B)*, 122 (1970).

134. J.-J. Delpuech and M. N. Deschamps, *Tetrahedron*, **26,** 2723 (1970).

135. R. D. Stolow, D. I. Lewis, and P. A. D'Angelo, *Tetrahedron*, **26**, 5831 (1970; M.-L. Stien, G. Chiurdoglu, R. Ottinger, J. Reisse, and H. Christol, *Tetrahedron*, **27**, 411 (1971).

135a. Y. Terui and K. Tori, *J.C.S. Perkin II*, 127 (1975).

135b. B. V. Unkovsky, S. V. Bogatkov, J. F. Malina, I. M. Studneva, T. D. Sokolova, and K. I. Romanova, *Tetrahedron*, **31**, 1321 (1975).

136. Y. Kawazoe, M. Tsuda, and M. Ohnishi, *Chem. and Pharm. Bull. (Japan)*, **15**, 51 (1967).

137. I. D. Blackburne, R. P. Duke, R. A. Y. Jones, A. R. Katritzky, and K. F. A. Record, *J.C.S. Perkin II*, 332 (1973).

138. M. J. Aroney, C-Y. Chen, R. J. W. Le Fèvre, and A. N. Singh, *J. Chem. Soc. (B)*, 98 (1966).

139. C-Y. Chen and R. J. W. Le Fèvre, *J. Chem. Soc.*, 3467 (1965).

140. N. L. Allinger and S. D. Jindal, *J. Org. Chem.*, **37**, 1042 (1972).

141. R. Haller and W. Ziriakus, *Arch. Pharm.*, **303**, 22 (1970).

142. R. Haller and W. Ziriakus, *Tetrahedron*, **28**, 2863 (1972); W. Hänsel and R. Haller, *Arch. Pharm.*, **302**, 147 (1969).

143. R. Haller and J. Ebersberg, *Arch. Pharm.*, **302**, 197 (1969).

144. E. A. Mistryukov and G. N. Smirnova, *Izvest. Akad. Nauk S.S.S.R., Ser. khim.*, 1525 (1970).

145. M. J. Cook, A. R. Katritzky, and M. Moreno Mañas, *J. Chem. Soc. (B)*, 1330 (1971).

146. Y. Shvo and E. D. Kaufman, *Tetrahedron*, **28**, 573 (1972).

147. R. E. Rolfe, K. D. Sales, and J. H. P. Utley, *J.C.S. Perkin II*, 1171 (1973).

148. F. Moll, *Tetrahedron Letters*, 5201 (1968).

149. C-Y. Chen and R. J. W. Le Fèvre, *Tetrahedron Letters*, 4057 (1965).

150. H. Dorn, A. R. Katritzky, and M. R. Nesbit, *J. Chem. Soc. (B)*, 501 (1967).

151. W. Barbieri and L. Bernardi, *Tetrahedron*, **21**, 2453 (1965).

152. M. Aroney and R. J. W. Le Fèvre, *J. Chem. Soc.*, 2161 (1960).

153. R. E. Lyle and L. N. Pridgen, *J. Org. Chem.*, **38**, 1618 (1973).

154. J. K. Becconsall, R. A. Y. Jones, and J. McKenna, *J. Chem. Soc.*, 1726 (1965).

155. R. F. Branch and A. F. Casy, *J. Chem. Soc. (B)*, 1087 (1968).

156. R. K. Harris and R. A. Spragg, *J. Chem. Soc. (B)*, 684 (1968).

157. H. Kessler and D. Leibfritz, *Tetrahedron Letters*, 4297 (1970).

158. J. M. Lehn and J. Wagner, *Chem. Comm.*, 414 (1970).

159. J. B. Lambert and W. L. Oliver, Jr., *Tetrahedron Letters*, 6187 (1968).

160. J. Cantacuzene and J. Leroy, *J. Amer. Chem. Soc.*, **93**, 5263 (1971).

161. G. A. Yousif and J. D. Roberts, *J. Amer. Chem. Soc.*, **90**, 6428 (1968).

162. R. R. Fraser and T. B. Grindley, *Tetrahedron Letters*, 4169 (1974).

163. Y. L. Chow, C. T. Colón, and J. N. S. Tam, *Canad. J. Chem.*, **46**, 2821 (1968).

164. R. A. Johnson, *J. Org. Chem.*, **33**, 3627 (1968).

165. J. M. Lehn and J. Wagner, *Chem. Comm.*, 1298 (1968).

166. T. P. Forrest, D. L. Hooper, and S. Ray, *J. Amer. Chem. Soc.*, **96**, 4286 (1974).

167. A. J. Jones, A. F. Casy, and K. M. J. McErlane, *J.C.S. Perkin I*, 2576 (1973).

168. I. Morishima, K. Yoshikawa, K. Oksada, T. Yonezawa, and K. Goto, *J. Amer. Chem. Soc.*, **95**, 165 (1973).

169. A. F. Casy and K. M. J. McErlane, *J.C.S. Perkin I*, 334 (1972).

170. A. J. Jones, C. P. Beeman, A. F. Casy, and K. M. J. McErlane, *Canad. J. Chem.*, **51**, 1790 (1973).

171. A. J. Jones, A. F. Casy, and K. M. J. McErlane, *Canad. J. Chem.*, **51**, 1782 (1973).

172. E. L. Eliel, W. F. Bailey, L. D. Kopp, R. L. Willer, D. M. Grant, R. Bertrand, K. A. Christensen, D. K. Dalling, M. W. Duch, E. Wenkert, F. M. Schell, and D. W. Cochran, *J. Amer. Chem. Soc.*, **97**, 322 (1975).

173. H. K. Hall, Jr., *J. Org. Chem.*, **29**, 3135 (1964).

174. C. G. Overberger, J. G. Lombardino, and R. G. Hiskey, *J. Amer. Chem. Soc.*, **80**, 3009 (1958).

175. S. Terashima, M. Wagatsuma, and S-i. Yamada, *Chem. and Pharm. Bull. (Japan)*, **18**, 1137 (1970); C. G. Overberger, N. P. Marullo, and R. G. Hiskey, *J. Amer. Chem. Soc.*, **83**, 1374 (1961).

176. E. A. Mistryukov, G. T. Katvalyan, and G. N. Smirnova, *Izvest. Akad. Nauk S.S.S.R.*, *Ser. khim.*, 1067 (1970).

177. E. A. Mistryukov and G. T. Katvalyan, *Izvest. Akad. Nauk S.S.S.R.*, *Ser. khim.*, 1298 (1970); A. F. Casy and W. K. Jeffrey, *Canad. J. Chem.*, **50**, 803 (1972).

178. E. A. Mistryukov, G. N. Smirnova, and N. I. Aronova, *Izvest. Akad. Nauk S.S.S.R.*, *Ser. khim.*, 1301 (1970).

179. N. J. Harper, A. H. Beckett, and A. D. J. Balon, *J. Chem. Soc.*, 2704 (1960).

180. A. F. Casy and K. M. J. McErlane, *J.C.S. Perkin I*, 726 (1972).

181. É. A. Mistryukov and N. I. Shvestov, *Izvest. Akad. Nauk S.S.S.R.*, *Ser. khim.*, 268 (1961).

182. A. F. Casy and P. Pocha, *J. Chem. Soc. (C)*, 979 (1967).

183. É. A. Mistryukov and V. F. Kucherov, *Izvest. Akad. Nauk S.S.S.R.*, *Ser. khim.*, 578 (1961).

184. A. F. Casy, A. H. Beckett, and M. A. Iorio, *Tetrahedron*, **23**, 1405 (1967).

185. K. M. J. McErlane and A. F. Casy, *J.C.S. Perkin I*, 339 (1972).

186. A. F. Casy, L. G. Chatten, and K. K. Khullar, *Canad. J. Chem.*, **48**, 2372 (1970).

187. D. V. Sokolov, *J. Gen. Chem. (U.S.S.R.)*, **30**, 839 (1960).

188. R. Haller and W. Hänsel, *Tetrahedron*, **26**, 2035 (1970).

189. E. A. Mistryukov and G. N. Smirnova, *Tetrahedron*, **27**, 375 (1971).

190. M. M. A. Hassan and A. F. Casy, *Tetrahedron*, **26**, 4517 (1970).

191. J. P. Li and J. H. Biel, *J. Org. Chem.*, **35**, 4100 (1970).

192. A. C. Cope and N. A. Le Bel, *J. Amer. Chem. Soc.*, **82**, 4656 (1960).

193. N. J. Leonard and S. Gelfand, *J. Amer. Chem. Soc.*, **77**, 3269, 3272 (1955).

194. N. J. Leonard, R. C. Sentz, and W. J. Middleton, *J. Amer. Chem. Soc.*, **75**, 1674 (1953).

195. J. McKenna, *Topics Stereochem.*, **5**, 275 (1970).

196. R. Brettle, D. R. Brown, J. McKenna, and J. M. McKenna, *Chem. Comm.*, 696 (1969).

197. W. Fedeli, F. Mazza, and A. Vaciago, *J. Chem. Soc. (B)*, 1218 (1970).

198. R. P. Duke, R. A. Y. Jones, A. R. Katritzky, J. R. Carruthers, W. Fedeli, F. Mazza, and A. Vaciago, *J.C.S. Chem. Comm.*, 455 (1972).

199. J. McKenna, J. M. McKenna, and J. White, *J. Chem. Soc.*, 1733 (1965).

200. D. R. Brown, J. McKenna, and J. M. McKenna, *J. Chem. Soc. (B)*, 1195 (1967).

201. J. McKenna, J. White, and A. Tulley, *Tetrahedron Letters*, 1097 (1962).

202. J. McKenna, J. M. McKenna, A. Tulley, and J. White, *J. Chem. Soc.*, 1711 (1965).

203. V. J. Baker, I. D. Blackburne, A. R. Katritzky, R. A. Kolinski, and Y. Takeuchi, *J.C.S. Perkin II*, 1563 (1974).

204. R. P. Duke, R. A. Y. Jones, and A. R. Katritzky, *J.C.S. Perkin II*, 1553 (1973).

205. V. J. Baker, I. D. Blackburne, and A. R. Katritzky, *J.C.S. Perkin II*, 1557 (1974).

206. T. M. Bare, N. D. Hershey, H. O. House, and C. G. Swain, *J. Org. Chem.*, **37**, 997 (1972).

207. J.-L. Imbach, A. R. Katritzky, and R. A. Kolinski, *J. Chem. Soc. (B)*, 556 (1966).

208. D. R. Brown, R. Lygo, J. McKenna, J. M. McKenna, and B. G. Hutley, *J. Chem. Soc. (B)*, 1184 (1967).

209. H. O. House, B. A. Tefertiller, and C. G. Pitt, *J. Org. Chem.*, **31**, 1073 (1966).

210. D. R. Brown, B. G. Hutley, J. McKenna, J. M. McKenna, *Chem. Comm.*, 719 (1966).

211. D. R. Brown, J. McKenna, J. M. McKenna, and J. M. Stuart, *Chem. Comm.*, 380 (1967).

212. A. F. Casy, A. H. Beckett, and M. A. Iorio, *Tetrahedron*, **22**, 2751 (1966).

213. R. A. Y. Jones, A. R. Katritzky, and P. G. Mente, *J. Chem. Soc. (B)*, 1210 (1970).

214. Y. Kawazoe and M. Tsuda, *Chem. and Pharm. Bull. (Japan)*, **15**, 1405 (1967).

215. J. McKenna, J. M. McKenna, and A. Tulley, *J. Chem. Soc.*, 5439 (1965).

216. M. A. Iorio and S. Chiavarelli, *Tetrahedron*, **25**, 5235 (1969).

217. R. V. Smith, F. W. Benz, and J. P. Long, *Canad. J. Chem.*, **51**, 171 (1973).

218. R. E. Lyle, E. W. Southwick, and J. J. Kaminski, *J. Amer. Chem. Soc.*, **94**, 1413 (1972).

219. B. G. Hutley, J. McKenna, and J. M. Stuart, *J. Chem. Soc. (B)*, 1199 (1967).

220. J. McKenna, B. G. Hutley, and J. White, *J. Chem. Soc.*, 1729 (1965).

221. J. McKenna and J. White, *Tetrahedron Letters*, 1493 (1963).

222. G. T. Katvalyan and E. A. Mistryukov, *Izvest. Akad. Nauk S.S.S.R., Ser. khim.*, 1868 (1969).

223. Y. Kawazoe, M. Tsuda, and M. Ohnishi, *Chem. and Pharm. Bull. (Japan)*, **15**, 214 (1967).

224. C. A. Grob in *Theoretical Organic Chemistry*, (Kekulé Symposium 1958), Butterworths, London, 1959, pp. 114–126.

225. R. D'Arcy, C. A. Grob, T. Kaffenberger, and V. Krasnobajew, *Helv. Chim. Acta*, **49**, 185 (1966).

225a. C. A. Grob, W. Kunz, and P. R. Marbet, *Tetrahedron Letters*, 2613 (1975).

225b. C. A. Grob, *Bull. Soc. chim. France*, 1360 (1960); *Angew. Chem. Internat. Edn.*, **8**, 535 (1969).

226. R. E. Lyle and W. E. Krueger, *J. Org. Chem.*, **30**, 394 (1965); **32**, 2873 (1967).

227. C. F. Hammer and S. R. Heller, *Tetrahedron*, **28**, 239 (1972).

228. E. M. Fry, *J. Org. Chem.*, **30**, 2058 (1965).

229. T. B. Zalucky, L. Malspeis, and G. Hite, *J. Org. Chem.*, **29**, 3143 (1964); T. B. Zalucky, S. Marathe, L. Malspeis, and G. Hite, *J. Org. Chem.*, **30**, 1324 (1965).

229a. M. Ferles and J. Pliml, *Adv. Heterocyclic Chem.*, **12**, 43 (1970).

230. M. Tišler and B. Stanovnik, *Adv. Heterocyclic Chem.*, **9**, 211 (1968).

231. K. R. Kopecky and J. Soler, *Canad. J. Chem.*, **52**, 2111 (1974).

232. J. A. Berson and S. S. Olin, *J. Amer. Chem. Soc.*, **92**, 1086 (1970).

233. E. L. Allred and J. C. Hinshaw, *J. Amer. Chem. Soc.*, **90**, 6885 (1968).

234. J. A. Berson and S. S. Olin, *J. Amer. Chem. Soc.*, **91**, 777 (1969).

235. D. C. Horwell and J. A. Deyrup, *J.C.S. Chem. Comm.*, 485 (1972).

236. B. M. Trost, R. M. Cory, P. H. Scudder, and H. B. Neubold, *J. Amer. Chem. Soc.*, **95**, 7813 (1973).

237. B. M. Trost and R. M. Cory, *J. Org. Chem.*, **37**, 1106 (1972).

238. P. C. Arora and D. Mackay, *Chem. Comm.*, 677 (1969).

239. Y. Kitahara, I. Murata, and T. Nitta, *Tetrahedron Letters*, 3003 (1967).

240. E. L. Allred and K. J. Voorhees, *J. Amer. Chem. Soc.*, **95**, 620 (1973).

241. V. T. Bandurco and J. P. Snyder, *Tetrahedron Letters*, 4643 (1969).

242. L. A. Paquette, *J. Amer. Chem. Soc.*, **92**, 5765 (1970).

243. D. A. Kleier, G. Binsch, A. Steigel, and J. Sauer, *J. Amer. Chem. Soc.*, **92**, 3787 (1970).

244. P. D. Bartlett and N. A. Porter *,J. Amer. Chem. Soc.*, **90**, 5317 (1968).

245. C. H. Hassall, R. B. Morton, Y. Ogihara, and W. A. Thomas, *Chem. Comm.*, 1079 (1969).

246. J. E. Anderson, *J. Amer. Chem. Soc.*, **91**, 6374 (1969).

247. J. E. Anderson and J. M. Lehn, *Bull. Soc. chim. France*, 2402 (1966).

248. B. J. Price, R. V. Smallman, and I. O. Sutherland, *Chem. Comm.*, 319 (1966).

249. B. Price, I. O. Sutherland, and F. G. Williamson, *Tetrahedron*, **22**, 3477 (1966).

250. J. C. Breliere and J. M. Lehn, *Chem. Comm.*, 426 (1965).

251. B. H. Korsch and N. V. Riggs, *Tetrahedron Letters*, 5897 (1966).

252. E. W. Bittner and J. T. Gerig, *J. Amer. Chem. Soc.*, **94**, 913 (1972).

253. J. E. Anderson and J. M. Lehn, *Tetrahedron*, **24**, 137 (1968).

254. C. H. Bushweller, *Chem. Comm.*, 80 (1966).

255. R. Daniels and K. A. Roseman, *Chem. Comm.*, 429 (1966).

256. J. Firl, *Chem. Ber.*, **102**, 2169, 2177 (1969).

257. J. E. Anderson and J. M. Lehn, *J. Amer. Chem. Soc.*, **89**, 81 (1967).

258. Y. Nomura, N. Masai, and T. Takeuchi, *J.C.S. Chem. Comm.*, 288 (1974) and 307 (1975).

259. J. Wagner, W. Wojnarowski, J. E. Anderson, and J. M. Lehn, *Tetrahedron*, **25**, 657 (1969).

260. J. E. Anderson and J. M. Lehn, *Tetrahedron*, **24**, 123 (1968).

261. J. M. Lehn and J. Wagner, *Tetrahedron*, **25**, 677 (1969).

262. R. A. Y. Jones, A. R. Katritzky, D. L. Ostercamp, K. A. F. Record, and A. C. Richards, *Chem. Comm.*, 644 (1971); R. A. Y. Jones, A. R. Katritzky, and R. Scattergood, *Chem. Comm.*, 644 (1971).

263. R. A. Y. Jones, A. R. Katritzky, D. L. Ostercamp, K. A. F. Record, and A. C. Richards, *J.C.S. Perkin II*, 34 (1972).

264. S. F. Nelsen, J. M. Buschek, and P. J. Hintz, *J. Amer. Chem. Soc.* **95**, 2013 (1973).

265. S. F. Nelsen and J. M. Buschek, *J. Amer. Chem. Soc.*, **96**, 6982 (1974).

266. S. F. Nelsen and J. M. Buschek, *J. Amer. Chem. Soc.*, **95**, 2011 (1973).

267. R. A. Y. Jones, A. R. Katritzky, K. A. F. Record, and R. Scattergood, *J.C.S. Perkin II*, 406 (1974).

268. W. Koch and M. Zollinger, *Helv. Chim. Acta*, **46**, 2697 (1963).

269. S. F. Nelsen and P. J. Hintz, *J. Amer. Chem. Soc.*, **93**, 7105 (1971).

270. K. R. Kopecky and S. Evani, *Canad. J. Chem.*, **47**, 4041 (1969).

271. S. G. Cohen, R. Zand, and C. Steel, *J. Amer. Chem. Soc.*, **83**, 2895 (1961).

272. H. Tanida, S. Teratake, Y. Hata, and M. Watanabe, *Tetrahedron Letters*, 5341 (1969).

273. S. G. Cohen and R. Zand, *J. Amer. Chem. Soc.*, **84**, 586 (1962).

274. B. S. Solomon, T. F. Thomas, and C. Steel, *J. Amer. Chem. Soc.*, **90**, 2249 (1968).

275. P. S. Engel, *J. Amer. Chem. Soc.*, **89**, 5731 (1967).

276. E. L. Allred and R. L. Smith, *J. Amer. Chem. Soc.*, **91**, 6766 (1969).

277. E. L. Allred and R. L. Smith, *J. Amer. Chem. Soc.*, **89**, 7133 (1967).

278. E. L. Allred and C. R. Flynn, *J. Amer. Chem. Soc.*, **92**, 1064 (1970).

279. J. C. Hinshaw and E. L. Allred, *Chem. Comm.*, **72**, (1969).

280. W. R. Roth and M. Martin, *Annalen*, **702**, 1 (1967).

281. E. L. Allred and J. C. Hinshaw, *Chem. Comm.*, 1021 (1969).

282. E. L. Allred and J. C. Hinshaw, *Tetrahedron Letters*, 387 (1972).

283. H. Schmidt, A. Schweig, B. M. Trost, H. B. Neubold, and P. H. Scudder, *J. Amer. Chem. Soc.*, **96**, 622 (1974).

284. E. L. Allred, J. C. Hinshaw, and A. L. Johnson, *J. Amer. Chem. Soc.*, **91**, 3382 (1969).

285. A. I. Meyers, D. M. Stout, and T. Takaya, *J.C.S. Chem. Comm.*, 1260 (1972).

285a. D. M. Stout, T. Takaya, and A. I. Meyers, *J. Org. Chem.*, **40**, 536 (1975).

286. D. J. Borwn, *The Pyrimidines*, Wiley-Interscience, New York, 1962 and Supplement I 1970.

287. O. Baudisch and D. Davidson, *J. Biol. Chem.*, **64**, 233 (1925).

288. P. Rouillier, J. Delmau, and C. Nofre, *Bull. Soc. chim. France*, 3515 (1966).

289. P. Rouillier, J. Delmau, J. Duplan, and C. Nofre, *Tetrahedron Letters*, 4189 (1966).

290. C. Nofre, M. Murat, and A. Cier, *Bull. Soc. chim. France*, 1749 (1965).

291. L. R. Subbaraman, J. Subbaraman, and E. J. Behrman, *J. Org. Chem.*, **38**, 1499 (1973).

292. B.-S. Hahn and S. Y. Wang, *J. Amer. Chem. Soc.*, **94**, 4764 (1972).

293. H. Hayatsu and S. Iida, *Tetrahedron Letters*, 1031 (1969).

293a. J. Cadet and R. Teoule, *Carbohydrate Res.*, **29**, 345 (1973).

293b. J. Cadet, J. Ulrich, and R. Teoule, *Tetrahedron*, **31**, 2057 (1975).

294. M. N. Khattak and J. H. Green, *Austral. J. Chem.*, **18**, 1847 (1965).

295. H. Hayatsu, Y. Wataya, and K. Kai, *J. Amer. Chem. Soc.*, **92**, 724 (1970).

296. R. Shapira, R. E. Servis, and M. Welcher, *J. Amer. Chem. Soc.*, **92**, 422 (1970).

297. K. Kai, Y. Wataya, and H. Hayatsu, *J. Amer. Chem. Soc.*, **93**, 2089 (1971).

298. K. Kirai, H. Matsuda, and Y. Kishida, *Chem. and Pharm. Bull. (Japan)*, **21**, 1090 (1973).

299. T. Kunieda and B. Witkop, *J. Amer. Chem. Soc.*, **93**, 3478 (1971).

300. F. I. Carroll and R. Meck, *J. Org. Chem.*, **34**, 2676 (1969); F. I. Carroll and G. N. Mitchell, *J. Medicin. Chem.*, **18**, 37 (1975).

301. E. C. Kleiderer and H. A. Shonle, *J. Amer. Chem. Soc.*, **56**, 1772 (1934).

302. C. E. Cook and C. R. Tallent, *J. Heterocyclic Chem.*, **6**, 203 (1969).

303. J. Knabe, H. Junginger, and W. Geismar, *Arch. Pharm.*, **304**, 1 (1971).

304. K. H. Dudley, I. J. Davis, D. K. Kim, and F. T. Ross, *J. Org. Chem.*, **35**, 147 (1970).

305. J. J. Hansen and A. Kjaer, *Acta Chem. Scand.*, **28B**, 418 (1974).

306. P. Blattmann and J. Rétey, *European J. Biochem.*, **30**, 130 (1972).

307. H. Hayashi and K. Nakanishi, *J. Amer. Chem. Soc.*, **95**, 4081 (1973).

308. F. I. Carroll and A. Sobti, *J. Amer. Chem. Soc.*, **95**, 8512 (1973).

309. W. J. Doran and H. A. Shonle, *J. Amer. Chem. Soc.*, **60**, 2880 (1938).

310. G. S. Skinner and R. deV. Huber, *J. Amer. Chem. Soc.*, **73**, 3321 (1951).

311. H. R. Henze and C. W. Gayler, *J .Amer. Chem. Soc.*, **74**, 3615 (1952).

312. D. A. Hahn and M. K. Seikel, *J. Amer. Chem. Soc.*, **58**, 647 (1936).

313. A. C. Cope and E. M. Hancock, *J. Amer. Chem. Soc.*, **61**, 776 (1939).

314. B. W. Bycroft, D. Cameron, L. R. Croft, and A. W. Johnson, *Chem. Comm.*, 1301 (1968).

315. B. W. Bycroft, D. Cameron, and A. W. Johnson, *J. Chem. Soc. (C)*, 3040 (1971).

316. B. W. Bycroft, L. R. Croft, A. W. Johnson, and T. Webb, *J.C.S. Perkin I*, 820 (1972).

317. B. W. Bycroft, D. Cameron, L. R. Croft, A. W. Johnson, T. Webb, and P. Coggan, *Tetrahedron Letters*, 2925 (1968).

318. G. Büchi and J. A. Raleigh, *J. Org. Chem.*, **36**, 873 (1971).

319. T. Wakamiya, T. Shiba, T. Kaneko, H. Sakakibara, T. Take, and J. Abe, *Tetrahedron Letters*, 3497 (1970).

320. I. H. Pitman, E. Shefter, and M. Ziser, *J. Amer. Chem. Soc.*, **92**, 3413 (1970).

321. F. G. Riddell, *J. Chem. Soc. (B)*, 560 (1967).

322. R. F. Farmer and J. Hamer, *Tetrahedron*, **24**, 829 (1968).

323. R. K. Harris and R. A. Spragg, *J. Chem. Soc. (B)*, 684 (1968).

324. R. O. Hutchins, L. D. Kopp, and E. L. Eliel, *J. Amer. Chem. Soc.*, **90**, 7174 (1968).

325. R. U. Lemieux in *Molecular Rearrangements*, P. De Mayo, Ed., Interscience, New York, 1974, p. 739; M. A. Kabayama and D. Patterson, *Canad. J. Chem.*, **36**, 563 (1958).

326. S. Wolfe, A. Rauk, L. M. Tel, and I. G. Csizmadia, *J. Chem. Soc. (B)*, 136 (1971).

327. H. Booth and R. U. Lemieux, *Canad. J. Chem.*, **49**, 777 (1971).

328. F. G. Riddell and D. A. R. Williams, *Tetrahedron Letters*, 2073 (1971).

329. R. A. Y. Jones, A. R. Katritzky, and M. Snarey, *J. Chem. Soc. (B)*, 131 (1970).

330. E. L. Eliel, L. D. Kopp, J. E. Dennis, and S. A. Evans, Jr., *Tetrahedron Letters*, 3409 (1971).

331. M. J. Cook, R. A. Y. Jones, A. R. Katritzky, M. M. Mañas, A. C. Richards, A. J. Sparrow, and D. L. Trepanier, *J.C.S. Perkin II*, 325 (1973).

332. F. P. Chen and R. G. Jesaitis, *Chem. Comm.*, 1533 (1970).

333. P. J. Halls, R. A. Y. Jones, A. R. Katritzky, M. Snarey, and D. L. Trepanier, *J. Chem. Soc. (B)*, 1320 (1971).

334. R. C. Cookson and T. A. Crabb, *Tetrahedron*, **24**, 2385 (1968).

335. P. G. Hall and G. S. Horsfall, *J.C.S. Perkin II*, 1280 (1973).

336. H. Piotrowska and T. Ubrasński, *J. Chem. Soc.*, 1942 (1962).

337. T. Tsuji and Y. Okamoto, *Chem. and Pharm. Bull. (Japan)*, **20**, 184 (1972).

338. T. Tsuji, *Chem. and Pharm. Bull. (Japan)*, **19**, 2551 (1971).

339. F. G. Riddell, *J. Chem. Soc. (B)*, 1028 (1971).

340. Y. Kondo and B. Witkop, *J. Amer. Chem. Soc.*, **90**, 764 (1968).

341. W. E. Cohen and D. G. Doherty, *J. Amer. Chem. Soc.*, **78**, 2863 (1956).

342. Y. Kondo and B. Witkop, *J. Amer. Chem. Soc.*, **90**, 3258 (1968).

343. Y. T. Pratt in *Heterocyclic Compounds*, Vol. 6, R. C. Elderfield, Ed., Wiley, New York, 1957, Ch. 9. p. 377.

344. G. W. H. Cheeseman and E. S. G. Werstiuk, *Adv. Heterocyclic Chem.*, **14**, 99 (1972).

345. C. Stoehr, *J. prakt. Chem.*, **51**, 449 (1895).

346. J. B. Conant and J. G. Aston, *J. Amer. Chem. Soc.*, **50**, 2783 (1928).

347. F. B. Kipping, *J. Chem. Soc.*, 368 (1937).

348. T. Hayashi, *Sci. Papers Inst. Phys. Chem. Res. (Tokyo)*, **38**, 466 (1941); *Chem. Abs.*, **41**, 5886 (1947).

349. T. Hayashi, *Nat. Sci. Rep. Ochanomizu Univ. Tokyo*, **1**, 64 (1951), *Chem. Abs.*, **49**, 1050 (1955).

350. F.B. Kipping and W. J. Pope, *J. Chem. Soc.*, 2396 (1924).

350a. G. Cignarella and G. G. Gallo, *J. Heterocyclic Chem.*, **11**, 985 (1974).

351. E. Fischer and K. Raske, *Ber.*, **39**, 3981 (1906); E. Fischer and A. H. Koelker, *Annalen*, **354**, 39 (1907).

352. J. P. Li and J. H. Biel, *J. Heterocyclic Chem.*, **5**, 703 (1968).

353. M. R. Harnden, *J. Chem. Soc. (C)*, 2341 (1967).

354. H. Poisel and U. Schmidt, *Chem. Ber.*, **106**, 3408 (1973).

355. J. Häusler and U. Schmidt, *Chem. Ber.*, **107**, 2804 (1974).

356. J. P. Greenstein and M. Winitz, *The Chemistry of the Amino Acids*, Vol. 2, Wiley, New York, 1961.

357. G. T. Morgan, W. J. Hickinbottom, and T. V. Barker, *Proc. Roy. Soc.*, **110A**, 502 (1926).

358. C. Stoehr, *J. prakt. Chem.*, **47**, 439 (1893).

359. F. B. Kipping and W. J. Pope, *J. Chem. Soc.*, 1076 (1926); W. J. Pope and J. Read, *J. Chem. Soc.*, 2325 (1912), 219 (1914).

360. C. Stoehr, *J. prakt. Chem.*, **47**, 439 (1893), **55**, 49 (1897).

361. F. B. Kipping, *J. Chem. Soc.*, 2889 (1929), 143 (1933).

362. F. B. Kipping, *J. Chem. Soc.*, 1160 (1931).

363. F. B. Kipping, *J. Chem. Soc.*, 1336 (1932).

364. R. K. Harris and N. Sheppard, *J. Chem. Soc. (B)*, 200 (1966).

365. W. E. Hanby and H. N. Rydon, *J. Chem. Soc.*, 833 (1945).

366. G. M. Bennett and E. Glynn, *J. Chem. Soc.*, 211 (1950).

367. B. Morosin and J. Howaston, *J.C.S. Perkin II*, 1087 (1972).

368. I. L. Karle, *J. Amer. Chem. Soc.*, **94**, 81 (1972).

369. J. W. Moncrief, *J. Amer. Chem. Soc.*, **90**, 6517 (1968).

370. R. Nagarajan, N. Neuss, and M. M. Marsh, *J. Amer. Chem. Soc.*, **90**, 6518 (1968); D. B. Cosulich, N. R. Nelson, and J. H. van den Hende, *J. Amer. Chem. Soc.*, **90**, 6519 (1968).

371. M. V. George and G. F. Wright, *Canad. J. Chem.*, **36**, 189 (1958).

372. M. V. George and G. F. Wright, *J. Amer. Chem. Soc.*, **80**, 1200 (1958).

373. J.-L. Imbach, R. A. Y. Jones, A. R. Katritzky, and R. J. Wyatt, *J. Chem. Soc. (B)*, 499 (1967).

373a. I. Horikoshi, M. Morii, and N. Takeguchi, *Chem. and Pharm. Bull. (Japan)*, **23**, 754 (1975).

373b. N. L. Allinger, J. G. D. Carpenter, and F. M. Karkowsky, *J. Amer. Chem. Soc.*, **87**, 1232 (1965).

374. R. K. Harris and R. A. Spragg, *Chem. Comm.*, 314 (1966).

375. R. A. Spragg, *J. Chem. Soc.*, *(B)* 1128 (1968).

376. G. G. Gallo and A. Vigevani, *J. Heterocyclic Chem.*, **2**, 418 (1965).

377. R. E. Lyle and J. J. Thomas, *Tetrahedron Letters*, 897 (1969).

378. J. J. Delpuech and Y. Martinet, *Tetrahedron*, **28**, 1759 (1972).

379. J. J. Delpuech and Y. Martinet, *Chem. Comm.*, 478 (1968).

380. J. J. Delpuech, Y. Martinet, and B. Petit, *J. Amer. Chem. Soc.*, **91**, 2158 (1969).

381. J. J. Delpuech and Y. Martinet, *Tetrahedron*, **27**, 2499 (1971).

382. J. L. Sudmeier and G. Occupati, *J. Amer. Chem. Soc.*, **90**, 154 (1968).

383. E. M. Popov, V. Z. Pletnev, S. L. Portnova, V. T. Ivanov, P. V. Kostetskii, and Y. A. Ovchinnikov, *J. Gen. Chem. (U.S.S.R.)*, **41**, 412 (1971).

384. M. Sakamoto and Y. Tomimatsu, *J. Pharm. Soc. Japan*, **90**, 544 (1970).

385. M. Sakamoto, Y. Tomimatsu, T. Momose, C. Iwata, and M. Hanaoka, *J. Pharm. Soc. Japan*, **92**, 1170 (1972).

386. J. W. Lown and M. H. Akhtar, *J.C.S. Chem. Comm.*, 511 (1973).

387. J. W. Lown and M. H. Akhtar, *Tetrahedron Letters*, 179 (1974).

388. M. Lamchen and T. W. Mittag, *J. Chem. Soc. (C)*, 1917 (1968).

389. F. G. Mann and A. Senior, *J. Chem. Soc.*, 4476 (1954).

390. H. Brockmann and H. Musso, *Chem. Ber.*, **89**, 241 (1956).

391. F. O. Lassen and C. H. Stammer, *J. Org. Chem.*, **36**, 2631 (1971).

392. C. Eguchi and A. Kakuta, *J. Amer. Chem. Soc.*, **96**, 3985 (1974).

393. C. Eguchi and A. Kakuta, *Bull. Chem. Soc. Japan*, **47**, 2277 (1974).

394. J. Vicar and K. Bláha, *Coll. Czech. Chem. Comm.*, **38**, 3307 (1973).

395. H. Poisel and U. Schmidt, *Chem. Ber.*, **105**, 625 (1972).

396. E. Öhler, F. Tataruch, and U. Schmidt, *Chem. Ber.* **106**, 165 (1973).

397. E. Öhler, H. Poisel, F. Tataruch, and U. Sdhmidt, *Chem. Ber.* **105**, 635 (1972).

398. E. Öhler, F. Tataruch, and U. Schmidt, *Chem. Ber.*, **105**, 3658 (1972).

398a. G. M. Strunz, M. Kakushima, and M. A. Stillwell, *Canad. J. Chem.*, **53**, 295 (1975).

399. Y. Kishi, S. Nakatsuka, T. Fukuyama, and T. Goto, *Tetrahedron Letters*, 4657 (1971).

400. F. G. Riddell and J. M. Lehn, *Chem. Comm.*, 375 (1966)

401. F. R. Farmer and J. Hamer, *Chem. Comm.*, 866 (1966).

402. G. B. Carter, M. V. McIvor, and R. G. J. Miller, *J. Chem. Soc. (C)*, 2591 (1968).

403. R. A. Y. Jones, A. R. Katritzky and M. Snarey, *J. Chem. Soc. (B)*, 135 (1970).

404. R. P. Duke, R. A. Y. Jones, A. R. Katritzky, R. Scattergood, and F. G. Riddell, *J.C.S. Perkin II*, 2109 (1973).

405. C. H. Bushweller, M. Z. Lourandos, and J. A. Brunelle, *J. Amer. Chem. Soc.*, **96**, 1591 (1974).

406. E. W. Lund, *Acta Chem. Scand.*, **5**, 678 (1951).

407. C. Schöpf, A. Komzak, F. Braun, E. Jacobi, M.-L. Bormuth, M. Bullnheimer, and I. Hagel, *Annalen*, **559**, 1 (1948).

408. C. Schöpf, H. Arm, and H. Krimm, *Chem. Ber.*, **84**, 690 (1951).

409. C. Schöpf and K. Otte, *Chem. Ber.*, **89**, 335 (1956).

410. C. Schöpf, H. Arm, and F. Braun, *Chem. Ber.*, **85**, 937 (1952).

411. W. L. F. Armarego and S. C. Sharma, *J. Chem. Soc. (C)*, 1600 (1970).

412. T. Kauffmann, G. Ruckelshauss, and J. Shulz, *Angew. Chem. Internat. Edn.*, **3**, 63 (1964).

413. J. E. Anderson and J. D. Roberts, *J. Amer. Chem. Soc.*, **90**, 4186 (1968).

414. R. A. Y. Jones, A. R. Katritzky, and A. C. Richards, *Chem. Comm.*, 708 (1969).

415. S. F. Nelsen and P. J. Hintz, *J. Amer. Chem. Soc.*, **94**, 3138 (1972).

416. R. A. Y. Jones, A. R. Katritzky, A. R. Martin, D. L. Ostercamp, A. C. Richards, and J. M. Sullivan, *J.C.S. Perkin II*, 948 (1974).

417. G. B. Ansell, J. L. Erickson, and D. W. Moore, *Chem. Comm.*, 446 (1970).

417a. G. B. Ansell and J. L. Erickson, *J.C.S. Perkin II*, 270 (1975).

418. S. Hammerum, *Acta Chem. Scand.*, **27**, 779 (1973).

419. A. D. Evnin and D. R. Arnold, *J. Amer. Chem. Soc.*, **90**, 5330 (1968).

420. D. R. Arnold, A. B. Evnin, and P. H. Kasai, *J. Amer. Chem. Soc.*, **91**, 784 (1969).

421. V. Prelog and S. Szpilfogel, *Helv. Chim. Acta*, **28**, 1684 (1945).

422. H. L. Lochte and A. G. Pittman, *J. Amer. Chem. Soc.* **82**, 469 (1960).

423. N. F. Albertson, *J. Amer. Chem. Soc.*, **72**, 2594 (1950).

424. I. N. Nazarov, G. A. Schwekhgeimer, and V. A. Rudneko, *Zhur. obshchei Khim.*, **24**, 319 (1954).

425. T. Henshall and E. W. Parnell, *J. Chem. Soc.*, 661 (1962).

426. E. A. Mistryukov, *Izvest. Akad. Nauk S.S.S.R.*, *Ser. khim.*, 1158 (1964).

427. I. N. Nazarov and E. A. Mistryukov, *Izvest. Akad. Nauk S.S.S.R.*, *Ser. khim.*, 565 (1958).

428. F. P. Guengerich, S. J. DiMari, and H. P. Broquist, *J. Amer. Chem. Soc.*, **95**, 2055 (1973).

429. V. Prelog and O. Metzler, *Helv. Chim. Acta*, **20**, 1170 (1946).

430. V. Prelog and O. Metzler, *Helv. Chim. Acta*, **20**, 1163 (1946).

431. A. I. Meyers, J. Schneller, and N. K. Rahlan, *J, Org. Chem.*, **28**, 2944 (1963).

432. G. G. Ayerst and K. Schofield, *J. Chem. Soc.*, 4097 (1958).

433. K. Jewers and J. McKenna, *J. Chem. Soc.*, 1575 (1960).

434. W. Holick, E. F. Jenny, and K. Heusler, *Tetrahedron Letters*, 3421 (1973).

435. A. McKillop, J. D. Hunt, E. C. Taylor, and F. Kienzle, *Tetrahedron Letters*, 5275 (1970).

436. K. Yamakawa, R. Sakaguchi, and K. Osumi, *Chem. and Pharm. Bull. (Japan)*, **22**, 576 (1974).

437. E. J. Eisenbraun, H. Auda, K. S. Schorno, G. R. Waller, and H. H. Appel, *J. Org. Chem.* **35**, 1346 (1970).

438. C. G. Casinovi, F. Delle Monache, G. Grandolini, G. B. Marini-Bettolò, and H. H. Appel, *Chem. and Ind.*, 984 (1963); G. W. K. Cavill and A. Zeitlin, *Austral. J. Chem.*, **20**, 349 (1967).

439. G. Jones, G. Ferguson, and W. C. Marsh, *Chem. Comm.*, 994 (1971).

440. J. H. Lister, *Fused Pyrimidines Part II. Purines*, D. J. Brown, Ed., Wiley-Interscience, New York, 1971.

441. J. Tafel and P. A. Houseman, *Ber.*, **40**, 3743 (1907).

442. J. Tafel, *Ber.*, **34**, 258 (1901).

443. J. Tafel, *Ber.*, **34**, 279 (1901).

444. W. L. F. Armarego and P. A. Reece, *Tetrahedron Letters*, 423 (1975).

445. W. L. F. Armarego and P. A. Reece, *J. C. S. Perkin I*, 1414 (1976).

446. J. L. Wong, R. Oesterlin, and H. Rapoport, *J. Amer. Chem. Soc.*, **93**, 7344 (1971), and earlier papers.

447. E. J. Schantz, V. E. Ghazarossian, H. K. Schnoes, F. M. Strong, J. P. Springer, J. O. Pezzanite, and J. Clardy, *J. Amer. Chem. Soc.*, **97**, 1238 (1975), and earlier papers; see also J. Bordner, W. E. Thiessen, H. A. Bates, and H. Rapoport, *J. Amer. Chem. Soc.*, **97**, 6008 (1975).

448. J. Thomas, *J. Chem. Soc.*, 725 (1912).

449. S. G. P. Plant and R. J. Rosser, *J. Chem. Soc.*, 2444 (1930).

450. T. Tokuyama, S. Senoh, T. Sakan, K. S. Brown, Jr., and B. Witkop, *J. Amer. Chem. Soc.*, **89**, 1017 (1967).

451. J. Roussel, J. J. Perie, J. P. Laval, and A. Lattes, *Tetrahedron*, **28**, 701 (1972).

452. I. Ninomiya, T. Kiguchi, and T. Naito, *J.C.S. Chem. Comm.*, 81 (1974).

453. J. Thesing and F. H. Funk, *Chem. Ber.*, **89**, 2498 (1956).

454. R. B. Roy and G. A. Swan, *Chem. Comm.*, 1446 (1968).

454a. J. M. Fayadh and G. A. Swan, *J. Chem. Soc. (C)*, 1781 (1969).

455. Y. Kobayashi, T. Kutsuma, Y. Hanzawa, Y. Iitaka, and H. Nakamura, *Tetrahedron Letters*, 5337 (1969).

456. P. G. Lehman, *Tetrahedron Letters*, 4863 (1972).

457. H. Leuchs, *Ber.*, **54**, 830 (1921).

458. E. E. Turner and M. M. Harris, *Quart. Rev.*, **1**, 299 (1947).

459. K. S. Brown, Jr., *J. Amer. Chem. Soc.*, **87**, 4202 (1965).

460. G. Kohl and H. Pracejus, *Annalen*, **694**, 128 (1966).

461. C. A. R. Baxter and H. C. Richards, *Tetrahedron Letters*, 3357 (1972).

462. G. Heller and A. Sourlis, *Ber.*, **41**, 2692 (1908); G. Heller, *Ber.*, **44**, 2106 (1911); G. Heller, *Ber.*, **47**, 2893 (1914).

463. H. Dunathan, I. W. Elliott, Jr., and P. Yates, *Tetrahedron Letters*, 781 (1961).

464. I. W. Elliott and J. Bordner, *Tetrahedron Letters*, 4481 (1971).

465. I. W. Elliott, W. T. Bowie, and D. L. T. Wong, *J. Heterocyclic Chem.*, **9**, 1425 (1972).

466. H. Booth, *J. Chem. Soc.*, 1841 (1964).

467. A. Einhorn, *Ber.*, **19**, 904 (1886); A. Einhorn and P. Sherman, *Annalen*, **287**, 26 (1895).

468. R. B. Woodward and E. C. Kornfeld, *J. Amer. Chem. Soc.*, **70**, 2508 (1948).

469. V. N. Gogte, M. A. Salama, and D. B. Tilak, *Tetrahedron*, **26**, 173 (1970).

470. K. G. Das, V. N. Gogte, M. Seetha, and B. D. Tilak, *Indian J. Chem.*, **10**, 924 (1972).

471. V. N. Gogte, K. M. More, and B. D. Tilak, *Indian J. Chem.*, **12**, 327 (1974).

472. T. P. Forrest, G. A. Dauphinee, and W. F. Miles, *Canad. J. Chem.*, **52**, 884 (1974).

473. W. Hückel and F. Stepf, *Annalen*, **453**, 163 (1927).

474. C. F. Bailey and S. M. McElvain, *J. Amer. Chem. Soc.*, **52**, 4013 (1930).

475. A. Skita and W. A. Meyer, *Ber.*, **45**, 3579 (1912).

476. L. Mascarelli, *Gazzetta*, **45**, 106, 127 (1915).

477. W. L. F. Armarego, *J. Chem. Soc.*, 377 (1967).

478. H. Booth and A. H. Bostock, *J.C.S. Perkin II*, 615 (1972).

479. A. Burger and L. R. Modlin, Jr., *J. Amer. Chem. Soc.*, **62**, 1079 (1940).

480. J. A. Hirsch and G. Schwartzkopf, *J. Org. Chem.*, **39**, 2044 (1974).

481. C. F. Koelsch and D. L. Ostercamp. *J. Org. Chem.*, **26**, 1104 (1961).

482. I. Inoue and Y. Ban, *J. Chem. Soc.* (*C*), 602 (1970).

483. F. E. King, T. Henshall, and R. L. St. D. Whitehead, *J. Chem. Soc.*, 1373 (1948).

484. Z-i. Horii, T. Watanabe, M. Ikeda, and Y. Tamura, *J. Pharm. Soc. Japan*, **83**, 930 (1963).

485. C. A. Grob and H. R. Kiefer, *Helv. Chim. Acta*, **48**, 799 (1965).

486. G. M. Badger, J. W. Cook, and T. Walker, *J. Chem. Soc.*, 2011 (1948); K. Hohenlohe-Oehringen, *Monatsh.*, **94**, 1217 (1963).

487. R. J. Sundberg and P. A. Bukowick, *J. Org. Chem.*, **33**, 4098 (1968).

488. D. V. Sokolov, G. S. Litvinenko, and K. I. Khludneva, *J. Gen. Chem.* (*U.S.S.R.*), **29**, 1083 (1959).

489. K. Nomura, J. Adachi, H. Manai, S. Nakayama, and K. Mitsuhashi, *Chem. and Pharm. Bull.* (*Japan*), **22**, 1386 (1974).

489a. S. S. Klioze and F. P. Darmory, *J. Org. Chem.*, **40**, 1588 (1975).

489b. W. Oppolzer and W. Fröstl, *Helv. Chim. Acta*, **58**, 590 (1975).

490. E. J. Moriconi and M. A. Stemniski, *J. Org. Chem.*, **37**, 2035 (1972).

491. T. Ibuka, Y. Inubushi, I. Saji, K. Tanaka, and N. Masaki, *Tetrahedron Letters*, 323 (1975).

492. G. DiMaio and P. Tardella, *Gazzetta*, **91**, 1345 (1961).

493. R. V. Stevens, R. K. Mehra, and R. L. Zimmerman, *Chem. Comm.*, 877 (1969).

494. R. V. Stevens, L. E. Du Pree, Jr., and P. L. Loewenstein, *J. Org. Chem.*, **37**, 977 (1972).

495. A. Popvici, C. F. Geschickter, E. L. May, and E. Mosettig, *J. Org. Chem.*, **21**, 1283 (1956).

496. R. A. Johnson, H C. Murray, L. M. Reineke, and G. S. Fonken, *J. Org. Chem.*, **33**, 3207 (1968).

497. W. C. Krueger, J. A. Johnson, and L. M. Pschigoda, *J. Amer. Chem.*, *Soc.* **93**, 4865 (1971).

498. R. A. Johnson, M. E. Herr, H. C. Murray, and G. S. Fonken, *J. Org. Chem.*, **33**, 3217 (1968).

499. H. Ripperger and K. Schreiber, *Tetrahedron*, **25**, 737 (1969).

500. H. Ripperger and K. Schreiber, *J. prakt. Chem.*, **313**, 825 (1971).

501. J. S. Roberts and C. Thomson, *J.C.S. Perkin II*, 2129 (1972).

502. J. Musher and R. E. Richards, *Proc. Chem. Soc.*, 230 (1958); J. T. Gerig and J. D. Roberts, *J. Amer. Chem. Soc.*, **88**, 2791 (1966); F. R. Jansen and B. H. Beck, *Tetrahedron Letters*, 4523 (1966).

503. H. Booth, *Tetrahedron*, **22**, 615 (1966).

504. H. Booth and A. H. Bostock, *Chem. Comm.*, 177 (1967).

505. H. Booth and D. V. Griffiths, *J.C.S. Chem. Comm.*, 666 (1973), and *J.C.S. Perkin II*, 111 (1975).

506. H. Booth and D. V. Griffiths, *J.C.S. Perkin II*, 842 (1973).

507. E. L. Eliel and F. W. Vierhapper, *J. Amer. Chem. Soc.*, **96**, 2257 (1974).

508. E. L. Eliel and F. W. Vierhapper, *J. Amer. Chem. Soc.*, **97**, 2424 (1975).

508a. K. Brown, A. R. Katritzky, and A. J. Waring, *J. Chem. Soc.* (*B*), 487 (1967), and *Proc. Chem. Soc.*, 257 (1964).

509. H. Booth and D. V. Griffiths, *J.C.S. Perkin II*, 2361 (1972).

510. V. S. Bogdanov, L. I. Ukhova, N. F. Uskova, A. P. Marochkin, A. M. Moissenkov, and A. A. Akhem, *Izvest. Akad. Nauk S.S.S.R.*, *Ser. khim.*, 2235 (1971).

511. P. N. Rylander, *Catalytic Hydrogenation over Platinum Metals*, Academic, New York, 1967.

512. M. Ehrenstein and W. Bunge, *Ber.*, **67**, 1715 (1934).

513. A. P. Gray, D. E. Heitmeier, and C. J. Cavallito, *J. Amer. Chem. Soc.*, **81**, 728 (1959).

514. G. Fodor, S.-Y. Abidi, and T. C. Carpenter, *J. Org. Chem.*, **39**, 1507 (1974).

515. G. Fodor and S. Abidi, *Tetrahedron Letters*, 1369 (1971).

516. S. Abidi, G. Fodor, C. S. Huber, I. Miura, and K. Nakanishi, *Tetrahedron Letters*, 355 (1972).

517. E. A. Mistryukov and V. F. Kucherov, *Izvest. Akad. Nauk. S.S.S.R.*, *Ser. khim.*, 1695 (1961).

518. E. A. Mistryukov, N. I. Aronov, and V. F. Kucherov, *Izvest. Akad. Nauk. S.S.S.R.*, *Ser. khim.*, 1514 (1962).

519. D. V. Sokolov, G. S. Litvinenko, and K. I. Khludneva, *J. Gen. Chem.* (*U.S.S.R.*,) **29**, 3518 (1959).

520. D. V. Sokolov, G. S. Litvinenko, and K. I. Khludneva, *J. Gen. Chem.* (*U.S.S.R.*), **30**, 844 (1960).

521. É. A. Mistryukov and V. F. Kucherov, *Izvest. Akad. Nauk. S.S.S.R.*, *Ser. khim.*, 1905 (1961).

522. O. V. Agashkin, G. S. Litvinenko, D. S. Sokolov, and S. S. Chasnikova, *J. Gen. Chem.* (*U.S.S.R.*), **31**, 794 (1961).

523. É. A. Mistryukov and V. F. Kucherov, *Izvest. Adak. Nauk S.S.S.R.*, *Ser. khim.*, 1702 (1961).

524. A. A. Akhrem, L. I. Ukhova, A. P. Marochkin, and G. V. Bludova, *Izvest. Akad. Nauk. S.S.S.R.*, *Ser. khim.*, 2192 (1972).

525. C. A. Grob, H. R. Kiefer, H. J. Lutz, and H. J. Wilkens, *Helv. Chem. Acta*, **50**, 416 (1967).

526. C. A. Grob, H. R. Kiefer, H. Lutz, and H. J. Wilkens, *Tetrahedron Letters*, 2901 (1964).

527. R. Gleiter, W. D. Stohrer, and R. Hoffmann, *Helv. Chim. Acta*, **55**, 893 (1972).

528. J. A. Marshall and J. H. Babler, *J. Org. Chem.*, **34**, 4186 (1969).

529. J. McKenna and A. Tully, *J. Chem. Soc.*, 945 (1960).

530. K. W. Bently, *The Isoquinoline Alkaloids*, Pergamon, New York, 1965; M. Shamma, *The Isoquinoline Alkaloids*, Academic, New York, 1972; T. Kametani, *The Isoquinoline Alkaloids*, Elsevier, Amsterdam, 1969.

531. W. M. Whaley and T. R. Govindachari in *Organic Reactions*, Vol. 6, R. Adams, Ed., Wiley, New York, 1951, Chap. 3, p. 151.

532. A. Brossi, A. Focella, and S. Teitel, *Helv. Chim. Acta*, **55**, 15 (1972).

533. E. A. Bell, J. R. Nulu, and C. Cone, *Phytochem.*, **10**, 2191 (1971).

534. M. E. Daxenbichler, R. Kleiman, D. Weisleder, C. H. VanEtten, and K. D. Carlson, *Tetrahedron Letters*, 1801 (1972).

535. T. Kametani, H. Sugi, H. Yagi, K. Fukumoto, and S. Shibuya, *J. Chem. Soc. (C)*, 2213 (1970).

536. M. A. Haimova, S. L. Spassov, S. I. Novkova, M. D. Palamareva, and B. J. Kurtev, *Chem., Ber.*, **104**, 2601 (1971).

537. M. A. Haimova, M. Palamareva, B. Kurtev, S. I. Novkova, and S. L. Spassov, *Chem. Ber.*, **103**, 1347 (1970).

538. W. M. Whaley and T. R. Govindachari in *Organic Reactions*, Vol. 6, R. Adams, Ed., Wiley, New York, 1951, Chap. 2, p. 74.

539. V. M. Potapov, V. M. Dem'yanovich, V. S. Soifer, and A. P. Terent'ev, *J. Gen. Chem. (U.S.S.R.)*, **37**, 2550 (1967).

540. G. Grethe, M. Uskokovic, T. Williams, and A. Brossi, *Helv. Chim. Acta*, **50**, 2397 (1967).

541. A. T. Nielsen, *J. Org. Chem.*, **35**, 2498 (1970).

542. P. Cerutti and H. Schmid, *Helv. Chim. Acta*, **47**, 203 (1964).

543. J. M. Bobbitt, I. Noguchi, H. Yagi, and K. H. Weisgraber, *J. Amer. Chem. Soc.*, **93**, 3551 (1971).

544. C. K. Bradsher, F. H. Day, A. T. McPhail, and P.-S. Wong, *Tetrahedron Letters*, 4205 (1971).

544a. C. K. Bradsher, F. H. Day, A. T. McPhail, and P.-S. Wong, *J.C.S. Chem. Comm.*, 156 (1973).

545. W. R. Schleigh, *J. Heterocyclic Chem.*, **9**, 675 (1972), and *Tetrahedron Letters*, 1405 (1969).

546. H. Böhme and K.-P. Stöcker, *Chem. Ber.*, **105**, 1578 (1972).

547. J.-L. Ferron and P. L'Ecuyer, *Canad. J. Chem.*, **33**, 352 (1955).

548. L. Marion and V. Grassie, *J. Amer. Chem. Soc.*, **66**, 1290 (1944).

549. R. H. F. Manske, E. H. Charlesworth, and W. R. Ashford, *J. Amer. Chem. Soc.*, **73**, 3751 (1951).

550. A. Brossi, J. O'Brien, and S. Teitel, *Helv. Chim. Acta*, **52**, 678 (1969).

551. H. Corrodi and E. Hardegger, *Helv. Chim. Acta*, **39**, 889 (1956).

552. J. F. Blount, V. Toome, S. Teitel, and A. Brossi, *Tetrahedron*, **29**, 31 (1973).

553. Y. L. Chow and C. J. Colón, *Canad. J. Chem.*, **46**, 2827 (1968).

554. M. Kajtár and L. Radics, *Chem. Comm.*, 784 (1967).

555. R. Bentley, *Molecular Asymmetry in Biology*, Vol. 2, Academic, New York, 1970, pp. 248–257.

556. T. C. Bruice and S. Benkovic, *Bioorganic Mechanisms*, Vol. 1, Benjamin, New York, 1966, pp. 212–258; D. M. Blow in *The Enzymes*, P. D. Boyer, Ed., Vol. III, Third Edition, Academic, New York, 1971, Chap. 6, p. 185, see also earlier edition, Vol. IV., 1960, Chaps. 1 and 5.

557. J. Knabe, R. Dörr, S. F. Dyke, and R. G. Kinsman, *Tetrahedron Letters*, 5373 (1972); R. G. Kinsman, A. W. C. White, and S. F. Dyke, *Tetrahedron*, **31**, 449 (1975).

557a. J. Knabe and H. Powilleit, *Arch. Pharm.*, **303**, 37 (1970).

557b. M. Sainsbury, S. F. Dyke, D. W. Brown, and R. G. Kinsman, *Tetrahedron*, **26**, 5265 (1970).

557c. R. G. Kinsman and S. F. Dyke, *Tetrahedron Letters*, 2231 (1975).

558. B. Witkop, *J. Amer. Chem. Soc.*, **70**, 2617 (1948).

559. A. P. Gray and D. E. Heitmeier, *J. Amer. Chem. Soc.*, **80**, 6274 (1958).

560. S. Kimoto, M. Okamoto, A. Watanabe, T. Baba, and I. Dobashi, *Chem. and Pharm. Bull. (Japan)*, **20**, 10 (1972).

561. V. Georgian, R. J. Harrisson, and L. L. Skaletzky, *J. Org. Chem.*, **27**, 4571 (1962).

562. M. Okamoto and M. Yamada, *Chem. and Pharm. Bull. (Japan)*, **11**, 554 (1963).

563. T. Tomioka, *J. Pharm. Soc. Japan*, **90**, 913 (1970).

564. R. T. Rapala, E. R. Lavagnino, E. R. Shepard, and E. Farkas, *J. Amer. Chem. Soc.*, **79**, 3770 (1957).

565. R. B. Woodward and W. E. Doering, *J. Amer. Chem. Soc.*, **66**, 849 (1944), and **67**, 860 (1945).

566. M. Okamoto, *Chem. and Pharm. Bull. (Japan)*, **15**, 168 (1967).

567. A. Marchant and A. R. Pinder, *J. Chem. Soc.*, 327 (1956).

568. B. Witkop, *J. Amer. Chem. Soc.*, **71**, 2559 (1949).

569. K. von Auwers, *Annalen*, **420**, 84 (1920); A. Skita, *Ber.*, **53**, 1792 (1920).

570. I. W. Mathison and R. C. Gueldner, *J. Org. Chem.*, **33**, 2510 (1968).

571. I. W. Mathison and P. H. Morgan, *J. Org. Chem.*, **39**, 3210 (1974).

572. S. M. McElvain and D. C. Remy, *J. Amer. Chem. Soc.*, **82**, 3960 (1960).

573. S. Dubé and P. Deslongchamps, *Tetrahedron Letters*, 101 (1970).

574. S. Durand-Henchoz and R. C. Moreau, *Bull. Soc. chim. France*, 3416 (1966).

575. S. M. McElvain and P. H. Parker, Jr., *J. Amer. Chem. Soc.*, **78**, 5312 (1956).

576. R. L. Augustine, *J. Org. Chem.*, **23**, 1853 (1958).

577. C. B. Clarke and A. R. Pinder, *J. Chem. Soc.*, 1967 (1958).

578. M. R. Uskoković, D. L. Pruess, C. W. Despreaux, S. Shiuey, G. Pizzolato, and J. Gutzwiller, *Helv. Chim. Acta*, **56**, 2834 (1973).

579. S. Heřmánek and J. Trojánek, *Coll. Czech. Chem. Comm.*, **22**, 1157 (1957).

580. R. A. Abramovitch and J. M. Muchowski, *Canad. J. Chem.*, **38**, 557 (1960).

581. R. A. Abramovitch and D. L. Struble, *Chem. Comm.*, 150 (1966).

582. R. A. Abramovitch and D. L. Struble, *Tetrahedron Letters*, 289 (1966).

583. R. A. Abramovitch and D. L. Struble, *Tetrahedron*, **24**, 705 (1968).

584. S. Kimoto, M. Okamoto, M. Uneo, S. Ohta, M. Nakamura, and T. Niiya, *Chem. and Pharm. Bull. (Japan)*, **18**, 2141 (1970).

585. S. Kimoto, M. Okamoto, M. Nakamura, and T. Baba, *J, Pharm. Soc. Japan*, **90**, 1538 (1970).

586. S. Kimoto, M. Okamoto, Y. Fumiwara, and K-i. Nakamura, *J. Pharm. Soc. Japan*, **93**, 939 (1973).

587. S. Kimoto, M. Okamoto, T. Mizumoto, and Y. Fujiwara, *Chem. and Pharm. Bull. (Japan)*, **16**, 2390 (1968).

588. S. Kimoto, M. Okamoto, T. Mizumoto, and Y. Fujiwara, *Chem. and Pharm. Bull. (Japan)*, **15**, 370 (1967).

589. M. Itoh, *Chem. and Pharm. Bull. (Japan)*, **16**, 455 (1968).

590. V. Boekelheide and W. M. Schilling, *J. Amer. Chem. Soc.*, **72**, 712 (1950).

591. R. Grewe and G. Winter, *Chem. Ber.*, **92**, 1092 (1959).

592. R. Grewe, H. Köpnick, and P. Roder, *Annalen*, **605**, 15 (1957).

593. R. Grewe and H. Jensen, *Chem. Ber.*, **92**, 1573 (1959).

594. C. A. Grob and R. A. Wohl, *Helv. Chim. Acta*, **49**, 2175 (1966).

595. C. A. Grob and R. A. Wohl, *Helv. Chim. Acta*, **49**, 2434 (1966).

596. W. Schneider, *Arch. Pharm.*, **293**, 838 (1960).

597. W. Schneider and R. Menzel, *Arch. Pharm.*, **297**, 65 (1964).

598. T. Omoto, *J. Pharm. Soc. Japan*, **80**, 132 (1960).

599. D. Perelman, S. Siscic, and Z. Welvart, *Tetrahedron Letters*, 103 (1970); G. Stork, A. Brizzolara, H. Landesman, J. Szmuskovicz, and R. Terrell, *J. Amer. Chem. Soc.*, 207 (1963).

600. H. G. O. Becker, *J. prakt. Chem.*, **23**, 259 (1964).

601. H. G. O. Becker and G. Landshulz, *J. prakt. Chem.*, **27**, 41 (1965).

602. N. Finch, L. Blanchard, R. T. Puckett, and L. H. Werner, *J. Org. Chem.*, **39**, 1118 (1974).

603. M. Onda, Y. Sugama, H. Yokoyama, and F. Tada, *Chem. and Pharm. Bull. (Japan)*, **21**, 2359 (1973).

604. S. Durand-Henchoz and R. C. Moreau, *Bull. Soc. chim. France*, 3428 and 3422 (1966).

605. S. Kimoto and M. Okamoto, *Chem. and Pharm. Bull. (Japan)*, **15**, 1045 (1967).

606. M. Geisel, C. A. Grob, and R. A. Wohl, *Helv. Chim. Acta*, **52**, 2206 (1969).

607. C. A. Grob and R. A. Wohl, *Helv. Chim. Acta*, **48**, 1610 (1965).

608. R. A. Wohl, *Helv. Chim. Acta*, **49**, 2162 (1966).

609. F. E. King and H. Booth, *J. Chem. Soc.*, 3798 (1954).

610. W. E. Bachmann, A. Ross, A. S. Dreiding, and P. A. S. Smith, *J. Org. Chem.*, **19**, 222 (1954).

611. S. Sicsic and N-T. Luong-Thi, *Tetrahedron, Letters*, 169 (1973).

612. K. Heusler, *Tetrahedron Letters*, 97 (1970).

613. H. Teufel, E. F. Jenny, and K. Heusler, *Tetrahedron Letters*, 3413 (1973).

614. H-C. Mez and G. Rihs, *Tetrahedron Letters*, 3417 (1973).

615. H-J. Trede, E. F. Jenny, and K. Heulser, *Tetrahedron Letters*, 3425 (1973).

616. M. Tichý, *Coll. Czech. Chem. Comm.*, **39**, 2673 (1974).

617. M. Tichý, *Tetrahedron Letters*, 2001 (1972).

618. M. Tichý, E. Duskova, and K. Bláha, *Tetrahedron Letters*, 237 (1974).

618a. R. W. Lockhart, K. Hanaya, F. W. B. Einstein, and Y. L. Chow, *J.C.S. Chem. Comm.*, 344 (1975).

619. B. Frydman, M. Los, and H. Rapoport, *J. Org. Chem.*, **36**, 450 (1971).

620. T. Takata, *Bull. Chem. Soc. Japan*, **35**, 1438 (1962).

621. H. Zondler and W. Pfleiderer, *Annalen*, **759**, 84 (1972).

622. F. Liberatore, A. Casini, V. Carelli, A. Arnone, and R. Mondelli, *Tetrahedron Letters*, 2381 (1971).

623. F. Liberatore, A. Casini, V. Carelli, A. Arnone, and R. Mondelli, *Tetrahedron Letters*, 3829 (1971).

624. F. Liberatore, A. Casini, V. Carelli, A. Arnone, and R. Mondelli, *J. Org. Chem.*, **40**, 559 (1975).

625. T. Kametani, K. Kigasawa, M. Hiiragi, and N. Wagatsuma, *Chem. and Pharm. Bull. (Japan,)* **16**, 296 (1968).

626. R. A. Carboni and R. V. Lindsey, Jr., *J. Amer. Chem. Soc.*, **81**, 4342 (1959).

627. W. S. Wilson and R. N. Warrener, *J.C.S. Chem. Comm.*, 211 (1972).

628. H. Koch and J. Pirsch, *Monatsh*, **93**, 661 (1962).

629. C. G. Overberger, G. Montaudo, J. Šebenda, and R. A. Veneski, *J. Amer. Chem. Soc.*, **91**, 1256 (1969).

630. A. Nakamura, *Chem. and Pharm. Bull. (Japan)*, **18**, 1426 (1970); A. Nakamura, and S. Kamiya, *Chem. and Pharm. Bull. (Japan)*, **20**, 69 (1972).

631. A. Nakamura and S. Kamiya, *J. Pharm. Soc. Japan*, **90**, 1069 (1970).

632. C. Mannich and O. Hieronimus, *Ber.*, **75**, 49 (1942).

633. T. Tsuda, S. Ikuma, M. Kawamura, R. Tachikawa, K. Sakai, C. Tamura, and O. Amakasu, *Chem. and Pharm. Bull. (Japan)*, **12**, 1357 (1964), and earlier papers.

634. T. Goto, Y. Kishi, S. Takahashi, and Y. Hirata, *Tetrahedron*, **21**, 2059 (1965), and references cited therein.

635. R. B. Woodward, *Pure and Appl. Chem.*, **9**, 49 (1964).

636. R. B. Woodward and J. Z. Gougoutas, *J. Amer. Chem. Soc.*, **86**, 5030 (1964).

637. T. Kishi, T. Fukuyama, M. Aratani, F. Nakatsubo, T. Goto, S. Inoue, H. Tanino, S. Sugiura, and H. Kakoi, *J. Amer. Chem. Soc.*, **94**, 9219 (1972), and earlier papers.

638. J. F. W. Keana, F. P. Mason, and J. S. Bland, *J. Org. Chem.*, **34**, 3705 (1969).

639. W. L. F. Armarego, *J. Chem. Soc. (C)*, 1812 (1971).

640. W. L. F. Armarego and P. A. Reece, *J.C.S. Perkin I*, 1740 (1975).

641. L. Bauer and C. N. V. Nambury, *J. Org. Chem.*, **26**, 1106 (1961).

642. W. L. F. Armarego and T. Kobayashi, *J. Chem. Soc. (C)*, 238 (1971).

643. W. L. F. Armarego and T. Kobayashi, *J. Chem. Soc. (C)*, 1635 (1969).

644. W. L. F. Armarego and T. Kobayashi, *J. Chem. Soc., (C)*, 1597 (1970).

645. W. L. F. Armarego and P. A. Reese, *J.C.S. Perkin I*, 2313 (1974).

646. G. Ziguener, V. Eisenreich, H. Weichsel, and W. Adam, *Monatsh.*, **101**, 1731 (1970).

647. G. Ziguener, V. Eisenreich, and W. Immel, *Monatsh.*, **101**, 1745 (1970).

648. W. Wendelin and A. Fuchsgruber, *Monatsh.*, **105**, 755 (1974).

649. G. Reck, E. Höhne, and G. Adam, *J. prakt. Chem.*, **316**, 496 (1974).

650. S. Shiotani and K. Mitsuhashi, *J. Pharm. Soc. Japan*, **84**, 656 (1964).

651. W. L. F. Armarego and T. Kobayashi, *J. Chem. Soc. (C)*, 2502 (1971).

652. W. L. F. Armarego, R. A. Y. Jones, A. R. Katritzky, and R. Scattergood, *Austral. J. Chem.*, **28**, 2323 (1975).

653. W. L. F. Armarego and T. Kobayashi, *J. Chem. Soc. (C)*, 3222 (1971).

654. A. R. Katritzky, M. R. Nesbit, B. J. Kurtev, M. Lyapova, and I. G. Pojarlieff, *Tetrahedron*, **25**, 3807 (1969).

655. W. L. F. Armarego, *Fused Pyrimidines Part I. Quinazolines*, D. J. Brown, Ed., Interscience, New York, 1967.

656. G. H. Fisher and H. P. Schultz, *J. Org. Chem.*, **39**, 631 (1974).

657. R. Aguilera, J. C. Duplan, and C. Nofre, *Bull. Soc. chim. France*, 4491 (1968).

658. R. A. Archer and H. S. Mosher, *J. Org. Chem.*, **23**, 1378 (1967).

659. M. J. Haddadin, H. N. Alkaysi, and S. E. Saheb, *Tetrahedron*, **26**, 1115 (1970).

660. R. C. DeSelms and H. S. Mosher, *J. Amer. Chem. Soc.*, **82**, 3762 (1960).

661. G. H. Fisher, P. J. Whitman, and H. P. Shultz, *J. Org. Chem.*, **35**, 2240 (1970).

662. G. H. Fisher and H. P. Schultz, *J. Org. Chem.*, **39**, 635 (1974).

663. M. Mousseron and G. Combes, *Bull. Soc. chim. France*, 82 (1947).

664. K. M. Beck, K. E. Hamlin, and A. W. Weston, *J. Amer. Chem. Soc.*, **74**, 605 (1952).

665. H. S. Broadbent, E. L. Allred, L. Pendleton, and C. W. Whittle, *J. Amer. Chem. Soc.*, **82**, 189 (1960).

666. S. Maffei and S. Pietra, *Gazzetta*, **88**, 556 (1958).

667. W. Christie, W. Rohde, and H. P. Schultz, *J. Org. Chem.*, **21**, 243 (1956).

668. E. Brill and H. P. Schultz, *J. Org. Chem.*, **28**, 1135 (1963).

669. E. Brill and H. P. Schultz, *J. Org. Chem.*, **29**, 579 (1964).

670. D. Gracian and H. P. Schultz, *J. Org. Chem.*, **36**, 3989 (1971).

671. G. Settimj, F. Fraschetti, and S. Chiavarelli, *Gazzetta*, **96**, 749 (1966).

672. G. W. H. Cheeseman, *Adv. Heterocyclic Chem.*, **2**, 204 (1963).

673. A. Albert and W. L. F. Armarego, *Adv. Heterocyclic Chem.*, **4**, 1 (1965).

674. D. D. Perrin, *Adv. Heterocyclic Chem.*, **4**, 43 (1965).

675. T. J. Batterham and J. A. Wunderlich, *J. Chem. Soc. (B)*, 489 (1969).

676. B. Evans and R. Wolfenden, *J. Amer. Chem. Soc.*, **94**, 5902 (1972), and *Biochemistry*, **12**, 392 (1973).

677. A. Albert and E. P. Serjeant, *J. Chem. Soc.*, 3357 (1964), and *Nature*, **199**, 1098 (1963).

678. H. Schlobach and W. Pfleiderer, *Helv. Chim. Acta*, **55**, 2518, 2525 (1972).

679. C. K. Mathews and F. M. Huennekens, *J. Biol. Chem.*, **235**, 3304 (1960); Y. Hatefi, P. T. Talbert, M. J. Osborn, and F. M. Huennekens, *Biochemical Preparations*, **7**, 89 (1960).

680. P. W. Boyle and M. T. Keating, *J.C.S. Chem. Comm.*, 375 (1974).

681. P. P. K. Ho and L. Jones, *Biochim. Biophys. Acta*, **148**, 622 (1967).

682. B. V. Ramasastri and R. L. Blakley, *Biochem. Biophys. Res. Comm.*, **12**, 478 (1963).

683. B. T. Kaufman, K. O. Donaldson, and J. C. Keresztesy, *J. Biol. Chem.*, **238**, 1498 (1963).

684. R. P. Leary, Y. Gaumont, and R. L. Kisliuk, *Biochem. Biophys. Res. Comm.*, **56**, 484 (1974).

685. P. R. Farina, L. J. Farina, and S. J. Benkovic, *J. Amer. Chem. Soc.*, **95**, 5409 (1973).

686. D. B. Cosulich, J. M. Smith, Jr., and H. P. Broquist, *J. Amer. Chem. Soc.*, **74**, 4215 (1952).

687. R. L. Blakley, *The Biochemistry of Folic Acid and Related Pteridines*, North-Holland, Amsterdam, 1969.

688. S. J. Benkovic and W. P. Bullard in *Progress in Bioorganic Chemistry*, Vol. 2, E. T. Kaiser and F. J. Kézdy, Eds., Wiley-Interscience, New York, 1973, pp. 133–175.

689. H. C. Chitwood and R. W. McNamee, U.S. Patent 2,345,237, *Chem. Abs.*, **38**, 4274 (1944); L. A. Cort and N. R. Francis, *J. Chem. Soc.*, 2799 (1964); H. Baganz, L. Domaschke, and G. Kirchner, *Chem. Ber.*, **94**, 2676 (1961).

690. J. M. Edwards, U. Weiss, R. D. Gilardi, and I. L. Karle, *Chem. Comm.*, 1649 (1968).

691. Z. Aizenshtat, E. Klein, H. Weiler-Feilshenfeld, and E. D. Bergmann, *Israel J. Chem.*, **10**, 753 (1972).

691a. G. A. Taylor and S. A. Proctor, *Chem. Comm.*, 1379 (1969).

692. W. H. Perkin, Jr., and W. G. Sedgwick, *J. Chem. Soc.*, 2437 (1924); W. H. Perkin, Jr. and S. G. P. Plant, *J. Chem. Soc.*, 2583 (1928).

692a. W. H. Perkin, Jr., and W. G. Sedgwick, *J. Chem. Soc.*, 438 (1926).

692b. M. M. Cartwright and S. G. P. Plant, *J. Chem. Soc.*, 1898 (1931).

693. H. Booth, *Tetrahedron*, **19**, 91 (1963).

694. D. A. Archer, H. Booth, P. C. Crisp, and J. Parrick, *J. Chem. Soc.*, 330 (1963).

695. H. Adkins and H. L. Coonradt, *J. Amer. Chem. Soc.*, **63**, 1563 (1941).

696. T. Masamune, M. Ohno, K. Takemura, and S. Ohuchi, *Bull. Chem. Soc. Japan*, **41**, 2458 (1968).

697. T. Masamune, *Chem. Comm.*, 244 (1968); T. Masamune and M. Takasugi, *Chem. Comm.*, 625 (1967).

698. N. Bărbulescu and F. Potmischil, *Annalen*, **735**, 132 (1970).

699. N. Bărbulescu and F. Potmischil, *Tetrahedron Letters*, 5279 (1969).

700. N. Bărbulescu and F. Potmischil, *Annalen*, **752**, 22 (1971).

701. A. Albert, *The Acridines*, E. Arnold, London, 1966; R. M. Acheson, *Acridines*, A. Weissberger, Ed., Interscience, New York, 1956.

702. B. R. Lowry and A. C. Huitric, *J. Org. Chem.*, **37**, 2697 (1972); B. R. Lowry, and A. C. Huitric, *J. Org. Chem.*, **37**, 1316 (1972).

703. A. C. Huitric, B. R. Lowry, and A. E. Weber, *J. Org. Chem.*, **40**, 965 (1975).

704. I. Ninomiya, T. Naito, and T. Kiguchi, *J.C.S. Perkin I*, 2257 (1973).

705. E. C. Taylor, Jr. and E. J. Strojny, *J. Amer. Chem. Soc.*, **78**, 5104 (1956).

706. T. Masamune, M. Takasugi, H. Suginome, and M. Yokoyama, *J. Org. Chem.*, **29**, 681 (1964).

707. T. Masamune, S. Ohuchi, S. Shimokawa, and H. Booth, *Tetrahedron*, **22**, 773 (1966).

708. W. N. Speckamp, U. K. Pandit, P. K. Korver, P. J. Van Der Haak, and H. O. Huisman, *Tetrahedron*, **22**, 2413 (1966).

709. Z-i. Horii, C. Iwata, and Y. Tamura, *Chem. and Pharm. Bull.* (*Japan*), **12**, 1493 (1964).

710. Z-i. Horii, T. Kurihara, S. Yamamoto, and I. Ninomiya, *Chem. and Pharm. Bull.* (*Japan*), **15**, 1641 (1967).

711. Z-i. Horii, T. Kurihara, and I. Ninomiya, *Cham. and Pharm. Bull.* (*Japan*), **16**, 668 (1968).

712. Z-i. Horii, T. Kurihara, S. Yamamoto, M-C. Hsü, C. Iwata, I. Ninomiya, and Y. Tamura, *Chem. and Pharm. Bull.* (*Japan*), **14**, 1227 (1966).

713. Z-i. Horii, T. Kurihara, and I. Ninomiya, *Chem. and Pharm. Bull.* (*Japan*), **17**, 1733 (1969).

714. F. Plénat, C. Montginoul, and H. Christol, *Bull. Soc. chim. France*, 697 (1973).

715. D. A. Walsh and E. E. Smissman, *J. Org. Chem.*, **39**, 3705 (1974).

716. W. Oppolzer, *Tetrahedron Letters*, 1001 (1974).

717. I. G. M. Campbell, C. G. Le Fèvre, R. J. W. Le Fèvre, and E. E. Turner, *J. Chem. Soc.*, 404 (1938).

718. G. R. Clemo and M. McIlwain, *J. Chem. Soc.*, 258 (1936).

719. S. Miyano and M. Nakao, *Chem. and Pharm. Bull.* (*Japan*), **20**, 1328 (1972).

720. R. M. Scott and M. Tomlinson, *J. Chem. Soc.*, 417 (1959).

721. M. Godchot and M. Mousseron, *Bull. Soc. chim. France*, 528 (1932).

722. G. R. Clemo and H. McIlwain, *J. Chem. Soc.*, 1698 (1936).

723. F. Hamaguchi, T. Nagasaka, and S. Ohki, *J. Pharm. Soc. Japan*, **94**, 351 (1974).

724. H. Akimoto, K. Okamura, M. Yui, T. Shioiri, M. Kuramoto, Y. Kikugawa, and S-i. Yamada, *Chem. and Pharm. Bull.* (*Japan*), **22**, 2614 (1974).

725. S-i. Yamada and T. Kunieda, *Chem. and Pharm. Bull.* (*Japan*), **15**, 499 (1967).

726. T. Kurihara, K. Nakamura, and H. Hirano, *Chem. and Pharm. Bull.* (*Japan*), **22**, 1839 (1974).

727. T. Tanaka, N. Taga, M. Miyazaki, and I. Iijima, *J.C.S. Perkin I*, 2110 (1974).

728. J. E. Anderson and A. C. Oehlschlager, *Chem. Comm.*, 284 (1ʼ 68).

728a. R. M. Moriarty, *Topics Stereochem.* **8**, 271 (1974).

729. K. E. Wilzbach and D. J. Rausch, *J. Amer. Chem. Soc.*, **92**, 2178 (1970).

730. R. C. DeSelms and W. R. Schleigh, *Tetrahedron Letters*, 3563 (1972).

731. M. G. Barlow, R. N. Haszeldine, and J. G. Dingwall, *J.C.S. Perkin I*, 1542 (1973).

732. F. W. Fowler, *J. Org. Chem.*, **37**, 1321 (1973).

733. U. Eisner, J. R. Williams, B. W. Matthews, and H. Ziffer, *Tetrahedron*, **26**, 899 (1970).

734. E. C. Taylor and W. W. Paudler, *Tetrahedron Letters*, No. 25, 1 (1960).

734a. E. C. Taylor, W. W. Paudler, and I. Kuntz, Jr., *J. Amer. Chem. Soc.*, **83**, 2967 (1961).

735. E. C. Taylor, R. O. Kan, and W. W. Paudler, *J. Amer. Chem. Soc.*, **83**, 4484 (1961).

736. E. C. Taylor and R. O. Kan, *J. Amer. Chem. Soc.*, **85**, 776 (1963).

737. G. Slomp, F. A. MacKellar, and L. A. Paquette, *J. Amer. Chem. Soc.*, **83**, 4472 (1961).

738. L. A. Paquette and G. Slomp, *J. Amer. Chem. Soc.*, **85**, 765 (1963).

739. W. A. Ayer, R. Hayatsu, P. deMayo, S. T. Reid, and J. B. Stothers, *Tetrahedron Letters*, 648 (1961).

740. L. J. Altman, M. F. Semmelhack, R. B. Hornby, and J. C. Vederas, *Chem. Comm.*, 686 (1968).

741. H. Igeta, T. Tsuchiya, and C. Kaneko, *Tetrahedron Letters*, 2883 (1971).

742. C. Kaneko, T. Tsuchiya, and H. Igeta, *Chem. and Pharm. Bull.* (*Japan*), **21**, 1764 (1973).

743. D. Elad, I. Rosenthal, and S. Sasson, *J. Chem. Soc.*, (*C*), 2053 (1971).

744. S. Sasson, I. Rosenthal, and D. Elad, *Tetrahedron Letters*, 4513 (1970).

745. J. C. Nnadi and S. Y. Wang, *Tetrahedron Letters*, 2211 (1969).

746. M. N. Khattak and S. Y. Wang, *Tetrahedron*, **28**, 945 (1972).

747. D. Suck and W. Saenger, *J. Amer. Chem. Soc.*, **94**, 6520 (1972).

748. P. J. Wagner and D. J. Bucheck, *J. Amer. Chem. Soc.*, **92**, 181 (1970).

749. S. Sasson and D. Elad, *J. Org. Chem.*, **37**, 3164 (1972).

750. J. A. Hyatt and J. S. Swenton, *J. Amer. Chem. Soc.*, **94**, 7605 (1972).

751. D. E. Bergstrom and W. C. Agosta, *Tetrahedron Letters*, 1087 (1974).

752. H. Morrison, A. Feeley, and R. Kleopfer, *Chem. Comm.*, 358 (1968).

753. R. Kleopfer and H. Morrison, *J. Amer. Chem. Soc.*, **94**, 255 (1972).

754. S. Y. Wang, J. C. Nnadi, and D. Greenfeld, *Tetrahedron*, **26**, 5913 (1970).

755. G. I. Birnbaum, J. M. Dunston, and A. G. Szabo, *Tetrahedron Letters*, 947 (1971).

756. G. M. Blackburn and R. J. H. Davies, *J. Chem. Soc.*, (*C*) 2239 (1966).

757. T. Kunieda and B. Witkop, *J. Amer. Chem. Soc.*, **93**, 3493 (1971).

758. G. M. Blackman and R. J. H. Davies, *Chem. Comm.*, 215 (1965).

759. R. Anet, *Tetrahedron Letters*, 3713 (1965).

760. N. Camerman and A. Camerman, *J. Amer. Chem. Soc.*, **92**, 2523 (1970).

761. S. Y. Wang and D. F. Rhoades, *J. Amer. Chem. Soc.*, **93**, 2554 (1971).

762. J. L. Flippen and I. L. Karle, *J. Amer. Chem. Soc.*, **93**, 2762 (1971).

763. S. Y. Wang, *J. Amer. Chem. Soc.*, **93**, 2762 (1971).

764. N. J. Leonard and R. L. Cundall, *J. Amer. Chem. Soc.*, **96**, 5904 (1974).

765. N. J. Leonard, K. Golankiewicz, R. S. McCredie, S. M. Johnson, and I. C. Paul, *J. Amer. Chem. Soc.*, **91**, 5855 (1969).

766. M. W. Logue and N. J. Leonard, *J. Amer. Chem. Soc.*, **94**, 2842 (1972).

767. J. K. Frank and I. C. Paul, *J. Amer. Chem. Soc.*, **95**, 2324 (1974).

767a. K. Mutai, B. A. Gruber, and N. J. Leonard, *J. Amer. Chem. Soc.*, **97**, 4095 (1975).

768. G. R. Evanega and D. L. Fabiny, *J. Org. Chem.*, **35**, 1757 (1970).

769. G. R. Evangea and D. L. Fabiny, *Tetrahedron Letters*, 2241 (1968).

770. A. I. Meyers and P. Singh, *Chem. Comm.*, 576 (1968).

771. J. N. Brown, R. L. L. Towns, and L. M. Trefonas, *J. Amer. Chem. Soc.*, **93**, 7012 (1971).

772. A. I. Meyers and P. Singh, *Tetrahedron Letters*, 4073 (1968).

773. A. I. Meyers and P. Singh, *J. Org. Chem.*, **35**, 3022 (1970).

774. Z. Koblicová and K. Weisner, *Tetrahedron Letters*, 2563 (1967).

775. R. Gault and A. I. Meyers, *Chem. Comm.*, 778 (1971).

776. G. R. Evanega and D. L. Fabiny, *Tetrahedron Letters*, 1749 (1971).

777. D. W. Jones, *J. Chem. Soc.*, (*C*) 1729 (1969).

778. H. C. Brown and B. Kanner, *J. Amer. Chem. Soc.*, **88**, 986 (1966).

779. J. M. Essery and K. Schofield, *J. Chem. Soc.*, 3939 (1961).

780. G. Gottarelli and B. Samori, *J.C.S. Perkin II*, 1462 (1974).

781. G. Gottarelli and B. Samori, *Tetrahedron Letters*, 2055 (1970).

782. H. E. Smith, L. J. Scnaad, R. B. Banks, C. J. Wiant, and C. F. Jordan, *J. Amer. Chem. Soc.*, **95**, 811 (1973).

783. W. von E. Doering and V. Z. Pasternak, *J. Amer. Chem. Soc.*, **72**, 143 (1950).

784. A. F. Bickel, *J. Amer. Chem. Soc.* **70**, 328 (1948).

785. R. Adams and J. L. Johnson, *J. Amer. Chem. Soc.*, **71**, 705 (1949).

786. N. K. Hart, S. R. Johns, and J. A. Lamberton, *Austral. J. Chem.*, **22**, 1283 (1969).

787. S. A. Monti, R. R. Schmidt III, B. A. Shoulders, and H. L. Lochte, *J. Org. Chem.*, **37**, 3834 (1972).

788. R. Kuhn and W. Münzing, *Chem. Ber.*, **85**, 29 (1952).

789. R. W. Faessinger and E. V. Brown, *J. Amer. Chem. Soc.*, **73**, 4606 (1951).

790. R. Bentley, *Molecular Asymmetry in Biology*, Vol. 2 Academic, New York, 1970, pp. 1–86.

791. T. C. Bruice and S. Benkovic, *Bioorganic Mechanisms*, Vol. 2 Benjamin, New York, 1966, pp. 301–349.

792. J. W. Cornforth, G. Ryback, G. Popják, C. Donninger, and G. Schroepfer, Jr., *Biochem. Biophys. Res. Comm.*, **9**, 371 (1962).

793. E. Späth and F. Kesztler, *Ber.*, **69**, 384 (1936); A.D. Rosenfeld and D. G. Kolesnikov, *Ber.*, **69**, 2022 (1936).

794. E. Späth, F. Kuffner, and N. Platzer, *Ber.*, **68**, 1384 (1935).

795. W. L. F. Armarego and J. I. C. Smith, *J. Chem. Soc.*, 5360 (1965).

796. D. J. Brown and K. Ienaga, *J.C.S. Perkin I*, 2182 (1975).

797. W. Merkel and W. Ried, *Chem. Ber.*, **106**, 471 (1973).

798. W. Merkel and W. Reid, *Chem. Ber.*, **106**, 956 (1973).

799. P. E. Fielding and R. J. W. Le Fèvre, *J. Chem. Soc.*, 1811 (1951).

800. J. E. Downes, *J. Chem. Soc.* (*C*), 1491 (1967).

801. T. McL. Spotswood and C. I. Tanzer, *Austral. J. Chem.* **20**, 1213 (1967).

802. I. C. Calder, T. McL. Spotswood, and C. I. Tanzer, *Austral. J. Chem.*, **20**, 1195 (1967).

803. W. L. F. Armarego and R. E. Willette, *J. Chem. Soc.*, 1258 (1965).

804. C. C. Steele and R. Adams, *J. Amer. Chem. Soc.*, **52**, 4528 (1930).

805. A. L. Stone and J. G. Breckenridge, *Canad. J. Chem.*, **30**, 725 (1952).

806. W. Theilacker and F. Baxmann, *Annalen*, **581**, 117 (1953).

807. M. Crawford and I. F. B. Smyth, *J. Chem. Soc.*, 3464 (1954).

807a. M. Crawford and I. F. B. Smyth, *J. Chem. Soc.*, 4133 (1952).

808. M. Crawford and R. B. Ingle, *J. Chem. Soc.* (*B*), 1907 (1971).

809. H. Mix, *Annalen*, **592**, 146 (1955).

810. D. L. Fields, T. H. Regan, and R. E. Graves, *J. Org. Chem.*, **36**, 2995 (1971).

811. D. L. Fields and T. H. Regan, *J. Org. Chem.*, **36**, 2991 (1971).

812. D. L. Fields, T. H. Regan, and D. P. Maier, *J. Org. Chem.*, **38,** 407 (1973).

813. D. L. Fields and T. H. Regan, *J. Org. Chem.*, **36,** 2986 (1971).

814. H. J. Teuber and L. Vogel, *Chem. Ber.*, **103,** 3319 (1970).

815. B. R. T. Keene and G. L. Turner, *Tetrahedron*, **27,** 3405 (1971).

816. D. L. Coffen and T. E. McEntee, Jr., *J. Org. Chem.*, **35,** 503 (1970).

817. T. P. Pitner, H. Sternglanz, C. E. Bugg, and J. D. Glickson, *J. Amer. Chem. Soc.*, **97,** 885 (1975).

818. S. Cohen "Jerusalem Symposium on Quantum Chemistry and Biochemistry," in E. D. Bergmann and B. Pullman, Eds., *The Purines. Theory and Experiment*, Academic, Jerusalem, 1972, pp. 539–547.

819. E. Chargaff and J. N. Davidson, *The Nucleic Acids*, Vols. I, II, and III, Academic, New York, 1955, 1955, and 1960; A. M. Michelson, *The Chemistry of Nucleosides and Nucleotides*, Academic, New York, 1963; P.O.P.T'so, *Basic Principles in Nucleic Acid Chemistry*, Vol. 1, Academic, New York, 1974.

820. N. K. Kochetkov and E. I. Budoviskii, Eds., *Organic Chemistry of Nucleic Acids*, Plenium Press, London, Part A, 1971 and Part B, 1972.

4 NITROGEN HETEROCYCLES: SEVEN- AND LARGER-MEMBERED RINGS, 1-AZABICYCLO[X.Y.0] ALKANES, *n*-AZABICYCLO [X.Y.Z] ALKANES AND MISCELLANEOUS CYCLIC AZA COMPOUNDS

I. INTRODUCTION

The present chapter is the third and last one on heterocycles containing only nitrogen atoms, and therefore includes miscellaneous compounds which were not previously discussed. The stereochemistry of seven-, eight-, nine-, and larger-membered rings is described together with the formation of bicyclic compounds by transannular reactions between NH and CO groups. Large rings including cyclophanes, catenanes, and cryptates are discussed briefly.

The basic stereochemistry of a series of 1-azabicyclo[*x.y.*0]alkanes (bicyclic systems) with a bridgehead nitrogen atom, for example, pyrrolizidines and quinolizidines, is presented. Their detailed chemistry, however, is excluded because it has been adequately reviewed in works on the alkaloids which contain the particular ring systems. A series of *n*-azabicyclo-[*x.y.z*]alkanes (tricyclic systems) is also reviewed insofar as to show the steric relationships of the atoms within the molecules. Several of these compounds may be symmetrical, but when considering them one has to know their three-dimensional structures. Their formulas have to be drawn in stereotype in order to define the molecules accurately and to understand many of their reactions. The stereochemistry of compounds related to Tröger base, propellanes, and azabullvalenes is included in the final sections.

II. SEVEN-MEMBERED RINGS

A. Azepines

The seven-membered ring is not planar in both the saturated and the unsaturated derivatives and has interesting conformational properties. Azepines are best synthesized by interesting reactions involving fused cyclopropanes or aziridines as intermediates. 4,6-Dimethoxycarbonyl-2,7-dimethyl-3H-azepine (1) together with its dimer is formed when 1,2-dihydro-3,5-dimethoxycarbonyl-2,6-dimethyl-2-tosyloxymethylpyridine is heated at 100°C. The latter is obtained by uv irradiation of 3,5-dimethoxycarbonyl-2,6-dimethylpyridine in methanolic solution followed by tosylation of the 1,2-methanol adduct.[1] No ring expansion occurs if the pyridine lacks the 2,6-dimethyl groups because they protect the adduct from undergoing further reactions, for example, addition or dimerization, and stabilize the azepine. An azanorcaradiene is postulated as the intermediate (Equation 1). Paquette and Haluska[2] reacted methoxycarbonyl-

(1)

nitrene with cycloheptatriene and obtained a 2.1:1 mixture of 2,3-homo-1H (2) and 4,5-homo-1H-azepine (3) which are most probably formed by valence-bond isomerization of the aziridine originally formed. Photolysis of the isomer 2 causes intramolecular cyclization to 5-methoxycarbonyl-5-aza-1α,2α,4α,6α-tricyclo[4.2.0.0^{2,4}]oct-7-ene (4), but 2-methoxycarbonyl-2-azabicyclo[3.3.0]octa-3,6-diene (5) is a minor by-product. This by-product was reduced to 2-methoxycarbonyl-2-azabicyclo[3.3.0]octane, which was independently synthesized by intramolecular cyclization of chloroaminoethylcyclopentane with $H_2SO_4/h\nu$: followed by reaction with ethyl chloroformate and triethylamine. A third synthesis involves a cycloaddition reaction between 2-phenylazirine and 2,5-dimethyl-3,4-diphenylcyclopen-

2   3   4   5

tadienone which is brought about by refluxing in benzene for long periods. The elements of carbon monoxide are lost and the intermediate 2,5-dimethyl-3,4,6-triphenyl-1-azabicyclo[4.1.0]heptene (6) is transformed first to 4,7-dimethyl-3,5,6-triphenyl-1H-azepine which then rearranges to the more stable 3H-azepine tautomer (7).[3] A 2:3 mixture of (d,l)-cis- and

(meso)-trans-3,6-dimethylperhydroazepines is obtained by reductive cyclization of a mixture of d,l- and meso-2,5-dicyanohexane or from 1,6-diamino-2,5-dimethylhexane. These are separable by distillation and their structures were assigned by ¹H nmr spectroscopy.[4]

X-Ray analysis of 1-p-bromobenzenesulfonyl-1H-azepine revealed that it exists in a "boat" conformation in the crystal, but there is no evidence that it exists in the azanorcaradiene (8) or the azahomoaromatic (9) forms.[5] ¹H nmr spectral data indicate that the azepine ring is conformationally labile involving two stable boat forms. The relative stabilities of these depend on the substituents, for example, 10 is more favored than 11,[6] and the equilibrium 7 ⇌ 12 is in favor of conformer 7, as would be expected

from consideration of nonbonded interactions.[3] The population of the lactam, 1,3-dihydro-3,5,7-trimethyl-2H-azepin-2-one (or thione) derived from 10, is also higher than that derived from 11.[6] The activation energies for these processes have been evaluated by variable temperature ¹H nmr from signal coalescence temperatures and signal band shapes and show no evidence of valence bond tautomerism. The $\Delta G^{\ddagger}_c$ values for ring inversion in 1, 4,6-dimethoxycarbonyl-6-methyl-2-styryl-1, and 2-anilino[7] 3H-azepines are 57.3, 59.4, and 42.7 kJ mol⁻¹ and are higher than those observed in 3H-azepin-2-one.[7] Perhydroazepines also exhibit conformational mobility and data suggest that they are more mobile than the unsaturated compounds. The $\Delta G^{\ddagger}_c$ values for 1-methyl- and 1-chloroperhydrazepines are 26.9 and 35.1 kJ mol⁻¹ respectively. Pseudorotation in these compounds is probably responsible for approximately 8–12 kJ mol^{-1} of the energy barrier.[8] Noe and Roberts[9] obtained, by ^{19}F nmr spectroscopy, a $\Delta G^{\ddagger}_c$ value for 4,4-difluoroperhydroazepin-2-one(γ,γ-difluoro-ϵ-caprolactam) of 43.5 kJ mol^{-1} and the spectra were best explained by presuming inverting "chair" conformations (e.g., **13**). The chiroptical properties of ($-$)-men-

13

R = Ac, CO_2Et, $COCH_2Cl$

14

thone lactam, 7-isopropyl-4-methylazepin-2-one and its *N*-methyl derivative also support a chair conformation for this system.[10] Three concerted conformational equilibria were observed in *N*-acyl-10,11-dihydrodibenz-[b,f]azepines (**14**), namely rotation of the ethane bridge, restricted rotation about the amide C–N bond, and inversion of the seven-membered ring.[11]

1-Alkoxycarbonyl- and 1-cyanoazepines are generally stable but slowly dimerize on long standing. The dimerization can be accelerated by heating and the [6 + 4]π cycloadducts, for example, 9,14-diazatricyclo[6.3.2.1^{2,7}]-tetradeca-3,5,10,12-tetraenes (**15**), are formed. When these are heated at higher temperatures they isomerize to the [6 + 6]π adducts, for example, 13,14-diazatricyclo[6.4.1.1^{2,7}]tetradeca-3,5,9,11-tetraenes (**16**).[12–16] These structures are supported by their chemical and physical data and were confirmed by X-ray analyses of their *N*-methyl derivatives.[17,18] The [6 + 6]π adducts (**16**) must be formed in a multistep reaction because the thermal cycloaddition is not allowed by orbital symmetry considerations.[14] The N_{13},N_{14}-diallyl and related compounds undergo intramolecular Diels-Alder reactions to form 19,20-diazaheptacyclo[6,4,2^{3,6}2,9,12 1^{1,8},1^{2,7},1^{13,19},1^{17,20}]-eicosanes (**17**) after catalytic reduction.[19] 1-Alkoxycarbonylazepine reacts with tetrachloroethylene faster than it does with itself and provides (4 + 2)π cycloadducts, 2-azabicyclo[3.2.2]nona-3,6-dienes, and not the (6 + 2)π adducts.[20,21] Similar adducts are formed with 4-methyl- and 4-bromo-1-methoxycarbonylazepines and the structure of the adduct from the bromo compound was confirmed by an X-ray analysis.[22] The adduct from *N*-

R = CO_2R', CN

15

16

17

ethoxycarbonylazepine gives the methoxybromo derivative **18** by reaction with bromine in methanol, and its structure was established by an X-ray study.[23] 1-Alkoxycarbonylazepines also undergo $(4 + 2)\pi$ cycloaddition reactions with 1,3-diphenylbenzoisofuran[24] and 4-phenyl-1,2,4-triazolin-3,5-dione,[25] but with *trans*-diethyl azodicarboxylate the $(6 + 2)\pi$ adduct **19** is obtained contrary to orbital symmetry rules.[26] A nonconcerted mechanism

18   **19**

must therefore be postulated. 1-Methylazepine-2,7-dione reacts as an enophile and yields endo and exo $(2 + 2)\pi$ adducts with cyclopentadiene.[27]

Valence isomerization occurs in 4,5-dihydro-2-ethoxy-4,4,6-trimethyl-3*H*-azepine when it is photolyzed in the presence of a photosensitizer, and 1-ethoxy-6-methoxy-7-azabicyclo[3.2.0]heptane is formed if methanolic sodium methoxide is used as solvent.[28] The 1*H*-azepine–benzene-1,2-imine valence isomerization, and the effect of electron-withdrawing substituents on the nitrogen atom, has been discussed in terms of results from extended Hückel calculations.[29]

Several dihydrodibenzazepines were prepared by Hall and co-workers[30–36] by bridging the 2,2'-positions in biphenyls, for example, by reaction of 2,2'-bis(bromomethyl)biphenyl with a primary amine. They prepared several optically active dihydrodibenzazepines (**20**) and azepinium salts by ring closure of the corresponding optically active 2,2'-disubstituted bi-

20

phenyls. They also studied the optical stabilities of azepines with various R^1 and R^2 substituents and found that they racemized faster than the respective 2,2'-nonbridged biphenyls.[31,32,34–37] An excellent review was written on this subject by Hall.[38] Mislow and collaborators[39] determined the absolute configuration of 6,6'-dinitro-2,2'-diphenic acid by reduction of the derived $(\pm)$-4',1''-dinitro-1,2:3,4-dibenzo-1,3-cycloheptadiene-6-one with a chiral alcohol of known absolute configuration in the presence of aluminum *t*-butoxide. They deduced the configuration from a knowledge of the mechanism of reduction. From this they correlated the absolute configuration of a large number of optically active biphenyls with their respective optical rotations and derived the optical displacement rule which states that "a symmetrically substituted hindered biaryl has the *S* (or *R*) configuration if the optical rotation becomes more positive (or negative) when the 2,2'-positions are bridged."[40-43] Thus the absolute configuration of several azepines with the general formula **20** is now known. The *S*(+)-*N*-methyl quaternary salt of **20** ($R^1 = R^2 = R^3 = Me$) rearranges with KNH_2/NH_3 into 9,10-dihydro-9-dimethylamino-4,5-dimethylphenanthrene (**21**) stereospecifically with the formation a chiral center at C–9. The absolute configuration at C–9 was shown to be *S* by ord and by ¹H nmr spectroscopy,[44] and it is consistent with the present views on the mechanism of the Stevens rearrangement.[45] The amine **21** undergoes an asym-

21

metric transformation whereby the configuration of the diphenyl changes from *S* to *R* without altering the chirality at C–9.[44]

Azepines related to the dibenzo compound **20** are conformationally mobile, as are the isomeric compounds **14**. This property has been examined by ¹H nmr spectroscopy.[46,47]

B. Diazepines

Pentasubstituted 1-ethoxycarbonylaminopyridine betaines cyclize to pentasubstituted 1-ethoxycarbonyl-1,2-diazepines.[48] α(o-Alkenylaryl)diazoalkanes, prepared from the tosyl hydrazones, yield 4,5-benzo-3*H*-1,2-diazepines which rearrange to the 5*H* isomers in the presence of alkali.[49] 3,6-Dimethyl- or 3,6-dimethoxycarbonyl-1,2,4,5-tetrazine and cyclopropanes produce $(4 + 2)\pi$ adducts which liberate a nitrogen molecule and yield 1,2-diazepines.[50] Like azepines, diazepines are not planar and there is X-ray analysis evidence that they have a "boat" structure.[51,52] Nmr data is best explained in terms of a conformational equilibrium between two boat forms, and in a multisubstituted derivative the proportion of one conformer may be higher than the other.[49,53] The coalescence temperatures of a few 3,7-substituted 4,5-benzo-5*H*-1,2-diazepines have been measured and gave $\triangle G^{\ddagger}_c$ values of the order of 79.5 kJ mol⁻¹.[49]

1,2-Diazepines, in contrast to azepines, are in equilibrium with appreciable amounts of the diazanorcaradienes which are energetically preferred.[50] The conformational equilibria and valence isomerizations are shown in Scheme 1.

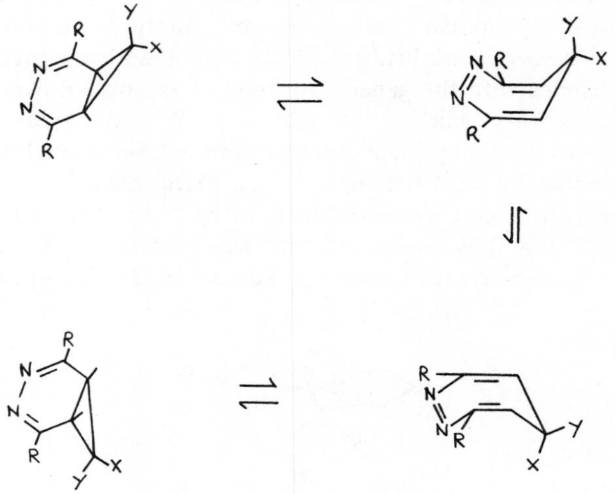

Scheme 1

A $(4 + 2)\pi$ dimer **22** is formed from 1-ethoxycarbonyl-1H-1,2-diazepine by acid catalysis,[54] and the $(4 + 2)\pi$ adduct **23** is obtained by condensation with tetracyanoethylene.[48] Intramolecular cyclizations in 1,2-diazepines can be achieved in various ways. Uv irradiation of 1-acetyl(or 1-ethoxycarbonyl)-3,5,7-triphenyl-1,2-diazepine yields *cis*-2-acetyl(or ethoxycarbonyl)-1,4,6-triphenyl-2,3-diazabicyclo[3.2.0]hepta-3,6-diene, whereas in the 2-methyl derivative the bicyclo compound is unstable and only 1-methyl-2,5-diphenylpyrazole is isolated.[55] Similar treatment of 6,7-dihydro-6-hydroxy-5-methyl-4-phenyl-1H(or acyl)-1,2-diazepines furnishes different

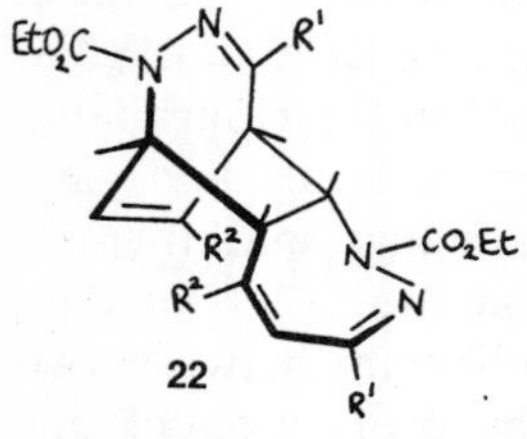

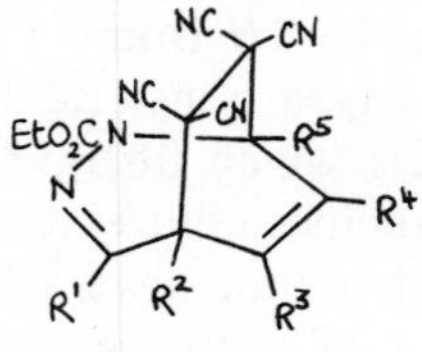

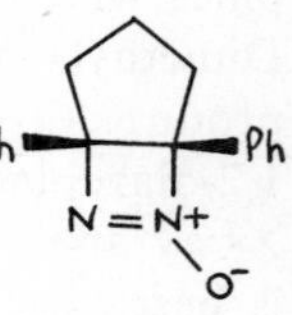

bicyclo compounds: *cis*-4-hydroxy-5-methyl-6-phenyl-1,2-diazabicyclo-[3.2.0]hept-6-enes (i.e., they contain a bridgehead nitrogen atom).[56] Such cyclizations, but with different arrangements of substituents, that is, the hydroxy group is now in position-6 and the phenyl group in position 4, are achieved by reaction with triethylamine,[57] and the 6-oxo-1,2-diazepines can also be made to undergo analogous[58] and other intramolecular cyclizations.[59]

Thermolysis of 3,4,5,6-tetrahydro-3,7-diphenyl-7H-1,2-diazepine leads to nitrogen extrusion and formation of cis- and trans-1,2-diphenylcyclopentanes without stereoselectivity.[60] Photolysis of the N-oxide of 4,5-dihydro-3,7-diphenyl-6H-1,2-diazepine furnishes the cis-diazabicyclo N-oxide (24) among other products.[61]

1,4- and 1,5-Benzodiazepines exhibit the same boat to boat conformational mobility of 1,2-diazepines, but some appear to be in a puckered-chair conformation. Temperature dependent ^{1}H nmr studies of several 2,3-benzfused derivatives were made and confirmed these equilibria.[7,62] The equilibria 25 ⇌ 26 were found to require larger activation energies than in the azepines (50–79.5 kJ mol^{-1}).[7,63] In contrast, cis-2,7-dimethylperhydro-2-oxo-1,4-diazepine, prepared from 2,4-diaminopentane and glyoxal bisulfite, was shown to assume the "chair" conformation 27.[64] Both cis- and trans-2,3,4,5-tetrahydro-2,4-dimethyl-1H-1,5-benzodiazepines were prepared by appropriate reduction and separation, from 2,4-dimethyl-1H-1,5-benzodiazepine.[65]

The important features of the effect of solvents on the cd curves of S(+)-1,11-dimethyl-6-phenyl-5H-dibenzo[d,f][1,3]diazepine (28) and S(+)-6,7-dihydro-1,11-dimethyl-6-oxo-5H-dibenzo[d,f][1,3]diazepine were described.[66]

Seven-membered ring nitrogen heterocycles were reviewed by Moore and Mitchell.[66a]

III. EIGHT-MEMBERED RINGS

A. Azocines

Lambert and Khan[67] recently examined the ^{1}H nmr (270 MHz) and ^{13}C nmr (22.6 MHz) spectra of perhydroazocine(azocane, azacyclooctane) and its 1-methyl and 1-chloro derivatives. The ^{1}H nmr signals of the three compounds collapsed at around $-95°C$ but the ^{13}C nmr signals were unaffected. On their evidence they excluded pseudorotation and interpreted the results in terms of skew boat-chair–chair-boat ring reversal while always assuming the puckered conformations 29–31.

29 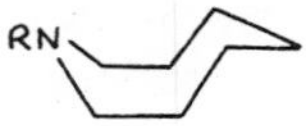   30 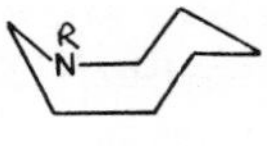   31

1,2,7,8-Tetrahydro-3,4:5,6-dibenzoazocine-1-spiro-1'-piperidinium io-dide (32) was resolved into its enantiomers by recrystallization of the (+) and (−)-α-bromocamphor-π-sulfonates. The (+)- and (−)-picrates race-mized rapidly in acetone at 23°C ($t_{1/2}$ = 1.4 min).[68] The N-alkyl derivatives of the isomeric ring system, 5,6-dihydro-7H,12H-dibenzo[c,f]azocine (32a), are also conformationally flexible and equilibria between rigid boat-chair and mobile twist-boat conformations are observable by nmr spectroscopy in CDCl$_3$. Protonation, that is, nmr in trifluoroacetic acid, raises the inter-conversion barriers, and the $\triangle G^{\ddagger}$ value for the N,N(6,6)-dimethyl iodide derivative of 32a is greater than 75.3 kJ mol^{-1}.[68a]

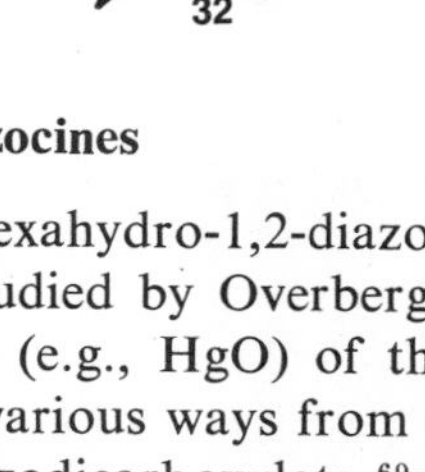

32

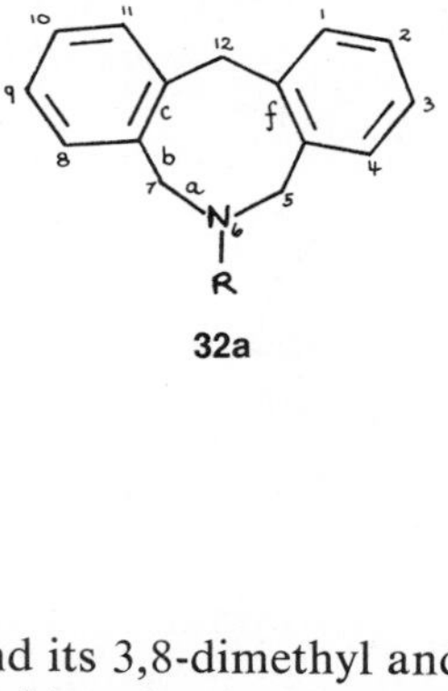

32a

B.   1,2-Diazocines

3,4,5,6,7,8-Hexahydro-1,2-diazocine and its 3,8-dimethyl and diaryl deriva-tives were studied by Overberger and his school. They synthesized these by oxidation (e.g., HgO) of the perhydrodiazocines which were in turn prepared in various ways from diketones or di-O-tosylates, and hydrazine or diethyl azodicarboxylate.[69–70] The structures of cis- and trans-3,8-diphenyl- and 3,8-di-p-methoxyphenyl-3,4,5,6,7,8-hexahydro-1,2-diazocines were determined from their dipole moments.[69] The N=N linkage of these diazocines were assumed to be cis (33), but Overberger and co-workers[71] recently irradiated the diazocine 33 with uv light and isomerized it to *trans*-3,4,5,6,7,8-hexahydro-1,2-diazocine (**34**) in 60% yield. They further showed that the previously recorded oxidation (HgO) of a mixture of *cis*- and *trans*-3,8-dimethylperhydro-1,2-diazocines[70] had in fact given *cis*-3,8-dimethyl-3,4,5,6,7,8-hexahydro-*cis*-1,2-diazocine and *trans*-3,8-dimethyl-3,4,5,6,7,8-hexahydro-*trans*-1,2-diazocine. These have been distinguished from each other by the differences in the ^{1}H nmr spectra of the protons α to the nitrogen atoms, by uv spectra (*cis*-azo isomers have uv maxima at longer wavelengths and have higher extinction coefficients than the trans isomers), and by the stronger binding of *cis*-hexahydrodiazocines with the shift reagent Eu(fod)$_3$ compared with that of the *trans*-hexahydrodiazocines because of the orientation of the nitrogen lone pair of electrons.[71] Pyrolysis and photolysis of cis and trans isomers of 3,8-diaryl-

33

34

3,4,5,6,7,8-hexahydro-1,2-diazocines gave *cis*- and *trans*-1,2-diphenylcyclohexanes, among other products, but in both cases the trans product predominated over the cis product.[69,70] These results should be reexamined in terms of the equilibrium **33** ⇌ **34**.

C. 1,4- and 1,5-Diazocines and 1,3,5,7-Tetraazocines

N,*N*'-Disubstituted 1,4-dihydro-1,4-diazocines (1,4-diazacycloocta-2,5,7-trienes) have recently been prepared by thermal 2σ → 2π isomerization of *N*,*N*'-disubstituted *cis*-diazabis-σ-homobenzenes. The latter were formed from *N*,*N*'-disubstituted *N*''-nitroso-*cis*-benzene trimine.[71a] Several 5,6,11,-12-tetrahydrodibenzo[*b*,*f*][1,4]diazocines were synthesized and examined by nmr techniques. They are conformationally mobile and the spectra can be interpreted in terms of two "chair-like" (e.g., **35**), two "boat-like" (**36**), and two intermediate "twist-boat" conformations making a total of twelve conformers.[72,73] The coalescence temperatures and free energy changes were evaluated and discussed.[72] The absolute stereochemistry S(+)-3,10-dicarboxy- (**37**, R^1=H,R^2=CO$_2$H) and S(+)-1,11-dimethyl-6,7-diphenyl-dibenzo[*e*,*g*][1,4]diazocines (**37**, R^1=Me,R^2=H) was deduced from a study of their chiroptical properties.[66]

Benzo[*b.f*][1,5]diazocines have also attracted attention and the 5,6,11,12-tetrahydro derivatives are conformationally mobile with chairlike and boatlike structures such as **35** and **36**, but with the nitrogen atoms in positions 1 and 5.[72,74-76] Hydroxyperhydro-1,5-diazocines (1,5-diazocanes) are readily prepared from 1,3-dibromopropan-2-ol[77] or 1-chloro-2-epoxy-2-methylpropane[78] and *p*-toluenesulfonamide anion. They undergo a num-

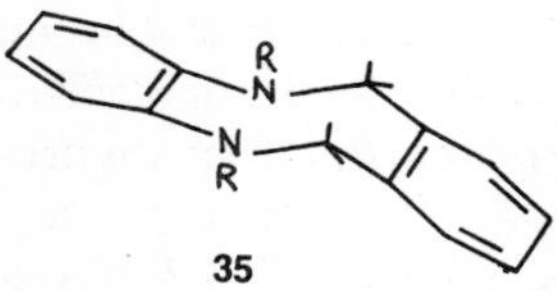

35

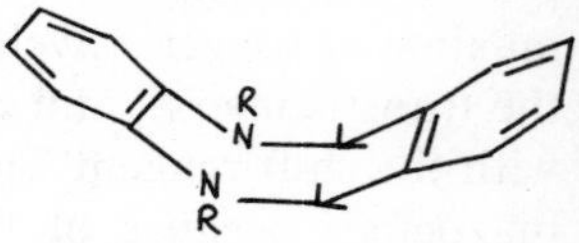

36

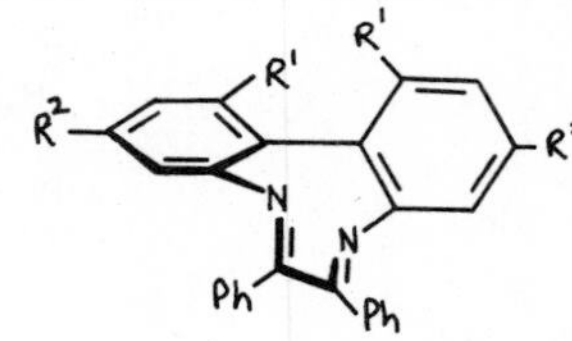

37

ber of intramolecular cyclizations which finally yield dialkyl piperazines, and the mechanisms are best explained by assuming a "crown" conformation for the diazocane.[79,80] They form interesting complexes with first row transition metals.[81] Dipole moment and ir measurements of *cis*- and *trans*-3,7-diethyl-3,7-dinitroperhydro-1,5-diazocine support the "crown" conformation (e.g., **38**, R = Et) for the trans isomer.[82] The crystal structure of cyclo-di-β-alanyl (2,6-dioxoperhydro-1,5-diazocine), however, revealed that the molecule is in the "boat" form **39**[83] (although empirical force-field calculations [gas phase] predict a "twist-boat" conformation),[83a] but the conformation may well be different in solution. The crystals of 1,3,5,7-tetranitro-1,3,5,7-tetraazaazocane are tetramorphic and the absorption spectra, dielectric constants, and X-ray diffraction patterns of three of them were studied by Wright and co-workers.[84] It was concluded that in the crystal they were in particular conformations bound by the lattice forces. None of the dipole moments calculated for the three conformers suggested were in agreement with the value measured in dioxane. It infers that in the absence of crystal lattice forces the molecules are conforma-

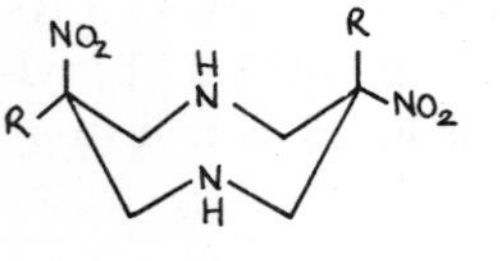

38

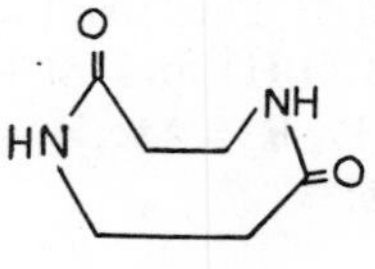

39

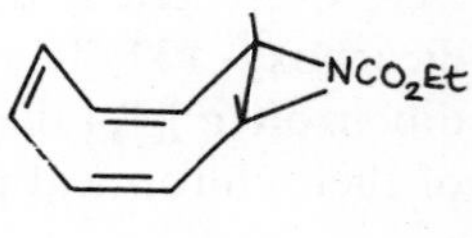

40

tionally mobile and have a dipole moment which is the statistical average of the moments of the conformers present.[85]

IV. NINE-MEMBERED RINGS : AZONINES

1-Ethoxycarbonylazonine (**41**) is obtained by photolysis[86,87] of the ethoxy-carbonyl nitrene adduct **40** of cyclooctatetraene or its thermal rearrangement product **42**.[88] Whereas the ^{1}H nmr spectra of oxonine and the

| 41 | 42 | 43 |

all-*cis*-cyclononatetraene are barely altered on cooling, the signals from 1-ethoxycarbonylazonine are temperature dependent. Masamune and co-workers[88] first explained these changes in terms of restricted rotation in the ethoxycarbonyl group ($\Delta G^{\ddagger} = 54.4$ kJ mol^{-1}), but Anastassiou and collaborators[89] thought it to be mainly due to the high barrier of nitrogen inversion. The results eliminated the planar geometry and supported the axisymmetric structure **43**. Hojo and Masamune[90] reexamined the data and demonstrated that their original interpretation was acceptable. The *N*-acylbicycloazonines related to **42** undergo Diels-Alder reactions and stereospecific intramolecular rearrangements to *cis*-azabicyclo systems.[91] The Diels-Alder products strongly suggest that *N*-acetyl-*cis,cis,trans,cis*-azonine is the intermediate involved.

[1]H nmr studies of 12-aza-12,13-dihydro-11*H*-dibenzo[*a.e*]cyclononene and related systems demonstrated that they are all conformationally labile, and activation energies for the conformational changes were evaluated.[92,93]

In a previous section (II.A), mention was made of how Mislow and co-workers[44] created an asymmetric center (**21**) by rearrangement of an optically active biphenyl. The reverse of this process, that is, conversion of a compound with asymmetric centers into an optically active biphenyl, was effected many years earlier. When thebaine was treated with phenyl magnesium bromide it underwent a rearrangement to a mixture of dia-stereomeric 4,5-(3-hydroxy-4-methoxybenzo)-6,7-(4-methoxybenzo)-1-methyl-2-phenyl-2,3:8,9-tetrahydroazonine. These differed in the chirality at C-2 because further degradation gave one optically active 2,2'-divinyl-biphenyl (Equation 2). From a knowledge of the absolute configuration of thebaine and the mechanism of the reaction, the absolute configuration (*R*) of the biphenyls was deduced and was the first unambiguous standard for the absolute configuration in these systems.[94]

(2)

Thermolysis of N,N',N''-trisubstituted *cis*-triazatris-σ-homobenzenes, for example, N,N',N''-trismethanesulfonyl *cis*-benzenetrimine, caused a $3\sigma \rightarrow 3\pi$ rearrangement into N,N',N''-trisubstituted 1,4,7-triazacyclonona-2,5,8-trienes.[94a]

V.   MISCELLANEOUS MEDIUM- AND LARGE-SIZED RINGS

A.   Medium-Sized Rings

1.   Transannular N,C_CO Interaction

The transannular interaction between the $\rightarrow$N: and C=O groups placed at 1,5- and 1,6-positions in azacycloalkanes was investigated extensively by Leonard and his school. This laudable work demonstrated how powerful physical methods could be when used judiciously. They showed how intricate chemical interactions were unraveled in an era when physical techniques were only just beginning to be appreciated as routine tools. They prepared several azacycloalkanones ranging from seven- to nineteen-membered rings by intramolecular Dieckmann[95] or acyloin cyclizations[96] of N-alkyl(or aryl)-N,N-bis(ω-alkoxycarbonylalkyl)amines (Equation 3). Only when the R_3N group was placed in the 1,5- or 1,6-position with respect to the C=O group (i.e., **44–46**) did they observe interaction between

$$(3)$$

them because of their steric proximity. The interaction is stronger in the 10-membered ring, that is, between the 1,6-positions (46), than in the eight-membered ring, that is, between the 1,5-positions (44).[95] Also the bulkier the group R (R = Me → t-Bu) the weaker the interaction because of F-strain.[97] However, this is not always entirely due to a steric effect since the electron withdrawing N-aryl or N-p-substituted aryl groups[98] (which

44   **45**   **46**

weaken the basicity of the N atom) and the electron withdrawing power of an N-cyclopropyl group[99] weaken this interaction. The interaction displays itself as an incipient transannular bond as shown in formula **47**

47   **48**   **49**

(although in the particular case drawn, it was not decided conclusively if the arrangement was cis or trans).[100] In the azacyclooctanones and aza-cyclodecanones incipient bonding and cyclization (see below) most probably leads to cis- and trans-fused systems respectively. Thus when the interaction occurred, the ir carbonyl stretching frequency was low (e.g., $\nu \sim 1666$ cm^{-1}), but when it was absent the ir frequency was normal (i.e., 1700 cm^{-1} and higher).[96—100] Protonation undoubtedly occurred on the oxygen atom and the salts (e.g., perchlorates of **48**, R^2 = H) had no carbonyl absorption, but showed OH stretching vibrations.[96—100] The pKa values measured in water were less than those measured in aqueous dimethyl formamide (e.g., 66%), suggesting that an acidic pKa was involved.[96—99] Uv spectral differences were observed but conclusive evidence from these alone could not be made at that time.[101] There was also evidence from dipole moment[102] and ord measurements.[103] In the latter case, the ord curve of (+)-1-(α-methylphenethyl)-1-azacyclooctan-5-one (**44**, R = CH$_3$–CH–CH$_2$Ph) was anomalous when compared with that of the open-chain analogue (+)-5-(N-ethyl-N-α-methylphenethylamino)pentan-2-one (**49**).

This was due to the induced chirality at the carbonyl group. Protonation caused a marked change in the ord spectrum.

A theoretical treatment by Yamabe, Kato, and Yonezawa[104] of these interactions in aza-, thia-, and oxacycloalkanones and cycloalkanones divulged a correlation between bond order and ir carbonyl absorption. Also the uv absorption near 221–235 nm was assigned to $\pi \rightarrow \pi^*$ transition due to intramolecular $N \rightarrow C\!=\!O$ charge transfer and the absorption near 217–190 nm was of the $n \rightarrow \sigma^*$ type.

An attempt by Leonard and Yethon[105] to obtain a similar interaction by replacing the $C\!=\!O$ group by an $S\!=\!O$ was unsuccessful but the salt **50** that they isolated was the first example in which intramolecular $N\!-\!H \cdots\cdots O\!=\!S$ hydrogen bonding across a ring was demonstrated.

Transannular N, C_{CO} interactions have been observed in certain natural products, for example, cryptopine and protopine in which cyclization to **51** (R = H) occurs on protonation, or to **51** (R = Me) on methylation.[106]

50   **51**   **52**

The site of microbiological hydroxylation of N-benzoylcyclooctane (**52**, $n=1, R^1 = R^2 = H_2$) or cyclononane (**52**, $n=2, R^1 = R^2 = H$) by *Sporotrichum sulphurescence* was shown to be at C–5 because oxidation of the products **52** ($n=1$ or 2, $R^1 = H, R^2 = OH$) gave the ketones **52** ($n=1$ or 2, $R^1, R^2 = O$) which demonstrated all the properties of N, C_{CO} transannular interaction.[107] This property was usefully applied to the synthesis, for example, of ($\pm$)-isoretronecanol (*trans*-1-hydroxymethylpyrrolizidine) from 1-benzyl-4-ethoxycarbonyl-5-oxo-1-azacyclooctane by transannular cyclization to the alcohol, and dehydration to the iminium salt followed by reduction.[108] Replacement of the $C\!=\!O$ in **52** ($R^1, R^2 = O$) by a $C\!=\!CH_2$ group does not alter the ability to form a transannular bond, and the substance behaves like a transannular enamine.[109] It forms mainly a salt such as **53**, but a little of the salt **54** is also produced.[110] Cleavage of the transannular bond in quaternary compounds such as **53** can be achieved by $NaNH_2$,[111,112] LAH, Li/NH_3,[113] or Hofmann elimination[114] as well as by basification. An interesting example is the 8-acetoxy-4-benzylpyrrolizidinium salt (**48**, $R^1 = CH_2Ph, R^2 = Ac$), prepared from the corresponding 8-hydroxy salt and ketene, which is a good acylating agent. It yields N-acetyl piperidine at

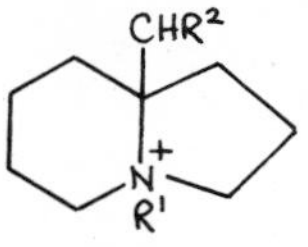

53

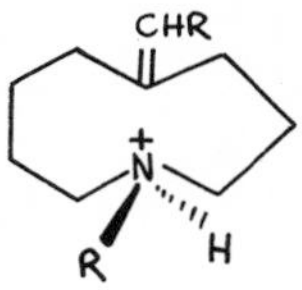

54

−20°C quantitatively in 2 min with piperidine, and will give acetic anhydride, also quantitatively, in 5 min when fused with potassium acetate.[115]

An irreversible transannular cyclization, not involving a carbonyl but taking advantage of the proximity of the 5-carbon atom in *N*-chloro-azacyclononane (**45**, R = Cl, and CH$_2$ replacing C=O) was achieved by treating this chloroamine in aqueous dioxane with silver ions. A nitrenium ion is apparently formed and indolizidine is obtained in 68% yield,[115a] (compare with Hofmann-Löffler-Freytag reactions).

2. *Azacyclophanes*

Azacyclophanes are medium-sized rings which incorporate one, two, and sometimes more carbocyclic and heterocyclic aromatic rings. A nomenclature for phanes was suggested by Vögtle and Neumann.[116] They are interesting stereochemically because they are conformationally mobile, the mobility is subject to intramolecular steric interactions and they may be chiral. The stereochemistry of [2.2]metacyclophanes was reviewed also by Vögtle and Neumann.[117]

Conformational mobility in these systems is revealed by their temperature dependent nmr spectra when, as the temperature is altered, some signals coalesce and then sharpen up again to give a different pattern of signals. From this data the free energy change for the equilibrium involved is determined. Thus the energy barrier for the diazaphane (**55**)[118] and *N*-cyclohexyl-2-aza[3,2]metacyclophane (**56**)[119] are 86.2 and 64.9 kJ mol^{-1} respectively. The $\Delta G^{\ddagger}_c$ values for the cyclophanes described below are given in brackets. The reason for noting these values is to show how much they can vary from one system to another. Steric crowding by the halogen atom is observed in **55,** and the mobility of [8] (2,5)pyrrolophanes (**57**) can

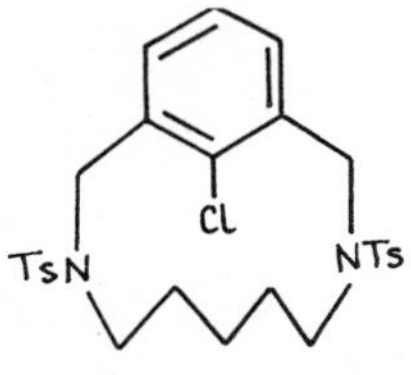

55

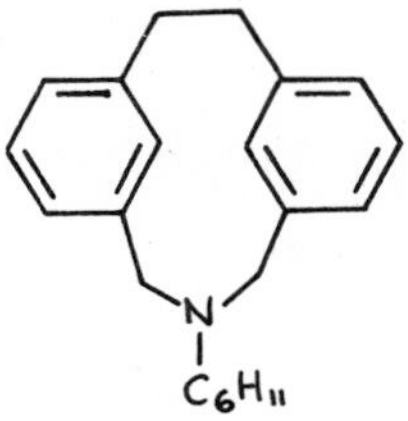

56

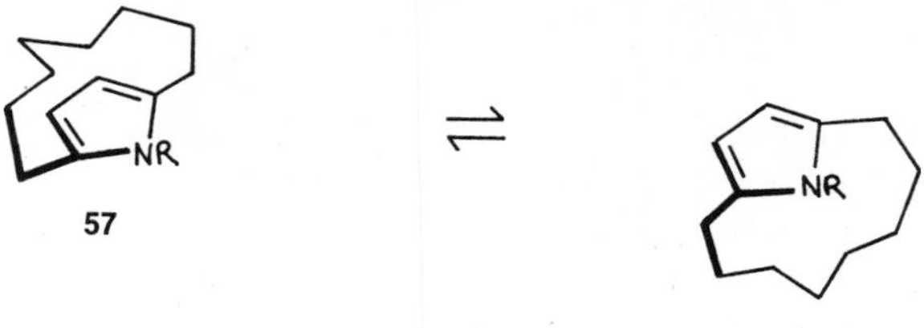

57   ⇌   57a

be seriously decreased by replacing the group R = Me by R = aryl. The pyrrolophanes are formed by a Paal-Knorr synthesis from cyclododecane-1,4-dione. The ultimate for restriction of movement is the novel type of *ansa* compound, *N,N'*-p-phenylene di[8](2,5)pyrrolophane (**58**) which is

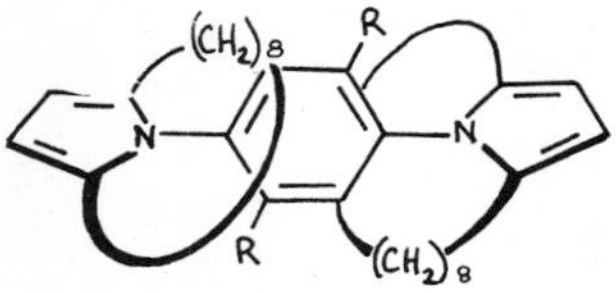

58 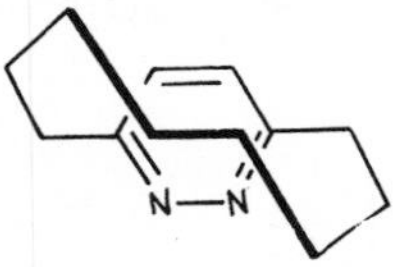   59 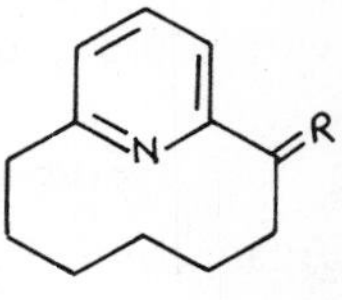   60

chiral[120] (see Part II, Chapter 2, Section XIV.C for definition of *ansa*). [8](3,6)-Pyridazinophane (**59**) (38.5 kJ mol^{-1}) was also obtained from cyclododecane-1,4-dione by Nozaki and co-workers.[121] The 1-methylene [7](2,6)pyridinophanes (**60,** R^2 = CH$_2$, 38.1 kJ mol^{-1}; **60,** R = CMe$_2$, 38.0 kJ mol^{-1}; **60,** R = CPh$_2$, 41.0 kJ mol^{-1}) were prepared from the pyridinophane (**60,** R = H$_2$) with *n*-BuLi/R$_2$C = O, or from the ketone (**60,** R = O) by means of a Wittig reaction. Their conformational mobility is similar to that of pyridazinophane (**59**).[122] Gerlach and Huber[123] synthesized the higher homologue [9](2,5)pyridinophane, among others, from the bis-chlorovinyl ketone (**61,** n = 8) and ammonia followed by reduction, and resolved it into its optical enantiomers by recrystallization of its (+)-2,2'-dihydroxy-1,1'-binaphthyl-3,3'-dicarboxylate salt. Parham and his school[124] synthesized benzochloro-[n](2,4)pyridinophanes (**62**) by the rearrangement

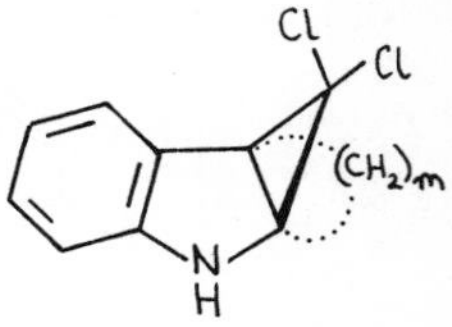

ClHC = CHCO-(CH$_2$)$_{m-1}$- COCH = CHCL

61

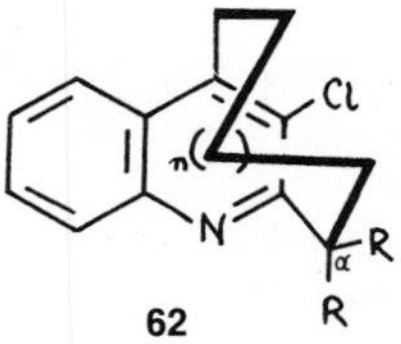

62

63

of the indole **63** obtained by addition of dichlorocarbene (from phenyl trichloromethyl mercury) to 2,3-polymethylene indoles. The limiting value of *m* in the indole **63** is 6; if *m* < 6 then no pyridinophanes are formed and the products are 4-chloro-2,3-polymethylene quinolines.[125] They prepared several pyridinophanes[124–126] related to **62** and their *N*-oxides. They found some stereoselectivity in the transformation of the *N*-oxides into α-substituted derivatives[127]—the position α is marked in formula **62,** and they studied the stereochemistry of substitution at this position.[128–131]

Sutherland and co-workers[132] compared the conformational equilibrium of [2,2](2,6)pyridinophane (**64** ⇌ **65,** 6.9 kJ mol⁻¹) with that of the benzene analogue (>113 kJ mol⁻¹) and found that ring inversion was faster in **64**

| **64** | **65** |

because the nonbonded interactions of the nitrogen lone pair of electrons were less than those of the hydrogen atoms in the benzene analogue. Vögtle and Effler[133] extended this study to several heterocyclic [2,2]metacyclophanes. Haley, Jr. and Keehn[133a] recently prepared pyrrolo[2.2](2,5)-pyrrolophane (the dipyrrole analogue of **64**) by a Paal-Knorr synthesis from cyclododecan-1,4,7,10-tetraone. Unlike the pyridine analogue, its nmr spectrum did not alter even on heating up to 190°C, indicating a ΔG‡ value for ring inversion greater than 113 kJ mol⁻¹. Several mixed [2,2](2,5)-pyrrolophanes were examined by dnmr and the equilibria in Scheme 2

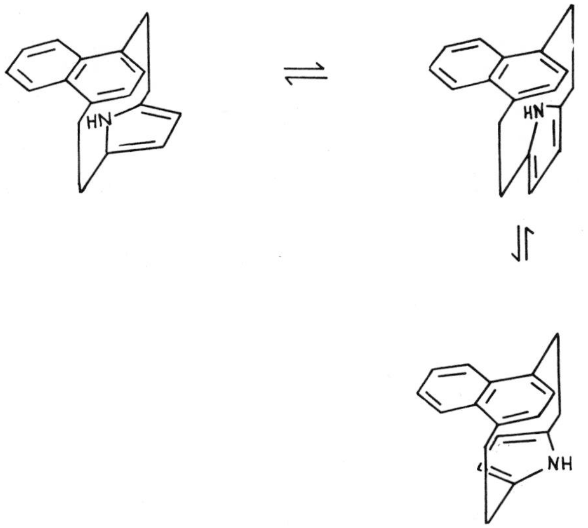

Scheme 2

were considered in order to explain the results obtained.[134] Vögtle and Neumann[135] prepared the "oligohetera" multibridged "phane," N,N',N''-tritosyl-2,11-20-triaza[3.3.3](1,3,5)cyclophane and the dnmr spectra were best interpreted in terms of the nitrogen and ring inversions shown in Scheme 3.

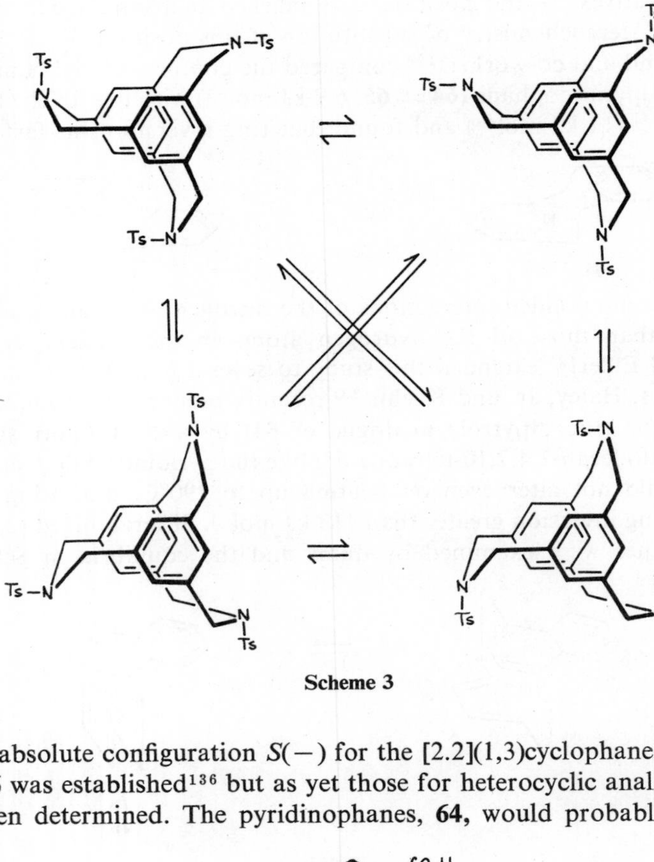

Scheme 3

The absolute configuration $S(-)$ for the [2.2](1,3)cyclophane carboxylic acid **66** was established[136] but as yet those for heterocyclic analogues have not been determined. The pyridinophanes, **64,** would probably racemize

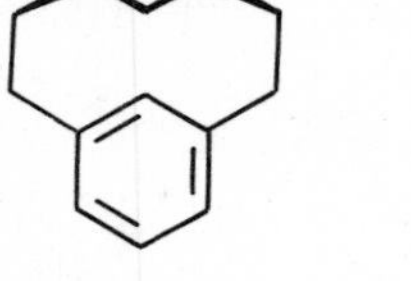

66

too rapidly for direct isolation of optically active forms. Oxygen and sulfur cyclophanes are discussed in Part II, Chapter 2, Section XIV.C, and Part II, Chapter 3, Section XII.D, respectively.

B. Large Rings

The arbitrary line between medium-sized and large rings is often difficult to define. In the present section rings with more than ten atoms will be considered as large although a few rings of about this size were discussed in the previous section. It might be more appropriate if the term "medium-large rings" were introduced but further classification is not warranted here.

Sicher and collaborators[137] methylated ten- and twelve-membered N-methylazacycloalkanes with *gem*-dimethyl groups at various positions around the rings. They studied the rates in methyl cyanide by polarography and found that the *gem*-methyl groups anchored the rings in certain preferred conformations and affected the relative rates accordingly. Similar biasing was also observed in *gem*-dimethyl 1-azacycloalkan-2-ones.[138] Feigenbaum and Lehn[139] found that 4,9-dihetero-*cis-cis*-deca-1,6-dienes were conformationally mobile and that the direction of the equilibria are as in Scheme 4 in which the boat form is disfavored.

$X = NH, O, S$

Scheme 4

Large cyclic amides have attracted much attention. The conformations of cyclic mono-[140] and dilactams[141] were studied and Dale[142] concluded that replacement of a CH_2 group in a macrocycle by O or N does not seriously alter the conformation, but insertion of multiple bonds drastically changes the geometry; for example, the cyclic lactone of ω-aminohexanoic acid has the structure **67** and not a completely zig-zag crown. Moriarty[143] measured the ^1H nmr spectra of several N-methyl cyclic lactams and observed that syn and anti isomerism (**68** $\rightleftharpoons$ **69**) about the amide bond oc-

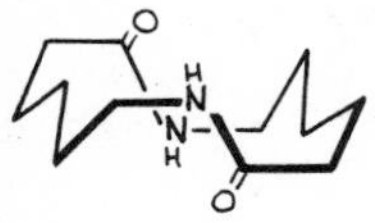

67

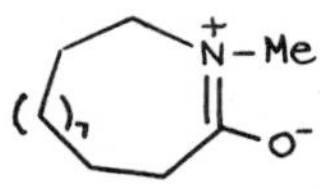

68

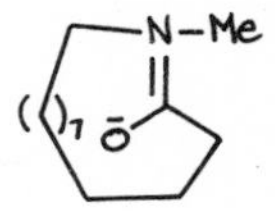

69

curred in rings larger than eleven-membered. The proportion of anti isomer increases with increase in the size of the ring.[144] In dilute CCl_4 the syn to anti ratios for the eleven-membered ($n = 5$) and thirteen-membered ($n = 7$) rings are approximately 55:45 and 40:60 respectively. Whereas in the cyclic lactams the anti form is favored at equilibrium, in the thio lactams the syn form of the thirteen-membered ring is thermodynamically more stable and crystallizes selectively in this form.[145] Several macrocyclic poly-peptide antibiotics are known (e.g., amidomycin[146] and polymyxin[147,148]) and a few iron-containing amide antibiotics (e.g., ferrichrysin and ferri-oxamine) are macrocycles and their structures have been elucidated.[149,150]

Rather ingenious syntheses of large rings have been devised such as the conversion of the twelve-membered N,N'-diphenyl-1,2-diazacycloalkane (70) to the twenty-membered ring 71 by a benzidine rearrangement,[151] and the intramolecular cyclization of m-12-methylaminododecylchloroben-

70

71

72

73

zene to a mixture of the seventeen- and sixteen-membered amines 72 and 73, which is mediated by phenyl lithium and involves a benzyne inter-mediate.[152] A third example worthy of mention is the preparation of the bicyclic amine 74 from intramolecular cyclization of N,N'-bis-10-bromo-decyl-p-phenylene diamine by Lüttringhaus and Simon.[153] Schill and co-

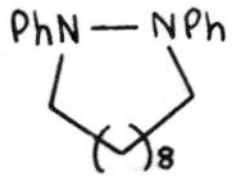

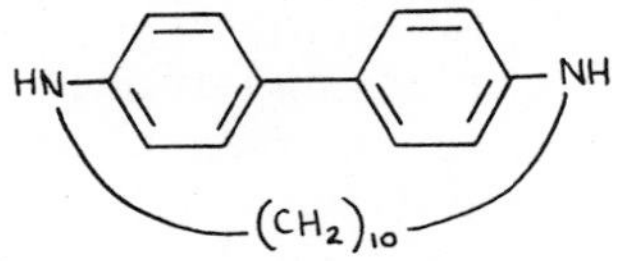

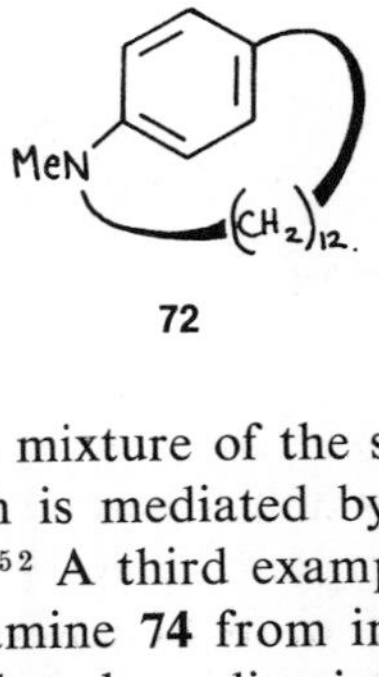

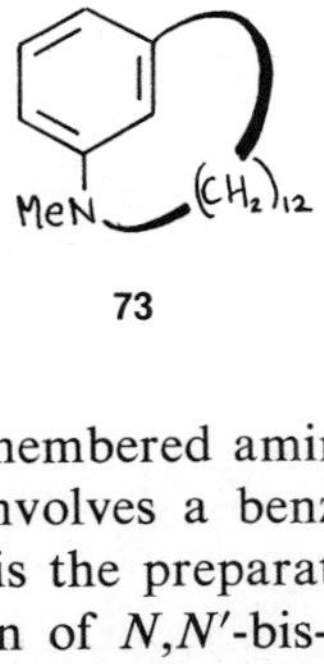

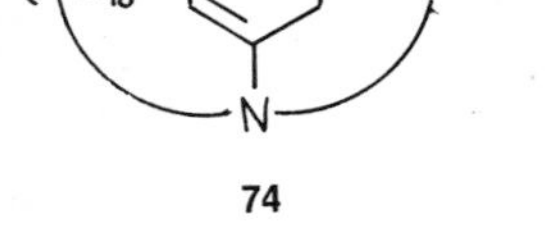

74

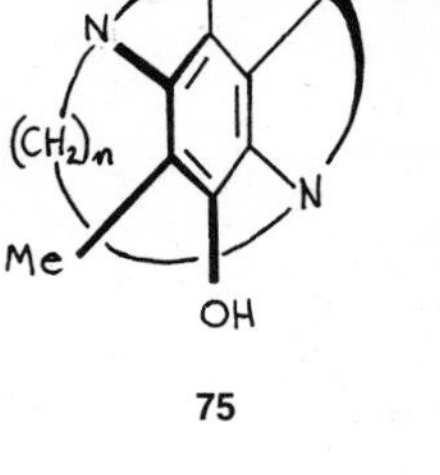

75

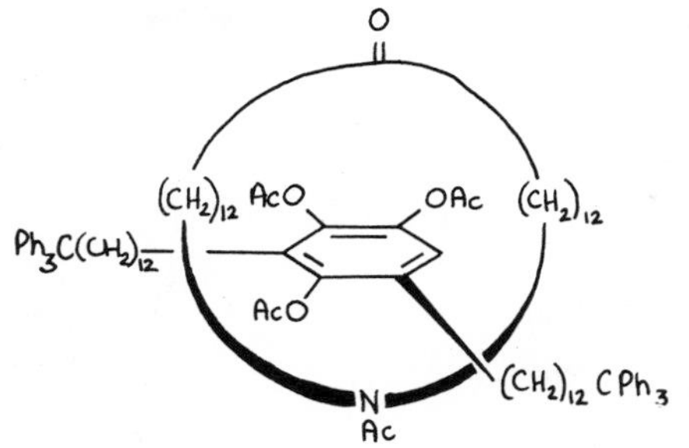

76

workers have extended this type of synthesis for the preparation of most intricate and interesting macrocycles. They prepared the rotaxanes, for example, **75,**[154,155] and the nonlinked rotaxane **76** in which the two parts of the molecule are held together by the triphenylmethyl groups.[156] The latter was prepared from the diansabromide (**77**). This approach was used

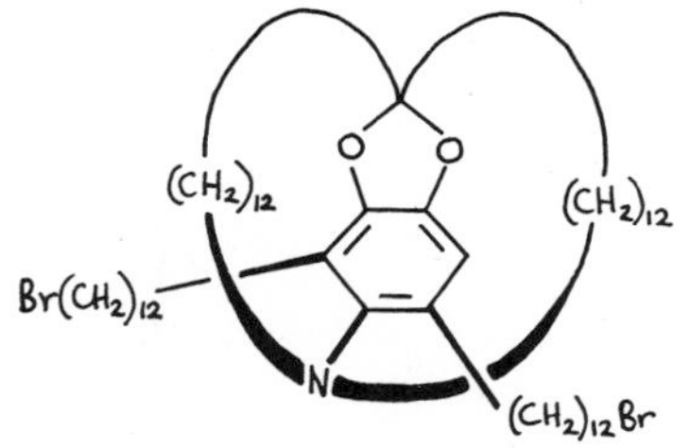

77

to prepare several catenanes,[157,158] and one in particular[159] is shown in Equation 4. The catenanes **78** (R = H) and **78** (R = Ac) were the first to be isolated in crystalline form. Schill and collaborators went even so far as to attempt to prepare a trefoil or knot such as **79.**[160] All the structures are amply supported by their physical data,[157] and in some cases cis and trans isomers were separated.[161] A monograph on catenanes, rotaxanes, and knots by Schill was published in 1971.[162] The "capped" porphyrin **80** was recently synthesized by Baldwin and co-workers.[163] The porphyrin gave an Fe^{II} complex which should stabilize a ferrous dioxygen complex by sterically inhibiting a bimolecular process.[164] This molecule is potentially capable of functioning as hemoglobin.

Nitrogen analogues of crown ethers, for example, **81,** and other examples in which the nitrogen atoms across the ring are linked by another carbon (or carbon and nitrogen) chain, are being actively studied[165] because of their usefulness as proton "cryptates." The proton finds a basic center in

$$(4)$$

78

79

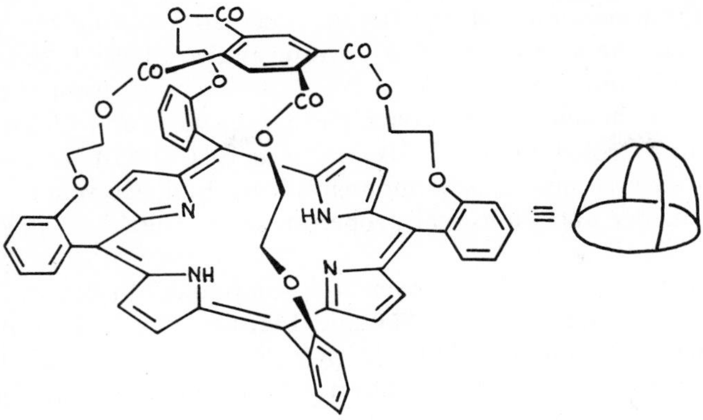

80

the cavity (crypt) of the molecule, and can be so effectively shielded that deprotonation is very slow even in a highly basic medium[166] (see Part II, Chapter 2, Section XIV.D). The ^{1}H nmr spectra of the condensation products of formaldehyde and ethylene diamine, o-phenylene diamine, and 3,4-diaminotoluene were interpreted by Volpp[167] in terms of the 1,4,6,9-tetraazatricyclo[4.4.1.^{1,6}1.^{4,9}]dodecane structure (**82**).

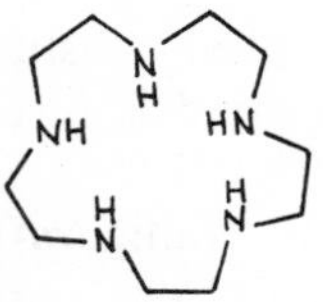

81

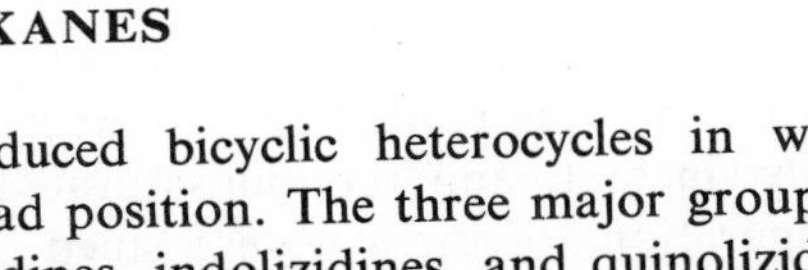

82

VI.   1-AZABICYCLO[x.y.0]ALKANES

1-Azabicyclo[x.y.0]alkanes are reduced bicyclic heterocycles in which the nitrogen atom is at a bridgehead position. The three major groups to be discussed here are the pyrrolizidines, indolizidines, and quinolizidines which are 5:5-, 5:6-, and 6:6-membered fused systems respectively. A small section on miscellaneous 1-azabicyclo[x.y.0]alkanes is also included. A comprehensive survey of heterocyclic systems with bridgehead nitrogen atoms has been published in two parts by Mosby.[168]

A.   1-Azabicyclo[3.3.0]octanes: Pyrrolizidines

Interest in the 1-azabicyclo[3.3.0]octanes came from the discovery of the pyrrolizidine alkaloids, and the ring system was unknown prior to this.

Much of the work done on this system is centered around the structures of the bicyclooctane, pyrrolizidine portion of the alkaloid. Indeed many of the pyrrolizidines are described by names derived from the original alkaloids. The chemistry of pyrrolizidines (Kochetkov and Linkhosherstov 1965)[169] and pyrrolizidine alkaloids (Leonard 1950, 1960)[170] was reviewed in detail, and a comprehensive monograph by Bull, Culvenor, and Dick on the chemistry and biological properties of pyrrolizidine alkaloids appeared in 1968.[171]

The higher stability of cis- over trans-bicyclo[3.3.0]octanes has been mentioned earlier (see formulas **150** and **151**, Chapter 2, Section III.E.3), and a similar situation prevails in the pyrrolizidines. These may be a little more mobile because the nitrogen atom is capable of inversion, but the cis juncture with respect to H–8 and the nitrogen lone pair of electrons is favored. (**83**). The parent substance is symmetrical, but the introduction of a substituent at the 1-, 2-, 3-, 5-, 6-, or 7-positions produces two chiral

83   **84**   **85**   **86**

centers and therefore four stereoisomers. Also, if a nonchiral substituent is placed at these positions (e.g., a =O) or if a double bond is inserted between two carbon atoms other than C–8, an asymmetric center at C–8 is created. Examples of these are known.[169–171] The absolute configurations of almost all the optically active pyrrolizidines obtained from natural sources have been determined by chemical correlation and/or X-ray methods.[171] Warren and von Klemperer determined the absolute configuration of heliotridane (**84**) as $1S,8S(-)$-1-methylpyrrolizidine by degradation to $S(+)$-methylheptane by three successive Hofmann reactions.[172] Because this was related to a number of other monohydroxy, dihydroxy, hydroxymethyl, and monounsaturated pyrrolizidines, then the absolute configurations of these were defined. Adams and Fleš[173] established the absolute configuration at C–1 of (−)-retronecanone (**85**), which had been previously prepared from (−)-3-methyl-5-aminovaleric acid,[174] as S by correlating the valeric acid with the known $S(-)$-methylsuccinic acid. Adams and Fleš[175] also determined the absolute configuration of C–8 in the pyrrolizidine moieties of the Senecio alkaloids by degradations in which the chiral C–8 atoms are undisturbed and converted to C–2 of S-proline. The relative configurations of substituents, and their relation to C–8, were determined chemically, for example, by formation of ethers such as **86** or cyclic sulfates.[176] Since the absolute configuration at C–8 is known then the absolute stereochemistry of the whole molecule is defined. Thus the configurations at C–8 of chloro isoheliotridine (**87**, $R^1 = Me, R^2 = H, R^3 = Cl$), retronecine (**87**, $R^1 = CH_2OH, R^2 = OH, R^3 = Cl$), retronecanol (**88**, $R^1 = Me, R^2 = R^4 = H, R^3 = OH$), and platynecine (**88**, $R^1 = CH_2OH, R^2 = R^4 = H, R^3 = $

87   88

OH) are *R*, and the configurations of heliotridane (**84**), isoretronecanol (**88**, $R^1 = CH_2OH, R^2 = R^3 = R^4 = H$), and supinidine (**87**, $R^1 = CH_2OH, R^2 = R^3 = H$) are *S*.[175] Optical resolutions of pyrrolizidines (e.g., of *d,l*-pseudo-heliotridane (the C–1 epimer of **84** with H–1 and H–8 trans)[177] and retronecine were achieved by recrystallization of diastereoisomeric salts of optically active acids,[178] for example, tartrates[177] and camphorates.[178] Most of the macrocyclic pyrrolizidine alkaloids have C–1 and C–7 joined by a chain of atoms of hydroxydicarboxylic acids (e.g., necic acid). The absolute configurations of Jacobine bromohydrin,[179] otonecine, and retusamine[180] were determined by X-ray analyses. The cd spectra of seventeen pyrrolizidine alkaloids were measured and the curves were discussed in terms of the pyrrolizidine nucleus, the chirality of the acid, and the macrocycle as a whole.[181]

The conformation of 1,2-dehydropyrrolizidines is slightly puckered with exo and endo conformations with structures **89** and **90** respectively, which

89   90

are slightly exaggerated in the formulas. Evidence for these came from X-ray data; for example, Jacobine bromohydrin is exo-buckled[179] and retusamine is endo-buckled to the extent of 30 and 40° respectively.[180] ¹H nmr data also provided evidence for the presence of these conformers, and generally retronecine derivatives (e.g., **87**, $R^1 = CH_2OH, R^2 = OH, R^3 = H$) are exo-buckled, and the epimeric heliotridine derivatives (e.g., **87**, $R^1 = CH_2OH, R^2 = H, R^3 = OH$) are an equilibrium mixture of both conformers. The reason why retronecine derivatives do not contain appreciable amounts of the endo-buckled conformation is because of the nonbonded interactions between the 7–OH and the substituent on C–1.[182] These conformational changes are, however, subtle and come into play when the substitution is unusual and when unusual internal strain prevails. Skvortsov and Elvidge[183] analyzed the temperature dependent ^{1}H nmr spectra of pyrrolizidine and its 3-methyl and methiodide derivatives. Pyrrolizidine, over the temperature range -70 to $190°C$, is mainly, if not entirely, in the cis-fused forms **91** and **92**, with pseudorotation occurring in each ring. The 3-*exo*-methyl derivative shows very little change in spectrum whereas the 3-*endo*-methyl isomer exists in the cis-fused form only below $-60°C$. Above this temperature an equilibrium mixture of cis and trans forms **(93)** is evident by the time-averaged signals. Pyrrolizidine cation and methiodide are in the cis configuration. 1- and 2-Hydroxypyrrolizidines are also cis-fused with weak intramolecular hydrogen bonding between the hydroxy groups and the bridgehead nitrogen atoms.[183a]

91   **92**   **93**

Several syntheses of pyrrolizidines (see the synthesis of **48** in Section V.A.1 of this chapter) were devised, but the more recent two step synthesis of 1-ethoxycarbonylpyrrolizidine, achieved by Pizzorno and Albonico,[184] from *N*-formyl *S*-proline is stereospecific yielding 80% of the thermodynamically more stable racemate (Equation 5). High stereospecificities are

$$
\begin{array}{ccc}
\text{(H, CO}_2\text{H, NCHO structure)} & \xrightarrow[\equiv \cdot CO_2Et]{Ac_2O} & \text{(CO}_2Et\text{ structure)} & \xrightarrow{10\% Pd-C|H} & \text{(H, CO}_2Et\text{ structure)}
\end{array} \tag{5}
$$

observed in the hydrogenation of retronecine (**87**, $R^1 = CH_2OH$, $R^2 = OH$, $R^3 = H$) to retronecanol (**88**, $R^1 = Me$, $R^2 = R^4 = H$, $R^3 = OH$) with $Pt–H_2O$ and to platynecine (**88**, $R^1 = CH_2OH$, $R^2 = R^4 = H$, $R^3 = OH$) with Raney nickel.[185] Similarly the 1-methylene compound **88** (R^1, $R^2 = CH_2$, $R^3 = OH$, $R^4 = H$) provides only retronecanol, and the epimeric 1-methylene compound **88** (R^1, $R^2 = CH_2$, $R^3 = H$, $R^4 = OH$) furnishes only 7β-hydroxy-1α-methyl-8β-pyrrolizidine (**88**, $R^1 = Me$, $R^2 = R^3 = H$, $R^4 = OH$).[186] However, this is not always the case because hydrogenation of 1-methylenepyrrolizidine (**88**, $R^1R^2 = CH_2$, $R^3 = R^4 = H$) with Raney nickel gives a mixture of heliotridane (**84**) and its epimer pseudoheliotridane.[187] Pyrrolizidine *N*-oxides are obtained by perbenzoic acid oxidation, and the configuration of the oxygen atom is cis with respect to H–8, for example, retronecine *N*-oxide (*N*-oxide of **87**, $R^1 = CH_2OH$, $R^2 = OH$, $R^3 = H$). 2,3-Epoxypyrrolizidines are generally reductively cleaved to the corresponding 3-hydroxy derivatives stereospecifically.[187a]

Intramolecular cyclization at two positions in a benzene ring occurred when *N*-chloroacetyl 3,4-dimethoxyphenethylamine was irradiated in ethanol or tetrahydrofuran. The structure of one arbitrarily chosen antipode, the racemic product, 1,2,5*a*,7*bβ*-tetrahydro-5*aβ*,5*bα*-dimethoxy-5*bH*-cyclobuta[1,4]cyclobuta[1,2,3-*g,h*]pyrrolizin-4(5*H*)-one (**94**), was established by X-ray analysis of a single optically active crystal picked out from the racemic conglomerate.[188]

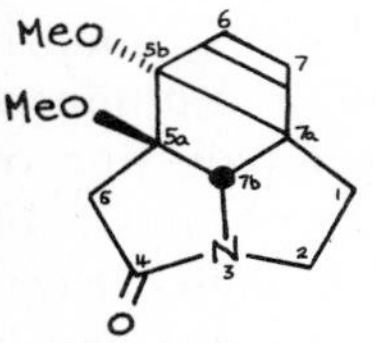

94 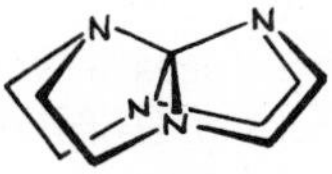   95 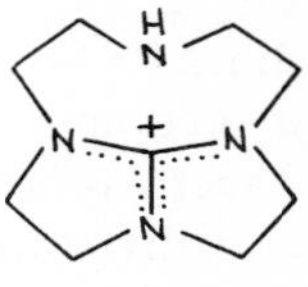   96

The highly symmetrical 1,4,7,10-tetraazatetracyclo[5.5.1.10.^{4,13}0^{10,13}]tridecane (**95**) is best discussed in this section. Its nmr spectrum was analyzed and the compound was shown to undergo degenerate prototropic and conformational equilibria. Ring-chain tautomerism, in the monocation, is possible between four identical tricyclic guanidinium cations (**96**) and four identical tetracyclic cations, that is, protonated (**95**), in minor amounts, and is supported by the physical data.[189]

B.   1-Azabicyclo[4.3.0]nonanes: Indolizidines

The indolizidine, 1-azabicyclo[4.3.0]nonane, nucleus (**97**) is present in a variety of alkaloids such as Strychnos,[190,191] Erythrina,[192] Galbulimima,[193] and Eleocarpus species[194] to name a few. Alkaline degradation of strychnine produced a base with the composition $C_{10}H_{11}N$ which was thought to be a reduced alkylindolizidine. Several alkyl indolizidines were prepared, in vain, in order to identify its reduction product.[195]

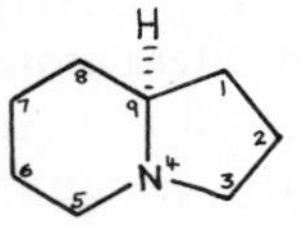

97 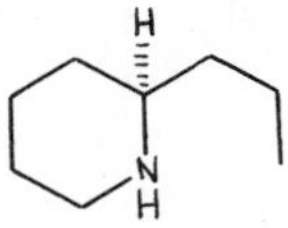   98 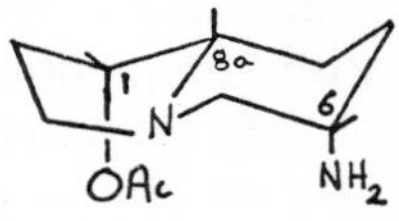   99

Unlike pyrrolizidine, the bridgehead carbon atom of indolizidine is asymmetric. Leonard and Middleton[196] resolved d,l-indolizidine into its enantiomers by recrystallization of its (+)-dibenzoyltartrate salt. The specific rotation of the laevo antipode was similar to that of δ-coniceine which Lellmann[197] had prepared many years previously from (+)-coniine (98) by N-bromination and cyclization in sulfuric acid. (+)-Coniine was correlated with R(+)-pipecolic acid of known absolute configuration, and because Lellmann's synthesis is known to proceed without upsetting the asymmetric center then the absolute configuration of (−)-indolizidine is R as shown in formula 97. The presence of substituents on the carbon atoms, other than C–9, introduce a second asymmetric center, and the substituent is related, cis or trans, to the bridgehead hydrogen atom at C–9. A mixture of two racemates is possible and these have been separated on occasions, for example, in 1-methyl-,[198] 3-methyl-,[198–200] and 2-hydroxyindolizidines.[201] Sometimes the method of synthesis is more specific and produces a predominance of one racemate.[203,203] The Wolff-Kishner reduction of 3-methyl-2-oxoindolizidine produces a 9:1 ratio of isomers A and B of 3-methylindolizidine;[199] whereas a Hofmann-Freytag-Löffler photocyclization of 1-chloro-2-propylpiperidine produces a 2:76 ratio of the A and B isomers respectively.[200] These racemates were separated by chromatography over alumina.[199] The absolute configuration of slaframine was revised to 1S,6S,8aS-1-acetoxy-6-aminoindolizidine (99)[204] and was confirmed by synthesis.[205] The geometrical isomers of 3-butyl-5-methylindolizidine were synthesized for the identification of insect trail pheromones.[205a]

The ability of the nitrogen atom to undergo atomic inversion in indolizidine introduces cis-trans isomerism in this system as in the 1-azabicyclo-[3.3.0]octanes (pyrrolizidines Section VI.A). In this case, however, the preference for the cis-fused form is less pronounced. Clemo and Ramage[206] prepared indolizidine by Clemmensen reduction of the 1-oxoindolizidine and Bouveault reduction of 3-oxoindolizidine. Two different picrates were isolated from the former but only one from the latter reduction product and they concluded that the different picrates were due to the cis and trans isomers. One of these is indeed indolizidine picrate, but it is doubtful whether the second picrate is a diastereomer, particularly as it is known that α-amino ketones can rearrange during a Clemmensen reduction.[207] (However, see the methylation below.) Direct methylation of indolizidine provides a 1:1 mixture of cis- and trans-fused N-methyl salts. The cis isomer can exist in two equilibrating conformers, 100 and 101, probably with 101 predominating because of decreased 1,3 nonbonded interactions of the methyl group. The trans isomer has the rigid structure 102.[208] An explanation for the higher field ^{1}H nmr resonance of the methyl group in the trans isomer with respect to the cis isomer was proposed.[183] The prepa-

100   **101**   **102**

ration of 4-methylindolizinium bromide by intramolecular cyclization of
N-methyl-2-(3-bromopropyl)piperidine or *N*-methyl-2-(4-bromobutyl)pyr-
rolidine yields exclusively the *cis*-methobromide.[208]

The cis and trans isomers of indolizidine are in equilibrium in favor of
the trans isomer with a $\triangle G°$ value of -10.0 kJ mol^{-1}.[209] This value is
closely similar to the one for quinolizidine ($\triangle G° = -10.04$ kJ mol^{-1}). In
contrast the free energy difference between *cis*- and *trans*-decalins (~ 11.3
kJ mol^{-1})[210] is higher than that of *cis*- and *trans*-hydrindanes (~ 1.3 kJ
mol^{-1})[211,212] with both being in favor of the trans isomer. It was suggested
from ^{1}H nmr, J_{gem} values evidence, that differences in the $\triangle G°$ values in
indolizidine and hydrindane were due to relief of strain in the former by
flattening of the bonds around the bridgehead nitrogen atom.[213] The
220-MHz ^{1}H nmr spectrum of 2,2-dideuterioindolizidine was measured by
Crabb and co-workers,[214] and the J_{gem} values for the C–5 protons (-11.0
Hz) were very close to those of the C–4 protons in quinolizidine, demonstrating that any strain in trans-fused indolizidine is not detectable by this
method. The presence of substituents can alter this balance of equilibrium
and the two most diagnostic means for establishing the configuration are
^{1}H nmr and ir spectroscopy. The magnitude of the coupling constants and
chemical shifts can assist in deciding between the three conformers such
as **100–102,** and the absence of Bohlmann bands (see Section VI.C) in
the ir is strong evidence for cis fusion.[215,216] The relative stereochemistry of a
hydroxy group and the ring junction in hydroxyindolizidines can be assigned by ir in solution and by examination for intramolecular hydrogen
bonding.[217,218] In these and other[216] substituted indolizidines the trans
form is preferred unless the nonbonded interactions of the substituents
are much larger in the trans-fused than in the cis-fused compounds. Crabb
and co-workers extended their studies in this system to diazaindolizidines,
that is, 1,8-diazabicyclo[4.3.0]nonanes,[219] and perhydropyrido[1,2-c]imid-
azoles and again found that unless steric interactions are severe the trans
stereochemistry is preferred at the ring junction.[220–222] Also in these systems 1,3- lone pair repulsion ("rabbit-ear" effect, see Chapter 3, Section
IV.C) can upset the trans-fused configuration. Thus 1,4-diazabicyclo[4.3.0]-
nonane prefers the trans configuration **103**; whereas in 1,8-diazabicyclo-
[4.3.0]nonane, the equilibrium between **104** and **105** is displaced in favor

103   104   105

of the cis form **105**.[223,224] The literature for 1970 and 1971 of indolizidines and related compounds[225] and the stereochemistry of bridgehead nitrogen compounds was reviewed by Crabb and co-workers.[226]

Armarego[227,228] examined the pKa values, uv and ¹H nmr spectra of indolizine (**106**, 1-azabicyclo[4.3.0]nona-2,5,7.9(1)-tetraene in present nomenclature) and several of its methyl derivatives. He found that the cations are protonated on C–1 (e.g., **107**) and/or C–3 (e.g., **108**). The cation of indolizine is protonated on C–3 (**108**, $R^1 = R^2 = H$), but by placing a methyl group at C–3 a 4:1 mixture of cations **107** ($R^1 = Me$, $R^2 = H$) and **108** ($R^1 = Me$, $R^2 = H$) is obtained. However, when a second methyl group is introduced at C–5 only the cation **108** ($R^1 = R^2 = Me$) is produced. This

106   107   108

is another demonstration of the fact that methyl groups in the peri positions, even in a 5:6- ring system, are crowding each other, and in this case the crowding is relieved by preferential protonation at C–3 which makes this an sp^3 hybridized atom (See Chapter 3, Sections XII.E and XIV.B).

C.   1-Azabicyclo[4.4.0]decanes: Quinolizidines

The quinolizidine ring system is present in a variety of alkaloids: lupinine,[229] yohimbine,[230–232] reserpine,[232] eburnamine, and vincadine,[233] and much of the chemistry and stereochemistry of this nucleus resulted from interest in the natural products. The references cited should be consulted for the detailed chemistry; only outlines of the stereochemical properties are discussed in this section.

Like pyrrolizidine and indolizidine, the cis and trans configurations of quinolizidine are interconvertible by nitrogen inversion. The cis structure is conformationally mobile (**109** and **110**) but is in equilibrium with the

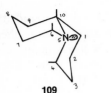

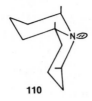

109   **110**   **111**

more rigid trans structure **111**. The equilibrium is strongly in favor of the trans isomer, and several estimates of the free energy difference have been made. By attributing the difference to 1,3-diaxial hydrogen interactions, Aaron,[234] from ir data, arrived at a $\triangle G$ value of 19.2 kJ mol^{-1}. This value was later[209] revised to 10.9 kJ mol^{-1} because he had not taken the energetically comparable 1,2-skew interactions into consideration. The latter value is closer to the one for cis-trans-decalin (11.3[210] and ~12.3[211] kJ mol^{-1}). Katritzky, Schofield, and collaborators[235] deduced values of 18.4 (from kinetics of quaternization) and 12.6 kJ mol^{-1} (from pKa measurements). For comparison with quinolizidine, they added 1.8 kJ mol^{-1} to the values for decalin in order to correct for the symmetry number, 2, of cis-decalin, and they obtained the values $\triangle G \sim$ 12.6–14.2 kJ mol^{-1} for the equilibrium involving one cis-decalin conformer. The similarity of these with the value obtained for the quinolizidine cation (i.e., from pKa data) is gratifying (see reference 212 of this chapter for further details).

   The configuration of quinolizidines at the bridgehead atoms, that is, cis or trans, has been assigned with great success from ir and nmr spectral data. In the ir spectra the presence of bands at 2700–2800 cm^{-1} is characteristic of trans-fused quinolizidines, and their absence or weak intensity is indicative of the cis configuration. These have been named "Bohlmann bands" and were first recognized by Wenkert and Roychauduri[236] in their studies of yohimbine and related alkaloids. Bohlmann[237–239] found that these bands were always present and intense in trans-quinolizidines but were absent or weak in the cis isomers, and he termed them "trans bands." By using simple quinolizidines and their deuterated derivatives he was able to assign these bands to C–H axial groups trans to a nitrogen lone pair of electrons.[240] Wiewiorowski and co-workers[241–243] made a quantitative study of these bands, which they called "T bands," using the lupin alkaloids[241] and deuterated sparteine systems,[242] and strengthened the previous assignments. They also defined the conditions for the appearance of these bands. One antiperiplanar axial C–H and nitrogen lone pair is sufficient to produce a band, but the complexity and intensity of these bands is roughly proportional to the number of axial C–H α to a nitrogen atom.

   Hamlow, Okuda, and Nakagawa[244] observed that the axial C–4 and C–6 protons in quinolizidine (but not its cation) absorbed at 0.93 ppm higher field than the corresponding equatorial protons. With this in mind, they rationalized the phenomenon of the Bohlmann bands and the large chemical shift difference in the nmr, by postulating a larger p-character to the atomic orbital of C-4 associated with the axial hydrogen than with the equatorial hydrogen atom. Also the bond energy of the axial C-H is lowered, with respect to the equatorial C-H, because of the interaction of the nitrogen lone pair with the antibonding orbital between C-4 and axial H-4. This will create a larger double bond character between N and C-4, make the axial H more electron dense (than equatorial H), and make the force constant of the C-H$_{ax}$ stretching frequency less than that of C-H$_{eq}$. The Bohlmann bands which appear in the same region of the ir as the N-CH$_3$ bands, which are due to Fermi resonance, are probably of the same origin. cis-Quinolizidines, **109** or **110**, have one αC-H$_{ax}$ whereas the trans isomer **111** has three such groups; hence the weaker and fewer bands in the cis compound. These bands have been used extensively for assigning configurations not only in quinolizidines[226,254,255] (see most of the references in this section) and condensed quinolizidines[226,245-250] but also in the indolizidines and related compounds (Section VI.B), and piperidines and related compounds (Chapter 3, Section II.B and C). In the latter case these bands were used in reverse, that is, to assign the orientation of the nitrogen lone pair with respect to the α axial C-H. A cautionary note must be made at this point because the absence of bands at 2700-2800 cm^{-1} is not necessarily evidence for the absence of a trans structure. Slight distortions in the molecule may cause changes in band frequencies. The bands for some reason may be at slightly higher frequencies (e.g., $trans$-4,10H-4-methylquinolizidine),[235,251] and data from systems other than quinolizidine should be treated with some reserve.[252]

^{1}H nmr spectroscopy was used extensively, in conjunction with ir spectroscopy, for assigning the configuration of quinolizidines and was used in almost all the references cited in this section. The chemical shift differences between the C-H$_{ax}$ and C-H$_{eq}$ α to the nitrogen atom were used to distinguish between cis and trans structures. Robinson[253] pointed out that the explanation of Hamlow and co-workers[244] (see above) is not entirely true because he showed that an adjacent methylene or methyl group normally deshields the vicinal axial hydrogen atom. Relative configurations of substituents were deduced in the usual ways from chemical shifts and coupling constants,[251,254] and in the case of hydroxy substituents the presence of intramolecular hydrogen bonding (e.g., **112**[255] and **113**[256] has assisted in the assignments. All the monohydroxy-,[257] 4-hydroxy-4-methyl-,[258] 2-hydroxy-10-methyl-,[259] all the 1-hydroxy-7-methyl-,[260] and 4-aryl-4-hydroxyquinolizidines[261,262] are predominantly (>95%) in the trans-fused configuration. 1-Hydroxy-1-phenylquinolizidine, on the other

112

113

114

115

116

117

hand, is a 40:60 mixture of the cis (114) and trans (115) isomers respectively.[209] Similarly lupinine (116) and epilupinine (117),[263–266] their 3-oxo derivatives[267] and the isomeric 3-hydroxymethylquinolizidines,[268,269] and 1-, 2-, 3-, and 4-methylquinolizidines[235,251,270] are all trans-fused. When 1,3-nonbonded interactions are large in the trans-fused structure then the equilibrium lies toward the cis isomer. Thus, whereas cis-4,10H-4-phenylquinolizidine is trans-fused, the trans-4,10H-4-phenyl isomer is cis-fused.[271] [13]C nmr spectroscopy was used to assign the stereochemistry of quinolizidine and its 8-methyl derivative, and the spectra were compared with those of the decalins.[272]

Quinolizidine is nonchiral but in its 1-, 2-, 3-, and 4-substituted quinolizidines, as in the methyl derivatives, a mixture of two racemates is possible in each case.[273,274] This is because substitution at C-1, C-2, C-3, and C-4 makes C-10 asymmetric. The absolute configuration of (−)-lupinine (116) was shown to be 1R,10R by a Hofmann degradation and conversion into R(−)-4-methylnonane,[275] in accord with earlier results.[276] Since the molecule is trans-fused (see above) then the nitrogen atom is chiral with the absolute configuration 5R. (−)-Lupinine was used to deduce the absolute configuration of several lupin alkaloids.[277] Mason, Schofield, and Wells[278] prepared 1-, 2-, and 3-oxoquinolizidines, resolved them, and examined their optical antipodes by ir, uv, and cd spectroscopy. The data showed that they existed in the trans-fused configuration (>90%) and that the absolute configurations were 10S(−)-1-oxo-, 10S(+)-2-oxo-, and 10R(−)-3-oxoquinolizidines. Yamada and co-workers[279,280] synthesized optically active 1-oxopyrrolidine, 1- and 8-oxoindolizidine, 1- and 2-oxoquinolizidine, and condensed systems and measured their ord spectra. They concluded that the octant rule was obeyed when the nitrogen atom was on the axis of the octant projection but was disobeyed when it was in the octant space. Their configuration for 1-oxoquinolizidine, however, was opposite to that reported by Mason and co-workers.[278] The absolute configuration of 1,2,3,6,7,11*b*-hexahydro-4*H*-pyrazino[2,1-*a*]isoquinoline (**118**) was derived from a comparison of its ord curves with those of *S*(−)-benzo[*a*]quinolizidine (**119**).[281]

Chemical elaboration of substituents on the quinolizidine nucleus has on many occasions been partially or completely stereospecific and use has been made of this in order to synthesize some alkaloids.[282–284] For other examples reference should be made to the reviews cited earlier.[229–233] One simple example of an alkaloid will suffice here—the catalytic reduction followed by reduction with LAH of 1-methoxycarbonyl-2-hydroxy-4-oxo-5*H*-6,7,8,9-tetrahydroquinolizine (**120**) yields (+)-lupinine (**116**) in 86% yield.[285] Other reactions of quinolizidines (non alkaloids) will be mentioned to demonstrate that stereospecificity can be achieved in this nucleus. The catalytic reduction of 1-oxoquinolizidine with PtO₂–AcOH furnished only the *cis*-1,10*H* alcohol **121**,[286,287] whereas reduction with LAH or sodium amalgam produced only the *trans*-1,10*H*-alcohol **122**.[287,288] Reduction of the 1,2,3,4,5,6,7,8,9-octahydro-4-methylquinolizinium ion yields exclusively *cis*-4,10*H*-4-methylquinolizidine (**123**), but reduction with NaBH₄ is less selective and gives a 25:75 ratio of cis and trans isomers.[270] Also,

catalytic reduction of 2,3,4,5,6,7,8,9-octahydro-5-methylquinolizinium salts (**124**) produces the cis-fused 5-methylquinolizidinium salts stereospecifically.[270] The specificity, however, can vary if the quinolizidine ring is fused to a benzene ring, and depends on the position of the double bond. that is, on whether the double bond is at C–10, N–5 or C–1, C–10.[289]

The quaternization of quinolizidines[235,251,258,264,270,290-292] and condensed quinolizidines[235,289,293,294] has attracted much attention, and quantitative studies were made by Katritzky, Schofield, and their co-workers.[235,295] Methylation of quinolizidine yields only one methiodide[291] which was shown to be the trans-fused isomer **125**.[251,270,290,292] *cis*-Quinolizidinium

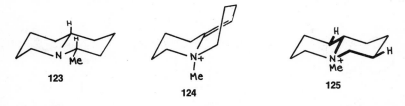

123   **124**   **125**

methiodide (**126**) was obtained exclusively by intramolecular cyclization of 2-ω-ethoxybutyl-1-methylpiperidine with hydriodic acid or by catalytic reduction of the quinolizidinium salt (**124**).[251,270,290,292] The methylation of 1-, 2-, 3-, and 4-methylquinolizidines was examined in great detail and the stereochemistry of the methiodides was consistent with what would be expected from the consideration of nonbonded interactions.[235,251,292] All the trans-fused bases with trans Me,10*H* relationships (i.e., Me equatorial) gave methiodides with retention of configuration. The bases with Me axial groups on the same side of the N lone pair, on the other hand, gave the cis-fused Me equatorial methiodides. The 2-methyl derivative with a cis Me,10*H* relationship and the 4-methyl derivative with a cis Me,10*H* relationship, both of which are in the trans-fused configuration, gave mixtures of cis- and trans-fused methiodides. Epilupinine (**117**) yields a methiodide with retention of configuration, whereas lupinine (**116**) forms the cis-fused methiodide.[264] The epimeric 1-hydroxy-1-phenyl- and 1-hydroxy-1-methyl-quinolizidines also gave methiodides which are consistent with the nonbonded interactions in the molecules.[258] Intramolecular quaternization occurs when the *O*-tosyl derivative of lupinine (**116**) is basified, and the azetidinium salt (**127**) is produced.[296] Methylation of quinolizidine, its 10-cyano, 10-methyl, 10-hydroxymethyl, and 10-nitromethyl derivatives with MeI gave trans to cis ratios of *N*-methylquinolizidinium iodides as

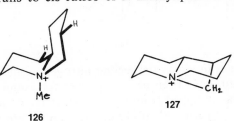

126   **127**

follows: 1:0, 17:1, 1:1, 1:5, and 0:1. Thus the proportion of cis-fused quaternary iodides was according to the size of the 10-substituent, that is, $CH_2NO_2 > CH_2OH > Me > CN > H.$[296a] Further studies revealed that the trans to cis ratios of N-methyl iodides from the quaternization, with MeI, of 10-ethyl-, 10-vinyl-, and 10-ethynylquinolizidines were 2:9, 4:9, and 15:2 respectively.[296b]

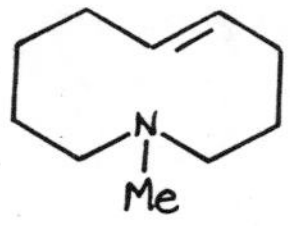

128 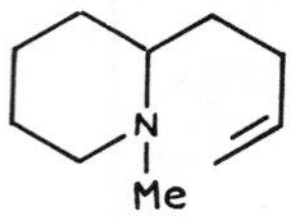   129    130

Hofmann degradation of *cis*- and *trans*-quinolizidinium methiodides gave 43:52 and 53:42 ratios of 1-methylcyclodec-5-ene (**128**) and 1-methyl-2-(but-3-en-1-yl)piperidine (**129**), respectively, together with 5% of quinolizidine.[270,290] The formation of **128** and **129** from the *cis*- and the *trans*-methiodides is clearly consistent with the known mechanism of the reaction (see Chapter 2, Section III.E.4 and Chapter 3, Sections VIII.B, IX.B.4.c, and X.B.3).

Grob fragmentation occurs when 1-acetoxyiminoquinolizidine (**130**) is treated with triethylamine in ethanol and N-(3-cyanopropyl)-3,4,5,6-tetrahydropyridinium cation (**131**) is produced. There is marked steric control in this fragmentation because it is 10^3 times faster than a similar fragmentation of 1-acetoxyiminoindolizidine (**132**). It must therefore proceed through the cis-fused isomer **130** because this conformation possesses the optimum stereoelectronic requirements for this fragmentation (see Chapter 3, Sections II.D.6, IX.B.4.c, and X.B.3). N-(2-Cyanoethyl)-3,4,5,6-tetrahydropiperidinium cation, which is formed from the indolizidine **132**, cyclizes spontaneously to 2-oxo-1-azaquinolizidine (**133**).[297]

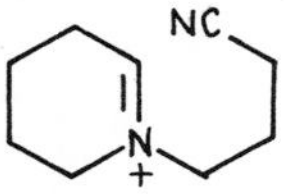

131 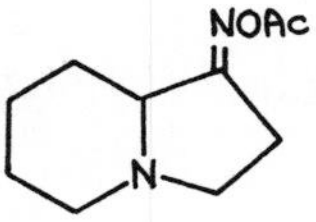   132 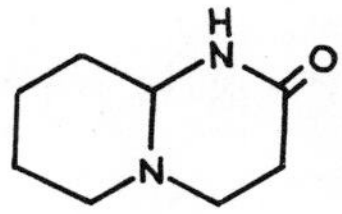   133

Azaquinolizidines in which the second nitrogen atom is α to the bridgehead nitrogen atom have been prepared,[298,299] for example, the polycyclic compound **134** was obtained from a Diels-Alder reaction between pyridazine and maleic anhydride.[299] 10-Azaquinolizidines are conformationally mobile,[300,301] and photoelectron spectroscopy revealed that the dihedral angle between the nitrogen lone pairs is about 180° (135).[302] Several other aza-[303-308] and condensed azaquinolizidines[309-314] were synthesized, and their configurations and conformations were deduced by ^{1}H nmr and ir spectroscopy. Their stereochemistry is not discussed in this monograph because Crabb and co-workers have reviewed this field recently.[225,226]

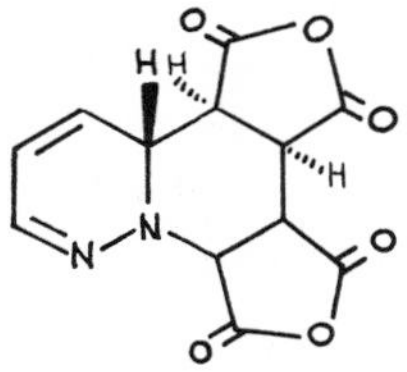

134

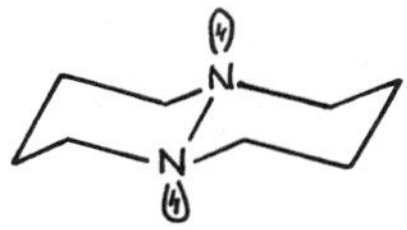

135

D.   Miscellaneous 1-Azabicyclo[x.y.0]alkanes

1-Azabicyclo[x.y.0]alkanes, other than those discussed in Sections A, B, and C, have not attracted much attention and only a few examples of stereochemical interest have been reported.

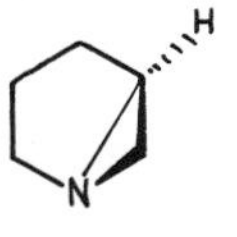

136

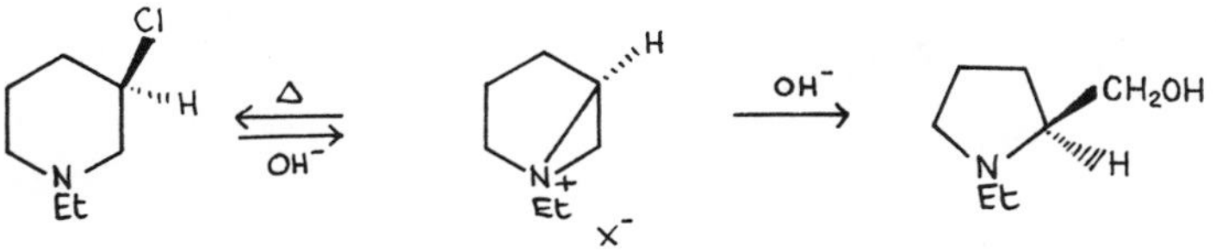

S(+)-1-Azabicyclo[3.1.0]hexane (136) was prepared by Buyle[315] by intramolecular cyclization of S-2-chloromethylpyrrolidine obtained from S-glutamic acid by way of S-prolinol. The free base polymerizes slowly on standing and rapidly in the presence of $BF_3 \cdot Et_2O$. The polymer has a strong levo rotation. The N-ethyl derivative was similarly prepared and can be obtained by direct alkylation of 136 with ethyl perchlorate.[316,317] The bicyclic ring is opened to yield a chiral 3-chloro-1-ethylpiperidine on heating, but it can be converted into N-ethylprolinol by a different bond cleavage (Equation 6). 1-Azabicyclo[3.1.0]hexanes are also formed by intramolecular thermal cyclization, with nitrogen elimination, of 5-azidopent-1-enes. 5-Azido-2-methylhex-1-ene gives a mixture of two isomeric 3,5-dimethylbicyclo[3.1.0]hexanes, which are separated by glc, together with 2,6-dimethyl-2,3,4,5-tetrahydropyridine and isopropyliminocyclobutane in the ratio of 4:5:1 (Equation 7).[318]

$$(7)$$

Interest in the 1-azabicyclo[3.2.0]heptanes arose because the nucleus is essentially that of the dethiapenicillins. The most effective synthesis adopted by Lowe and Ridley[319,320] involves a photochemical Wolff rearrangement of 2-diazo-1,3-dioxopyrrolizidines. Photolysis is carried out in the presence of 1-methyl-1-phenylethoxycarbonylhydrazine, and a separable mixture of cis- and trans-7-oxo-1-azabicyclo[3.2.0]heptanes is formed (Equation 8). The three-dimensional structure of this ring system was confirmed by an X-ray study of 1-methyl-1-azoniabicyclo[3.2.0]heptane chloride.[321]

$$(8)$$

R = CO₂CMe₂Ph

The 1-azabicyclo[4.2.0]octane ring system, called conidine by Löffler, was found in the alkaloid coniceine.[322] Toy and Price[323] synthesized (+)- and (−)-conidine by intramolecular cyclization of optically active 2-(2-chloroethyl)piperidine in the presence of a base. The chloro compound was obtained from the corresponding 2-(2-hydroxyethyl)piperidine which was resolved by way of the (+)-10-camphorsulfonate salt. The ring closure took place with inversion of optical rotation, and treatment with BF_3Et_2O causes polymerization with an inversion in the sign of rotation and increased specific rotation (compare with 1-azabicyclo[3.1.0]hexane above). 2-Methylconidine exists in four optically active forms, that is, two racemates, one of which is the naturally occurring ε-coniceine. Löffler[324] obtained two isomers in optically active forms, and Fodor and Cooke[325] showed that inversion of configuration at the chiral center in the side chain of *threo-O*-mesylsedridine occurs in the cyclization (Equation 9). Löffler[326] also synthesized 3-methylconidine, separated the two racemates, and resolved one of them by recrystallization of the (+)-tartrate salts. Apparently only one 3-phenylconidine was formed when 2-(2-chloro-1-phenylethyl)piperidine hydrochloride was basified.[327] The 1-azabicyclo-[4.2.0]octane ring system is present in dethiacephalosporins and syntheses with this in mind were devised by Lowe and Parker.[328] They photolyzed *N*-(*t*-butoxycarbonyl)diazoacetylpiperidine and obtained a 1:2 mixture of *cis*- and *trans*-7-*t*-butoxycarbonyl-8-oxo-1-azabicyclo[4.2.0]octane in 80% yield. This cyclization, which was originally developed by Corey and Felix,[329] was adopted for the synthesis of optically active nuclear analogues of dethiacephalosporins by Lowe and co-workers (Equation 10).[330]

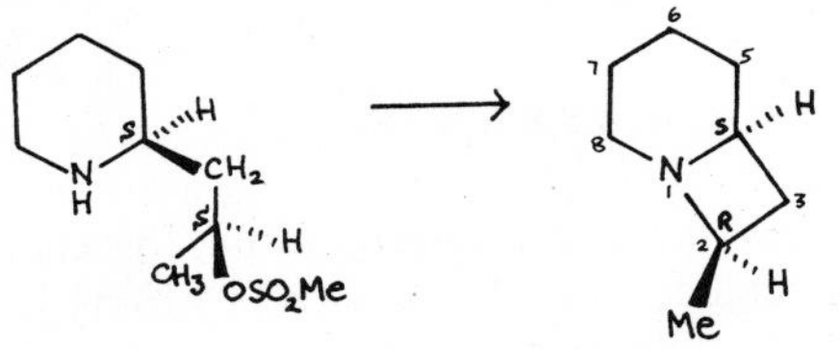

(9)

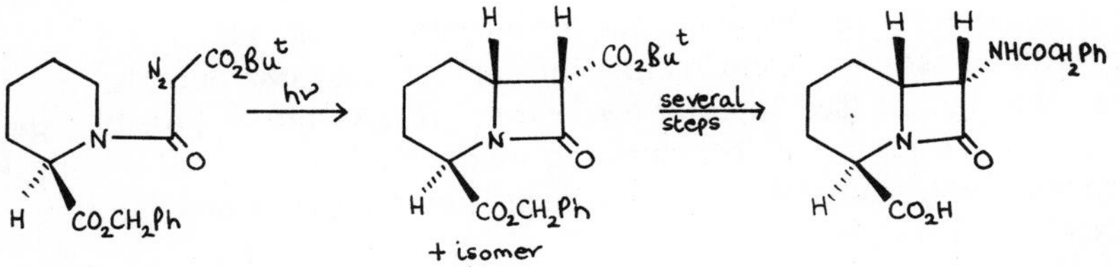

(10)

Meyers and Rahlan[331] prepared 2,2-dimethylcyclopenteno[*d*]-1-azabicyclo[4.2.0]octene (137), 1-azabicyclo[4.3.0]nonene (138), and 1-azabicyclo-[4.4.0]decene (139) by intramolecular cyclization of the corresponding 1-(*ω*-chloroalkyl)-3,3-dimethyl-1,2,3,4,6,7-hexahydro-5*H*-2-pyridines at pH 8–9. The stereochemistry indicated in the formulas was proposed from spectroscopic evidence.

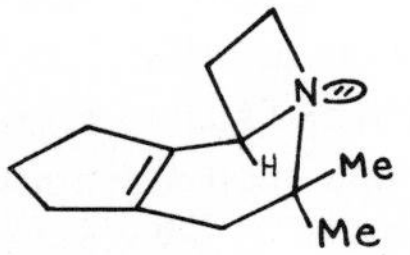

137

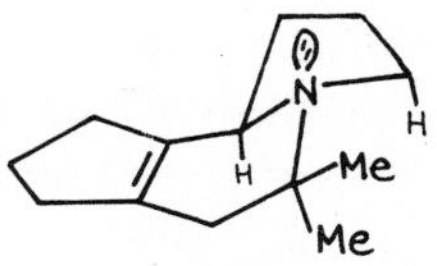

138

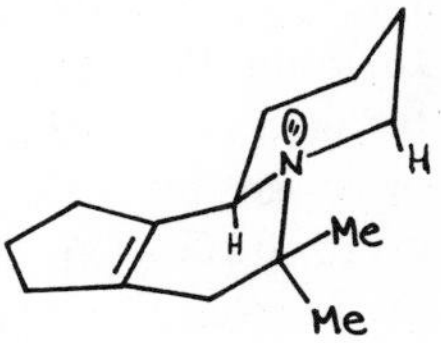

139

Intramolecular cyclization was once again the method of choice for preparing 6-oxo-1-azabicyclo[5.4.0]undecane from 2-ethoxycarbonyl-1-(4-cyanobutyl)piperidine. Addition of aryl magnesium bromides to this ketone provided the equatorial-aryl: axial hydroxy derivative stereoselectively as evidenced by intramolecular hydrogen bonding. Dehydration of these alcohols followed by reduction of the olefin formed provided the 6-equatorial-aryl-trans-fused 1-azabicyclo[5.4.0]undecanes (Equation 11).[332]

$$Ar = 2\text{-MeOC}_6\text{H}_4\text{-}, \quad 3\text{-MeOC}_6\text{H}_4\text{-},$$
$$4\text{-MeOC}_6\text{H}_4\text{-}, \quad 3,4\text{(MeO)}_2\text{C}_6\text{H}_4\text{-}$$

VII. *n*-AZABICYCLO[*x.y.z*]ALKANES

The *n*-azabicyclo[*x.y.z*]alkanes are numerous and for each bicycloalkane the nitrogen atom can replace any of the carbon atoms in the molecules. Only the ones that were studied more frequently are discussed here, but an attempt is made to mention at least one example of each ring system; hence a subsection on miscellaneous azabicycloalkanes. The von Baeyer system is the most convenient method of naming these compounds and is adopted throughout (see Chapter 1, Section III). As far as possible, the ring systems are discussed in increasing number of atoms, and within a ring system compounds with the nitrogen atoms in positions 1,2,3, . . ., are described in turn.

A. Azabicyclo[2.2.1]heptanes

1-Azabicyclo[2.2.1]heptanes (**140**) are readily formed by intramolecular cyclization of 4-chloromethylpiperidines.[333] The absolute configuration of 2,3-benzo fused derivatives of **140**, for example, 4-*S*-(−)-2,3-benzo-6,6-dimethyl-1-azabicyclo[2.2.1]hept-2-ene, which were prepared in a similar manner was discussed previously (see Chapter 3, Section IX.A.2 and formula **171**).[334] Another reaction which was very useful for making the bridged compounds is the Hofmann-Löffler-Freytag reaction in which the nitrogen atom of an *N*-haloheterocycle in concentrated sulfuric acid cyclized on to an alkyl side chain. The reaction has sometimes been assisted by uv irradiation. Lukeš and Ferles[335] found that irradiation of 1-chloro-4-ethylpiperidine in sulfuric acid gave solely 7-methyl-1-azabicyclo[2.2.1]heptane. Corey and Hertler[336] examined this reaction and observed that the attack of the nitrogen atom on to the carbon atom was not stereospecific and that a free radical mechanism was involved. The reaction has been called the Hofmann-Löffler and the Löffler-Freytag reaction, but the three names together will be used here.

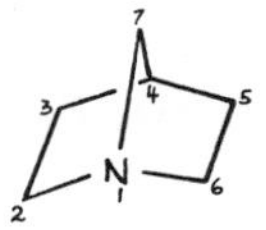

140

Gassman and co-workers[337-339] noted that methanolysis of *N*-chloro-4,7,7-trimethyl-2-azabicyclo[2.2.1]heptane causes a deep seated rearrangement with the formation of 2-*exo*-methoxy-3,3,4-trimethyl-1-azabicyclo-[2.2.1]heptane (**141**). The rate of this rearrangement is enhanced by a factor of 2 × 10³ in the presence of silver ions, and the mechanism suggested is shown in Equation 12. *N*-Bromo-,[339] *N*-*p*-nitrobenzoyloxy-,[340] *N*-tosyloxy-,[341] and *N*-amino-2-azabicyclo[2.2.1]heptanes[342] will rearrange in a similar manner on solvolysis, except for the *N*-amino compounds in which the reaction is mediated by nitrous acid. The reaction is not always clean, that is, other products are formed, and the last one named suggests that a nitrenium intermediate may be involved. Gassman and co-workers[343,344] also studied the solvolysis of 2-substituted bicyclo[2.2.1]heptanes and observed that the presence of a bridgehead nitrogen atom causes a rate acceleration of the order of 10³. The driving force must obviously arise from the nitrogen lone pair overlapping and stabilizing the incipient carbenium ion. These results must be borne in mind when considering the limitations of Bredt's rule.[345]

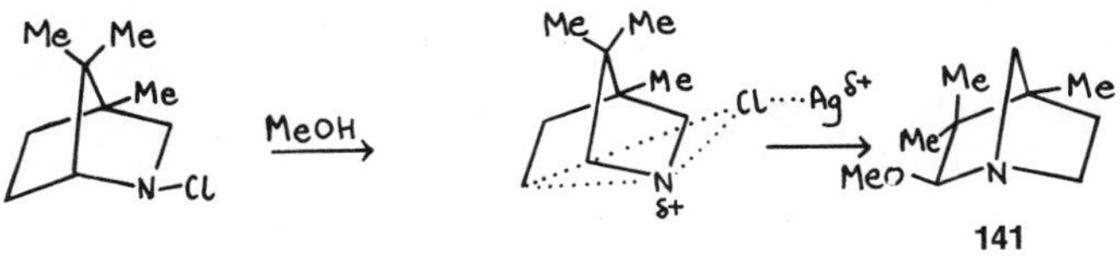

(12)

141

Oppolzer[346] came up with yet another ingeneous cycloaddition reaction in the synthesis of 7-acetyl-1,7-diazabicyclo[2.2.1]heptane (**142**) from *N*-acetyl-*N'*-but-3-en-1-ylhydrazine (**143**) and formaldehyde. This reaction was also extended to the preparation of 1,7-diaza- and 1,8-diazabicyclo-[3.2.1]octanes.

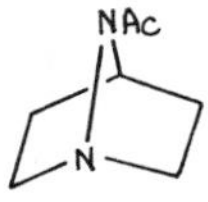

142

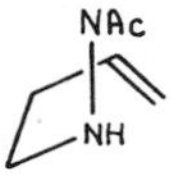

143

Portoghese and Sepp[347] obtained 6-alkoxy-2-azabicyclo[2.2.1]heptanes from 2-azatricyclo[2.2.1.0^{1,6}]heptane by reaction of its methiodide with a nucleophile (Equation 13). The tricycloheptane was formed from lead tetraacetate oxidation of 4-aminomethylcyclopent-1-ene (compare with Equation 3 in Chapter 2.I.1.a). The ratio of *exo-* and *endo-*alcohols formed (R = H) is similar to that observed in norborneol and it was suggested as evidence for the closely similar steric demands of the nitrogen lone pair compared with a C–H.[347,348]

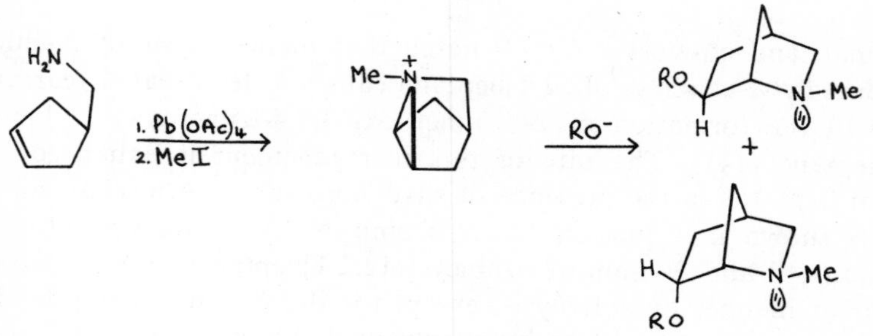

(13)

7-Azabicyclo[2.2.1]hepta-2,5-dienes are readily accessible by [4 + 2]π Diels-Alder addition of dienophiles to 1,3,4-trisubstituted pyrroles. A variety of acetylenes have been used[349,350] and benzyne yields the corresponding 2,3-benzo derivative (Equation 14).[351,352] The relatively unstable isoindoles have been trapped by forming 7-azabicyclo[2.2.1]heptenes with benzyne[353] or N-phenylphthalimide.[354] The reaction can be reversed by heating, yielding either the original pyrrole or a new pyrrole with its 3- and 4-substituents altered. Photolysis of some of these 7-azabicyclo[2.2.1]-hepta-2,5-dienes furnishes quadricyclanes which can be isolated and characterized, for example, 3-(p-toluenesulfonyl)-1,5-bis(trifluoromethyl)-3-azatetracyclo[3.2.0.0.^{2,7}0^{4,6}]heptane (**144**). When heated in chloroform these rearrange to the isomeric azepines.[349–351] The 2,3-benzo derivatives rearrange more readily to the 3-benzazepines.[351,355] Nitrogen inversion in

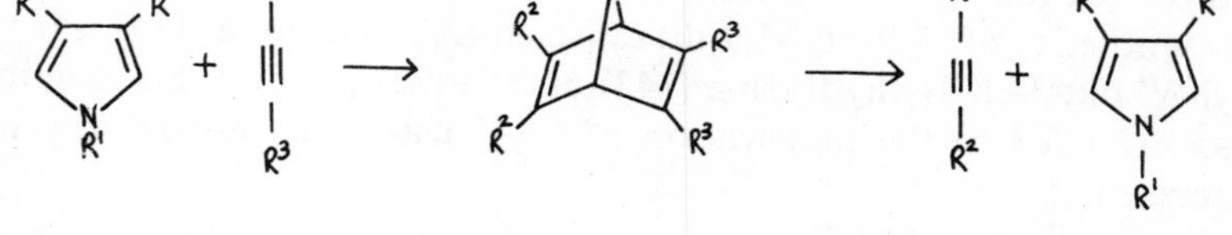

(14)

2,3-dimethoxycarbonyl-7-methyl-7-azabicyclo[2.2.1]hepta-2,5-diene is slow enough at 36°C that the two invertomers **145** and **146** are clearly discernible by ^{1}H nmr spectroscopy.[356] The ratio of syn (**145**) to anti (**146**) compounds is 1:2. The nitrogen atom in 2,3-(tetrachloro- and tetrafluorobenzo)-7-methyl-7-azabicyclo[2.2.1]hepta-2,5-diene inverts more rapidly ($\triangle F^{\ddagger} = 58.6$ kJ mol^{-1}), but on cooling to about $-10°C$ the two invertomers can be observed and the equilibrium, also in this example, is in favor of the *anti*-methyl conformer. Stereospecific orientation of the nitroxide group in 2,3-benzo-7-oxyl-7-azabicyclo[2.2.1]hept-2-ene (**147**) and 2,3-benzo-7-oxyl-7-azabicyclo[2.2.1]hepta-2,5-diene (**148**) as indicated in the formulas explains the esr spectra that are produced.[357]

144   **145**   **146**

147   **148**

B. Azabicyclo[2.2.2]octanes

1. *1-Azabicyclo[2.2.2]octanes: Quinuclidines*

The 1-azabicyclo[2.2.0]octane ring system has been known for many years because it occurs in a large number of alkaloids which include the Cinchona,[358] Iboga, Voacanga,[359] Ajmaline, Sarpagine,[360] and the 2,2′-indolylquinuclidine alkaloids.[360a] The chemistry and stereochemistry were admirably discussed in the reviews cited and recently by Yakhontov.[361] It is not the intention to recapitulate all these, but some basic and more recent data are described here. A reference, however, must be given for the ingenious synthesis of quinine by Woodward and Doering[362] in 1944. Interest in Cinchona alkaloids was revived recently by some novel stereoselective and stereospecific syntheses by Taylor,[363] by Uskoković,[364] and by Coffen,[365] and their co-workers, but because these eventually lead to the synthesis of alkaloids a discussion of these is beyond the scope of this monograph.

The key reactions for the formation of the quinuclidine ring system include the intramolecular cyclization of 4-carboxymethylpiperidines followed by reduction of the amide carbonyl,[366] of 4-haloethylpiperidine,[334,364] and Hofmann-Löffler-Freytag cyclization of *N*-halo-4-ethylpiperidines.[335,367]

Wittig and Steinhoff[368] effected an unusual Diels-Alder reaction whereby 1,4-dihydronaphthalene-1,4-oxide added across the 9,10-positions of acridine-1-oxide with loss of the *N*-oxide group. The adduct lost a second oxygen atom when it was heated at 80°C in polyphosphoric acid and gave a 35% yield of 2,3:5,6-dibenzo-7,8-naphtho-1-azabicyclo[2.2.2]octa-2,5,7-triene, benzoazatriptycene (**149**). Intramolecular alkylation of *N*-chloroethylpiperazine yields 1,4-diazabicyclo[2.2.2]octane.[369] The X-ray crystal structures of this diazabicycle,[370] which is also called triethylene diamine, and of *R*(−)-3-acetoxyquinuclidine methiodide (**150**) (i.e., 3-*R*-3-acetoxy-1-methyl-1-azabicyclo[2.2.2]octane) were reported. In the latter case the study revealed the absolute configuration depicted in **150**.[371]

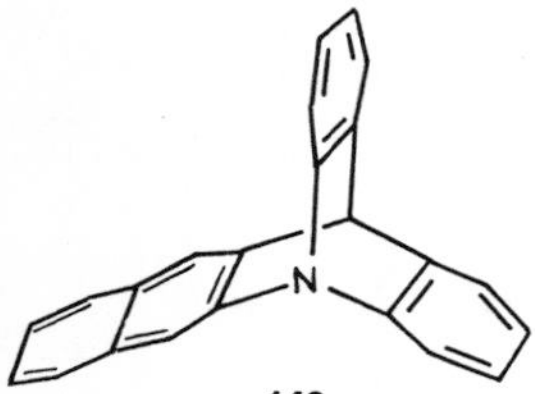

149 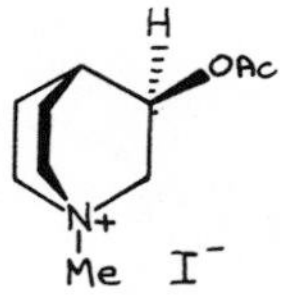   150 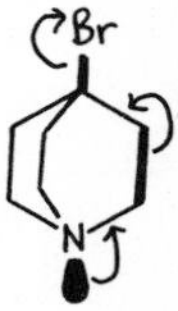   151

The halogen atom in 4-bromo-1-azabicyclo[2.2.2]octane (**151**) is ideally placed, stereoelectronically, for synchronous fragmentation. It solvolyzes 5.5×10^4 times faster than the corresponding carbocyclic compound[372] (see Chapter 3, Sections II.D.6, IX.B.4.c, and X.B.3; and Chapter 4,

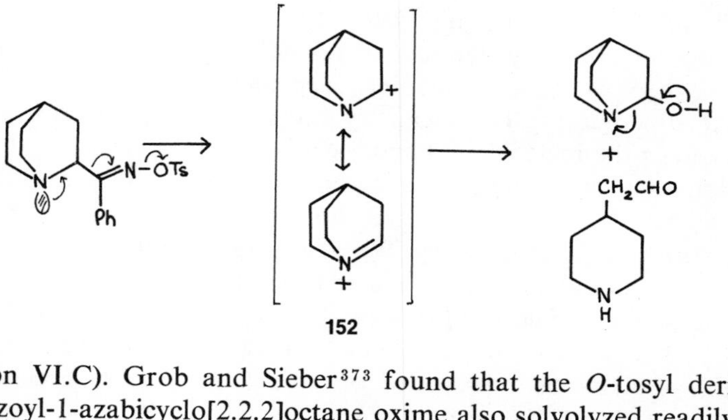

(15)

152

Section VI.C). Grob and Sieber[373] found that the *O*-tosyl derivative of 2-benzoyl-1-azabicyclo[2.2.2]octane oxime also solvolyzed readily in aqueous ethanol or tetrahydrofuran with synchronous fragmentation. The products of this decomposition are benzonitrile, 2-hydroxy-1-azabicyclo-[2.2.2]octane, and piperidin-4-ylacetaldehyde (Equation 15). The evidence indicates that the quinuclidine cation (**152**) is more stable than the corresponding carbocyclic bicyclo[2.2.2]octane cation and that the positive charge in **152** is delocalized. Thus Bredt's rule is not strictly applicable when the carbonium ion is stabilized by a nitrogen atom. This is to be compared with the data of Gassman and Fox[374] on the electrolytic decarboxylation of 2-carboxy-1-azabicyclo[2.2.2]octane (see Section VII.A). The *p*K*a* values and rates of methylation of a large number of 4-substituted 1-azabicyclo[2.2.2]octanes were measured and excellent Hammett-type correlations were deduced.[375] The results reveal that the polar effects are ideally relayed from the substituent to the nitrogen atom, and when a comparison of the *p*K*a* was made with the corresponding 4-substituted pyridines the polar and resonance effects of substituents could be adequately separated.[376]

2.   *2-Azabicyclo[2.2.2]octanes: Isoquinuclidines*

A simple synthesis of 2-azabicyclo[2.2.2]octanes is the intramolecular cyclization of *cis*-1-amino-4-carboxycyclohexanes,[377–379] and the stereochemistry of the substituents is dictated by their relative configurations in the original substituted cyclohexanes.[377]

(16)

In a second synthesis 1-substituted 1,2-dihydropyridines undergo stereospecific [4 + 2]π cycloaddition reactions with dienophiles to form 5-substituted 5-azabicyclo[2.2.2]oct-2-enes (Equation 16).[380–384] The dihydropyridines are formed by reduction, e.g. with $NaBH_4$, of the 1-substituted pyridinium salts. Some of these cycloadditions are highly regiospecific and the same effect can be achieved by placing the heteroatom on the dienophile. For example, cycloaddition between iminocarbamates ($R_2C{=}NCO_2R$) and cyclohexa-1,3-diene yields 5-alkoxycarbonyl-5-azabicyclo[2.2.2]oct-2-enes.[385] 2-Pyridones react similarly with dienophiles and produce 6-oxo-5-azabicyclo[2.2.2]oct-2-enes stereo- and regiospecifically,[386–391] as do 2-pyrazinones.[392] Bradsher and his school[393–395] have added olefins across the 1,4-positions of 3-alkyl-2-methylisoquinolinium salts and found the reaction to be highly regiospecific. In the case with methyl vinyl ether, the adduct **153**, with the methoxy group over the benzene ring and anti to the nitrogen

153

atom, is formed and was confirmed by X-ray analysis.[396] The specificity is as would be expected from a consideration of the repulsive forces in the transition state (see Chapter 3, Section X.A.1).[394,395] This method of forming 2-azabicyclo[2.2.2]octa-2,5-dienes has been extended to condensed systems such as the 4α-azoniaanthracene cations and their benzologues.[397] The dienophiles used were cyclopentene,[398] phenylacetylene,[399] N-arylmaleimides,[400] maleic and fumaric esters,[401,402] and 1,1-diethoxyethylenes.[403] The addition of azodicarboxylic esters to anthracene (and tetracene) yields 7,8 - bisalkoxycarbonyl - 2,3,5,6 - dibenzo - 7,8 - diazabicyclo[2.2.2]octa - 2,5-dienes.[404]

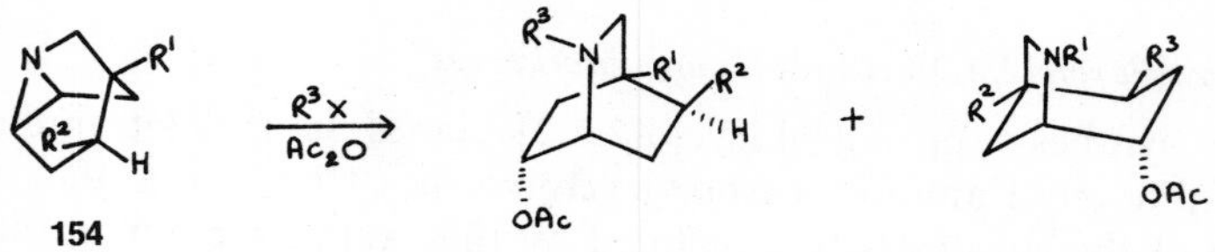

(17)

154

Among the miscellaneous, but interesting, methods by which this ring system is formed in a stereospecific manner is the alkylation followed by a nucleophilic attack of the aziridine **154** which produces a 4:1 mixture of a 2-azabicyclo[2.2.2]- and 6-azabicyclo[3.2.1]octanes (Equation 17).[405] In a similar way, a catalyzed reaction of 3-N-chloro-N-methylaminomethyl-1-methoxycyclohex-1-ene gives a mixture of these two bicyclo compounds in varying proportions depending on the catalyst used. The catalysts were $AlCl_3$, $AlBr_3$, BF_3, and BF_3Et_2O.[406]

A synthesis which involves another rearrangement is the decomposition of 1-azidobicyclo[2.2.1]heptane in methanol or propan-2-ol. The minor product is 1-methoxy- or 1-isopropoxy-2-azabicyclo[2.2.2]octane, but the major product is 1-methoxy- or isopropoxy-2-azabicyclo[3.2.1]octane (Equation 18). The intermediates in the formation of these products must

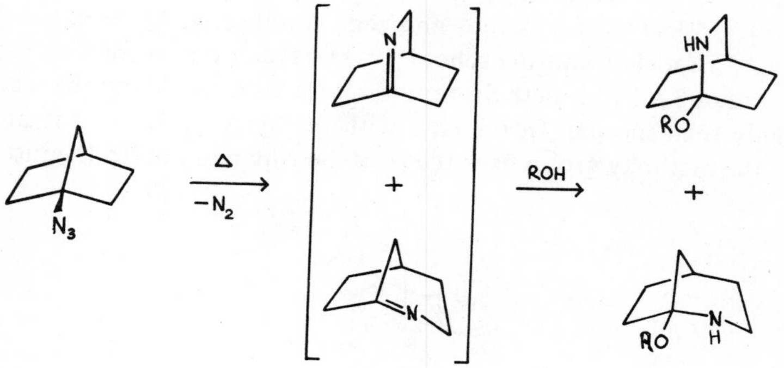

(18)

contain a double bond between the nitrogen atom and the bridgehead carbon atom, or a mesomeric form of it; thus providing another example in which Bredt's rule is violated (see Sections VII.A and B.1).[407] A rearrangement with elimination of chlorine, which is characteristic of β-chloro bicyclic compounds in which a cis nitrogen bridge is present, for example, 2-β-chloro-8-methyl-8-azabicyclo[3.2.1]octane produces the 2-azoniabicyclo[2.2.2]oct-2-ene cation (155) in boiling aqueous acetone.[408]

The methylation of 5-methyl-5-azabicyclo[2.2.2]oct-2-ene (156) and -octane (157) proceeded with low stereospecificity indicating that the double bond has little effect on the direction of alkylation.[409]

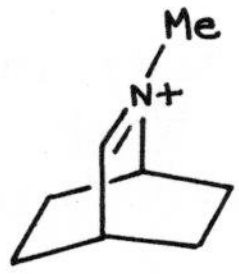

155    156    157

C. Azabicyclo[3.2.1]octanes

1. 1-Azabicyclo[3.2.1]octanes

1-Azabicyclo[3.2.1]octanes are readily prepared by intramolecular cyclization of 3-(2-haloethyl)piperidines,[410,411] 3-ethoxycarbonyl-1-ethoxycarbonylmethylpiperidines,[412] and 1,3-bis(cyanomethyl)- or 3-ethoxycarbonyl-1-(2-ethoxycarbonylethyl)-pyrrolidines.[413] Gassman and Fox[414,415] found that, as in related systems (see Section VII.A), *N*-chloro-2-azabicyclo[2.2.2]octane rearranged in methanol containing silver nitrate to 2-methoxy-1-azabicyclo[3.2.1]octane. Here again a carbonium ion such as 158 may be involved (see Sections VII.A.1 and 2, VII.B.1 and 2). A Hofmann reaction on 1-methyl-1-azabicyclo[3.2.1]octane cations proceeds as expected with formation of the corresponding 1-methyl-3-(prop-2-en-1-yl)-pyrrolidine. However, when the cation is heated with water in a sealed tube, fragmentation takes a different course and the product is 3-(2-hydroxyethyl)-1-methylpiperidine together with some di[2-(piperidin-3-yl)ethyl]ether (Equation 19).[411,412,416,417]

2-(2-Bromoethyl)-1,2,3,4-tetrahydroquinoxaline cyclized as readily as the above to (±)-2,3-benzo[*b*]-1,4-diazabicyclo[3.2.1]oct-2-ene (159).[418] All the

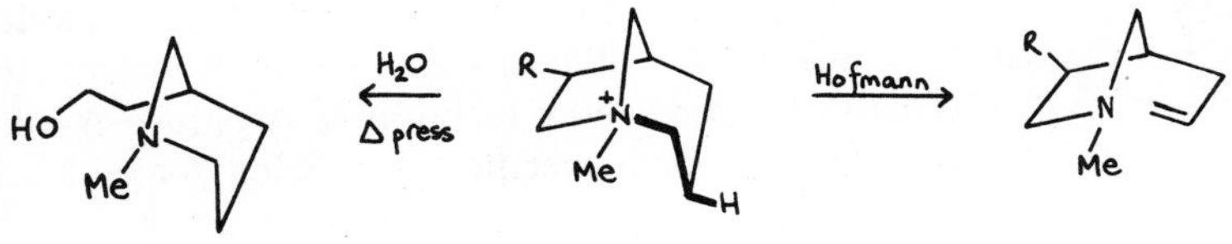

(19)

158   159

^{1}H nmr chemical shifts and coupling constants were assigned for this compound by using the "shift reagent" Eu(fod)$_3$ and are consistent with the structure indicated.[419]

2.   2-, 3-, and 6-Azabicyclo[3.2.1.]octanes

exo-2-Chloronorbornane (exo-2-chlorobicyclo[2.2.1]heptane) reacts with trichloramine to form the exo-2-dichloroamino derivative which rearranges to 2-azabicyclo[3.2.1]oct-2-ene.[420] Reduction of this product yields the octane.

Camphidine, 1,8,8-trimethyl-3-azabicyclo[3.2.1]octane, was the chosen example in the studies on this system. ^{1}H nmr spectroscopy of N-nitrosocamphidine at 220 MHz revealed that it was a 1:1 mixture of syn and anti isomers. The "shift reagent" Eu(dpm)$_3$ was used to assign the resonances.[421] The evidence on quaternization of N-methyl camphidine by primary alkylating agents is that axial attack is preferred.[422,423] Dealkylation of the quaternary salts derived from N-methyl- and N-ethylcamphidines with ^{14}C-methyl iodide, by reaction with sodium phenoxide, removes the alkyl group that was introduced by the alkylation reaction more readily than the alkyl group originally present.[424] The quaternization of camphidine was discussed in a recent review on the quaternization of piperidines by McKenna.[425]

Most of the syntheses of 6-azabicyclo[3.2.1]octanes were brought about by intramolecular cyclization reactions or by molecular rearrangement of some reactive intermediate generated in the reaction. Intramolecular addition of a nitrogen atom onto a double bond is exemplified in the photolysis of 4-N-nitroso-N-methylcarbamylcyclohex-1-ene,[426] or on the treatment of 4-N-methylaminomethylcyclohex-1-ene[426–428] with lead tetraacetate or silver salts. Although, theoretically, 2-azabicyclo[2.2.2]octane and 6-azabicyclo[3.2.1]octane are possible products by addition of nitrogen to C–1 and C–2 respectively, usually the latter is formed regiospecifically (Equation 20). The reaction involves free radicals, and attack at C–4 occurs from both the exo and endo sides. Similarly, photolysis of 4-N-chloro-N-methylamidocyclohex-1-enes furnishes 6-methyl-6-azabicyclo[3.2.1]octan-7-ones in moderate yields.[429] The Hofmann-Löffler-Fretag reaction of N-chloro-N-alkylaminomethylcyclohexanes is also regiospecific and yields 6-azabicyclo-[3.2.1]nonanes.[430–432]

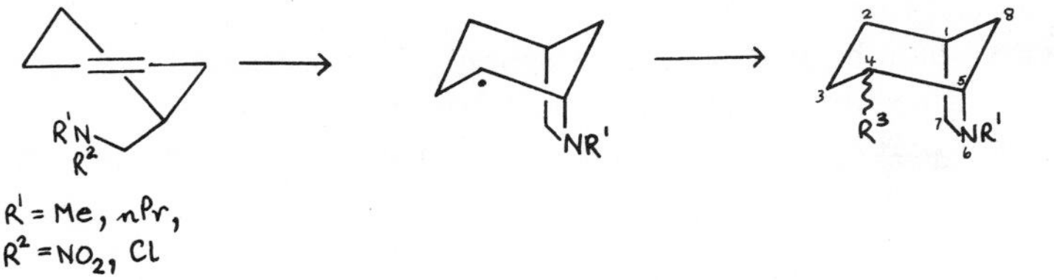

R^1 = Me, *n*Pr,
R^2 = NO_2, Cl

(20)

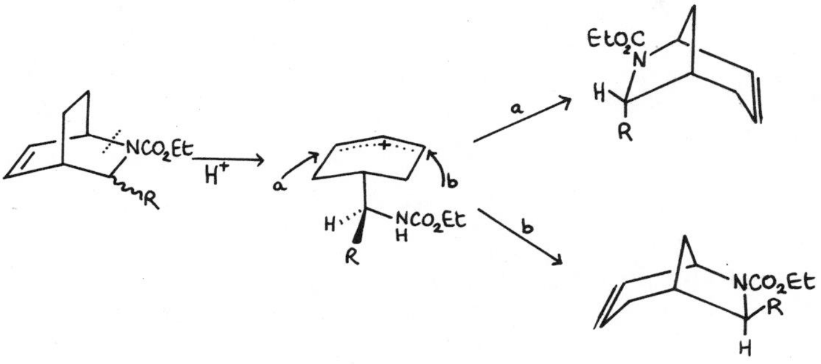

(21)

5-Ethoxycarbonyl-5-azabicyclo[2.2.2]oct-2-ene rearranges in benzene in the presence of *p*-toluenesulfonic acid monohydrate to form 7-ethoxycarbonyl-7-azabicyclo[3.2.1]oct-2-ene. It should be noted that the presence of the double bond alters the numbering of the heteroatom from N–6 to N–7. Irrespective of whether the substituent on C–6 in the [2.2.2]octene is endo or exo oriented, the ratio of 6-*endo*- and 6-*exo*-substituted 7-azabicyclo-[3.2.1]oct-2-ene always remains the same. This is strong evidence for the participation of the double bond in the rearrangement (Equation 21).[433]

In solution *cis*- and *trans*-8a-carbamoyl-2-tetralones (**160**) are in equilibrium with the corresponding 6-azabicyclo[3.2.1]octan-7-ones (**161**).[434]

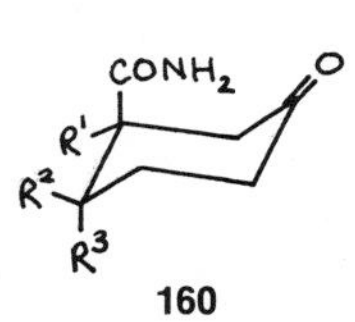

160 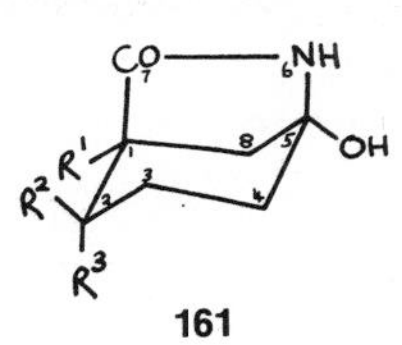   **161**

3. 8-Azabicyclo[3.2.1]octanes: Tropanes

Two restrictions are placed on the material to be included in this subsection. The first is that most of the work on 8-azabicyclo[3.2.1]octanes arose from interest on tropane alkaloids—which does not fall within the scope of this monograph. The second is that reviews on tropane alkaloids[435,436] and the stereochemistry of tropane[436] alkaloids and tropanols[437] have discussed this matter. The series of papers on the stereochemistry of tropane alkaloids by Fodor and his school[438] should also be consulted for detailed information of certain aspects of tropanes. The stereochemical features of 8-azabicyclo[3.2.1]octanes which were reported after these reviews will be treated here.

162   163   164

Dipole moment measurements of tropanes by several workers[439–441] showed that undoubtedly the six-membered ring is in the chair form and that the methyl group is predominantly equatorially disposed as in formula **162**. However, substituents in the α position of C–3 distort the six-membered ring and probably the skeleton as a whole.[439,441,442] Studies of hydrogen bonding between an OH and the nitrogen atom in tropanes have been very useful in assigning structures for the intramolecularly hydrogen bonded molecules.[443] C–3 can come close to N–8 by hydrogen bonding between a 3-β-OH group, but this would require that the six-membered ring should assume a boat conformation. This is certainly true when C–3 carries a large α substituent, and also in the transfer of an acyl group from N–8 to the C–3 oxygen atom.[444,445] An interesting case is the addition of N–8 to the carbonyl of 3-α-phenyl-3-β-benzoyl-8-azabicyclo[3.2.1]octane and its 8-methyl derivative which exists as the hemi ketal **163**. The uv spectrum indicated lack of conjugation between the carbonyl and benzene ring of the benzoyl group, and the structure is allowed to take this form because of otherwise strong nonbonded interactions between the 3-α-phenyl group and the C–6 and C–7 hydrogen atoms. This is supported by the evidence that 1-methyl-4-phenyl-4-benzoyl-piperidine does not behave similarly.[446]

Tropane is a symmetrical molecule but insertion of a substituent at C–1 (5), C–2 (4), or C–6 (7) makes the molecule asymmetric. Thus all four optical isomers of 2-hydroxytropane were prepared. The 2-α-isomer was resolved with (+)- and (−)-tartaric acids and the optically active 2-β-isomers were obtained by epimerization of the (+)- and (−)-α-isomers.[447] Asymmetry can be introduced into the tropane molecule by cis/trans isomerism and was demonstrated by the isolation of (−)-tropinone oxime **(164)** from the recrystallization of the (+)-10-camphorsulfonate salt.[448]

The specificity of reactions at C–3 are still attracting attention. The catalytic reduction of 8-alkyl-3-oxo-8-azabicyclo[3.2.1]octanes yields essentially the 3α-alcohol, whereas reduction of the quaternary salts produces smaller amounts of the 3α-alcohol with increasing size of the *N* substituents.[449] The steric course of nucleophilic substitution at C–3 depends on the nature of the leaving group. A 3α-methanesulfonyloxy group is displaced by a phenoxide ion to form stereospecifically the β-phenyl ether. A 3α-chloro group, on the other hand, leads to the α-phenyl ether.[450,451] When (+)-2α-hydroxy-8-azabicyclo[3.2.1]octane, in acetic anhydride containing a drop of perchloric acid, is kept at 25°C it is converted into the 2α-acetate, but when it is boiled complete racemization is observed. The (−)-2β- isomer, on the other hand, is acetylated without racemization under

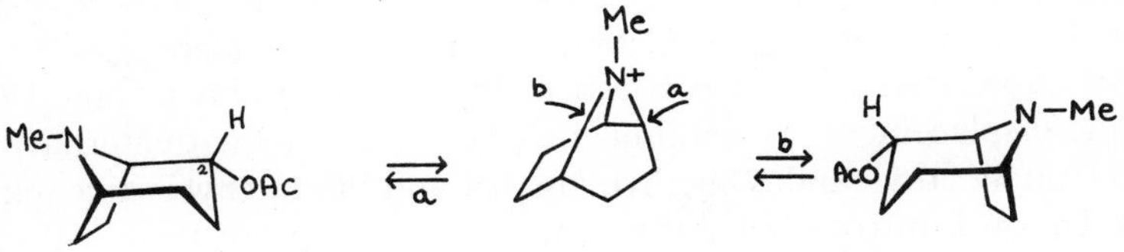
Me
N+
b
a
Me–N
H
OAc
H
N–Me
AcO

(22)

these conditions. A symmetrical intermediate would account for racemization in the former case (Equation 22).[452] If the acid esterifying the 3α-OH group is optically active (e.g., in scopolamine **165**) then the anisotropy that is introduced makes H–6 and H–7 magnetically nonequivalent in the ^{1}H nmr spectrum.[453]

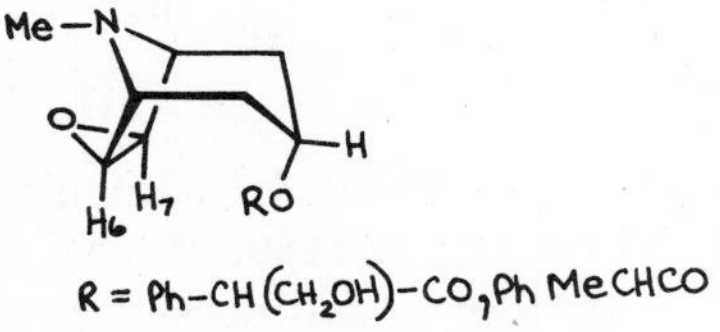
Me—N
H
H7
H6
RO
R = Ph–CH(CH2OH)–CO, Ph MeCHCO

165

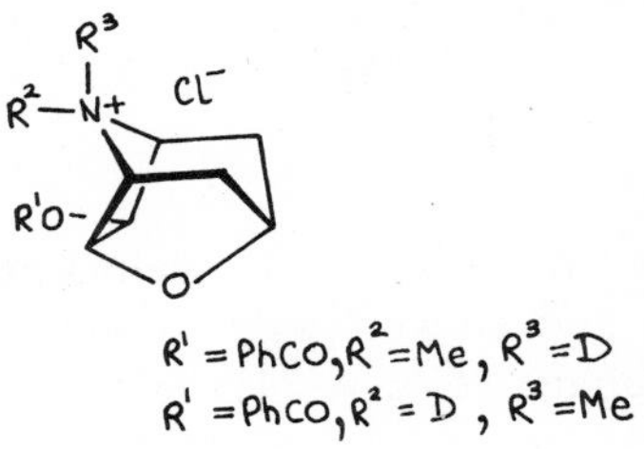
R3
R2—N+
Cl−
R'O–
R' = PhCO, R2 = Me, R3 = D
R' = PhCO, R2 = D, R3 = Me

166

8-Methyl-8-azabicyclo[3.2.1]octanes reveal one *N*-methyl signal in the ^{1}H nmr spectrum and the methiodides reveal two signals of equal intensity. The cations, however, give two *N*-methyl groups which can be assigned to axial and equatorial *N*-methyl cations. The ratio of these signals varies with the substituent. When there is very little nonbonded interaction, the signals are of equal intensities, that is, equal amounts of axial and equatorial *N*-methyl cations as in benzoyl scopoline deuterochloride (**166**).[454]

The quaternization of 8-azabicyclo[3.2.1]octane was examined in some detail and was discussed in a recent review.[425] Quaternization studies of N-methyl and N-isopropyl nortropane suggest that it occurs by way of an axial approach[422,455,456] with methyl iodide. Many of the results are confusing apparently because the data is not clear-cut and the interpretations of structures are not necessarily correct.[455,457,458] Fodor and co-workers examined the alkylation of several tropane derivatives and in certain examples confirmed their structures by X-ray analyses. They found that consistently, the direction of ethylation, hydroxyethylation, chloroethylation, and alkoxycarbonylmethylation in this series is equatorial (**167**).[459] In contrast, in the quaternization of tropidines (**168**) the axial approach is preferred probably because it is less hindered due to the flattening of the six-membered ring.[460] Thermal demethylation of 3α-hydroxy-N-^{14}C-methyl-N-methyl-8-azoniabicyclo[3.2.1]octane chloride is not stereospecific with the label equally distributed between methyl chloride and tropine.[461] Nucleophilic displacement, by sodium thiophenoxide, of the quaternary salts derived from N-ethyl- and N-methyltropanes and ^{14}C-methyl iodide removes the alkyl group, introduced during the alkylation, more readily than the original N-alkyl group.[424] N-Oxidation of tropine is preferentially axial and the isomeric N-oxides, among other tropane derivatives, were identified by their 220-MHz ^{1}H nmr spectra.[462]

167   **168**   **169**

The tropane system is ideally suited for a study of the Grob fragmentation. Grob and co-workers[463] solvolyzed 3β-chlorotropane (**169**, R = Me) and nortropane (**169**, R = H) in 80% aqueous ethanol and observed rate enhancements of 1.35×10^4 and 1.35×10^3 with respect to the deaza analogue exo-3-chlorobicyclo[3.2.1]octane. 3α-Chlorotropane did not undergo fragmentation. (Compare with Chapter 3, Sections VI.C, II.D.6, IX.B.4.c, and X.B.3.)

Katritzky and co-workers found that 1-methyl-,[464,465] 1-phenyl-,[464–466] 1-(2,4-dinitrophenyl),[466,467] and 1-(3-nitro-2-pyridyl)-3-oxidopyridinium betaines[468] react as 1,3- dipoles, and form Diels-Alder adducts with acrylonitrile, methyl acrylate,[464,467,468] N-phenyl maleimide,[465] 2,3-dimethylbutadiene, maleic and fumaric esters,[468] fulvenes,[468a] and benzyne[465,466] by addition across the 2- and 6-positions of the pyridine ring. The products are

8-substituted 4-oxo-8-azabicyclo[3.2.1]oct-2-enes, and in the case of acrylonitrile and methyl acrylate both the *exo*- and *endo*-7-cyano- and -7-methoxy-
carbonyl derivatives are obtained regiospecifically. (Equation 23.) The last
mentioned betaine forms the dimer **170** reversibly and is in a convenient
form for reaction with the olefinic dipolarophiles.[468]

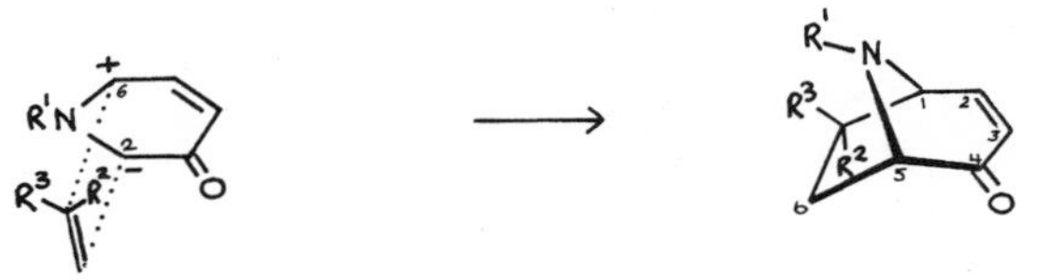

(23)

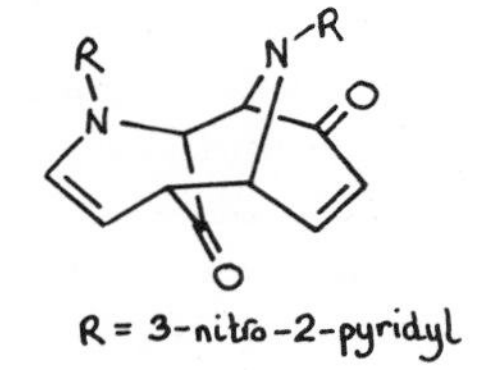

R = 3-nitro-2-pyridyl

170

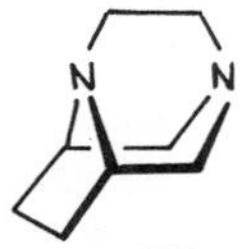

171

3,8-Diazabicyclo[3.2.1]octanes have been described, and it was possible
to link N–3 and N–8 by an ethylene bridge as in 3,8-diazatricyclo[3.2.1.2³,⁸]-
decane (**171**).[469]

D.   Azabicyclo[3.3.1]nonanes

1.   3-Azabicyclo[3.3.1]nonanes

3-Substituted 3-azabicyclo[3.3.1]nonanes are conveniently made from cyclohexanones by an intramolecular Mannich reaction with two molecules
of formaldehyde and a primary amine.[470,471] An alternative synthesis is
from the enamines derived from *N*-methyl-4-piperidones and acrolein.[472]
Both these methods provide the 9-oxo derivative, but the latter method
provides a mixture of epimeric 7-alcohols. A reaction that takes advantage
of the proximity of syn-axial substituents is the photolysis of *cis*- and
trans-1,1-dimethyldecalin-10-acylazides (**172**) in cyclohexane which yields
cis- and *trans*-1-methyl-8,9-trimethylene-4-oxo-3-azabicyclo[3.3.1]nonanes
(**173**), among other products.[473] The rigidity of the systems, compared
with 1,3-disubstituted cyclohexane, is an important factor in the cyclization
reaction.

The two six-membered rings in 3-azabicyclo[3.3.1]nonanes prefer to be
in the double-chair conformation bringing N–3 and C–7 quite close to
each other. Dobler and Dunitz[474] determined the X-ray structure of

$$\underline{cis} \quad R', R^3 = -(CH_2)_3-, \ R^2 = H$$
$$\underline{trans} \quad R', R^2 = -(CH_2)_3-, \ R^3 = H$$

172

$$\underline{cis} \quad R', R^3 = -(CH_2)_3-, \ R^2 = H$$
$$\underline{trans} \quad R', R^2 = -(CH_2)_3-, \ R^3 = H$$

173

174

3-azabicyclo[3.3.1]nonane hydrobromide and although they confirmed the double-chair conformation (**174**) they found that the bond angles were slightly different from the ideal values. Pumphrey and Robinson[475] examined this system in order to obtain information regarding the steric requirements of the nitrogen lone pair of electrons, a hydrogen atom, and a methyl group. They found that 7-*exo-t*-butyl-3-azabicyclo[3.3.1]nonane and its 3-methyl derivative (**175**) exist in the conformation in which the lone pair is endo and the hydrogen atom and methyl group exo. The evidence is from the ABX pattern of signals in the ^{1}H nmr spectrum for H–1 and H–2 and from the presence of Bohlmann bands at 2700–2800 cm^{-1} in the ir. The latter data infers a trans diaxial arrangement of the lone pair and axial H–2 and H–4. This is conclusive evidence that the lone pair on a nitrogen atom is less sterically demanding than a hydrogen atom. McKenna and co-workers[476] also concluded that the 3-methyl-3-azabicyclo[3.3.1]nonane hydrochloride is in the double-chair conformation with the *endo*-hydrogen and *exo*-methyl group similar to the free base and the parent hydrobromide **174**, but its methiodide is almost exclusively in the boat-chair conformation **176**. Methylation of 3-benzyl- and 3-ethyl-3-azabicyclo[3.3.1]nonanes gives an approximately 9:1 mixture of two quaternary salts implying that a boat-axial attack by the reagent is preferred. There was no evidence that the system was twisted. House and Tefertiller[477] examined the quaternization of 9-oxo- (**177**, n = 2), *anti*-9-hydroxy- (**178**, n = 2), and *syn*-9-hydroxy-3-methyl-3-azabicyclo[3.3.1]nonanes (**179**, n = 2) with methyl bromoacetate and showed the 9-substituent influenced the direction of attack by the alkylating agent. They obtained a 1:1 mixture of *endo*- and *exo*-methoxy-carbonylmethyl, *exo*-methoxycarbonylmethyl, and a 1:3 mixture of *endo*- and *exo*-methoxycarbonylmethyl salts respectively. The last compound was isolated as the cyclic ester **180**. House and Pitt[478] made a comparative study of the alkylation of **177**, **178**, and **179**, in which n = 1, 2, and 3, with trideuteromethyl *p*-toluenesulfonate. The preferred direction of methylation in each example was syn to the oxygen atom, that is, from the exo side (see **174** and **177**). House and co-workers[479,480] also studied the reduction and the reactions with Grignard and organo lithium reagents with the ketones **177**, $n = 1$, $n = 2$, and $n = 3$. The stereoselectivity was different for each of these systems but results for the reduction and addition reactions were similar. In the octane series (i.e., $n = 1$) attack was predominantly from side α; in the nonane series (i.e., $n = 2$) a $1:2$ ratio of α to β side attack was observed and in the decane series β side attack was the main pathway of reaction (see formula **177**). Stereoselectivity is apparent in the catalytic reduction of *N*-methyl 7,8-benzo-1-methyl-9-oximino-3(and 2)aza-bicyclo[3.3.1]non-7-ene. The *syn*- and *anti*-9-acetamido derivatives are formed when platinum in acetic acid and platinum in acetic acid containing sulfuric acid are used respectively.[481]

No unusual effects were observed in the ir and uv spectra of the ketones **177**, $n = 1$, $n = 2$, and $n = 3$ suggesting the absence of interaction between N–3 and the 9-carbonyl group.[482] The proximity of N–3 and C–9, however, shows up again in the different behavior of the *endo*- and the *exo*-9-ethoxy-carbonyl-3-tosyl derivatives upon electron impact. In the exo series, there is successive loss of EtOH and tosyl radicals apparently by anchimeric assistance from H–7; whereas there is no loss of EtOH in the endo series, and expulsion of the tosyl radical is probably assisted by a carbonyl radical.[483] Intramolecular carbene insertion was found by Oida and Ohki[484] to be sensitive to ring size. The carbene generated from 3-methyl-9-tosylhydrazono-3-azabicyclo[3.3.1]nonane (**181**, $n = 2$) attacks C–6 in the carbocyclic ring and yields 3-methyl-3-azatricyclo[6.1.0.0^{5,9}]nonane (**182**), while the carbene generated from the bicyclooctane (**181**, $n = 1$) attacks C–5 of the piperidine ring and forms 2-methyl-2-azatricyclo[5.1.0.0^{4,8}]octane (**183**). The structures of the tricyclo compounds were confirmed by Hofmann degradations.

181   182   183

2. *9-Azabicyclo[3.3.1]nonanes*

The 9-azabicyclo[3.3.1]nonane ring system is present in the alkaloids from pomegranate bark and reference to reviews on these alkaloids should be made for the chemistry and early work.[485] ψ-Pelletierine, 9-methyl-3-oxo-9-azabicyclo[3.3.1]nonane exists in the chair-chair conformation **184**. [1]H nmr data by Chen and Le Fèvre[486] further demonstrated that 3β-granatanol, 3-*exo*-hydroxy-9-methyl-9-azabicyclo[3.3.1]nonane, also is in the chair-chair conformation but 3α-granatanol, the 3-*endo*-hydroxy isomer, is in the chair-boat form. The nonbonded interaction between the bulky 3-*endo* substituent, and H–7 is relieved by conversion of that ring into a boat conformation. A similar chair-boat conformation for 9-benzoyl-3α-bromo-2-oxo(and 2-β-hydroxy)-9-azabicyclo[3.3.1]nonane (**185**) was confirmed by an X-ray study.[487] Dipole moment measurements of the 6,8-ethano derivative of ψ-pelletierine (**184**) demonstrated that it was in the chair-chair conformation with the *N*-methyl group axially oriented.[488]

184   185   186

A most convenient synthesis of 9-azabicyclo[3.3.1]nonanes is the addition of a nitrogen atom across the double bonds of cycloocta-1,5-diene regiospecifically. Thus *N,N* dibromo *p*-toluenesulfonamide yields 31% of 2,6-dibromo-9-tosyl-9-azabicyclo[3.3.1]nonane (**186**, R = Br). Similarly 1,2:5,6-diepoxycyclooctane and ammonia give a 45% yield of the corresponding

2,6-dihydroxy derivative (**186**, R = OH).[488a] A second synthesis involving intramolecular addition of a nitrogen atom to a double bond is the reaction of 5-*N*-chloro-*N*-methylaminocyclooct-1-ene in acetone containing silver perchlorate. The cyclization is not completely regiospecific because 2*α*-chloro-9-methyl-9-azabicyclo[3.3.1]nonane (**187**) and 2*β*-chloro-9-methyl-9-azabicyclo[4.2.1]nonane are obtained in proportions depending on the conditions used, but the chlorine atom is always cis with respect to the nitrogen bridge.[489] Findlay[490] prepared racemic 2-methoxycarbonyl-9-methyl-3-oxo-9-azabicyclo[3.3.1]nonane by a Robinson-Schöpf synthesis and resolved it by way of its tartrate salt. Tursch and co-workers[491] synthesized adaline, which is present in lady bugs from the genus adalia (*Olesplera coccinellidae*), starting from 1-octen-3-ol. The absolute configuration for the natural product, 1*R*-3-oxo-1-*n*-pentyl-9-azabicyclo[3.3.1]nonane was deduced from the application of the octant rule to the ord curves.

The proximity of C–3 and C–7 is demonstrated in the pyrolysis of the sodium salt of the tosylhydrazone of *ψ*-pelletierine in diglyme at 165°C. An 80% yield of 9-methyl-9-azatricyclo[3.3.1.0³,⁷]nonane (**188**) is formed, that is, C–3 and C–7 are directly bonded.[492] Hofmann degradation of the methiodide caused cleavage in the expected manner and the all-*cis*-7-dimethylaminobicyclo[3.3.0]oct-2-ene (**189**) is obtained.[493]

1-Hydroxy-9-methyl-9-azabicyclo[3.3.1]nonane, prepared from the corresponding 9-oxa analogue, is dehydrated to 9-methyl-9-azabicyclo[3.3.1]-non-1-ene (**190**) by boiling in benzene with potassium *t*-amyloxide. This is yet another example which violates Bredt's rule. Ir and nmr evidence indicate no interaction of the nitrogen lone pair and the double bond. The nitrogen atom, however, participates most strongly in the solvolysis of 1-chloro-9-methyl-9-azabicyclo[3.3.1]nonane (**191**, X = NMe) as evidenced by the relative rates of **191** (X = NMe): **191** (X = CH₂): **191** (X = O): **191**

(X = S) which are 7.5 × 10⁸ : 93 : 30 : 1 respectively and obviously stabilizes the positive charge at C–1 (compare with Chapter 4, Sections VII.C.1 and D.2; Part II, Chapter 2, Section XIV.E; and Part II, Chapter 3, Section XIII).[494]

3. Miscellaneous Diazabicyclo[3.3.1]nonanes

The synthesis of 1,2-,[495] 1,3-,[496–499] 1,4-,[500] 1,5-,[501,502] 2,3-,[503] 2,7-,[504] and 3,9-diazabicyclo[3.3.1]nonanes[505,506] were reported and most evidence is consistent with a "chair-chair" conformation for the two six-membered rings. The facile formation of 1,5-diazabicyclo[3.3.1]nonanes from 1,5-diazacyclooctanes and aldehydes in 54–83% yields demonstrates that the nitrogen atoms are quite close together in space.[501] 3-Isopropyl-2-oxo-1,3-diazabicyclo[3.3.1]nonane (**192**), readily obtained from 3-(N-isopropylaminomethyl)piperidine and phosgene, is a relatively stable compound to hydrolysis, with a normal amide carbonyl ($\nu_{max} = 1650$ cm⁻¹)[499]—another violation of Bredt's rule. In contrast, 2-oxo-1-azabicyclo[2.2.2]octane (**193**) polymerizes readily on hydrolysis[507] and the carbonyl stretching frequency is high ($\nu = 1750$ cm⁻¹).[508,509] Careful evaluation of the coupling constants of 1,5-diazabicyclo[3.3.1]nonanes and 1,5-diazabicyclo[3.2.1]octanes indicates that although the hexahydropyrimidine rings are in the chair conformation there is substantial flattening of the rings.[510]

Several syntheses of 3,7-diazabicyclo[3.3.1]nonanes (bispidines) were described,[511,512] but the most common is by way of an intramolecular Mannich reaction between 4-piperidones, a primary amine, and formaldehyde.[513–516] The dipole moments and ¹H nmr spectra of bispidines are consistent with a slightly flattened chair-chair conformation.[517,518] The pKa values of bispidines, which titrate as monoamines, are much larger (> 1.5 units) than those of related piperidines. This is readily explained by the proximity of the nitrogen atoms and formation of a stable adamantane type of structure (**194**).[518,519]

The bispidine ring system is present in the lupin alkaloids, sparteine, ψ-sparteine, isosparteine, and related natural products.[229,519–524]

4. Azaadamantanes

Adamantane (**195**) is a symmetrical, rigid, and cagelike molecule but can undergo a variety of rearrangements which were reviewed recently.[525] Re-

192   **193**   **194**

placement of a methylene group or bridgehead carbon atom by a hetero-atom does not upset this symmetry. Heterocycles of the adamantane type with nitrogen, sulfur, oxygen, phosphorus, and mixed heteroatoms are known.[526] Although adamantane is itself not very interesting from the stereochemical point of view, reactions which lead to its synthesis give information about the proximity of groups in the intermediates. Thus the formation of azaadamantanes by some form of intramolecular cyclization of 3-aminobicyclo[3.3.1]nonanes[527–529] and aza derivatives[530,531] is evidence of this proximity between C–3 and C–7. The nitrogen atoms in bispidines (see Section VII.C.3 of this chapter) can be linked with aldehydes to form diazaadamantanes.[532,533] There is strong evidence from uv, nmr, pKa, and photoelectron spectroscopy that through-σ-bond interaction occurs between the nitrogen lone pair and the carbonyl π system in 6-oxo-1,3-diazaada-mantanes (**196**).[532–534] This is strictly due to the stereochemical arrange-

195

$R = CO_2Et, Ph$

196

ment of the nitrogen atoms with respect to the carbonyl groups imposed by the rigid structure. The carbonyl group does not behave in the usual way and the uv spectrum of **196** (R = CO$_2$Et) has $\lambda_{max} = 262$ nm ($\epsilon = 3600$). By comparison, 3,5-bisethoxycarbonyl-4-piperidone and 1,5-bisethoxycar-bonyl-9-oxo-bispidine have no absorption maximum in that region.[534] The system therefore behaves as a chromophore.

Photolysis of 1-azidoadamantane causes an exocyclic nitrogen atom to be inserted into a ring, and the azahomoadamantane (4-azatricyclo[4.3.1.1^{3,8}]-undecane) is formed. The intermediate which violates Bredt's rule must be formed initially and then reacted with the protic medium to yield the hy-droxy or alkoxy azahomoadamantanes. In nonprotic media (e.g., cyclohex-ane) it gives a dimer (4,13-diazaheptacyclo[13.3.1.1.^{3,17}1.^{6,10}1.^{8,11}0.^{3,13}0^{4,12}]-docosane), but it readily decomposes to 3-hydroxy-4-azatricyclo[4.3.1.1^{3,8}]-undecane. The latter can be obtained by intramolecular cyclization of *endo*-7-(aminomethyl)bicyclo[3.3.1]nonan-3-one (Scheme 5).[534a]

Hexamine, hexamethylene tetramine, is 1,3,5,7-tetraazaadamantane[535,536] and can be easily alkylated at room temperature. It was alkylated with chiral agents, for example, α-phenyl ethyl bromide or ethyl α-bromopro-pionate, and the ^{1}H nmr spectra of the quaternary salts demonstrated the intrinsic diastereotopic nature of the geminal methylene protons.[536]

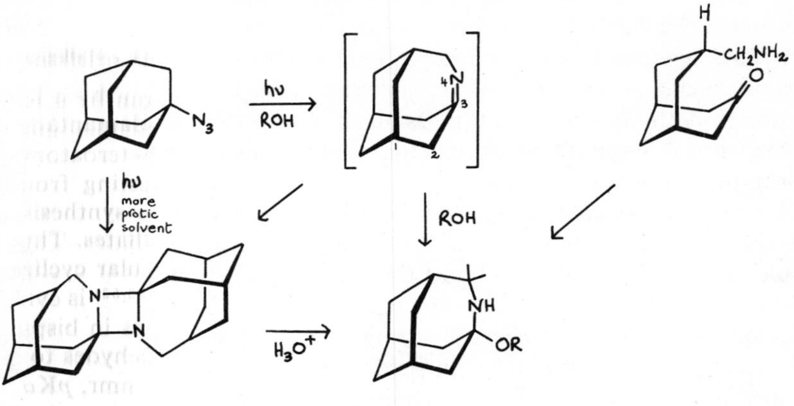

Scheme 5

E.   Miscellaneous Azabicyclononanes, -decanes, and -undecanes

9-Azabicyclo[4.2.1]nonenes are obtained by the 1,4-addition of cyanonitrene (from cyanogen azide) to cyclooctatetraene. The 1,2-adduct **197** is also formed but is less stable than the 1,4-adduct 9-cyano-9-azabicyclo-[4.2.1]nona-2,4,7-triene (**198**).[537] Oxidation of the triene (**198**) with peracetic acid demonstrated some stereoselectivity and gave a 1:3 mixture of 7-*exo*-epoxy-9-cyano-9-azabicyclo[4.2.1]nona-2,4-diene and 2-*exo*-epoxy-9-cyano-9-aza[4.2.1]nona-4,7-diene.[538] Hofmann elimination in 9,9-dimethyl-4-oxo-3-aza-9-azonia[4.2.1]nonane (**199**) proceeds with formation of 7-dimethylamino-2-oxo-1-azaoct-3-ene (**200**), a process which is reversed by transannular cyclization with perchloric acid. Similar transformations are exhibited by 10,10-dimethyl-4-oxo-3-aza-10-azoniabicyclo[4.3.1]decane and 11,11-dimethyl-4-oxo-3-aza-11-azoniabicyclo[4.4.1]undecane and suggest that the conformation of the bicyclic system is not very different from that of the monocyclic unsaturated lactam.[539]

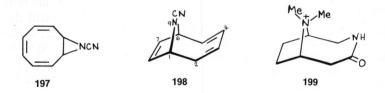

| 197 | 198 | 199 |

The uniparticulate electrophile chlorosulfonyl isocyanate adds across the 1,2- double bond of 1-methyl-, 1-methoxy-, and 1-phenylcyclooctatetraenes from the endo side of the boat to form the intermediate homotropylium cation (201) which cyclizes in two ways (at C–3 and C–5) when treated with sodium hydroxide in acetone. It affords, after alkylation with $Me_3O^+BF_4^-$, the 2- (202, $R^2 = H$) and 9-substituted (202, $R^1 = H$) 8-methoxy-7-azabicyclo[4.2.2]deca-2,4,7,9-tetraenes in varying proportions. When R is methyl, methoxy, or phenyl the ratios of 202 ($R^2 = H$, $R^1 = Me$, OMe or Ph) to 202 ($R^1 = H$, $R^2 = Me$, OMe, or Ph) are 65:35, 83:17, or 80:20 respectively.[540]

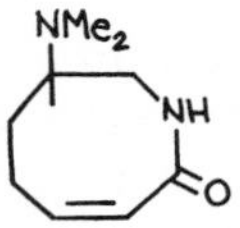

200 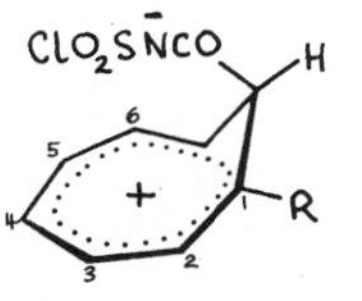   201 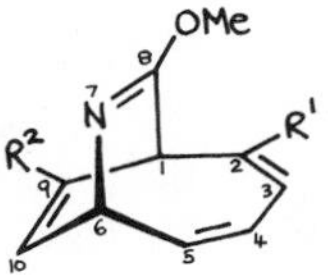   202

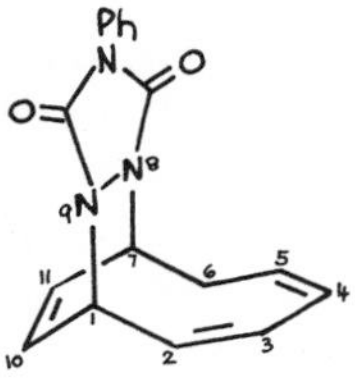

203 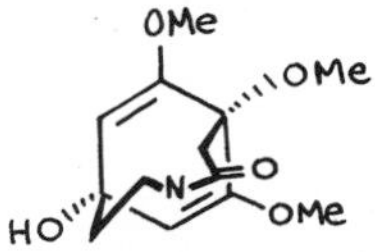   204 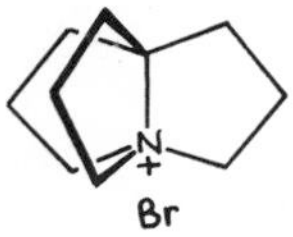   205

The cycloaddition of 4-phenyl-1,2,4-triazolin-3,5-dione to *cis,cis,cis,cis*-1,3,5,7-cyclononatetraene at −78°C is a (4 + 2)π and not a (6 + 2)π process and gives a 60% yield of 8,9-*N*-phenyliminodicarbonyl-8,9-diazabicyclo[5.2.2]undeca-2,4,10-triene (203).[541] Another bicycloundecane was obtained by photocyclization of *N*-chloroacetyl β-3,4,5-trimethoxyphenyl-ethylamine in aqueous ethanol. The three-dimensional structure of this 7-hydroxy-1,9,10-trimethoxy-3-oxo-4-azabicyclo[5.2.2]undeca-8,10-diene (204) was established by X-ray analysis.[542]

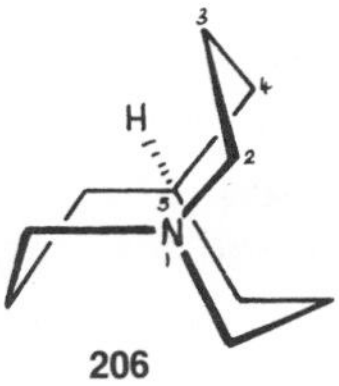

206

Leonard and co-workers[543] cleaved the bond between the bridgehead atoms of 1-azoniatricyclo[3.3.3.0]undecane bromide (205) with sodium in liquid ammonia and obtained the novel 1-azabicyclo[3.3.3]undecane (206) which they named "manaxine" after the emblem of the Isle of Man. The The carbocyclic analogue was called "manaxane."[543,544] This molecule showed interesting physical features which were attributed to flattening of the bridgehead regions which affected the hybridization of the nitrogen atom. The ^{1}H nmr spectra indicated that the molecule was flipping rapidly, the chemical shift of the bridgehead H–5 was at rather low field ($\delta = 2.57$ ppm), and ^{13}C nmr suggested that there was unusual hydridization around H–5 due to the proximity of the nitrogen atom. The pKa value (9.9 in H_2O) was rather low for a saturated amine (compare pKa of quinuclidine which is 10.9 in H_2O) and it absorbed in the uv with a λ_{max} at 240 nm ($\epsilon = 2935$ in Et_2O). The flattening of the structure due to strain in the molecule at the bridgehead atoms was confirmed by X-ray analysis of the hydrochloride.[545] A trinitodiaza derivative of manaxine, 3,7,10-triethyl-3,7,10-trinitro-1,5-diazabicyclo[3.3.3]undecane, was prepared from nitropropane, formaldehyde, and ammonia,[546] and the double-chair structure was confirmed by dipole moment measurements.[82]

207   208   209

VIII. TRÖGER BASE AND RELATED COMPOUNDS

Tröger base, 1,2′-methylene-3-(4′-tolyl)-6-methyl-1,2,3,4-tetrahydroquinazoline, was resolved into its optical antipodes by Prelog and Wieland[547] in 1944 by passing it through an activated d-lactose column. It was the first classical demonstration that optical activity was due to an asymmetric trivalent nitrogen atom. The specific rotations of the enantiomers 207 and 208 were $[\alpha]_D^{17} + (287 \pm 7°)$ (c 0.281, hexane) and $[\alpha]_D^{16.5} - (272° \pm 8°)$ (c 0.275, hexane), but varied with concentration. It racemizes slowly in $0.1N$ alcoholic hydrochloric acid probably because of cleavage of the –NCH$_2$N– group, but it is recovered unchanged on vacuum sublimation. Červinka, Fábryová, and Novák[548] measured the ord spectra of (+)-Tröger base and (–)-argemonine (209) and from the empirical comparison deduced that the absolute configuration in these was $1S,5S$. Mason, Schofield, and co-workers[549,550] analyzed the cd curves of $(+)$- and $(-)$-Tröger base in detail and reversed the configuration. Thus $(+)$-Tröger base has the $1R,5R$ configuration 207. This showed that the empirical approach was of limited value for stereochemical correlations. The $1S,5S$ (209) configuration of $(-)$-argemonine was confirmed by Baker and Battersby[551] by degradation to di-n-propyl $S(-)$-N,N-dimethylaspartate of known absolute configuration. For further details on the chemistry of Tröger base reference should be made to the monograph by Armarego.[552]

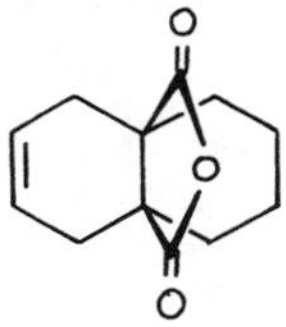

210

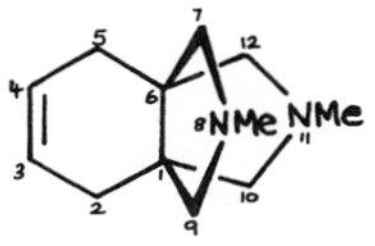

211

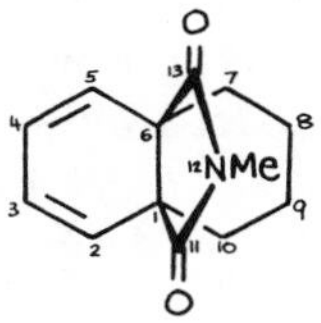

212

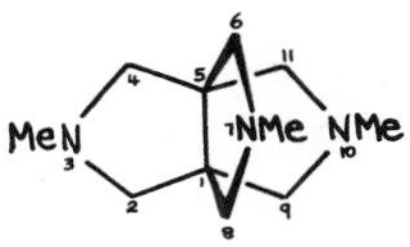

213

IX. PROPELLANES

A brief mention of heterocyclic analogues of propellanes should be made because they have interesting three-dimensional structures. One such compound, 205, has already been mentioned. Propellanes were explored almost entirely by Ginsburg and co-workers who have synthesized them from $4a,8a$-disubstituted hydronaphthalenes such as the anhydride 210. By using quite conventional syntheses they prepared 8,11-dimethyl-8,11-diazatricyclo[4.3.3.0^{1,6}]dodec-3-ene (211),[553] 12-methyl-11,13-dioxo-12-azatricyclo[4.4.3.0^{1,6}]trideca-2,4-diene (212),[554] and 3,7,10-trimethyl-3,7,10-triazatricyclo[3.3.3.0^{1,5}]undecane (213)[555] to mention only a few examples. They also prepared several propellanes with oxygen (see Part II, Chapter 2, Section XII.C), sulfur (see Part II, Chapter 3, Section XIV.E), and mixed heteroatoms in much the same way.[553] Unsaturated propellanes like 212 are capable of undergoing Diels-Alder reactions which are stereospecific.[554,556] Ginsburg reviewed the chemistry of propellanes in 1969[557] and in 1974.[557a]

X. AZABULLVALENES AND RELATED COMPOUNDS

Azabullvalenes and related compounds are described briefly here in order to demonstrate their structure. An awareness of the three-dimensional cage structure is necessary to understand its chemistry. We are indebted to Paquette and his school for much of our present knowledge on azabullvalene which, like bullvalene,[558] is in a state of rapid flux between valence-bond isomers.[559—561]

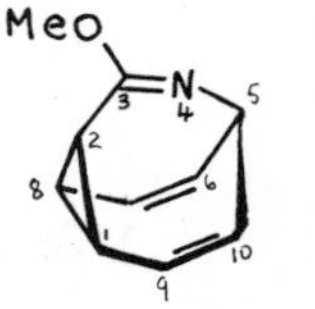

214

215

3-Methoxyazabullvalene, 3-methoxy-4-azatricyclo[3.3.2.0^{2,8}]deca-3,6,9-triene (**214**), is obtained by uv irradiation of 8-methoxy-7-azabicyclo[4.2.2]-deca-2,4,7,9-tetraene (**202**, $R^1 = R^2 = H$) or by methylation of the 3,4-dihydro-3-oxo derivative of **214** which was in turn prepared by irradiation of the 7,8-dihydro-8-oxo derivative of **202** ($R^1 = R^2 = H$).[560,562] Benzazabullvalene was similarly synthesized from the 2,3-benzo derivative of **202** which was prepared from benzocyclooctatetraene and chlorosulfonyl isocyanate followed by methylation.[563,564] Variable temperature ^1H nmr demonstrated that the bonds are in a state of flux with the three-membered ring moving around the molecule as in bullvalene. There is no evidence of such valence-bond tautomerism in the 3,4-dihydro derivative (**215**).[562]

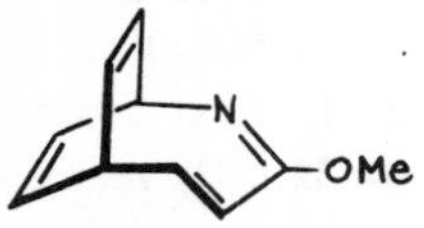

216

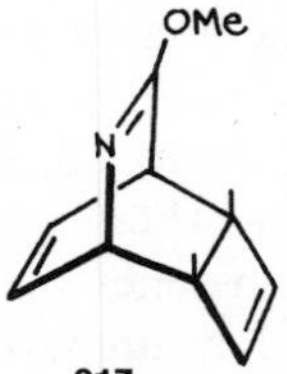

217

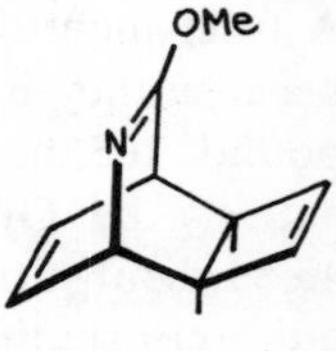

218

The pyrolysis of methoxyazabullvalenes at 600°C produces methoxy-quinolines and isoquinolines in much the same ratio as in the pyrolysis of the bicyclo[4.2.2]decatetraene (**202**).[565,566] Photolysis of 3-methoxy-4-azabullvalene (**214**) yields two bicycloazadecatetraenes, **202** ($R^1 = R^2 = H$) and **216**, and two tricycloazadecatrienes, **217** and **218**, in varying proportions depending on the nature of the solvent.[567] Reduction of the 3,4-dihydro-3-oxo derivative of **214** with LAH caused a rearrangement as well, and 7-azabicyclo[3.3.2]deca-2,9-diene (**219**) was formed.[564]

219   220   221

A compound related to azabullvalene but short of two carbon atoms (220) has also been reported. Semiazabullvalene (220) was prepared by Paquette and Krow[568] in three steps from Dewar benzene (bicyclo[2.2.0]-hexa-2,5-diene) and chlorosulfonyl isocyanate. Anastassiou and co-workers[569] prepared 9-cyano-9-azatricyclo[3.3.1.0^{2,8}]nona-3,6-diene (221), which is one carbon short of bullvalene, by photolysis of 9-cyano-9-azabicyclo-[4.2.1]nona-2,4,7-triene and called it 9-azabarbaralane. The nmr spectrum of tropilium azide is consistent with the sandwich ion structure 222.[570] Theoretical calculations on 3,7-diazasemibullvalene, 3,7-diazatricyclo[3.3.0.0^{2,8}]-octa-3,6-diene, predict that the classical structure should be unstable and that it should exist as a nonclassical bishomopyrazine, for example, 223.[571]

222   223

XI.   REFERENCES

1. T. J. van Bergen and R. M. Kellogg, *J. Org. Chem.*, **36**, 978 (1971).

2. L. A. Paquette and R. J. Haluska, *J. Org. Chem.*, **35**, 132 (1970).

3. V. Nair, *J. Org. Chem.*, **37**, 802 (1972).

4. E. Niemers, W. Neagele, and D. Wendisch, *Tetrahedron Letters*, 3791 (1970).

5. I. C. Paul, S. M. Johnson, L. A. Paquette, J. H. Barrett, and R. J. Haluska, *J. Amer. Chem. Soc.*, **90**, 5023 (1968).

6. L. A. Paquette, *J. Amer. Chem. Soc.*, **86**, 4096 (1964).

7. A. Mannschreck, G. Rissman, F. Vögtle, and D. Wild, *Chem. Ber.*, **100**, 335 (1967).

8. J. M. Lehn and J. Wagner, *Chem. Comm.*, 414 (1970).

9. E. A. Noe and J. D. Roberts, *J. Amer. Chem. Soc.*, **93**, 7261 (1971).

10. H. Ogura, H. Takayanagi, K. Kubo, and K. Furuhata, *J. Amer. Chem. Soc.*, **95**, 8056 (1973).

11. R. J. Abraham, L. J. Kricka, and A. Ledwith, *J.C.S. Chem. Comm.*, 282 (1973).

12. A. L. Johnson and H. E. Simmons, *J. Amer. Chem. Soc.*, **89**, 3191 (1967).

13. A. L. Johnsón and H. E. Simmons, *J. Amer. Chem. Soc.*, **88**, 2591 (1966).

14. L. A. Paquette and J. H. Barrett, *J. Amer. Chem. Soc.*, **88**, 2590 (1966).

15. L. A. Paquette, J. H. Barrett, and D. E. Kuhla, *J. Amer. Chem. Soc.*, **91**, 3616 (1969)

16. I. C. Paul, S. M. Johnson, J. H. Barrett, and L. A. Paquette, *Chem. Comm.*, 6 (1969).

17. S. M. Johnson and I. C. Paul, *J. Chem. Soc.* (*B*), 1244 (1969).

18. G. Habermehl and S. Göttlicher, *Angew. Chem. Internat. Edn.*, **6**, 805 (1967).

19. A. L. Johnson and H. E. Simmons, *J. Org. Chem.*, **34**, 1139 (1969).

20. J. E. Baldwin and R. A. Smith, *J. Amer. Chem. Soc.*, **87**, 4819 (1965).

21. A. S. Kende, P .T. Izzo, and J. E. Lancaster, *J. Amer. Chem. Soc.*, **87**, 5044 (1965).

22. R. A. Smith, J. E. Baldwin, and I. C. Paul, *J. Chem. Soc.* (*B*), 112 (1967).

23. J. H. van den Hende and A. S. Kende, *Chem. Comm.*, 384 (1965).

24. L. A. Paquette, D. E. Kuhla, J. H. Barrett, and L. M. Leichter, *J. Org. Chem.*, **34**, 2888 (1969).

25. T. Saski, K. Kanematsu, and K. Hayakawa, *J. Chem. Soc.* (*C*), 2142 (1971).

26. W. S. Murphy and J. P. McCarthy, *Chem. Comm.*, 1129 (1970).

27. R. Shapiro and S. Nesnow, *J. Org. Chem.*, **34**, 1695 (1969).

28. T. H. Koch and D. A. Brown, *J. Org. Chem.*, **36**, 1934 (1971).

29. W-D. Stohrer, *Chem. Ber.*, **106**, 970 (1973).

30. D. M. Hall and E. E. Turner, *J. Chem. Soc.*, 1242 (1955).

31. S. R. Ahmed and D. M. Hall, *J. Chem. Soc.*, 3043 (1958).

32. D. M. Hall and M. M. Harris, *J. Chem. Soc.*, 490 (1960).

33. S. R. Ahmed and D. M. Hall, *J. Chem. Soc.*, 4165 (1960).

34. D. M. Hall and T. M. Poole, *J. Chem. Soc.* (*B*), 1034 (1966).

35. D. M. Hall and J. M. Insole, *J.C.S. Perkin II*, 1164 (1972).

36. P. A. Browne and D. M. Hall, *J.C.S. Perkin I*, 2717 (1972).

37. J. M. Insole, *J.C.S. Perkin II*, 1168 (1972).

38. D. Muriel Hall, *Progr. Stereochem.* **4**, 1 (1968).

39. P. Newman, P. Rutkin, and K. Mislow, *J. Amer. Chem. Soc.*, **80**, 465 (1958).

40. M. Siegel and K. Mislow, *J. Amer. Chem. Soc.*, **80**, 473 (1958).

41. F. A. McGinn, A. K. Lazarus, M. Siegel, J. E. Ricci, and K. Mislow, *J. Amer. Chem. Soc.*, **80**, 476 (1958).

42. D. D. Fitts, M. Siegel, and K. Mislow, *J. Amer. Chem. Soc.*, **80**, 480 (1958).

43. K. Mislow, *Angew. Chem.*, **70**, 683 (1958).

44. H. Joshua, R. Gans, and K. Mislow, *J. Amer. Chem. Soc.*, **90**, 4884 (1968).

45. A. R. Lepley and A. G. Giumanini in *Mechanisms of Molecular Migrations*, Vol. 3, B. S. Thyagarajan, Ed., Wiley-Interscience, New York, 1971, p. 297.

46. I. O. Sutherland and M. V. J. Ramsay, *Tetrahedron*, **21**, 3401 (1965).

47. K. Mislow, M. A. W. Glass, H. B. Hopps, E. Simon, and G. H. Wahl, Jr., *J. Amer. Chem. Soc.*, **86**, 1710 (1964); N. V. Riggs and S. M. Verma, *Austral. J. Chem.*, **23**, 1913 (1970).

48. T. Sasaki, K. Kanematsu, A. Kakehi, I. Ichikawa, and K. Hayakawa, *J. Org. Chem.*, **35**, 426 (1970).

49. A. A. Reid, J. T. Sharp, H. R. Sood, and P. B. Thorogood, *J.C.S. Perkin I*, 2543 (1973).

50. A. Steigel, J. Sauer, D. A. Kleier, and G. Binsch, *J. Amer. Chem. Soc.*, **94**, 2770 (1972).

51. R. Gerdil, *Helv. Chim. Acta*, **55**, 2159 (1972).

52. J. N. Brown, R. L. Towns, and L. M. Trefonas, *J. Amer. Chem. Soc.*, **92**, 7436 (1970).

53. A. Neszmélyi, E. Gács-Baitz, G. Horváth, T. Láng, and J. Körösi, *Chem. Ber.*, **107**, 3894 (1974).

54. B. Willig and J. Streith, *Tetrahedron Letters*, 4167 (1973).

55. G. Kan, M. T. Thomas, and V. Snieckus, *Chem. Comm.*, 1022 (1971).

56. J-L. Derocque, W. T. Theuer, and J. A. Moore, *J. Org. Chem.*, **33**, 4381 (1968).

57. S. M. Rosen and J. A. Moore, *J. Org. Chem.*, **37**, 3770 (1972).

58. J. A. Moore, F. J. Marascia, R. W. Medeiros, and R. L. Wineholt, *J. Org. Chem.*, **31**, 34 (1966).

59. O. S. Rothenberger, R. T. Taylor, D. L. Dalrymple, and J. A. Moore, *J. Org. Chem.*, **37**, 2640 (1972).

60. C. G. Overberger and J. G. Lombardino, *J. Amer. Chem. Soc.*, **80**, 2317 (1958).

61. W. R. Dolbier, Jr., and W. M. Williams, *J. Amer. Chem. Soc.*, **91**, 2818 (1969).

62. P. W. W. Hunter, and G. A Webb, *Tetrahedron*, **29**, 147 (1973).

63. P. Lincheid and J. M. Lehn, *Bull. Soc. chim. France*, 992 (1967).

64. R. H. McDougall and S. H. Malik, *J. Chem. Soc. (C)*, 2044 (1969).

65. J. A. Baltrop, C. G. Richards, and D. M. Russell, *J. Chem. Soc.*, 1423 (1959).

66. J. M. Insole, *J. Chem. Soc. (C)*, 1712 (1971).

66a. J. A. Moore and E. Mitchell in *Heterocyclic Compounds*, R. C. Elderfield, Ed., Vol. 9, Wiley, New York, 1967, p. 224.

67. J. S. Lambert and S. A. Khan, *J. Org. Chem.*, **40**, 369 (1975).

68. S. R. Ahmed and D. M. Hall, *J. Chem. Soc.*, 3383 (1959).

68a. R. R. Frazer, M. A. Raza, R. N. Renaud, and R. B. Layton, *Canad. J. Chem.*, **53**, 167 (1975).

69. C. G. Overberger, J. W. Stoddard, C. Yaroslavsky, H. Katz, and J-P. Anselme, *J. Amer. Chem. Soc.*, **91**, 3226 (1969); C. G. Overberger and I. Tashlick, *J. Amer. Chem. Soc.*, **81**, 217 (1959).

70. C. G. Overberger and J. W. Stoddard, *J. Amer. Chem. Soc.*, **92**, 4922 (1970).

71. C. G. Overberger, M. S. Chi, D. G. Pucci, and J. A. Barry, *Tetrahedron Letters*, 4565 (1972).

71a. H. Prinzbach, M. Breuninger, B. Gallenkamp, R. Schwesinger, and D. Hunkler, *Angew. Chem. Internat. Edn.*, **14**, 348 (1975).

72. R. Crossley, A. P. Downing, M. Nógrádi, A. Braga de Oliveira, W. D. Ollis, and O. I. Sutherland, *J.C.S. Perkin I*, 205 (1973).

73. A. Saunders and J. M. Sprake, *J.C.S. Perkin II*, 1660 (1972).

74. W. Metlesics, R. Taveres, and L. H. Sternbach, *J. Org. Chem.*, **31**, 3356 (1966).

75. W. D. Ollis, J. F. Stoddart, and I. O. Sutherland, *Tetrahedron*, **30**, 1903 (1974).

76. A. Albert and H. Yamamoto, *J. Chem. Soc. (C)*, 1944 (1968).

77. W. W. Paudler, G. R. Gapski, and J. M. Barton, *J. Org. Chem.*, **31**, 277 (1966).

78. W. W. Paudler and A. G. Zeiler, *J. Org. Chem.*, **32**, 2425 (1967).

79. W. W. Paudler, A. G. Zeiler, and G. R. Gapski, *J. Org. Chem.*, **34**, 1001 (1969).

80. D. A. Nelson, J. J. Worman, and B. Keen, *J. Org. Chem.*, **36**, 3361 (1971).

81. W. P. Kensen, J. J. Worman, and O. D. Filbey, *J. Heterocyclic Chem.*, **9**, 145 (1972).

82. R. Kolínski, H. Piotrowska, and T. Ubrański, *J. Chem. Soc.*, 2319 (1958).

83. D. N. J. White and J. D. Dunitz, *Israel J. Chem.*, **10**, 249 (1972).

83a. D. N. J. White and M. P. H. Guy, *J.C.S. Perkin II*, 43 (1975).

84. M. Bedard, H. Huber, J. L. Myers, and G. F. Wright, *Canad. J. Chem.*, **40**, 2278 (1962).

85. P. G. Hall and G. S. Horsfall, *J.C.S. Perkin II*, 1280 (1973).

86. S. Masamune, K. Hojo, and S. Takada, *Chem. Comm.*, 1204 (1969).

87. A. G. Anastassiou and J. H. Gebrian, *J. Amer. Chem. Soc.*, **91**, 4011 (1969).

88. S. Masamune and N. T. Castellucci, *Angew. Chem. Internat. Edn.*, **3**, 582 (1964).

89. A. G. Anastassiou, R. P. Cellura, and H. J. Gebrian, *Chem. Comm.*, 375 (1970).

90. K. Hojo and S. Masamune, *J. Amer. Chem. Soc.*, **92**, 6690 (1970).

91. A. G. Anastassiou, R. L. Elliott, H. W. Wright, and J. Clardy, *J. Org. Chem.*, **38**, 1959 (1973).

92. W. D. Ollis and J. F. Stoddart, *Angew. Chem. Internat. Edn.*, **13**, 728 (1974).

93. D. J. Brickwood, W. D. Ollis, and J. F. Stoddart, *Angew. Chem. Internat. Edn.*, **13**, 731 (1974).

94. J. A. Berson, *J. Amer. Chem. Soc.*, **78**, 4170 (1956).

94a. H. Prinzbach, R. Schwesinger, M. Breuninger, B. Gallenkamp, and D. Hunkler, *Angew. Chem. Internat. Edn.*, **14**, 347 (1975).

95. N. J. Leonard, M. Ōki, and S. Chiavarelli, *J. Amer. Chem. Soc.*, **77**, 6234 (1955).

96. N. J. Leonard, R. C. Fox, and M. Ōki, *J. Amer. Chem. Soc.*, **76**, 5708 (1954).

97. N. J. Leonard and M. Ōki, *J. Amer. Chem. Soc.*, **76**, 3463 (1954).

98. N. J. Leonard and M. Ōki, *J. Amer. Chem. Soc.*, **77**, 6241 (1955).

99. N. J. Leonard and M. Ōki, *J. Amer. Chem. Soc.*, **77**, 6245 (1955).

100. N. J. Leonard, M. Ōki, J. Brader, and H. Boaz, *J. Amer. Chem. Soc.*, **77**, 6237 (1955).

101. N. J. Leonard and M. Ōki, *J. Amer. Chem. Soc.*, **77**, 6239 (1955).

102. N. J. Leonard, D. F. Morrow, and M. T. Rogers, *J. Amer. Chem. Soc.*, **79**, 5476 (1957).

103. N. J. Leonard, J. A. Adamcik, C. Djerassi, and O. Halpern, *J. Amer. Chem. Soc.*, **80**, 4858 (1958).

104. H. Yamabe, H. Kato, and T. Yonezawa, *Bull. Chem. Soc. Japan*, **44**, 611 (1971).

105. N. J. Leonard and A. E. Yethon, *Tetrahedron Letters*, 4259 (1965).

106. E. H. Mottus, H. Schwarz, and L. Marion, *Canad. J. Chem.*, **31**, 1144 (1953).

107. R. A. Johnson, M. E. Herr, H. C. Murray, and G. S. Fonken, *J. Org. Chem.*, **33**, 3187 (1968).

108. N. J. Leonard and T. Sato, *J. Org. Chem.*, **34**, 1066 (1969).

109. R. A. Johnson, *J. Org. Chem.*, **37**, 312 (1972).

110. M. G. Reinecke, L. R. Kray, and R. F. Francis, *J. Org. Chem.*, **37**, 3489 (1972).

111. M. G. Reinecke and R. F. Francis, *J. Org. Chem.*, **37**, 3494 (1972).

112. M. G. Reinecke and R. G. Daubert, *J. Org. Chem.*, **38**, 3281 (1973).

113. Y. Arata, S. Yoshifuji, and Y. Yasuda, *Chem. and Pharm. Bull.* (*Japan*), **16**, 569 (1968).

114. E. W. Warnoff and P. Reynolds-Warnhoff, *Canad. J. Chem.*, **51**, 2338 (1973).

115. N. J. Leonard and J. A. Klainer, *J. Org. Chem.*, **33**, 4269 (1968).

115a. O. E. Edwards, D. Vocelle, J. W. ApSimon, and F. Haque, *J. Amer. Chem. Soc.*, **87**, 678 (1965).

116. F. Vögtle and P. Neumann, *Tetrahedron*, **26**, 5847 (1970).

117. F. Vögtle and P. Neumann, *Angew. Chem. Internat. Edn.*, **11**, 73 (1972).

118. F. Vögtle and P. Neumann, *Tetrahedron Letters*, 115 (1970).

119. A. Maquestiau, Y. Van Haverbeke, R. Flammang, and C. DuBray, *Tetrahedron Letters*, 3645 (1970).

120. H. Nozaki, T. Koyama, and T. Mori, *Tetrahedron*, **25**, 5357 (1969).

121. T. Hiyama, S. Hirano, and H. Nozaki, *J. Amer. Chem. Soc.*, **96**, 5287 (1974).

122. S. Fujita, K. Imamura, and H. Nozaki, *Bull. Chem. Soc. Japan*, **45**, 1881 (1972).

123. H. Gerlach and E. Huber, *Helv. Chim. Acta*, **51**, 2027 (1968).

124. W. E. Parham, R. W. Davenport, and J. B. Biasotti, *Tetrahedron Letters*, 557 (1969).

125. W. E. Parham, R. W. Davenport, and J. B. Biasotti, *J. Org. Chem.*, **35**, 3775 (1970).

126. W. E. Parham, D. C. Egberg, and S. S. Salgar, *J. Org. Chem.*, **37**, 3248 (1972).

127. W. E. Parham, K. B. Sloan, K. R. Reddy, and P. E. Olson, *J. Org. Chem.*, **38**, 927 (1973).

128. W. E. Parham, P. E. Olson, K. R. Reddy, and K. B. Sloan, *J. Org. Chem.*, **39**, 172 (1974).

129. W. E. Parham, P. E. Olson, and K. R. Reddy, *J. Org. Chem.*, **39**, 2432 (1974).

130. W. E. Parham, and P. E. Olson, *J. Org. Chem.*, **39**, 2916 (1974).

131. W. E. Parham and P. E. Olson, *J. Org. Chem.*, **39**, 3407 (1974).

132. I. Gault, B. J. Price, and I. O. Sutherland, *Chem. Comm.*, 540, (1967).

133. F. Vögtle and A. H. Effler, *Chem. Ber.*, **102**, 3071 (1969).

133a. J. F. Haley, Jr., and P. M. Keehn, *Tetrahedron Letters*, 1675 (1975).

134. S. Rosenfeld and P. M. Keehn, *Tetrahedron Letters*, 4021 (1973).

135. F. Vögtle and P. Neumann, *Chem. Comm.*, 1464 (1970).

136. B. Kainradl, E. Langer, H. Lehner and K. Schlögl, *Annalen*, **766**, 16 (1972).

137. M. Havel, J. Krupička, J. Sicher, and M. Svoboda, *Tetrahedron Letters*, 4009 (1967).

138. J. Smolíková, M. Havel, S. Vašičková, A. Vítek, M. Svoboda, and K. Bláha, *Coll. Czech. Chem. Comm.*, **39**, 459 (1974).

139. A. Feigenbaum and J. M. Lehn, *Bull. Soc. chim. France*, 198 (1973).

140. J. Smolíkova, M. Havel, S. Vašičková, A. Vítek, M. Svoboda, and K. Bláha, *Coll. Czech. Chem. Comm.*, **39**, 293 (1974).

141. J. Dale and R. Coulon, *J. Chem. Soc.*, 182 (1964).

142. J. Dale, *J. Chem. Soc.*, 93 (1963).

143. R. M. Moriarty, *J. Org. Chem.*, **29**, 2748 (1964).

144. R. M. Moriarty and J. M. Kleigman, *J. Org. Chem.*, **31**, 3007 (1966).

145. R. M. Moriarty, E-L. Yeh, V. A. Curtis, C-L. Yeh, J. L. Flippen, J. Karle, and K. C. Ramey, *J. Amer. Chem. Soc.*, **94**, 6871 (1972).

146. L. C. Vining and W. A. Taber, *Canad. J. Chem.*, **35**, 1109 (1957).

147. K. Vogler, R. O. Studer, W. Lergier, and P. Lanz, *Helv. Chem. Acta*, **43**, 1751 (1960); R. O. Studer, K. Vogler, and W. Lergier, *Helv. Chim. Acta*, **44**, 131 (1961).

148. K. Vogler, R. O. Studer, P. Lanz, W. Lergier, and E. Böhni, *Helv. Chim. Acta*, **48**, 1161 (1965) and **48**, 1371 (1965).

149. W. Keller-Schierlein and A. Deér, *Helv. Chim. Acta*, **46**, 1907 (1963); W. Keller-Schierlein, *Helv. Chim. Acta*, **46**, 1920 (1963).

150. W. Keller-Schierlein, P. Mertens, V. Prelog, and A. Walser, *Helv. Chim. Acta*, **48**, 710 (1965).

151. G. Wittig and J. E. Grolig, *Chem. Ber.*, **94**, 2148 (1961).

152. R. Huisgen, H. König, and A. R. Lepley, *Chem. Ber.*, **93**, 1496 (1960).

153. A. Lüttringhaus and H. Simon, *Annalen*, **557**, 120 (1947).

154. G. Schill and H. Neubauer, *Annalen*, **750**, 76 (1971).

155. G. Schill and H. Zollenkopf, *Annalen*, **721**, 53 (1969).

156. G. Schill, C. Zürcher, and W. Vetter, *Chem. Ber.*, **106**, 228 (1973).

157. G. Schill, *Chem. Ber.*, **100**, 2021 (1967); W. Vetter and G. Schill, *Tetrahedron*, **23**, 3079 (1967).

158. G. Schill, K. Murjahn, and W. Vetter, *Annalen*, **740**, 18 (1970).

159. G. Schill, E. Logemann, and W. Vetter, *Angew. Chem. Internat. Edn.*, **11**, 1089 (1972).

160. G. Schill, R. Henschel, and J. Boeckmann, *Annalen*, 709 (1974).

161. G. Schill and W. Vetter, *Chem. Ber.*, **104**, 3582 (1971).

162. G. Schill, *Catenanes, Rotaxanes and Knots*, Academic, New York, 1971.

163. J. Almog, J. E. Baldwin, R. L. Dyer, and M. Peters, *J. Amer. Chem. Soc.*, **97**, 226 (1975).

164. J. Almog, J. E. Baldwin, and J. Huff, *J. Amer. Chem. Soc.*, **97**, 227 (1975).

165. J. E. Richman and T. J. Atkins, *J. Amer. Chem. Soc.*, **96**, 2268 (1974).

166. J. Cheney and J. M. Lehn, *J.C.S. Chem. Comm.*, 487 (1972).

167. G. Volpp, *Chem. Ber.*, **95**, 1493 (1962).

168. W. L. Mosby, *Heterocyclic Systems with Bridgehead Nitrogen Atoms*, Parts 1 and 2, Interscience, New York, 1961.

169. N. K. Kochetkov and A. M. Likhosherstov, *Adv. Heterocyclic Chem.*, **5**, 315 (1965).

170. N. J. Leonard in *The Alkaloids*, Vol. 1, R. H. F. Manske and H. L. Holmes, Eds., Academic, New York, 1950, Ch. 4, p. 107; N. J. Leonard in *The Alkaloids*, Vol. VI, Academic, New York, 1960, Ch. 3, p. 35.

171. L. B. Bull, C. C. J. Culvenor, and A. T. Dick, *The Pyrrolizidine Alkaloids*, North-Holland, Amsterdam, 1968.

172. F. L. Warren and M. E. von Klemperer, *J. Chem. Soc.*, 4574 (1958).

173. R. Adams and D. Fleš, *J. Amer. Chem. Soc.*, **81**, 4946 (1959).

174. R. Adams and N. J. Leonard, *J. Amer. Chem. Soc.*, **66**, 257 (1944).

175. R. Adams and D. Fleš, *J. Amer. Chem. Soc.*, **81**, 5803 (1959).

176. R. Adams and B. L. ban Duuren, *J. Amer. Chem. Soc.*, **76**, 6379 (1954).

177. N. J. Leonard and D. L. Felley, *J. Amer. Chem. Soc.*, **72**, 2537 (1950).

178. T. A. Geissman and A. C. Waiss, Jr., *J. Org. Chem.*, **27**, 139 (1962).

179. J. Fridrichsons, A. McL. Matheison, and D. J. Sutor, *Tetrahedron Letters*, No. 23, 35 (1960).

180. J. A. Wunderlich, *Chem. and Ind.*, 2089 (1962).
181. C. C. J. Culvenor, D. H. G. Crout, W. Klyne, W. P. Mose, J. D. Renwick, and P. M. Scopes, *J. Chem. Soc.* (*C*), 3653 (1971).
182. C. C. J. Culvenor, M. L. Heffernan, and W. G. Woods, *Austral. J. Chem.*, **18**, 1605 (1965).
183. I. M. Skvortsov and J. A. Elvidge, *J. Chem. Soc.* (*B*), 1589 (1968).
183a. H. S. Aaron, C. P. Rader, Jr., and G. E. Wicks, *J. Org. Chem.*, **31**, 3502 (1966).
184. M. T. Pizzorno and S. M. Albonico, *J. Org. Chem.*, **39**, 731 (1974).
185. R. Adams and E. F. Rogers, *J. Amer. Chem. Soc.*, **63**, 537 (1941).
186. C. C. J. Culvenor and L. W. Smith, *Austral. J. Chem.*, **14**, 284 (1961).
187. C. C. J. Culvenor and L. W. Smith, *Austral. J. Chem.*, **12**, 255 (1959).
187a. L. J. Dry, M. J. Koekemoer, and F. L. Warren, *J. Chem. Soc.*, 59 (1955).
188. O. Yonemitsu, Y. Okuno, Y. Kanaoka, I. L. Karle, and B. Witkop, *J. Amer. Chem. Soc.*, **90**, 6522 (1968).
189. J. E. Richman and H. E. Simmons, *Tetrahedron*, **30**, 1769 (1974).
190. H. L. Holmes in *The Alkaloids*, Vol. II, R. H. F. Manske and H. L. Holmes, Eds., Academic, New York, 1952, Ch. 15, p. 513.
191. G. F. Smith in *The Alkaloids*, Vol. VIII, R. H. F. Manske, Ed., Academic, New York, 1965, Ch. 17, p. 591.
192. L. Marion in *The Alkaloids*, Vol. II, R. H. F. Manske and H. L. Holmes, Eds., Academic, New York, 1952, Ch. 14, p. 499.
193. L. N. Mander, R. H. Prager, M. Rasmussen, E. Ritchie, and W. C. Taylor, *Austral. J. Chem.*, **20**, 1473 (1967).
194. N. K. Hart, S. R. Johns, and J. A. Lamberton, *Chem. Comm.*, 460 (1971).
195. G. R. Clemo, N. Fletcher, G. R. Fulton, and R. Raper, *J. Chem. Soc.*, 1140 (1950).
196. N. J. Leonard and W. J. Middleton, *J. Amer. Chem. Soc.*, **74**, 5776 (1952).
197. E. Lellmann, *Ber.*, **23**, 2141 (1890).
198. A. Crabtree, A. W. Johnson, and J. C. Tebby, *J. Chem. Soc.*, 3497 (1961).
199. N. J. Leonard and S. H. Pines, *J. Amer. Chem. Soc.*, **72**, 4931 (1950).
200. M. F. Grundon and B. E. Reynolds, *J. Chem. Soc.*, 3898 (1963).
201. G. R. Clemo and T. P. Metcalfe, *J. Chem. Soc.*, 1518 (1937).
202. G. R. Clemo, W. McG. Morgan, and R. Raper, *J. Chem. Soc.*, 1743 (1935).
203. G. R. Clemo, T. P. Metcalfe, and R. Raper, *J. Chem. Soc.*, 1429 (1936).
204. R. A. Gardiner, K. L. Rinehart, Jr., J. J. Snyder, and H. P. Broquist, *J. Amer. Chem. Soc.*, **90**, 5639 (1968).
205. W. J. Gensler and M. W. Hu, *J. Org. Chem.*, **38**, 3848 (1973).
205a. P. E. Sonnet and J. E. Oliver, *J. Heterocyclic Chem.*, **12**, 289 (1975).
206. G. R. Clemo and G. R. Ramage, *J. Chem. Soc.*, 2969 (1932).
207. V. Prelog and R. Seiwerth, *Ber.*, **72**, 1638 (1939); N. J. Leonard and W. C. Wildman, *J. Amer. Chem. Soc.*, **71**, 3089 (1949); N. J. Leonard and E. Barthel, *J. Amer. Chem. Soc.*, **71**, 3098 (1949).
208. W. L. Meyer and N. Sapianchiay, *J. Amer. Chem. Soc.*, **86**, 3343 (1964).
209. H. S. Aaron and C. P. Ferguson, *Tetrahedron Letters*, 6191 (1968).
210. D. M. Speros and F. D. Rossini, *J. Phys. Chem.*, **64**, 1723 (1960).

211. N. L. Allinger and J. L. Coke, *J. Amer. Chem. Soc.*, **82**, 2553 (1960).

212. E. L. Eliel, N. L. Allinger, S. J. Angyal, and G. A. Morrison, *Conformational Analysis*, Interscience, New York, 1965, p. 230.

213. T. A. Crabb and R. F. Newton, *Tetrahedron Letters*, 1551 (1970).

214. R. Cahill, T. A. Crabb, and R. F. Newton, *Org. Magn. Resonance*, **3**, 263 (1971).

215. A. E. Theobald and R. G. Lingard, *Spectochim. Acta*, **24**, 1245 (1968).

216. B. Lüning and C. Lundin, *Acta Chem. Scand.*, **21**, 2136 (1967).

217. C. P. Rader, R. L. Young, Jr., and H. S. Aaron, *J. Org. Chem.*, **30**, 1536 (1965); E. P. Hibbett and J. Sam, *J. Heterocyclic Chem.*, **7**, 857 (1970).

218. H. S. Aaron, C. P. Rader, and G. E. Wicks, Jr., *J. Org. Chem.*, **31**, 3502 (1966).

219. T. A. Crabb and R. F. Newton, *J. Heterocyclic Chem.*, **6**, 301 (1969).

220. P. J. Chivers, T. A. Crabb, and R. O. Williams, *Tetrahedron*, **24**, 6625 (1968).

221. P. J. Chivers, T. A. Crabb, and R. O. Williams, *Tetrahedron*, **25**, 2921 (1969).

222. P. J. Chivers and T. A. Crabb, *Tetrahedron*, **26**, 3389 (1970).

223. M. E. Freed and A. R. Day, *J. Org. Chem.*, **25**, 2108 (1960).

224. T. A. Crabb and J. Mitchell, *J. Heterocyclic Chem.*, **8**, 721 (1971).

225. T. A. Crabb in M.T.P. *International Review of Science*, Vol. 4, *Heterocyclic Compounds*, K. Schofield, Ed., Butterworths, London, 1973, pp. 291–317.

226. T. A. Crabb, R. F. Newton, and D. Jackson, *Chem. Rev.*, **71**, 109 (1971).

227. W. L. F. Armarego, *J. Chem. Soc.*, 4226 (1964).

228. W. L. F. Armarego, *J. Chem. Soc.*, (B) 191 (1966).

229. N. J. Leonard in *The Alkaloids*, R. H. F. Manske, Ed., Academic, New York, 1953, Vol. III, Ch. 19, p. 119; 1959, Vol. VII, Ch. 14, p. 253.

230. A. R. Battersby and H. F. Hodson in *The Alkaloids*, Vol. VIII, R. H. F. Manske, Ed., Academic, New York, 1965, Ch. 15, p. 515.

231. G. A. Morrison, *Fortschr. Chem. org. Naturstoffe.*, **25**, 269 (1967).

232. E. Schlittler in *The Alkaloids*, Vol. VIII, R. H. F. Manske, Ed., Academic, New York, 1965, Ch. 13, p. 287.

233. W. I. Taylor in *The Alkaloids*, Vol. VIII, R. H. F. Manske, Ed., Academic, New York, 1965, Ch. 11, p. 249.

234. H. S. Aaron, *Chem. and Ind.*, 1338 (1965).

235. C. D. Johnson, R. A. Y. Jones, A. R. Katritzky, C. R. Palmer, K. Schofield, and R. J. Wells, *J. Chem. Soc.*, 6797 (1965).

236. E. Wenkert and D. K. Roychaudhuri, *J. Amer. Chem. Soc.*, **78**, 6417 (1956).

237. F. Bohlmann, *Angew. Chem.*, **69**, 641 (1957).

238. F. Bohlmann, *Chem. Ber.*, **92**, 1798 (1959).

239. F. Bohlmann, *Chem. Ber.*, **91**, 2157 (1958).

240. F. Bohlmann, D. Schumann, and M. Schulz, *Tetrahedron Letters*, 173 (1965).

241. M. Wiewiorowski and J. Skolik, *Bull. Acad. polon. Sci., Ser. Sci. chim.*, **10**, 1 (1962), *Chem. Abs.*, **57**, 13313 (1962).

242. J. Skolik, P. J. Krueger, and M. Wiewiorowski, *Tetrahedron*, **24**, 5439 (1968).

243. M. Wiewiorowski, O. E. Edwards, and M. D. Bratek-Wiewiorowska, *Canad. J. Chem.*, **45**, 1447 (1967).

244. H. P. Hamlow, S. Okuda, and N. Nakagawa, *Tetrahedron Letters*, 2553 (1964).

245. F. Bohlmann, E. Winterfeldt, and G. Boroschewski, *Chem. Ber.*, **94**, 3174 (1961).

246. F. Bohlmann, H-J. Müller, and D. Schumann, *Chem. Ber.*, **106**, 3026 (1973).

247. F. Bohlmann and C. Arndt, *Chem. Ber.*, **91**, 2167 (1958).

248. T. Masamune and M. Takasugi, *Chem. Comm.*, 625 (1967).

249. P. J. Chivers and T. A. Crabb, *Tetrahedron*, **26**, 3369 (1970).

250. P. J. Chivers, T. A. Crabb, and R. O. Williams, *Tetrahedron*, **24**, 4411 (1968).

251. T. M. Moynehan, K. Schofield, R. A. Y. Jones, and A. R. Katritzky, *J. Chem. Soc.*, 2637 (1962).

252. C-Y. Chen and R. J. W. Le Fèvre, *Tetrahedron Letters*, 1611 (1965).

253. M. J. T. Robinson, *Tetrahedron Letters*, 1153 (1968).

254. E. Toromanoff, *Bull. Soc. chim. France*, 3357 (1966).

255. H. Möhrle, C. Karl, and U. Scheidegger, *Tetrahedron*, **24**, 6813 (1968).

256. J. Sam, J. D. England, and D. Temple, *J. Medicin. Chem.*, **12**, 144 (1969).

257. H. S. Aaron, G. E. Wicks, Jr., C. P. Rader, *J. Org. Chem.*, **29**, 2248 (1964).

258. D. Temple and J. Sam, *J. Heterocyclic Chem.*, **5**, 441 (1968).

259. F. Bohlmann, E. Winterfeldt, O. Schmidt, and W. Reusche, *Chem. Ber.*, **94**, 1767 (1961).

260. H. Möhrle and C. Karl, *Arch. Pharm. (Weinheim)*, **301**, 530 (1968).

261. D. L. Temple and J. Sam, *J. Heterocyclic Chem.*, **7**, 847 (1970).

262. J. D. England and J. Sam, *J. Heterocyclic Chem.*, **3**, 482 (1966).

263. F. Bohlmann, E. Winterfeldt, P. Studt, H. Laurent, G. Boroschewski, and K-M. Kleine, *Chem. Ber.*, **94**, 3151 (1961).

264. W. D. Crow, *Austral. J. Chem.*, **11**, 366 (1958).

265. F. Galinovsky and H. Nesvadba, *Monatsh.*, **85**, 1300 (1954).

266. A. F. Thomas, H. J. Vipond, and L. Marion, *Canad. J. Chem.*, **33**, 1290 (1955).

267. F. Bohlmann, E. Winterfeldt, D. Schumann, U. Zarnack, and P. Wandrey, *Chem. Ber.*, **95**, 2365 (1962).

268. I. Matou, *J. Pharm. Soc. Japan*, **81**, 1078 (1961).

269. S. Ohki and Y. Noike, *Chem. and Pharm. Bull. (Japan)*, **7**, 708 (1959).

270. K. Schofield and R. J. Wells, *Austral. J. Chem.*, **18**, 1423 (1965).

271. F. Bohlmann, D. Schumann, and C. Arndt, *Tetrahedron Letters*, 2705 (1965).

272. R. T. Lalonde and T. N. Donvito, *Canad. J. Chem.*, **52**, 3778 (1974).

273. N. J. Leonard, R. W. Fulmer, and A. S. Hay, *J. Amer. Chem. Soc.*, **78**, 3457 (1956); N. J. Leonard and E. D. Nicolaides, *J. Amer. Chem. Soc.*, **73**, 5210 (1951); C. Schöpf and O. Thomä, *Annalen*, **98**, 465 (1928); V. Boekelheide and J. M. Ross, *J. Amer. Chem. Soc.*, **77**, 5691 (1955).

274. I. Matsuo, K. Sugimoto, and S. Ohki, *Chem. and Pharm. Bull. (Japan)*, **16**, 1680 (1968).

275. S. Okuda, H. Kataoka, and K. Tsuda, *Chem. and Pharm. Bull. (Japan)*, **13**, 487 (1965).

276. R. C. Cookson, *Chem. and Ind.*, 337 (1953); P. Karrer, F. Canal, K. Zohner, and R. Widmer, *Helv. Chim. Acta*, **11**, 1062 (1928).

277. S. Okuda, H. Kataoka, and K. Tsuda, *Chem. and Pharm. Bull. (Japan)*, **13**, 491 (1965).

278. S. F. Mason, K. Schofield, and R. J. Wells, *J. Chem. Soc.* (*C*), 626 (1967).

279. S-i. Yamada and K. Kunieda, *Chem. and Pharm. Bull.* (*Japan*), **15**, 490 (1967).

280. T. Kunieda, K. Koga, and S-i. Yamada, *Chem. and Pharm. Bull.* (*Japan*), **15**, 337 (1967).

281. H. Ripperger and K. Schreiber, *Chem. Ber.*, **100**, 1383 (1967).

282. Y. Arata, S. Yasuda, and K. Yamanouchi, *Chem. and Pharm. Bull.* (*Japan*), **16**, 2074 (1968).

283. S. I. Goldberg and A. H. Lipkin, *J. Org. Chem.*, **37**, 1823 (1972).

284. Y. Kobayashi, *Chem. and Pharm. Bull.* (*Japan*), **7**, 472 (1959).

285. T. Kappe, *Monatsh.*, **98**, 1852 (1967).

286. N. J. Leonard and R. R. Sauers, *J. Amer. Chem. Soc.*, **79**, 6210 (1957).

287. G. A. Swan, *J. Chem. Soc.*, 2051 (1958).

288. N. J. Leonard, A. S. Hay, R. W. Fulmer, and V. W. Gash, *J. Amer. Chem. Soc.*, **77**, 439 (1955).

289. R. Fujii, M. Nohara, M. Mitsukuchi, M. Ohba, K. Shikata, S. Yoshifuji, and S. Ikegami, *Chem. and Pharm. Bull.* (*Japan*), **23**, 144 (1975).

290. K. Schofield and R. J. Wells, *Chem. and Ind.*, 572 (1963).

291. N. J. Leonard and W. C. Wildman, *J. Amer. Chem. Soc.*, **71**, 3100 (1949).

292. T. M. Moynehan, K. Schofield, R. A. Y. Jones, and A. R. Katritzky, *Proc. Chem. Soc.*, 218 (1961).

293. K. Schofield and R. J. Wells, *J. Chem. Soc.* (*C*), 621 (1967).

294. S. Saeki, *Chem. and Pharm. Bull.* (*Japan*), **9**, 226 (1961).

295. J-L. Imbach, A. R. Katritzky, and R. A. Kolinski, *J. Chem. Soc.* (*B*), 556 (1966).

296. O. E. Edwards, G. Fodor, and L. Marion, *Canad. J. Chem.*, **44**, 13 (1966).

296a. Y. Arata, T. Aoki, M. Hanaoka, and M. Kamei, *Chem. and Pharm. Bull.* (*Japan*), **23**, 333 (1975).

296b. H. Arata, M. Hanaoka, and S. K. Kim, *Chem. and Pharm. Bull.* (*Japan*), **23**, 1142 (1975).

297. C. A. Grob, H. P. Fischer, H. Link, and E. Renk, *Helv. Chim. Acta*, **46**, 1190 (1963).

298. H. Hjeds and I. K. Larsen, *Acta Chem. Scand.*, **25**, 2394 (1971).

299. R. C. Cookson and N. S. Isaacs, *Tetrahedron*, **19**, 1237 (1963).

300. B. Price, I. O. Sutherland, and F. G. Williamson, *Tetrahedron*, **22**, 3477 (1966).

301. J. P. Kintzinger, J. M. Lehn, and J. Wagner, *Chem. Comm.*, 206 (1967).

302. S. F. Nelsen and J. M. Buschek, *J. Amer. Chem. Soc.*, **95**, 2011 (1973).

303. Y. Arata and Y. Nakagawa, *Chem. and Pharm. Bull.* (*Japan*), **33**, 1248 (1973).

304. T. A. Crabb and R. F. Newton, *J.C.S. Perkin II*, 1920 (1972).

305. R. Cahill and T. A. Crabb, *J.C.S. Perkin II*, 1374 (1972).

306. T. A. Crabb and R. F. Newton, *Tetrahedron*, **26**, 701 (1970).

307. T. A. Crabb and R. F. Newton, *Tetrahedron*, **24**, 6327 (1968).

308. P. Aeberli and W. J. Houlihan, *J. Org. Chem.*, **34**, 2720 (1969).

309. H. B. Sullivan and A. R. Day, *J. Org. Chem.*, **29**, 326 (1964).

310. L. Birkofer and H. D. Engels, *Chem. Ber.*, **95**, 2212 (1962).

311. S. Ohki, M. Akiba, H. Shimada, and K. Kunihiro, *Chem. and Pharm. Bull.* (*Japan*), **16**, 1889 (1968).

312. Y. Arata, Y. Nagakawa, Y. Arakawa, and M. Hanaoka, *Chem. and Pharm. Bull.* (*Japan*), **22**, 157 (1974).

313. F. Bohlmann, D. Habeck, E. Poetsch, and D. Schumann, *Chem. Ber.*, **100**, 2742 (1967).

314. F. Bohlmann and O. Schmidt, *Chem. Ber.*, **97**, 1354 (1964).

315. R. Buyle, *Chem. and Ind.*, 195 (1966).

316. C. F. Hammer, S. R. Heller, and J. H. Craig, *Tetrahedron*, **28**, 239 (1972).

317. C. F. Hammer and S. R. Heller, *Chem. Comm.*, 919 (1966).

318. A. L. Logothetis, *J. Amer. Chem. Soc.*, **87**, 749 (1965).

319. G. Lowe and D. D. Ridley, *J.C.S. Chem. Comm.*, 328 (1973).

320. G. Lowe and D. D. Ridley, *J.C.S. Perkin I*, 2024 (1973).

321. R. Majeste and L. M. Trefonas, *J. Heterocyclic Chem.*, **5**, 663 (1968).

322. L. Marion in *The Alkaloids*, R. H. F. Manske, Ed., Academic, New York, 1950, Vol. I, p. 224 and 1960, Vol. VI, p. 123.

323. M. S. Toy and C. C. Price, *J. Amer. Chem. Soc.*, **82**, 2613 (1960).

324. K. Löffler, *Ber.*, **42**, 948 (1909).

325. G. Fodor and G. A. Cooke, *Tetrahedron*, **22**, suppl. 8, 113 (1966).

326. K. Löffler and A. Grosse, *Ber.*, **40**, 1325 (1907).

327. I. Weisz and A. Dudás, *Monatsh.*, **91**, 840 (1960).

328. G. Lowe and J. Parker, *Chem. Comm.*, 577 (1971).

329. E. J. Corey and A. M. Felix, *J. Amer. Chem. Soc.*, **87**, 2518 (1965).

330. D. M. Brunwin, G. Lowe, and J. Parker, *J. Chem. Soc.* (*C*), 3756 (1971).

331. A. I. Meyers and N. K. Ralhan, *J. Org. Chem.*, **28**, 2950 (1963).

332. N. N. Abdul-Enein and J. Sam, *J. Heterocyclic Chem.*, **8**, 7 (1971).

333. G. R. Clemo and T. P. Metcalfe, *J. Chem. Soc.*, 1523 (1937).

334. G. Kohl and H. Pracejus, *Annalen*, **694**, 128 (1966).

335. R. Lukeš and M. Ferles, *Coll. Czech. Chem. Comm.*, **20**, 1227 (1955).

336. E. J. Corey and W. R. Hertler, *J. Amer. Chem. Soc.*, **82**, 1657 (1960).

337. P. G. Gassman and R. L. Cryberg, *J. Amer. Chem. Soc.*, **90**, 1355 (1968).

338. P. G. Gassman and R. L. Cryberg, *J. Amer. Chem. Soc.*, **91**, 2047 (1969).

339. P. G. Gassman, K. Shudo, R. L. Cryberg, and A. Battisti, *Tetrahedron Letters*, 875 (1972).

340. P. G. Gassman and G. D. Hartman, *J. Amer. Chem. Soc.*, **95**, 449 (1973).

341. J-P. Fleury, J-M. Biehler, and M. Desbois, *Tetrahedron Letters*, 4091 (1969).

342. P. G. Gassman and K. Shudo, *J. Amer. Chem. Soc.*, **93**, 5899 (1971).

343. P. G. Gassman, A. J. Battitsi, and K. Shudo, *Tetrahedron Letters*, 3773 (1972).

344. P. G. Gassman, R. L. Cryberg, and K. Shudo, *J. Amer. Chem. Soc.*, **94**, 7600 (1972).

345. E. L. Eliel, *Stereochemistry of Carbon Compounds*, McGraw-Hill, New York, 1962, pp. 298–302.

346. W. Oppolzer, *Tetrahedron Letters*, 1707 (1972).

347. P. S. Portoghese and D. T. Sepp, *Tetrahedron*, **29**, 2253 (1973).

348. P. S. Portoghese and D. L. Lattin, *J. Heterocyclic Chem.*, **9**, 395 (1972).

349. H. Prinzbach, G. Kaupp, R. Fuchs, M. Joyeux, R. Kitzing, and J. Markert, *Chem. Ber.*, **106**, 3824 (1973).

350. R. C. Bansal, A. W. McCulloch, and A. G. McInnes, *Canad. J. Chem.*, **47**, 2391 (1969).

351. G. Kaupp, J. Perreten, R. Leute, and H. Prinzbach, *Chem. Ber.*, **103**, 2288 (1970).

352. G. W. Gribble, N. R. Easton, Jr., and J. T. Eaton, *Tetrahedron Letters*, 1075 (1970).

353. G. Wittig, E. Knauss, and K. Niethammer, *Annalen*, **630**, 10 (1960).

354. R. Bonnett, R. F. C. Brown, and R. G. Smith, *J.C.S. Perkin I*, 1432 (1973).

355. P. D. Rosso, J. Oberdier, and J. S. Swenton, *Tetrahedron Letters*, 3947 (1971).

356. R. Kitzing, R. Fuchs, M. Joyeux, and H. Prinzbach, *Helv. Chim. Acta*, **51**, 888 (1968).

357. I. Morishima, K. Yoshikawa, K. Bekki, M. Kohno, and K. Arita, *J. Amer. Chem. Soc.*, **95**, 5815 (1973).

358. R. B. Woodward and R. B. Turner in *The Alkaloids*, Vol. III, R. H. F. Manske and H. L. Holmes, Eds., Academic, New York, 1953, Ch. 16, p. 1.

359. W. I. Taylor in *The Alkaloids*, Vol. VIII, R. H. F. Manske, Ed., Academic, New York, 1965, Ch. 9, p. 203.

360. W. I. Taylor in *The Alkaloids*, Vol. VIII, R. H. F. Manske, Ed., Academic, New York, 1965, Ch. 22, p. 785, Ch. 10, p. 237.

361. L. N. Yakhontov, *Adv. Heterocyclic Chem.*, **11**, 473 (1970).

362. R. B. Woodward and W. E. Doering, *J. Amer. Chem. Soc.*, **67**, 860 (1945).

363. E. C. Taylor and S. F. Martin, *J. Amer. Chem. Soc.*, **94**, 6218 (1972), *J. Amer. Chem. Soc.*, **96**, 8095 (1974).

364. G. Grethe, J. Gutzwiller, H. L. Lee, and M. R. Uskoković, *Helv. Chim. Acta*, **55**, 1044 (1972); G. Grethe, H. L. Lee, T. Mitt, and M. R. Uskoković, *J. Amer. Chem. Soc.*, **93**, 5904 (1971); J. Gutzwiller and M. R. Uskoković, *J. Amer. Chem.*, *Soc.* **92**, 204 (1970).

365. D. L. Coffen and T. E. McEntee, Jr., *Chem. Comm.*, 539 (1971).

366. H. Pracejus, *Chem. Ber.*, **98**, 2897 (1965).

367. S. Wawzoneck, M. F. Nelson, Jr., and P. J. Thelen, *J. Amer. Chem. Soc.*, **73**, 2806 (1951); R. Lukěs and M. Ferles, *Coll. Czech. Chem. Comm.*, **16**, 416 (1951).

368. G. Wittig and G. Steinhoff, *Chem. Ber.*, **95**, 203 (1962).

369. F. G. Mann and D. P. Mukherjee, *J. Chem. Soc.*, 2298 (1949).

370. T. Wada, E. Kishida, Y. Tomiie, H. Suga, S. Seki, and I. Nitta, *Bull. Chem. Soc. Japan*, **33**, 1317 (1960).

371. R. W. Baker and P. J. Pauling, *J.C.S. Perkin II*, 2340 (1972).

372. P. Brenneisen, C. A. Grob, R. A. Jackson, and M. Ohta, *Helv. Chim. Acta*, **48**, 146 (1965).

373. C. A. Grob and A. Sieber, *Helv. Chem. Acta*, **50**, 2531 (1967).

374. P. G. Gassman and B. L. Fox, *J. Org. Chem.*, **32**, 480 (1967).

375. J. Palaček and J. Hlavatý, *Coll. Czech. Chem. Comm.*, **38**, 1985 (1973).

376. R. W. Taft and C. A. Grob, *J. Amer. Chem. Soc.*, **96**, 1236 (1974).

377. V. A. Snieckus, T. Onouchi, and V. Boekelheide, *J. Org. Chem.*, **37**, 2845 (1972); J. Witte and V. Boekelheide, *J. Org. Chem.*, **37**, 2849 (1972).

378. J. I. DeGraw and J. G. Kennedy, *J. Hetreocyclic Chem.*, **4**, 251 (1967).

379. G. N. Walker and D. Alkalay, *J. Org. Chem.*, **32**, 2213 (1967).

380. M. Saunders and E. H. Gold, *J. Org. Chem.*, **27**, 1439 (1962).

381. Y. Ban, T. Oishi, M. Ochiai, T. Wakamatsu, and Y. Fujimoto, *Tetrahedron Letters*, 6385 (1966).

382. E. E. Knaus, F. M. Pasutto, and C. S. Giam, *J. Heterocyclic Chem.*, **11**, 843 (1974).

383. G. Krow, E. Michener, and K. C. Ramey, *Tetrahedron Letters*, 3653 (1971).

384. H. Greuter and H. Schmid, *Helv. Chim. Acta*, **57**, 1204 (1974).

385. G. Krow, R. Rodebaugh, R. Carmosin, W. Figures, H. Pannella, G. DeVicaris, and M. Grippi, *J. Amer. Chem. Soc.*, **95**, 5273 (1973).

386. H. Tomisawa and H. Hongo, *Tetrahedron Letters*, 2465 (1969).

387. E. B. Sheinin, G. E. Wright, C. L. Bell, and L. Bauer, *J. Heterocyclic Chem.*, **5**, 859 (1968).

388. H. Tomisawa, R. Fujita, K. Noguchi, and H. Hongo, *Chem. and Pharm. Bull. (Japan)*, **18**, 941 (1970).

389. H. Hongo, *Chem. and Pharm. Bull. (Japan)*, **20**, 226 (1972).

390. H. Tomiaswa and H. Hongo, *Chem. and Pharm. Bull. (Japan)*, **18**, 925 (1970).

391. N. J. Mruk and H. Tieckelmann, *Tetrahedron Letters*, 1290 (1970).

392. P. J. Machin, A. E. A. Porter, and P. G. Sammes, *J.C.S. Perkin I*, 404 (1973).

393. C. K. Bradsher and F. H. Day, *Tetrahedron Letters*, 409 (1971).

394. C. K. Bradsher, F. H. Day, A. T. McPhail, and P-S. Wong, *J.C.S. Chem. Comm.*, 156 (1973).

395. C. K. Bradsher and F. H. Day, *J. Heterocyclic Chem.*, **11**, 23 (1974); F. H. Day, C. K. Bradsher, and T-H. Chen, *J. Org. Chem.*, **40**, 1195 (1975).

396. C. K. Bradsher, F. H. Day, A. T. McPhail, and P-S. Wong, *Tetrahedron Letters*, 4205 1971.

397. D. L. Fields, T. H. Regan, and J. C. Dignan, *J. Org, Chem.*. **33**, 390 (1968).

398. C. K. Bradsher and F. H. Day, *J. Heterocyclic Chem.*, **10**, 1031 (1973).

399. W. S. Burnham and C. K. Bradsher, *J. Org. Chem.*, **37**, 355 (1972).

400. C. K. Bradsher and D. J. Harvan, *J. Org. Chem.*, **36**, 3778 (1971).

401. C. K. Bradsher and J. A. Stone, *J. Org. Chem.*, **33**, 519 (1968).

402. C. K. Bradsher and J. A. Stone, *J. Org. Chem.*, **34**, 1700 (1969).

403. D. L. Fields and T. H. Regan, *J. Org. Chem.*, **35**, 1870 (1970).

404. B. J. Price and I. O. Sutherland, *J. Chem. Soc. (B)*, 573 (1967).

405. W. Nagata, S. Hirai, K. Kawata, and T. Okamura, *J. Amer. Chem. Soc.*, **89**, 5046 (1967); W. Nagata, S. Hirai, K. Kawata, T. Okamura, and T. Aoki, *J. Amer. Chem. Soc.*, **89**, 5045 (1967).

406. A. Heumann, R. Furstoss, and B. Waegell, *Tetrahedron Letters*, 993 (1972).

407. J. O. Reed and W. Lwowski, *J. Org. Chem.*, **36**, 2864 (1971).

408. J. D. Hobson and W. D. Riddell, *Chem. Comm.*, 1180 (1968).

409. D. R. Brown and J. McKenna, *J. Chem. Soc.*, *(B)* 570 (1969).

410. V. Prelog, S. Heimbach, and E. Cerkovnikov, *J. Chem. Soc.*, 677 (1939).

411. A. D. Yanina and M. V. Rubtsov, *J. Gen. Chem.*, *(U.S.S.R.)*, **32**, 1774 (1962).

412. A. D. Yanina and M. V. Rubtsov, *J. Gen. Chem.* *(U.S.S.R.)*, **30**, 2527 (1960).

413. B. P. Thill and H. S. Aaron, *J. Org. Chem.*, **33**, 4376 (1968).

414. P. G. Gassman and B. L. Fox, *Chem. Comm.*, 153 (1966).

415. P. G. Gassman and B. L. Fox, *J. Amer. Chem. Soc.*, **69**, 338 (1967).

416. A. D. Yanina and M. V. Rubtsov, *J. Gen. Chem. (U.S.S.R.)*, **32**, 3866 (1962).

417. A. D. Yanina and M. V. Rubtsov, *J. Gen. Chem.*, *(U.S.S.R.)*, **30**, 3098 (1962).

418. H. C. Cunningham and A. R. Day, *J. Org. Chem.*, **38**, 1225 (1973).

419. L. W. Morgan and M. C. Bourlas, *Tetrahedron Letters*, 2631 (1972).

420. P. Kovacic, M. K. Lowery, and P. D. Roskos, *Tetrahedron*, **26**, 529 (1970).

421. J. W. ApSimon and J. D. Cooney, *Canad. J. Chem.*, **49**, 2377 (1971).

422. D. R. Brown, J. McKenna, and J. M. McKenna, *J. Chem. Soc. (B)*, 1195 (1967).

423. J. McKenna, J. White, and A. Tulley, *Tetrahedron Letters*, 1097 (1962).

424. B. G. Hutley, J. McKenna, and J. M. Stuart, *J. Chem. Soc. (B)*, 1199 (1967).

425. J. McKenna, *Topics Stereochem.*, **5**, 275 (1970).

426. Y. L. Chow and R. A. Perry, *Tetrahedron Letters*, 531 (1972).

427. J-M. Surzur, L. Stella, and R. Nouguier, *Tetrahedron Letters*, 903 (1971).

428. O. E. Edwards, G. Bernath, J. Dixon, J. M. Paton, and D. Vocelle, *Canad. J. Chem.*, **52**, 2123 (1974).

429. M. E. Kuehne and D. A. Horne, *J. Org. Chem.*, **40**, 1287 (1975).

430. G. Esposito, R. Furstoss, and B. Waegell, *Tetrahedron Letters*, 895 (1971); P. G. Gassman and B. L. Fox, *J. Org. Chem.*, **32**, 3679 (1967).

431. W. J. Gensler, C. D. Gatsonis, and Q. A. Ahmed, *J. Org. Chem.*, **33**, 2968 (1968).

432. G. Esposito, R. Furstoss, and B. Waegell, *Tetrahedron Letters*, 899 (1971).

433. G. R. Krow, R. Rodebaugh, C. Hyndman, R. Carmosin, and G. DeVicaris, *Tetrahedron Letters*, 2175 (1973).

434. W. Nagata, I. Kikkawa, and M. Fujimoto, *Chem. and Pharm. Bull. (Japan)*, **11**, 226 (1963).

435. H. L. Holmes in *The Alkaloids*, Vol. I, H. L. Holmes and R. H. F. Manske, Eds., Academic, New York, 1950, Ch. 6, p. 271.

436. G. Fodor in *The Alkaloids*, Vol. VI, R. H. F. Manske, Ed., Academic, New York, 1960, Ch. 5, p. 145.

437. A. Sekera, *Ann. Chim. (France)*, **7**, 537 (1962).

438. G. Fodor and F. Sóti, *J. Chem. Soc.*, 6830 (1965) and earlier papers.

439. R. J. Bishop, G. Fodor, A. R. Katritzky, F. Sóti, L. E. Sutton, and F. J. Swinbourne, *J. Chem. Soc. (C)*, **74**, (1966).

440. J. M. Eckert, R. J. W. Le Fèvre, *J. Chem. Soc.*, 3991 (1962).

441. P. Scheiber, G. Kraiss, K. Nádor, and A. Neszmélyi, *J. Chem. Soc. (B)*, 2149 (1971).

442. P. Scheiber, G. Kraiss, and K. Nádor, *J. Chem. Soc. (B)*, 1366 (1970).

443. F. Uchimaru, *Chem. and Pharm. Bull. (Japan)*, **10**, 238 (1962).

444. G. Fodor and K. Nádor, *Nature*, **169**, 462 (1952), *J. Chem. Soc.*, 721 (1953).

445. A. Nickon and L. F. Fieser, *J. Amer. Chem. Soc.*, **74**, 5566 (1952).

446. M. R. Bell and S. Archer, *J. Amer. Chem. Soc.*, **82**, 151 (1960).

447. E. R. Atkinson and D. D. McRitchie, *J. Org. Chem.*, **36**, 3240 (1971).

448. H. Singh and B. Razdan, *Tetrahedron Letters*, 3243 (1966).

449. M. Shimizu, F. Uchimaru, and B. Kurihara, *Chem. and Pharm. Bull. (Japan)*, **10**, 204 (1962).

450. G. Kraiss, P. Scheiber, and K. Nádor, *J. Org. Chem.*, **33**, 2601 (1968).

451. G. Kraiss, P. Scheiber, and K. Nádor, *J. Chem. Soc. (B)*, 2145 (1971).

452. S. Archer, T. R. Lewis, M. R. Bell, and J. W. Schulenberg, *J. Amer. Chem. Soc.*, **83**, 2386 (1961).

453. S. R. Johns and J. A. Lamberton, *Chem. Comm.*, 458 (1965).

454. G. L. Closs, *J. Amer. Chem. Soc.*, **81**, 5456 (1959).

455. D. R. Brown, R. Lygo, J. McKenna, J. M. McKenna, and B. G. Hutley, *J. Chem. Soc. (B)*, 1184 (1967).

456. C. C. Thut and A. T. Bottini, *J. Amer. Chem. Soc.*, **90**, 4752 (1968).

457. G. Fodor, J. D. Medina, and N. Mandava, *Chem. Comm.*, 581 (1968).

458. D. R. Brown, J. McKenna, J. M. McKenna, and J. M. Stuart, *Chem. Comm.*, 380 (1967).

459. G. Fodor, R. V. Chastain, Jr., D. Frehel, M. J. Cooper, N. Mandava, E. L. Gooden, *J. Amer. Chem. Soc.*, **93**, 403 (1971).

460. J. H. Supple and E. Eklum, *J. Amer. Chem. Soc.*, **93**, 6684 (1971).

461. K. Schmid, W. von Philipsborn, H. Schmid, and P. Karrer, *Helv. Chim. Acta*, **39**, 394 (1956).

462. K. Bachmann and W. von Philipsborn, *Helv. Chim. Acta*, **55**, 637 (1972).

463. A. T. Bottini, C. A. Grob, E. Schumacher, and J. Zergenyi, *Helv. Chim. Acta*, **49**, 2516 (1966).

464. A. R. Katritzky and Y. Takeuchi, *J. Amer. Chem. Soc.*, **92**, 4134 (1970).

465. N. Dennis, A. R. Katritzky, T. Matsuo, S. K. Parton, and Y. Takeuchi, *J.C.S. Perkin I*, 746 (1974).

466. N. Dennis, A. R. Katritzky, S. K. Parton, and Y. Takeuchi, *J.C.S. Chem. Comm.*, 707 (1972).

467. N. Dennis, B. Ibrahim, A. R. Katritzky, I. G. Taulov, and Y. Takeuchi, *J.C.S. Perkin I*, 1883 (1974).

468. N. Dennis, B. Ibrahim, and A. R. Katritzky, *J.C.S. Chem. Comm.*, 500 (1974).

468a. N. Dennis, B. Ibrahim, and A. R. Katritzky, *J.C.S. Chem. Comm.*, 425 (1975).

469. G. Cignarella, E. Occelli, G. G. Gallo, and E. Testa, *Gazzetta*, **98**, 848 (1968); G. Cignarella, G. Maffii, and E. Testa, *Gazzetta*, **93**, 226 (1963).

470. B. Shimizu, A. Ogiso, and I. Iwai, *Chem. and Pharm. Bull. (Japan)*, **11**, 766 (1963).

471. A. Ogiso, B. Shimizu, and I. Iwai, *Chem. and Pharm. Bull. (Japan)*, **11**, 770 (1963).

472. A. Z. Britten and J. O'Sullivan, *Tetrahedron*, **29**, 1331 (1973).

473. R. F. C. Brown, *Austral. J. Chem.*, **17**, 47 (1964).

474. M. Dobler and J. D. Dunitz, *Helv. Chim. Acta*, **47**, 695 (1964).

475. M. W. J. Pumphrey and M. J. T. Robinson, *Chem. and Ind.*, 1903 (1963).

476. R. Lygo, J. McKenna, and I. O. Sutherland, *Chem. Comm.*, 356 (1965).

477. H. O. House and B. A. Tefertiller, *J. Org. Chem.*, **31**, 1068 (1966).

478. H. O. House and C. G. Pitt, *J. Org. Chem.*, **31**, 1062 (1966).

479. H. O. House and W. M. Bryant III, *J. Org. Chem.*, **30**, 3634 (1965).
480. H. O. House, H. C. Müller, C. G. Pitt, and P. P. Wickham, *J. Org. Chem.*, **28**, 2407 (1963).
481. S. Shiotani and K. Mitsuhashi, *J. Pharm. Soc. Japan*, **92**, 97 (1972).
482. H. O. House, P. P. Wickham, and H. C. Müller, *J. Amer. Chem. Soc.*, **84**, 3139 (1962).
483. A. W. J. D. Dekkers, N. M. M. Nibbering, and W. N. Speckamp, *Tetrahedron*, **28**, 1829 (1972).
484. S. Oida and E. Ohki, *Chem. and Pharm. Bull. (Japan)*, **16**, 654 (1968).
485. L. Marion in *The Alkaloids*, R. H. F. Manske and H. L. Holmes, Eds., Academic, New York, 1950, Vol. I, Ch. 5, p. 165; and R. H. F. Manske, Ed., 1960, Vol. VI, Ch. 4, p. 123.
486. C-Y. Chen and R. J. W. Le Fèvre, *J. Chem. Soc. (B)*, 539 (1966).
487. P. D. Cradwick and G. A. Sim, *J. Chem. Soc. (B)*, 2218 (1971); C. Tamura and G. A. Sim, *J. Chem. Soc. (B)*, 1241 (1968).
488. L. A. Paquette and J. W. Heimaster, *J. Amer. Chem. Soc.*, **88**, 763 (1966).
488a. H. Stetter and K. Heckel, *Tetrahedron Letters*, 801 (1972).
489. J. D. Hobson and W. D. Riddell, *Chem. Comm.*, 1178 (1968).
490. S. P. Findlay, *J. Org. Chem.*, **23**, 391 (1958).
491. B. Tursch, C. Chome, J. C. Braekman, and D. Daloze, *Bull. Soc. chim. belges*, **82**, 699 (1973).
492. T. Sasaki, S. Eguchi, and T. Kiriyama, *J. Amer. Chem. Soc.*, **91**, 212 (1969).
493. T. Sasaki, S. Eguchi, and T. Kiriyama, *Tetrahedron*, **27**, 893 (1971).
494. H. O. Krabbenhoft, J. R. Wiseman, and C. B. Quinn, *J. Amer. Chem. Soc.*, **96**, 258 (1974).
495. S. Shiotani, T. Hori, and K. Mitsuhashi, *Chem. and Pharm. Bull. (Japan)*, **15**, 88 (1967).
496. S. Shiotani and K. Mitsuhashi, *Chem. and Pharm. Bull. (Japan)*, **12**, 647 (1964).
497. T. Kametani and K. Kigasawa, *Chem. and Pharm. Bull. (Japan)*, **14**, 566 (1966).
498. E. E. Smissman, R. A. Robinson, J. B. Carr, and A. J. B. Matuszak, *J. Org. Chem.*, **35**, 3821 (1970).
499. H. K. Hall, Jr., and R. C. Johnson, *J. Org. Chem.*, **37**, 697 (1972).
500. T. Kametani, K. Kigasawa, M. Hiiragi, and H. Ishimaru, *Chem. and Pharm. Bull. (Japan)*, **13**, 295 (1965).
501. J. H. Billman and L. C. Dorman, *J. Org. Chem.*, **27**, 2419 (1962).
502. S. Shiotani and K. Mitsuhashi, *Chem. and Pharm. Bull. (Japan)*, **14**, 608 (1966).
503. S. Shiotani and K. Mitsuhashi, *Chem. and Pharm. Bull. (Japan)*, **15**, 761 (1967).
504. T. Kametani, K. Kigasawa, and T. Hayasaka, *Chem. and Pharm. Bull. (Japan)*, **13**, 300 (1965).
505. S. Shiotani and K. Mitsuhashi, *J. Pharm. Soc. Japan*, **84**, 1032 (1964).
506. T. Sasaki, S. Eguchi, and T. Kiriyama, *J. Org. Chem.*, **36**, 2061 (1971).
507. R. Lukeš, *Coll. Czech. Chem. Comm.*, **10**, 148 (1938).
508. H. Pracejus, *Chem. Ber.*, **92**, 988 (1959).
509. H. Pracejus, M. Kehlen, H. Kehlen, and H. Matschiner, *Tetrahedron*, **21**, 2257 (1965).
510. S. F. Nelsen, P. J. Hintz, and R. T. Landis II, *J. Amer. Chem. Soc.*, **94**, 7105 (1972).

511. E. A. Steck, L. T. Fletcher, and R. P. Brundage, *J. Org. Chem.*, **28**, 2233 (1963).

512. F. Galinovsky and H. Langer, *Monatsh.*, **86**, 449 (1955); F. Bohlmann, N. Ottawa, and R. Keller, *Annalen*, **587**, 162 (1954).

513. G. Settimj, R. L. Vittory, F. Delle Monache, and S. Chiavarelli, *Gazzatta*, **96**, 331 (1966).

514. G. Settimj, R. L. Vittory, F. Gatta, N. Sarti, and S. Chiavarelli, *Gazetta*, **96**, 604 (1966).

515. C. W. Thornber, *J.C.S. Chem. Comm.*, 238 (1973).

516. S. Chiavarelli, F. Toffler, L. Gramiccioni, and G. P Valsecchi, *Gazzatta*, **98**, 1126 (1968).

517. M. R. Chakrabarty, R. L. Ellis, and J. L. Roberts, *J. Org. Chem.*, **35**, 541 (1970).

518. J. E. Douglass and T. B. Ratliff, *J. Org. Chem.*, **33**, 355 (1968).

519. N. J. Leonard, P. D. Thomas, and V. W. Gash, *J. Amer. Chem. Soc.*, **77**, 1552 (1955).

520. E. Leete and L. Marion, *Canad. J. Chem.*, **30**, 563 (1952).

521. B. P. Moore and L. Marion, *Canad. J. Chem.*, **31**, 187 (1953).

522. N. J. Leonard and R. E. Beyler, *J. Amer. Chem. Soc.*, **72**, 1316 (1950).

523. F. Galinovsky, P. Knoth, and W. Fischer, *Monatsh.*, **86**, 1014 (1955).

524. B. Eda, K. Tsuda, and M. Kubo, *J. Amer. Chem. Soc.*, **80**, 2426 (1958).

525. M. A. McKervey, *Chem. Soc. Rev.* **3**, 479 (1974).

526. G. Gelbard, *Ann. Chim. (France)*, **4**, 331 (1969).

527. W. H. Staas and L. A. Spurlock, *J. Org. Chem.*, **39**, 3822 (1974).

528. R. E. Portmann and C. Ganter, *Helv. Chim. Acta*,. **56**, 1986 (1973).

529. R. J. Schultz, W. H. Staas, and L. A. Spurlock, *J. Org. Chem.*, **38**, 3091 (1973).

530. H. Stetter and K. Heckel, *Tetrahedron Letters*, 1907 (1972).

531. R-M. Dupeyre and A. Rassat, *Tetrahedron Letters*, 2699 (1973).

532. T. Sasaki, S. Eguchi, T. Kiriyama, and Y. Sakito, *J. Org. Chem.*, **38**, 1648 (1973).

533. T. Sasaki, S. Eguchi, T. Kiriyama, Y. Sakito, and H. Kato, *J.C.S. Chem. Comm.*, 725 (1974).

534. R. C. Cookson, J. Henstock, and J. Hudec, *J. Amer. Chem. Soc.*, **88**, 1060 (1966); J. Kuthan and J. Paleček, *Coll. Czech. Chem. Comm.*, **28**, 2260 (1963).

534a. H. Quast and P. Eckert, *Annalen*, 1727 (1974).

535. M. E. Foss, E. L. Hirst, J. K. N. Jones, H. D. Springall, A. T. Thomas, and T. Ubrański, *J. Chem. Soc.*, 1691 (1950).

536. J. McKenna, J. M. McKenna, and B. A. Wesby, *Chem. Comm.*, 867 (1970).

537. A. G. Anastassiou, *J. Amer. Chem. Soc.*, **90**, 1527 (1968).

538. A. G. Anastassiou and R. P. Cellura, *J. Org. Chem.*, **37**, 3126 (1972).

539. L. A. Paquette and L. D. Wise, *J. Amer. Chem. Soc.*, **87**, 1561 (1965).

540. L. A. Paquette, J. R. Malpass, and T. J. Barton, *J. Amer. Chem. Soc.*, **91**, 4714 (1969).

541. A. G. Anastassiou and R. P. Cellura, *Tetrahedron Letters*, 911 (1970).

542. O. Yonemitsu, K. Nakai, Y. Kanaoka, I. L. Karle, and B. Witkop, *J. Amer. Chem. Soc.*, **91**, 4591 (1969).

543. N. J. Leonard and J. C. Coll, *J. Amer. Chem. Soc.*, **92**, 6685 (1970); J. C. Coll, DeL.R. Crist, M. delC. G. Barrio, and N. J. Leonard, *J. Amer. Chem. Soc.*, **94**, 7092 (1972).

544. M. Doyle, W. Parker, P. A. Gunn, J. Martin, and D. D. Macnicol, *Tetrahedron Letters*, 3619 (1970).

545. N. J. Leonard, J. C. Coll, A. H-J. Wang, R. J. Missavage, and I. C. Paul, *J. Amer. Chem. Soc.*, **93**, 4628 (1971); A. H-J. Wang, R. J. Missavage, S. R. Byrn, and I. C. Paul, *J. Amer. Chem. Soc.*, **94**, 7100 (1972).

546. J. K. N. Jones, R. Koliński, H. Piotrowska, and T. Ubrański, *Bull. Acad. polon. Sci.*, *Cl. III*, **4**, 521 (1956), *Chem. Abs.*, **51**, 8765 (1957), *Roszniki Chem.*, **31**, 101 (1957), and *Chem. Abs.*, **51**, 14718 (1957).

547. V. Prelog and P. Wieland, *Helv. Chim. Acta*, **27**, 1127 (1944).

548. O. Červinka, A. Fábryová, and V. Novák, *Tetrahedron Letters*, 5375 (1966).

549. S. F. Mason, G. W. Vane, K. Schofield, R. J. Wells, and J. S. Whitehurst, *J. Chem. Soc.* (*B*), 553 (1967).

550. S. F. Mason, K. Schofield, R. J. Wells, J. S. Whitehurst, and G. W. Vane, *Tetrahedron Letters*, 137 (1967).

551. A. C. Baker and A. R. Battersby, *Tetrahedron Letters*, 135 (1967).

552. W. L. F. Armarego, *Fused Pyrimidines Part I. Quinazolines*, D. J. Brown, Ed., Interscience, New York, 1967, pp. 406–409.

553. J. Altman, E. Babad, J. Pucknat, N. Reshef, and D. Ginsburg, *Tetrahedron*, **24**, 975 (1968).

554. C. Amith, J. Kalo, B. E. North, and D. Ginsburg, *Tetrahedron*, **30**, 479 (1974); J. M. Ben-Bassat and D. Ginsburg, *Tetrahedron*, **30**, 483 (1974).

555. J. Altman, E. Babad, J. Itzchaki, and D. Ginsburg, *Tetrahedron*, **22**, suppl. 8, 279 (1966).

556. C. Amith and D. Ginsburg, *Tetrahedron*, **30**, 3415 (1974); M. Korat, D. Tatarsky, and D. Ginsburg, *Tetrahedron*, **28**, 2315 (1972).

557. D. Ginsburg, *Accounts Chem. Res.*, **2**, 121 (1969).

557a. D. Ginsburg, *Tetrahedron*, **30**, 1487 (1974).

558. G. Schröder, J. F. M. Oth, and R. Merényi, *Angew. Chem. Internat. Edn.*, **4**, 752 (1965).

559. L. A. Paquette, J. P. Malpass, and G. R. Krow, *J. Amer. Chem. Soc.*, **92**, 1980 (1970).

560. L. A. Paquette and T. J. Barton, *J. Amer. Chem. Soc.*, **89**, 5480 (1967).

561. L. A. Paquette, T. J. Barton, and E. B. Whipple, *J. Amer. Chem. Soc.*, **89**, 5481 (1967).

562. G. R. Krow and J. Reilly, *Tetrahedron Letters*, 3075 (1973).

563. L. A. Paquette and J. R. Malpass, *J. Amer. Chem. Soc.*, **90**, 7151 (1968).

564. L. A. Paquette, J. R. Malpass, G. R. Krow, and T. J. Barton, *J. Amer. Chem. Soc.*, **91**, 5296 (1969).

565. L. A. Paquette, G. R. Krow, and J. R. Malpass, *J. Amer. Chem. Soc.*, **91**, 5522 (1969).

566. L. A. Paquette, G. R. Krow, J. R. Malpass, and T. J. Barton, *J. Amer. Chem. Soc.*, **80**, 3600 (1968).

567. L. A. Paquette and G. R. Krow, *J. Amer. Chem. Soc.*, **90**, 7149 (1968).

568. L. A. Paquette and G. R. Krow, *J. Amer. Chem. Soc.*, **91**, 6107 (1969).

569. A. G. Anastassiou, A. E. Winston, and E. Riechmanis, *J.C.S. Chem. Comm.*, 779 (1973).

570. D. S. Wulfman and J. J. Ward, *Chem. Comm.*, 276 (1967).

571. M. J. S. Dewar, Z. Náhlovská, and B. D. Náhlovský, *Chem. Comm.*, 1377 (1971).

INDEX

This index contains most of the names of heterocyclic and other compounds of stereochemical interest, but not all the compounds in the text. Authors' names mentioned in the text are indexed.